Anthocyanins in Health and Disease

Anthocyanins in Health and Disease

Edited by
Taylor C. Wallace
M. Monica Giusti

CRC Press
Taylor & Francis Group
Boca Raton London New York

CRC Press is an imprint of the
Taylor & Francis Group, an **informa** business

CRC Press
Taylor & Francis Group
6000 Broken Sound Parkway NW, Suite 300
Boca Raton, FL 33487-2742

Printed on acid-free paper
Version Date: 20130730

International Standard Book Number-13: 978-1-4398-9471-2 (Hardback)

Library of Congress Cataloging-in-Publication Data

Anthocyanins in health and disease / editors, Taylor C. Wallace, M. Monica Giusti.
     pages cm
   "A CRC title."
   Includes bibliographical references and index.
   ISBN 978-1-4398-9471-2 (hardcover : alk. paper) 1. Anthocyanins. 2. Flavonoids. 3. Vegetarianism--Health aspects. I. Wallace, Taylor C., editor of compilation. II. Giusti, M. Monica, editor of compilation.

QK898.A55A64 2014
613.2'62--dc23

2013028854

Visit the Taylor & Francis Web site at
http://www.taylorandfrancis.com

and the CRC Press Web site at
http://www.crcpress.com

Contents

Preface

It is well accepted that diets rich in colorful fruits and vegetables are linked to health, longevity, and a reduced risk for the development of chronic diseases. In an endeavor to identify active health-promoting ingredients, many researchers have focused on the properties of the flavonoids, a large class of polyphenolic compounds that are abundant in such foods. Most prominent among the flavonoids are anthocyanins—secondary plant metabolites responsible for the orange-red to blue-violet hues evident in many fruits, vegetables, cereal grains, and flowers. Represented by over 700 molecular structures identified in nature to date, anthocyanins are of particular interest to the industry because of their potential health-promoting properties, as well as their unique ability to impart vibrant colors to a variety of products. Recent studies using purified anthocyanins or anthocyanin-rich extracts on *in vitro* experimental systems have confirmed the potential of these pigments to positively effect cell homeostasis and function. Mechanistic data from these studies as well as the information obtained from *in vivo* animal experiments and human clinical trials demonstrate efficacy and biological activity of anthocyanins far beyond the GI tract. Many questions remain whether these apparent health benefits stem from anthocyanins alone, from their degradation products, or from their synergistic interactions with other phenolic compounds and plant constituents present in foods and anthocyanin-rich extracts. Imperative to the study of these compounds is a greater understanding of their fate in the GI tract upon fermentation and metabolism by the gut microbiota and whether the food matrix and diet have the ability to enhance the bioavailability of anthocyanins and their derived breakdown/metabolic products.

Demonstrable benefits of a diet rich in anthocyanins may include protection against the development of many age- and obesity-related chronic diseases such as but not limited to certain types of cancers, type-2 diabetes, metabolic syndrome and cardiovascular disease in addition to positive effects in the eye and on cognitive functions and exercise-induced immunity. Nevertheless, anthocyanins are considered nonessential but "nutritive" in the sense that their intake has been associated with a variety of beneficial health outcomes. Advances in scientific studies relating to the role of anthocyanins in human health and disease prevention make compilation of the data into book form an important stride toward advancing research and enhancing communication of these value-added functional ingredients. Research on the health effects of anthocyanins drastically expanded during the past decade but is still in its early stages from a public policy standpoint. The literature overwhelmingly suggests that anthocyanins like many other bioactives prevalent in the diet are safe and beneficial when consumed at levels within the normal nutritional range. A shift in research paradigm from the traditional evidence-based medicine approach toward a more evidence-based nutrition or evidence-informed nutrition method of studying nonnutrient dietary components dramatically affects the way scientists study compounds such as anthocyanins on a long term. Although anthocyanins may have a marked effect on human health, it should be recognized that they are not pharmaceuticals and their intended use should be in helping consumers to achieve healthier lifestyles and maintain or extend normal physiological functions during the process of aging.

This book has the purpose of effectively communicating modern-day research to a large group of scientific audiences ranging from university classrooms to industry product developers and basic researchers in the field of food/nutrition science, or related fields. The search for an international assortment of scientists working on the health benefits of anthocyanins who were qualified to contribute to this book was a major editorial challenge in editing this book. The interdisciplinary range of content that is covered by the different authors made this work particularly challenging, but is one of the key aspects of this comprehensive book. We would like to personally thank each contributing author for his or her dedication and involvement in the generation of this book. The unique expertise of each distinguished scientist in his or her particular field, covering a wide array of specialized knowledge, makes the book an authoritative and leading source of information for the industry, academia, and governments across the globe.

This book will strengthen the position of anthocyanins and anthocyanin-rich products in the future by facilitating access to the total body of information available to date and encouraging the increased consumption of anthocyanin-rich foods—in hopes of contributing to a healthier population.

Taylor C. Wallace, PhD, CFS, FACN
Council for Responsible Nutrition
Developing Solutions, LLC

M. Monica Giusti, PhD
Department of Food Science and Technology
The Ohio State University

Peer Review Team

The editors would like to recognize the meticulous contributions of the diverse team of experts who helped ensure that this book represents the utmost quality up-to-date scientific research.

Lindsay Brown
Biological and Physical Sciences
University of Southern Queensland,
Australia

Oliver Chen
Antioxidants Research Laboratory
Jean Mayer UDSA Human Nutrition
Research Center on Ageing, USA

Jian He
Food and Beverages
Unilever, China

Wilhelmina Kalt
Food Chemistry/Chimiste Alimentaire
Agriculture and Agri-Food
Canada/Agriculture et Agroalimentaire
Canada

Tia M. Rains
Biofortis-Provident Clinical Research,
USA

Luis E. Rodriguez-Saona
Food Science and Technology
Department
The Ohio State University, USA

Joseph Scheerens
Department of Horticulture & Crop
Science
The Ohio State University, USA

Andrea Tarozzi
Department of Pharmacology
University of Bologna

Takanori Tsuda
College of Bioscience and Biotechnology
Chubu University, Japan

Editors

Taylor C. Wallace, PhD, CFS, FACN, is the senior director of Science, Policy, and Government Affairs at the National Osteoporosis Foundation (NOF) and the senior director of Scientific and Clinical Programs at the National Bone Health Alliance (NBHA), a public–private partnership, managed and operated by the NOF. Dr. Wallace is responsible for ensuring that NOF's scientific, legislative, and policy program is broad-based, comprehensive, and evidence-based, aimed at strengthening bone health, decreasing the prevalence for osteoporosis by working with various key government agencies and scientific societies toward improving tests and therapies associated with the prevention, diagnosis, and treatment of osteoporosis. In addition, Dr. Wallace provides scientific leadership, content expertise, and project management leadership in support of NBHA projects and activities such as the Secondary Fracture Prevention Initiative, a recently piloted NBHA Fracture Liaison Service (FLS) program that coordinates post-fracture care through an FLS coordinator (a nurse or other allied health professional) who ensures that individuals who suffer fractures receive appropriate diagnosis, treatment, and support. He also manages several NOF and NBHA scientific initiatives in areas such as bone turnover standardization and rare bone diseases. Before joining NOF and NBHA, Dr. Wallace served as the senior director of Scientific & Regulatory Affairs at the Council for Responsible Nutrition (CRN), where he was responsible for providing scientific and regulatory expertise for evaluating scientific research, ensuring legislative and that policy positions were based on credible science rationale, while developing new scientific reviews and original research for the peer-reviewed literature. His academic background includes a PhD and an MS in food science and nutrition from the Ohio State University and a BS in food science and technology from the University of Kentucky. Dr. Wallace currently serves as a trustee and the acting treasurer of Feeding Tomorrow, the Foundation of the Institute of Food Technologists. Dr. Wallace was recently elected as a fellow of the American College of Nutrition, and he is also a member of the editorial board for the *Journal of the American College of Nutrition*. He has produced over 20 peer-reviewed publications and book chapters. Dr. Wallace is the coeditor of *Anthocyanins in Health and Disease* and the editor of *Dietary Supplements in Health and Disease Prevention* to be published in 2014. He is an active member of the Institute of Food Technologists, the American Society of Nutrition, the American College of Nutrition, and the Phi Tau Sigma Honor Society for Food Science and Technology.

Photo by Ken Chamberlain, OSU CommTech.

M. Mónica Giusti is an associate professor and graduate studies chair at the Food Science and Technology Department, the Ohio State University (OSU), Columbus, Ohio and a visiting faculty of the Facultad de Industrias Alimentarias, Universidad Nacional Agraria, La Molina, Peru. Her research has focused on the chemistry and functionality of flavonoids, with strong emphasis on natural colorants and functional foods.

Dr. Giusti's interest on anthocyanins started at graduate school, where she explored the use of anthocyanins as alternatives to the use of synthetic dyes under the guidance of Dr. Wrolstad, a highly renowned expert in the field. Since then she has investigated different aspects of anthocyanins including their incidence and concentration in plants, stability and interactions with food matrices, novel analytical procedures, and the bioavailability, bio-transformations, and potential bioactivity of these wonderful plant pigments. Dr. Giusti has become a leading researcher in the field of anthocyanins with her work being cited over 1500 times in a plethora of scientific publications.

Dr. Giusti has 50 peer-reviewed publications and 20 book chapters, and has presented her research around the world at more than 100 national and international meetings, conferences, and workshops. For her innovative work on anthocyanins and their food applications, she was conferred the 2010 Ohio Agricultural Research and Development Center Director's Innovator of the Year, and the 2011 TechColumbus Outstanding Woman in Technology. Dr. Giusti was granted a patent on the chemoprotective effects of anthocyanin-rich extracts, and has three additional patents pending in the field of anthocyanins. Dr. Giusti is also an entrepreneurial scholar for the OSU College of Food, Agriculture and Environmental Sciences, working with the OSU Office of Technology Commercialization and Knowledge Transfer to help translate research into impactful commercial outcomes. Her patent-pending technology for isolation of anthocyanin mixtures was the foundation for the start-up company AnthoScyantific, LCC, where she served as chief scientist.

Dr. Giusti is a member of the American Chemical Society and the Institute of Food Technologists (IFT), where she served on the executive board of the IFT Fruit and Vegetable division and as officer for the Ohio Valley section of the IFT.

Before joining the OSU, Dr. Giusti was a faculty member at the Department of Nutrition and Food Science at the University of Maryland. Dr. Giusti, born in Lima, Peru, received a food engineer degree from the Universidad Nacional Agraria, La Molina, Peru and master's and doctorate degrees in food science from Oregon State University, Corvallis, Oregon.

Contributors

Øyvind M. Andersen
Department of Chemistry
University of Bergen
Bergen, Norway

Xiuli Bi
School of Life Sciences
Liaoning University
Shenyang, China

Steven Carmella
Masonic Cancer Center
University of Minnesota
Minneapolis, Minnesota

Diana M. Cheng
School of Environmental
    and Biological Sciences
Rutgers, The State University
    of New Jersey
New Brunswick, New Jersey

Bertold Fridlender
School of Environmental
    and Biological Sciences
Rutgers, The State University
    of New Jersey
New Brunswick, New Jersey
and
Nutrasorb LLC
North Brunswick, New Jersey

M. Monica Giusti
Department of Food Science
    and Technology
The Ohio State University
Columbus, Ohio

Brittany Graf
School of Environmental
    and Biological Sciences
Rutgers, The State University
    of New Jersey
New Brunswick, New Jersey

Stephen Hecht
Masonic Cancer Center
University of Minnesota
Minneapolis, Minnesota

Tim H.-M. Huang
Department of Molecular Medicine/
    Institute of Biotechnology
University of Texas Health Science
    Center
San Antonio, Texas

Yi-Wen Huang
Department of Obstetrics
    and Gynecology
Medical College of Wisconsin
Milwaukee, Wisconsin

Roger Donald Hurst
Food and Wellness Group
The New Zealand Institute for Plant
    and Food Research Ltd.
Palmerston North, New Zealand

Suzanne Maria Hurst
Food and Wellness Group
The New Zealand Institute for Plant
    and Food Research Ltd.
Palmerston North, New Zealand

Pu Jing
Department of Food Science
and Engineering
Shanghai Jiao Tong University
Shanghai, People's Republic of China

Monica Jordheim
Department of Chemistry
University of Bergen
Bergen, Norway

Wilhelmina Kalt
Agriculture and Agri-Food Canada
Atlantic Food and Horticulture
Research Centre
Kentville, Nova Scotia, Canada

Chieh-Ti Kuo
Department of Medicine
Medical College of Wisconsin
Milwaukee, Wisconsin

Daniel A. Linseman
Department of Biological Sciences
Eleanor Roosevelt Institute
University of Denver
and
Research Service, Veterans Affairs
Medical Center
Denver, Colorado

Zhongfa Liu
Division of Pharmaceutic
Ohio State University
Columbus, Ohio

Bernadene A. Magnuson
Department of Nutritional Science
University of Toronto
Toronto, Ontario, Canada

Tony K. McGhie
Biological Chemistry and Bioactives
Group
The New Zealand Institute for Plant
and Food Research Ltd
Palmerston North, New Zealand

Maria Fernanda Nunez
Department of Nutritional Science
University of Toronto
Toronto, Ontario, Canada

Dan Peiffer
Department of Medicine
Medical College of Wisconsin
Milwaukee, Wisconsin

Ilya Raskin
School of Environmental
and Biological Sciences
Rutgers, The State University
of New Jersey
New Brunswick, New Jersey

David Ribnicky
School of Environmental
and Biological Sciences
Rutgers, The State University
of New Jersey
New Brunswick, New Jersey

Leonel E. Rojo
School of Environmental
and Biological Sciences
Rutgers, The State University
of New Jersey
New Brunswick, New Jersey

and

Universidad Arturo Prat
Iquique, Chile

Diana E. Roopchand
School of Environmental
and Biological Sciences
Rutgers, The State University
of New Jersey
New Brunswick, New Jersey

Daniel Rosenberg
Department of Genetics
and Developmental Biology
University of Connecticut
Farmington, Connecticut

Erika K. Ross
Department of Biological Sciences
Eleanor Roosevelt Institute
University of Denver
and
Research Service, Veterans Affairs
    Medical Center
Denver, Colorado

Nita Salzman
Department of Pediatrics
    and Gastroenterology
Medical College of Wisconsin
Milwaukee, Wisconsin

Claire Seguin
Comprehensive Cancer Center
Ohio State University
Columbus, Ohio

David E. Stevenson
Food and Wellness Group
The New Zealand Institute for Plant
    and Food Research Ltd
Palmerston North, New Zealand

Gary Stoner
Department of Medicine
Medical College of Wisconsin
Milwaukee, Wisconsin

Kristen Stoner
Comprehensive Cancer Center
Ohio State University
Columbus, Ohio

Francois Tremblay
Department of Ophthalmology
    and Visual Sciences
Dalhousie University
IWK Health Centre
Halifax, Nova Scotia, Canada

Taylor C. Wallace
Council for Responsible Nutrition
Washington, DC

Li-Shu Wang
Department of Medicine
Medical College of Wisconsin
Milwaukee, Wisconsin

Aimee N. Winter
Department of Biological Sciences
Eleanor Roosevelt Institute
University of Denver
Denver, Colorado

Xianli Wu
Hershey Center for Health
    and Nutrition
The Hershey Company
Hershey, Pennsylvania

Guang-Yu Yang
Department of Pathology
Northwestern University
Chicago, Illinois

Wencai Yang
Department of Pathology
University of Illinois at Chicago, Illinois
Chicago, Illinois

1 Anthocyanins in Health and Disease Prevention

Maria Fernanda Nunez and
Bernadene A. Magnuson

CONTENTS

With a growing body of evidence on the health benefits of fruits and vegetables, the old adage "An apple a day keeps the doctor away" no doubt sounds promising in today's climate of rising rates of diet-related chronic diseases and healthcare costs. Regular intake of colorful fruits and vegetables is an important component of a healthy lifestyle that can confer protection against chronic diseases—and with good reason. Low intake of this vibrant food group accounts for an estimated 1.7 million deaths globally including those caused by gastrointestinal cancer (14%), ischemic heart disease (11%), and stroke (9%) annually (World Health Organization, 2004). In support, epidemiological surveys suggest associations between high intakes of fruits and vegetables and lower incidences of chronic diseases such as cancer (Riboli and Norat, 2003) and cardiovascular disease (Liu et al., 2000; Bazzano et al., 2002). Such staggering figures have shaped dietary guidelines around the world recommending a daily intake between 5 and 10 servings of fruits and vegetables or 400 g (World Health Organization, 2004).

Although suboptimal fruit and vegetable intake is a global public health concern (Hall et al., 2009), consumers are steadily becoming more nutrition- and health-conscious (Agriculture and Agri-Food Canada, 2011), seeking foods that may provide them with beneficial physiological benefits beyond those related to energy and essential nutrients. With their reputation as nutritional cure-alls for a spectrum of diet-related diseases (Cooke et al., 2005), fruits and vegetables are at the heart of an active research field aimed at discovering the components in their phytochemical profiles that render them "superfoods." As researchers studying various chronic diseases investigate the impact of phytochemicals, it is becoming apparent that one class of compounds, the anthocyanins, are consistently reported to be beneficial for very diverse health conditions. These water-soluble compounds impart the characteristic red–blue color of many fruits and vegetables, including berries, grapes, red cabbage, and purple potatoes.

A fascinating tale of the wide scope and diversity of the biological properties of anthocyanins has begun to emerge in the literature, and will be presented in detail in the chapters that are to follow. Thus, the new version of the old adage may soon be "Binge on berries every day to keep the doctor away"! This introductory chapter provides a very brief taste of what the smorgasbord that follows will offer.

1.1 WHAT AND WHERE ARE ANTHOCYANINS? (CHAPTERS 1 AND 2)

Anthocyanins are among the most interesting and vigorously studied plant compounds, representing a large class of polyphenolic pigments within the flavonoid family that exist ubiquitously in the human diet (Williams and Grayer, 2004). As will be discussed in detail in Chapter 2, by Andersen and Jordheim, the chemistry of anthocyanins is not trivial. In addition to their multiple phenyl groups, anthocyanins are characterized by a sugar moiety typically conjugated to the C3 hydroxyl group in ring C, making them glycosides (Williams and Grayer, 2004). They exist in this form almost exclusively; their nonsugar (aglycon) counterparts known as anthocyanidins are poorly stable and thus rarely found in nature. To date, over 700 structurally distinct anthocyanin molecules have been identified (Andersen, 2012) but only six account for ~90% of those found in nature: cyanidin (50%), delphinidin (12%), malvidin (12%), pelargonidin (12%), peonidin (7%), and petunidin (7%) occurring as glycosides (Kong et al., 2003).

Anthocyanins are found in copious amounts in Rubus berries (red and black raspberries, blackberries), Vaccinium species (cranberries, blueberries), strawberries, cherries, and grapes (muscadines, Concord cultivar) (Wu et al., 2006; de Pascual-Teresa and Sanchez-Ballesta, 2008). Other dietary sources include red wine, and some leafy and root vegetables such as red cabbage, purple potatoes, and radish. Nearly four decades ago, the average daily intake of anthocyanins was estimated to range between 180 and 215 mg/day in the United States (Kuhnau, 1976), yet recent calculations state a much lower figure of approximately 12.5 mg daily (Wu et al., 2006) depending on dietary sources. Murphy and colleagues (2012) recently reported estimates of approximately 10 mg anthocyanidins/day for adults not meeting the recommended intakes of fruits and vegetables, as compared to 34.4–36.0 mg/day for men and women consuming recommended amounts of fruits and vegetables. Additional information on the sources of anthocyanins in the diet will also be discussed in Chapter 2.

Anthocyanins are present in different plant components, and their content varies between and among fruits of the same type (de Pascual-Teresa and Sanchez-Ballesta, 2008). Differences in growth environments, genetic factors, samples, cultivars, preparation, and extraction methods contribute to anthocyanin variability (de Pascual-Teresa and Sanchez-Ballesta, 2008), which may make it challenging to accurately determine their content and human intake. These water-soluble compounds are likely produced by plants to help protect them against oxidative stress, UV light, and additional environmental threats.

1.1.1 ANTHOCYANINS IN THE BODY (CHAPTERS 3 THROUGH 5)

McGhie and Stevenson present current understanding of the absorption, digestion, metabolism, and excretion of anthocyanins in Chapter 3. Overall, the degree of absorption of anthocyanins is low, but is highly dependent upon structure (Novotny et al., 2012). Unlike many glycosides that are completely hydrolyzed in the gut to eliminate their sugar moieties, thereby producing aglycones prior to absorption, studies have shown that intact anthocyanins are absorbed into the blood (Prior and Wu, 2006). Nonetheless, some studies suggest that some of the anthocyanins are hydrolyzed in the gut to anthocyanidins; either by gut flora or intestinal enzymes and that such microbial-derived anthocyanin metabolites are more stable and may be biologically active (Keppler and Humpf, 2005) than their intact parent molecules. Thus, is it the parent or the metabolite mediating the physiological effects observed *in vivo* following the consumption of anthocyanins?

To accurately determine the fate of anthocyanins in the body, validated methods for the detection, identification, and quantification of intact anthocyanins and their metabolites in tissues must be developed and employed in pharmacokinetic studies. Chapter 4 by Giusti will describe the current methodology for the assessment of anthocyanins in the blood, plasma, urine, and tissues.

As will be discussed by Wu in Chapter 5, anthocyanins are regarded as potent antioxidants and noted free radical scavengers with the potential to prevent or delay protein, lipid, or DNA oxidation (Wang et al., 1997) induced by reactive oxygen species (ROS) that otherwise overwhelm cells' natural antioxidant defenses and result in apoptosis. Under controlled conditions, particularly *in vitro* models, anthocyanin-rich foods and their extracts can counteract harmful oxidative effects induced by exposure to environmental stressors including UV radiation. For example, recently, a significant reduction in DNA damage and increased cell survival was observed in human fibroblasts exposed to UV-A radiation when preincubated with a strawberry anthocyanin-rich extract compared with control incubations (Giampieri et al., 2012). Structure–antioxidant activity patterns indicate that such potent antioxidant activity is associated with hydroxyl groups (Rice-Evans et al., 1996).

1.1.2 ANTHOCYANINS AND HEALTH

Given the biomedicinal reputation of other flavonoid groups (Erdman et al., 2007), anthocyanins have also garnered intense research interest to reveal their potential as preventative and therapeutic plant agents. Historically, anthocyanin-rich foods have been used in traditional medicines in many cultures. In "Anthocyanins in Health and Disease," a plethora of observational and clinical studies will be reviewed that highlight the health effects of anthocyanins in humans, supported by *in vitro* and animal studies revealing the potential cellular processes and gene regulatory changes responsible for these benefits. Anthocyanins are implicated in a number of biological pathways linked with protection against cardiovascular disease, metabolic syndrome, and type II diabetes, various cancers including in the gastrointestinal tract, vision problems, neurodegenerative diseases, aging skin, inflammation, and other medical conditions.

Anthocyanins are promising candidates accounting for many of the health benefits of fruits and vegetables observed in epidemiological studies but the diversity of their reported benefits is impressive, which begs the question: how can a single class of phytochemicals provide such robust protection against so many various illnesses? As the following chapters will show, the development of the aforementioned chronic diseases involves changes affecting a cluster of overlapping pathways. In turn, *in vitro* studies suggest several target mechanisms by which anthocyanin-rich extracts or whole foods can protect against disease, many of which are centered on the basic principle of oxidative stress and inflammation common in the etiology of major chronic diseases (Rosa et al., 2012).

1.1.3 ANTHOCYANINS AND DISEASE PREVENTION (CHAPTERS 6 THROUGH 8)

The anti-inflammatory and antioxidant effects of anthocyanins play a critical role in protection against cardiovascular disease as detailed by Wallace in Chapter 6 "Anthocyanins in Cardiovascular Disease." A large body of evidence supports the role of anthocyanins in scavenging free radicals, which would potentially prevent the oxidation of LDL cholesterol, reduce foam cell formation, and minimize vascular dysfunction. Anthocyanins can additionally scavenge peroxynitrite, a product of the signaling molecule nitric oxide that normally promotes vasodilation and exerts antiatherogenic actions; however, peroxynitrite can be toxic if present in excess amounts (Kim et al., 2004). Anthocyanins, which have been found to accumulate in the tissue of pigs after long-term intake (Kalt et al., 2008), thereby affect several mechanisms related to the development of cardiovascular disease, including the prevention of ischemic damage observed in animals provided a high-anthocyanin diet (Ahmet et al., 2009). It must be noted, however, that although results from cell and animal studies provide evidence in support of benefits, it is challenging to replicate in humans.

Inflammation is also associated with several diseases including obesity and type II diabetes as discussed by Raskin and colleagues in Chapter 7 "Anthocyanins in the Prevention of Metabolic Syndrome and Type-2 Diabetes." Inflammation, a natural and important component of human innate immunity, is in fact a protective response that helps create a barrier against harmful stimuli and initiate recovery processes to heal tissues. However, when excessive or uncontrolled, proinflammatory activities can damage normal biological systems. Adipose tissue is a highly active organ that secretes a range of peptides like the proinflammatory adipokines, TNFα, and C-reactive protein (CRP), into circulation, which in turn creates a low-grade chronic inflammatory environment that promotes impaired glucose tolerance, insulin resistance, and atherogenic events. In addition to reducing markers of lipid oxidation, anthocyanin-rich fruits like strawberries and chokeberries have been found to decrease CRP, IL-6, and TNFα in animal studies. Raskin and coauthors (Chapter 7) will review the intriguing animal studies that have shown that supplementation of purified anthocyanins, but not anthocyanin-rich foods, can be beneficial for body weight maintenance. Clearly, with the rates of obesity and diabetes escalating worldwide, this research area has significant public health implications.

The potential for anthocyanins and anthocyanin-rich extracts to prevent cancer development has been studied for over 30 years, and hundreds of publications can be found in the literature. Anthocyanins, anthocyanidins, and anthocyanin-rich extracts from a variety of sources have been shown to prevent the growth of human cancer cells from leukemia, breast, prostate, uterine, lung, vulva, stomach, melanoma, and colon cancer (Cooke et al., 2005). Animal studies using chemically and genetically induced tumors of the breast, skin, stomach, esophagus, and colon have all shown evidence of prevention by anthocyanins (Kocic et al., 2011). These chemopreventive effects may result from the modulation of cell signaling pathways implicated in the proliferation, differentiation, and survival of cells (e.g., downregulation of NF-kB activation) (Hou et al., 2004); and perhaps through the physical integration of anthocyanins in cellular components (e.g., membrane) (Giampieri et al., 2012), which helps protect against oxidative stress in endothelial cells (Youdim et al., 2002).

Among the most promising are studies of the prevention of cancers of the gastrointestinal (GI) tract, including oral, stomach, esophagus, and colon. Evidence for, and the molecular pathways involved in, the prevention of the cancers in the GI tract is the focus of Wang and colleagues in Chapter 8 "Anthocyanins, Anthocyanin Derivatives, and Colorectal Cancer." Mechanisms include promoting/accelerating apoptosis, inhibiting angiogenesis, interference of cell cycle progression, and minimizing cancer-induced DNA damage, as evidenced by changes such as a reduction in urinary levels of 8-hydroxy-2'-deoxyguanosine.

In addition to prevention of cancer, a role of anthocyanins in the treatment of existing cancers as an adjunct to chemotherapy has promise. Following studies showing that anthocyanin-rich extracts inhibited the growth (Zhao et al., 2004) and cell cycle (Malik et al., 2003) of colon cancer, but not normal colon cells, and inhibition of numerous biomarkers of colon cancer in a rat model (Lala et al., 2006), we investigated whether pretreatment of rats with anthocyanin-rich extracts would offer protection against the known adverse side effects of 5-flurouracil (5-Fu) used for cancer treatments. We observed that chokeberry anthocyanin-rich extract enhanced the growth inhibition by 5-Fu in colon cancer cell lines HT29 and SW620, but had little effect on normal colon cells at the same concentration. Doses of 200 mg/kg 5-Fu resulted in severe gastrointestinal toxicity and myleotoxicity in the rats. Prior treatment with the extract partially reduced the gastrointestinal damage caused by 5-Fu, suggesting that anthocyanin-rich extracts may be a beneficial adjuvant for 5-Fu chemotherapy (Su and Magnuson, unpublished). Similar findings have been reported with bilberry anthocyanin-rich extract and 5-Fu (Choi et al., 2007). Recently, however, a reduced number of doxorubicin-mediated DNA strand breaks in colon cancer cells during coincubation with anthocyanins was considered as a possible interference of chemotherapy efficacy by anthocyanins (Esselen et al., 2011). Clearly, additional studies are required in tumor-bearing animals to determine if the feasibility of improvement of chemotherapy through the reduction of adverse side effects by anthocyanins.

1.2 ANTHOCYANINS AND HEALTH PROMOTION (CHAPTERS 9 THROUGH 12)

Consumers are looking not only to prevent disease, but also to achieve and maintain optimal performance of mental and physical state, as they move into older age and have the desire to have active and enjoyable lifestyles.

In Chapter 9, "Anthocyanins in Visual Performance and Ocular Diseases," Tremblay and Kalt reveal data, which suggest that anthocyanins cross the blood–brain barrier after long-term feeding giving them the opportunity to play an active role in maintaining normal vision and preventing ocular pathologies. Similarly, in Chapter 10, "Effects of Anthocyanins on Neuronal and Cognitive Brain Functions," Linseman highlights how anthocyanins can help protect against neurodegenerative diseases, which also share common elements of oxidative stress, excitotoxicity, neuroinflammation, and cellular apoptosis. Improvement in memory and potential protection against Alzheimer's and Parkinson's diseases with anthocyanins are also discussed.

Rojo and colleagues address additional protective effects against skin aging in Chapter 11, "Anthocyanins in Cosmeceuticals and Skin Health," an area of study that has grown to prominence and reflects consumers' increasing concern with conserving their dermatological health. A significant catalyst is growing demand for natural and efficacious cosmetic ingredients that can mitigate the appearance of aging skin (Baumann et al., 2009) and, more critically, help confer protection against cutaneous malignancies that account for up to 40% of newly diagnosed cancers in the United States alone (Afaq et al., 2005).

As examined by Hurst and Hurst in Chapter 12 "Anthocyanins, Innate Immunity, and Exercise," exercise also triggers an acute inflammatory response and oxidative stress through the production of ROS and reactive nitrogen species (RNS). Compared with sedentary individuals, people who regularly engage in exercise have stronger innate immune responses, which include decreased susceptibility to common infections and reduced recovery time from tissue damage. However, these health benefits are affected by the duration, type, and intensity of exercise and have been found to be hormetic, with strenuous exercise in fact reversing the positive effects on immunity and increasing susceptibility to pathogens during a "window of infection risk" 3–72 h postexercise. Given their antioxidant dependent and independent properties, anthocyanins can therefore play a pivotal role in facilitating immune recovery by thwarting the oxidative stress produced after intense exercise. The effects of anthocyanin consumption on innate immunity postexercise are not fully known but studies suggest that intake of berries before exercise attenuates markers of oxidative stress and increases anti-inflammatory markers following workouts. Furthermore, as the authors will explore in more detail, anthocyanins may complement the health benefits associated with frequent moderate exercise by upregulating cytoprotective signaling pathways.

1.3 BEYOND ANTIOXIDANT ACTIVITY

An exciting emerging trend is that anthocyanins can also exert effects that alter specific pathways of certain diseases and heath conditions that are not just limited to antioxidant activities. For instance, anthocyanins can prevent fluctuations in intracellular calcium concentrations during aging, which is critical for suppressing excitotoxicity and preventing neuronal apoptosis in relevant neurodegenerative diseases (Ward et al., 2000). Bilberry anthocyanins may help improve insulin sensitivity by increasing AMPK (AMP-activated protein kinase) phosphorylation, which is an important step in cellular energy homeostasis that leads to the activation of catabolic pathways (i.e., fatty acid oxidation) and inactivation of anabolic pathways (i.e., gluconeogenesis, lipogenesis) (Takikawa et al., 2010). In adults with type II diabetes, sour cherries can also exert antidiabetic effects including decreasing HbA1C, a clinical biomarker of long-term glucose control (Mulabagal et al., 2009), although the exact mechanisms have yet to be fully determined. In studies on the chemopreventive effects of anthocyanins, multiple cellular targets have been identified including: inhibition of cyclooxygenase enzymes (COX); activation of IKK and NFkB; inhibition of epidermal growth factor receptor (EGFR) tyrosine kinase activity; inhibition of activator protein 1 (AP1); and inhibition of ERK and JNK phosphorylation (Cooke et al., 2005). As this volume will illustrate, there is much more to anthocyanins than antioxidant activity.

1.4 CONSIDERATIONS AND QUESTIONS

Studies have frequently assessed the antioxidant capacity of anthocyanins *in vitro*; the available *in vivo* studies generally measured total antioxidant capacity in plasma after consumption of a source rich in anthocyanins. Although a positive relationship between total serum anthocyanin concentration and antioxidant capacity in humans has been reported in some studies, it is not yet established if the same dosages and extracts tested in cell culture studies can exert comparable effects *in vivo,* which may challenge the relevance of testing such high doses of anthocyanins in human studies. Moreover, during the evaluation of potential health claims based on antioxidant activity, it has been questioned if high antioxidant capacity in plasma is sufficient to predict protection against oxidative damage and thereby support a cause and effect relationship between the intake of a food rich in antioxidants including anthocyanins and beneficial health outcomes (EFSA Panel on Dietetic Products, Nutrition and Allergies, 2010).

Validation of biomarkers of anthocyanin intake and distribution is a critical need to further our understanding of the role of these compounds in health and disease. Ten different anthocyanin metabolites have been detected in human circulation after intake of anthocyanin-rich extracts, with the most common ones being phenolic acids and aldehydes (Kay et al., 2005; Forester and Waterhouse, 2010). Anthocyanin conjugates have also been found including sulfated, methylated, glucuronidated, and glycosylated forms (Gonthier et al., 2003; Manach et al., 2005). Anthocyanins have been detected at low concentrations in circulation (McGhie and Walton, 2007), yet they have been found to accumulate in animal tissues after long-term feeding. Further research is needed to determine if they accumulate in specific sites (i.e., cardiac or vascular tissues) and thereby prevent disease compared with acting while in circulation. As will be discussed in future chapters, although expensive, studies with 13C or 14C isotope-labeled anthocyanins can help elucidate the bioavailability, metabolism, tissue distribution, and mode of action of these compounds.

Because of their varying chemical structures and sources, different types of anthocyanins may have varying degrees of effectiveness. As discussed in Chapter 1, the chemical structure of anthocyanins including the position, number, and types of substitutions influences their biological activities. Antiproliferative activity is related to the anthocyanin structure, such that activity is more potent with the presence of certain moieties on ring B of the molecule (i.e., hydroxyl or methoxyl groups). Hydroxyl groups on anthocyanidins conferred higher proapoptotic activity in human leukemia cells. Anthocyanidins may have greater potency than glycosylated versions, to inhibit the growth of embryonic fibroblasts and of malignant human cancer cells (Hou et al., 2004). Delphinidin possessed the highest grown-inhibitory activity, which gives a blue–red color to Cabernet Sauvignon grapes, cranberries, Concord grapes, and pomegranates and is characterized by two "OH groups" (C3 and C5 of ring B).

As will be illustrated in the many studies to be discussed in the following chapters, delineating the active compound or combination of compounds in extracts and foods can be challenging, as, sometimes, beneficial effects are greater with purified compounds and/or extracts; yet, often consumers are eating the whole food. The presence of other polyphenols and compounds in whole foods or anthocyanin-rich compounds may obscure the effects of anthocyanins; thus, studies utilizing highly purified anthocyanins can be preferred. Different anthocyanins may act synergistically between themselves or with other polyphenols present in fruits and vegetables. These studies are not feasible, however, until economical sources of large amounts of purified anthocyanins become available. Emerging work includes the development of chemically stable and clinically effective anthocyanin-rich formulations for a variety of therapeutic applications. Roopchand and colleagues have found that anthocyanins and other polyphenols can be efficiently separated from highly polar carbohydrates, bound, concentrated, and stabilized into protein-rich, food matrixes, such as defatted soybean flour (DSF) and soy protein isolate (SPI), while preserving their pharmacological effects (Roopchand et al., 2012). Stabilizing anthocyanins by electrostatically binding them to protein matrices may provide another strategy for protecting their structural integrity, function, and color.

Last, human clinical trials remain the gold standard for the demonstration of efficacy of treatments, for either health promotion or disease prevention. Demonstration of efficacy must consider the characteristics of the test population as compared to the general population, the duration of treatment and effects, the strength, consistency, specificity, and dose–response of the effect, the biological plausibility of relationship, and, finally, the source and characteristics of the test compound (purified compound, extracts, or whole foods). Based on the exciting developments to be presented in the chapters of "Anthocyanins in Health and Disease Prevention," more clinical trials are most likely.

In conclusion, the world of anthocyanin research continues to grow and diversify to all aspects of health and disease, encompassing a wide array of diverse foods. The new version of the old adage may well soon be *"Binge on berries, pop in purple potatoes, revel with radishes, grab the grapes, and wind down with red wine."* Enjoy!

REFERENCES

Afaq, F., Adhami, V.M., and Mukhtar, H. 2005. Photochemoprevention of ultraviolet B signaling and photocarcinogenesis. *Mutation Research/Fundamental and Molecular Mechanisms of Mutagenesis* **571**(1–2): 153–173. PMID:15748645.

Agriculture and Agri-Food Canada. 2011. Agri-Food Trade Service: Health and wellness trends for Canada and the world [online]. Available from http://www.ats-sea.agr.gc.ca/inter/4367-eng.htm [accessed 21 May 2012].

Ahmet, I., Spangler, E., Shukitt-Hale, B., Juhaszova, M., Sollott, S.J., Joseph, J.A., Ingram, D.K., and Talan, M. 2009. Blueberry-enriched diet protects rat heart from ischemic damage. *PLoS One* **4**(6): e5954. doi:10.1371/journal.pone.0005954. PMID:19536295

Andersen, Ø. M. 2012. Personal database on anthocyanins.

Baumann, L., Woolery-Lloyd, H., and Friedman, A. 2009. "Natural" ingredients in cosmetic dermatology. *J. Drugs Dermatol.* **8**(6 Suppl): s5–s9. PMID:19562883.

Bazzano, L.A., He, J., Ogden, L.G., Loria, C.M., Vupputuri, S., Myers, L., and Whelton, P.K. 2002. Fruit and vegetable intake and risk of cardiovascular disease in US adults: The first National Health and Nutrition Examination Survey Epidemiologic Follow-up Study. *Am. J. Clin. Nutr.* **76**(1): 93–99. PMID:12081821.

Choi, E.H., Ok, H.E., Yoon, Y., Magnuson, B.A., Kim, M.K., and Chun, H.S. 2007. Protective effect of anthocyanin-rich extract from bilberry (*Vaccinium myrtillus* L.) against myelotoxicity induced by 5-fluorouracil. *Biofactors* **29**(1): 55–65. PubMed PMID: 17611294.

Cooke, D., Steward, W.P., Gescher, A.J., and Marczylo, T. 2005. Anthocyans from fruits and vegetables—Does bright colour signal cancer chemopreventive activity? *Eur. J. Cancer* **41**(13): 1931–1940. doi:10.1016/j.ejca.2005.06.009. PMID:16084717.

de Pascual-Teresa, S. and Sanchez-Ballesta, M.T. 2008. Anthocyanins: From plant to health. *Phytochem. Rev.* **7**: 281–299.

EFSA Panel on Dietetic Products, Nutrition and Allergies. 2010. Scientific opinion on the substantiation of health claims related to various food(s)/food constituent(s) and protection of cells from premature aging, antioxidant activity, antioxidant content and antioxidant properties, and protection of DNA, proteins and lipids from oxidative damage pursuant to Article 13(1) of Regulation(EC) No 1924/2006. *EFSA Journal* **8**(2): 1489–1552.

Erdman, J.W., Jr, Balentine, D., Arab, L., Beecher, G., Dwyer, J.T., Folts, J., Harnly, J. et al. 2007. Flavonoids and heart health. *Proceedings of the ILSI North America Flavonoids Workshop*, May 31–June 1, 2005, Washington, DC. *J. Nutr.* **137**(3 Suppl 1): 718S–737S. PMID:17311968.

Esselen, M., Fritz, J., Hutter, M., Teller, N., Baechler, S., Boettler, U., Marczylo, T.H., Gescher, A.J., and Marko, D. 2011. Anthocyanin-rich extracts suppress the DNA-damaging effects of topoisomerase poisons in human colon cancer cells. *Mol. Nutr. Food Res.* **55**(Suppl 1): S143–S153. doi: 10.1002/mnfr.201000315. Epub 2011 Jan 31. PubMed PMID: 21280204.

Forester, S.C. and Waterhouse, A.L. 2010. Gut metabolites of anthocyanins, gallic acid, 3-O-methylgallic acid, and 2,4,6-trihydroxybenzaldehyde, inhibit cell proliferation of Caco-2 cells. *J. Agric. Food Chem.* **58**(9): 5320–5327. doi:10.1021/jf9040172. PMID:20373763.

Giampieri, F., Alvarez-Suarez, J.M., Tulipani, S., Gonzales-Paramas, A.M., Santos-Buelga, C., Bompadre, S., Quiles, J.L., Mezzetti, B., and Battino, M. 2012. Photoprotective potential of strawberry (*Fragaria × ananassa*) extract against UV-A irradiation damage on human fibroblasts. *J. Agric. Food Chem.* **60**(9): 2322–2327. doi:10.1021/jf205065x. PMID:22304566.

Gonthier, M.P., Cheynier, V., Donovan, J.L., Manach, C., Morand, C., Mila, I., Lapierre, C., Remesy, C., and Scalbert, A. 2003. Microbial aromatic acid metabolites formed in the gut account for a major fraction of the polyphenols excreted in urine of rats fed red wine polyphenols. *J. Nutr.* **133**(2): 461–467. PMID:12566484.

Hall, J.N., Moore, S., Harper, S.B., and Lynch, J.W. 2009. Global variability in fruit and vegetable consumption. *Am. J. Prev. Med.* **36**(5): 402–409.e5. doi:10.1016/j.amepre.2009.01.029. PMID:19362694.

Hou, D.X., Fujii, M., Terahara, N., and Yoshimoto, M. 2004. Molecular mechanisms behind the chemopreventive effects of anthocyanidins. *J. Biomed. Biotechnol.* **2004**(5): 321–325. doi:10.1155/S1110724304403040. PMID:15577196.

Kalt, W., Blumberg, J.B., McDonald, J.E., Vinqvist-Tymchuk, M.R., Fillmore, S.A., Graf, B.A., O'Leary, J.M., and Milbury, P.E. 2008. Identification of anthocyanins in the liver, eye, and brain of blueberry-fed pigs. *J. Agric. Food Chem.* **56**(3): 705–712. doi:10.1021/jf071998l. PMID:18211026.

Kay, C.D., Mazza, G.J., and Holub, B.J. 2005. Anthocyanins exist in the circulation primarily as metabolites in adult men. *J. Nutr.* **135**(11): 2582–2588. PMID:16251615.

Keppler, K. and Humpf, H.U. 2005. Metabolism of anthocyanins and their phenolic degradation products by the intestinal microflora. *Bioorg. Med. Chem.* **13**(17): 5195–5205. PMID:15963727.

Kim, J.Y., Jung, K.J., Choi, J.S., and Chung, H.Y. 2004. Hesperetin: A potent antioxidant against peroxynitrite. *Free Radic. Res.* **38**(7): 761–769. PMID:15453641.

Kocic, B., Filipovic, S., Nikolic, M., and Petrovic, B. 2011. Effects of anthocyanins and anthocyanin-rich extracts on the risk for cancers of the gastrointestinal tract. *J. Boun.* **16**(4): 602–8. PubMed PMID: 22331709.

Kong, J.M., Chia, L.S., Goh, N.K., Chia, T.F., and Brouillard, R. 2003. Analysis and biological activities of anthocyanins. *Phytochemistry* **64**(5): 923–933. PMID:14561507.

Kuhnau, J. 1976. The flavonoids. A class of semi-essential food components: Their role in human nutrition. *World Rev. Nutr. Diet.* **24**: 117–191. PMID:790781.

Lala, G., Malik, M., Zhao, C., He, J., Kwon, Y., Giusti, M.M., and Magnuson, B.A. 2006. Anthocyanin-rich extracts inhibit multiple biomarkers of colon cancer in rats. *Nutr. Cancer* **54**(1): 84–93. PubMed PMID: 16800776.

Liu, S., Manson, J.E., Lee, I.M., Cole, S.R., Hennekens, C.H., Willett, W.C., and Buring, J.E. 2000. Fruit and vegetable intake and risk of cardiovascular disease: The Women's Health Study. *Am. J. Clin. Nutr.* **72**(4): 922–928. PMID:11010932.

Malik, M., Zhao, C., Schoene, N., Guisti, M.M., Moyer, M.P., and Magnuson, B.A. 2003. Anthocyanin-rich extract from *Aronia meloncarpa* E induces a cell cycle block in colon cancer but not normal colonic cells. *Nutr. Cancer* **46**(2): 186–96. PubMed PMID: 14690795.

Manach, C., Williamson, G., Morand, C., Scalbert, A., and Remesy, C. 2005. Bioavailability and bioefficacy of polyphenols in humans. I. Review of 97 bioavailability studies. *Am. J. Clin. Nutr.* **81**(1 Suppl): 230S–242S. PMID:15640486.

McGhie, T.K. and Walton, M.C. 2007. The bioavailability and absorption of anthocyanins: Towards a better understanding. *Mol. Nutr. Food Res.* **51**(6): 702–713. doi:10.1002/mnfr.200700092. PMID:17533653.

Mulabagal, V., Lang, G.A., DeWitt, D.L., Dalavoy, S.S., and Nair, M.G. 2009. Anthocyanin content, lipid peroxidation and cyclooxygenase enzyme inhibitory activities of sweet and sour cherries. *J. Agric. Food Chem.* **57**(4): 1239–1246. doi:10.1021/jf8032039. PMID:19199585.

Murphy, M.M., Barraj, L.M., Herman, D., Bi, X., Cheatham, R., Randolph, R.K. 2012. Phytonutrient intake by adults in the United States in relation to fruit and vegetable consumption. *J. Acad. Nutr. Diet* **112**(2): 222–229.

Novotny, J.A., Beverly, A., Clevidenc, B.A., and Kurilich, A.C. 2012. Anthocyanin kinetics are dependent on anthocyanin structure. *Br. J. Nutr.* **107**: 504–509.

Prior, R.L. and Wu, X. 2006. Anthocyanins: Structural characteristics that result in unique metabolic patterns and biological activities. *Free Radic. Res.* **40**(10): 1014–28.

Riboli, E. and Norat, T. 2003. Epidemiologic evidence of the protective effect of fruit and vegetables on cancer risk. *Am. J. Clin. Nutr.* **78**(3 Suppl): 559S–569S. PMID:12936950.

Rice-Evans, C.A., Miller, N.J., and Paganga, G. 1996. Structure-antioxidant activity relationships of flavonoids and phenolic acids. *Free Radic. Biol. Med.* **20**(7): 933–956. PMID:8743980.

Roopchand, D.E., Grace, M.H., Kuhn, P., Cheng, D.M., Plundrich, N., Poulev, A., Howell, A., Fridlender, B., Lila, M.A., and Raskin, I. 2012. Efficient sorption of polyphenols to soybean flour enables natural fortification of foods. *Food Chem.* **131**(4): 1193–1200.

Rosa, F.T., Zulet, M.A., Marchini, J.S., and Martinez, J.A. 2012. Bioactive compounds with effects on inflammation markers in humans. *Int. J. Food Sci. Nutr.* **63**(6): 749–765. doi: 10.3109/09637486.2011.649250. PMID:22248031.

Takikawa, M., Inoue, S., Horio, F., and Tsuda, T. 2010. Dietary anthocyanin-rich bilberry extract ameliorates hyperglycemia and insulin sensitivity via activation of AMP-activated protein kinase in diabetic mice. *J. Nutr.* **140**(3): 527–533. doi:10.3945/jn.109.118216. PMID:20089785.

Wang, H., Cao, G., and Prior, R.L. 1997. Oxygen radical absorbing capacity of anthocyanins. *J. Agric. Food. Chem.* **45**(2): 304–309.

Ward, M.W., Rego, A.C., Frenguelli, B.G., and Nicholls, D.G. 2000. Mitochondrial membrane potential and glutamate excitotoxicity in cultured cerebellar granule cells. *J. Neurosci.* **20**(19): 7208–7219. PMID:11007877.

Williams, C.A. and Grayer, R.J. 2004. Anthocyanins and other flavonoids. *Nat. Prod. Rep.* **21**(4): 539–573. doi:10.1039/b311404j. PMID:15282635.

World Health Organization. 2004. Promoting fruit and vegetable consumption around the world. *Global Strategy on Diet, Physical Activity and Health.* [online]. Available from http://www.who.int/dietphysicalactivity/fruit/en/index2.html [accessed 21 May 2012].

Wu, X., Beecher, G.R., Holden, J.M., Haytowitz, D.B., Gebhardt, S.E., and Prior, R.L. 2006. Concentrations of anthocyanins in common foods in the United States and estimation of normal consumption. *J. Agric. Food Chem.* **54**(11): 4069–4075. doi:10.1021/jf060300l. PMID:16719536.

Youdim, K.A., Spencer, J.P., Schroeter, H., and Rice-Evans, C. 2002. Dietary flavonoids as potential neuroprotectants. *Biol. Chem.* **383**(3–4): 503–519. doi:10.1515/BC.2002.052. PMID:12033439.

Zhao, C., Giusti, M.M., Malik, M., Moyer, M.P., and Magnuson, B.A. 2004. Effects of commercial anthocyanin-rich extracts on colonic cancer and nontumorigenic colonic cell growth. *J Agric Food Chem.* **52**(20): 6122–6128. PubMed PMID: 15453676.

2 Basic Anthocyanin Chemistry and Dietary Sources

Øyvind M. Andersen and Monica Jordheim

CONTENTS

2.1   INTRODUCTION

The last two decades have witnessed in several areas a renaissance in research activities on anthocyanins, mainly related to their potential health-promoting properties, their use as natural food colorants, and their appearance in cultivars and plant mutants with new colors and color patterns. Today, there exists convincing scientific evidence in support of the association between diet and chronic diseases. As a consequence, dietary guidelines have been formulated around the world for general prevention of cancer, cardiovascular diseases, diabetes, osteoporosis, and so on, incorporating the recommendation to increase the consumption of plant-based foods that are good sources of biologically active phytochemicals, also including anthocyanins and other flavonoids. So, *if* intake of anthocyanins has positive health effect(s), and *if* the various anthocyanins or their derivatives in the human body have different properties, then of course both the qualitative and quantitative anthocyanin content of our food as well as the *individual chemistry* of these compounds should be more closely considered.

The total number of anthocyanin structures identified after isolation from plant extracts is 702 (Andersen 2012). In addition, nearly 200 different anthocyanins have been presented with tentative structures. Most of the literatures written in the English language reporting major anthocyanins and quantitative anthocyanin content of fruits and vegetables used in the human diet have been summarized in Tables 2.3 and 2.4. The dietary sources presented here are thus most representative for typical consumption in Europe, America, and Australia, and to a lesser degree for people living in Asia and Africa. As seen in Tables 2.3 and 2.4, the range of anthocyanins in the diet is limited to altogether 96 different structures. There exists a very noticeable difference between the qualitative anthocyanin content of fruits and vegetables (Section 2.3.2), which might influence how we design our diet in the future!

This chapter will to some extent (Section 2.4.1) describe how genetic relationship influences the anthocyanin content of some fruits and vegetables, both on family, genus, cultivar, and population level. It will further indicate how environmental factors and agricultural practice influence the anthocyanin content of the various sources used in our diet (Section 2.4.2). Many studies have focused on elucidating these factors with the purpose of obtaining increased yield of anthocyanins in food products. In Section 2.6, the importance of careful interpretation of experimental anthocyanin data is stressed.

The anthocyanins, which constitute a major flavonoid group, are water-soluble pigments synthesized by plants. They consist of an aglycone (anthocyanidin), sugar unit(s) and in many cases acyl group(s). A description of the structural units of the various types of anthocyanins is presented in Section 2.2, with main focus on anthocyanin structures found in fruits and vegetables of the human diet. As basis for understanding possible chemical transformations and properties of anthocyanins, the reactivity of the various parts of the anthocyanins is treated in some detail in Section 2.2.5. In this context, some anthocyanidin derivatives like pyranoanthocyanins and heterodimer anthocyanins formed during storage and processing in plant-derived foods including wines have been described, together with some considerations related to stability of anthocyanins. The exact mechanisms for degradation and derivative formation of anthocyanins in natural systems, whether it is intact plants, processed plant products, or human metabolism, have been difficult to establish due to the complexity of the chemical reactions taking place. In Section 2.5.1, three different thermal degradation pathways of anthocyanins under pH influence in model systems have been summarized, while anthocyanin degradation products and phase II metabolites detected in humans and animals have been described in Sections 2.5.4 and 2.5.3, respectively.

Anthocyanins are outstanding compounds by the way their aglycone parts (anthocyanidins) are involved in a series of equilibriums, which are particularly dependent on pH, giving rise to several forms (secondary structures). These forms have different chemical and possibly nutraceutical properties, however, their exact nature is poorly known, with the exception of the flavylium cation form. For instance, complete structural assignments of the quinonoidal form(s) have been incomplete for all anthocyanins. In Section 2.2.3, the secondary anthocyanidin structures are treated with some details.

The major aim of this chapter is to show how the anthocyanins in fruits, vegetables, and products thereof vary substantially with respect to structures and quantities, with serious consequences for anthocyanin reactivity, stability, and bioavailability, including the formation of anthocyanin degradation products and phase II metabolites. With the stage set by a probable difference in distribution of secondary anthocyanidin forms under *in vivo* conditions, depending on the type of anthocyanin in the diet, the chemistry of individual anthocyanins within the human body is extraordinary, largely unknown, however, likely to be prosperous when understood!

2.2 ANTHOCYANIN STRUCTURES, STABILITY, AND REACTIVITY

2.2.1 ANTHOCYANIN OCCURRENCES

Anthocyanin compounds are the major reason for most cyanic colors ranging from salmon, pink through red, and violet to dark blue of most fruits, vegetables, and flowers among angiosperms (Andersen and Jordheim 2010). They are sometimes present in other plant tissues, and are also found in various gymnosperms, ferns, and some bryophytes. In ferns and bryophytes they occur as deoxyanthocyanins and deoxyanthocyanidins (Table 2.1), which are located in the cell walls. In fruits, the anthocyanins are most prevalent in the epidermal and hypodermal layers of the skin,

TABLE 2.1

Structures of Naturally Occurring Anthocyanidins Based on the Same C$_{15}$-Skeleton, 2-Phenylbenzopyrylium[a]

Anthocyanidins	Substitution Pattern						
	3	5	6	7	3'	4'	5'
Common Anthocyanidins							
Pelargonidin (Pg)	OH	OH	H	OH	H	OH	H
Cyanidin (Cy)	OH	OH	H	OH	H	OH	H
Delphinidin (Dp)	OH	OH	H	OH	OH	OH	OH
Peonidin (Pn)	OH	OH	H	OH	OMe	OH	H
Petunidin (Pt)	OH	OH	H	OH	OMe	OH	OH
Malvidin (Mv)	OH	OH	H	OH	OMe	OH	OMe
A-Ring Methoxylated Anthocyanidins							
5-O-MethylCy	OH	OMe	H	OH	OH	OH	H
7-O-MethylCy	OH	OH	H	OMe	OH	OH	H
7-O-MethylPn (Rosinidin)	OH	OH	H	OMe	OMe	OH	H
7-O-MethylDp	OH	OH	H	OMe	OH	OH	OH
7-O-MethylPt	OH	OH	H	OMe	OMe	OH	OH
7-O-MethylMv (Hirsutidin)	OH	OH	H	OMe	OMe	OH	OMe
5,7-Di-O-methylDp (Pulchellidin)[b]	OH	OMe	H	OMe	OH	OH	OH
5,7- Di-O-methylPt (Europinidin)[b]	OH	OMe	H	OMe	OMe	OH	OH
5,7- Di-O-methylMv (Capensinidin)[b]	OH	OMe	H	OMe	OMe	OH	OMe
6-Hydroxylated Anthocyanidins							
6-HydroxyPg	OH	OH	OH	OH	H	OH	H
6-HydroxyCy	OH	OH	OH	OH	OH	OH	H
6-HydroxyDp	OH	OH	OH	OH	OH	OH	OH
3-Deoxyanthocyanidins							
Apigeninidin (Ap)	H	OH	H	OH	H	OH	H
Luteolinidin (Lt)	H	OH	H	OH	OH	OH	H
Tricetinidin (Tr)	H	OH	H	OH	OH	OH	OH
7-O-MethylAp	H	OH	H	OMe	H	OH	H
5-O-MethylLt	H	OMe	H	OH	OH	OH	H
6-Hydroxy-5-O-methylAp (Carajurone)	H	OMe	OH	OH	H	OH	H
6-Hydroxy-5,4'-di-O-methylAp (Carajurin)	H	OMe	OH	OH	H	OMe	H
6-Hydroxy-5-O-methylLt	H	OMe	OH	OH	OH	OH	H
6-Hydroxy-5,4'-di-O-methylLt	H	OMe	OH	OH	OH	OMe	H

[a] See Figure 2.1 for anthocyanidins with extended C15-skeletons; Riccionidins A and B, sphagnoru-
     bins A-C, pyranoanthocyanidins including rosacyanins A1, A2, and B and heterodimers.

[b] Revised structure (Skaar et al. 2012).

or throughout the fruit as in many berries. In leaves, the anthocyanins may be found in upper or lower epidermis, and in palisade or spongy mesophyll.

Anthocyanins are within the cells often found dissolved uniformly in vacuolar solutions, typically seen in anthocyanin-rich berries like raspberries. At the subcellular level in maize (*Zea mays*), it has been found that light induces an alteration in the way the anthocyanins were distributed within vacuolar compartments (Irani and Grotewold 2005). Some anthocyanins have been reported to occur in intensively colored intravascular bodies called AVIs (anthocyanic vacuolar inclusions), mainly in flowers (Markham et al. 2000, Zhang et al. 2006). However, similar structures have been found for instance in leaves of red cabbage (Small and Pecket 1982, Nozzolillo et al. 1995), in grapevine in cell cultures (Cormier et al. 1997, Conn et al. 2010), and in the tubers of sweet potato in suspension culture (Nozue et al. 1997). Anatomical observations of anthocyanin-rich cells in apple skin carried out by light and electron microscopy showed that the skin with fully developed red color had more layers of anthocyanin-containing epidermal cells than those of green skin (Bae et al. 2006). The anthocyanins were frequently found in clusters or in agglomerations that were round in shape in the epidermal cells of the red skin, however, there was no distinct envelope membrane on the anthocyanin granule in the vacuoles. It has generally been indicated that the anthocyanin-containing globular inclusions may be protein matrices, and that they possess neither a membrane boundary nor an internal structure. However, in a recent study, no unique protein component was detected in AVI's in grape cell suspension cultures (Conn et al. 2010). In two lines of cell suspension cultures of grapevine (*Vitis vinifera*), AVIs appeared as dark red to purple spheres of various sizes in vacuoles, due to their interaction with anthocyanins (Conn et al. 2003). Compared with the total anthocyanin profile, the profile of the AVI-bounded anthocyanins showed an increase of approximately 28–29% in acylated (*p*-coumarylated) anthocyanins in both lines. In sorghum (*Sorghum bicolor*) 3-deoxyanthocyanidins accumulate as inclusions in leaf cells under fungal attack, and function as phytoalexins by inhibiting infection in a site-specific response (Snyder and Nicholson 1990).

Although it is generally accepted that anthocyanins as other flavonoids are synthesized on the cytoplasmic surface of the endoplasmic reticulum membrane, the mechanisms for transportation and anthocyanin accumulation in the cells are more indecisive and under elucidation (Pourcel et al. 2012).

2.2.2 NATURAL ANTHOCYANIDINS (PRIMARY STRUCTURES)

Altogether 27 different anthocyanidins based on the same carbon-C_{15} skeleton have been isolated from plants (Table 2.1). In addition, at least 19 anthocyanidins with extended C15 skeletons, such as riccionidins A (**1**) and B, sphagnorubins A–C (**2–4**), flavanol-anthocyanidin heterodimers (**5–8**), and pyranoanthocyanidins (**9–14**), including rosacyanins A1, A2, and B (**15–17**), have been reported to occur in plants (Figure 2.1).

Most anthocyanins found in plants (89.8%) are based on Cy, Dp, Pg, Pn, Mv, and Pt, which have different substitutions on their B-rings. These anthocyanidins are therefore referred to as the *common anthocyanidins*. With just a few exceptions, the

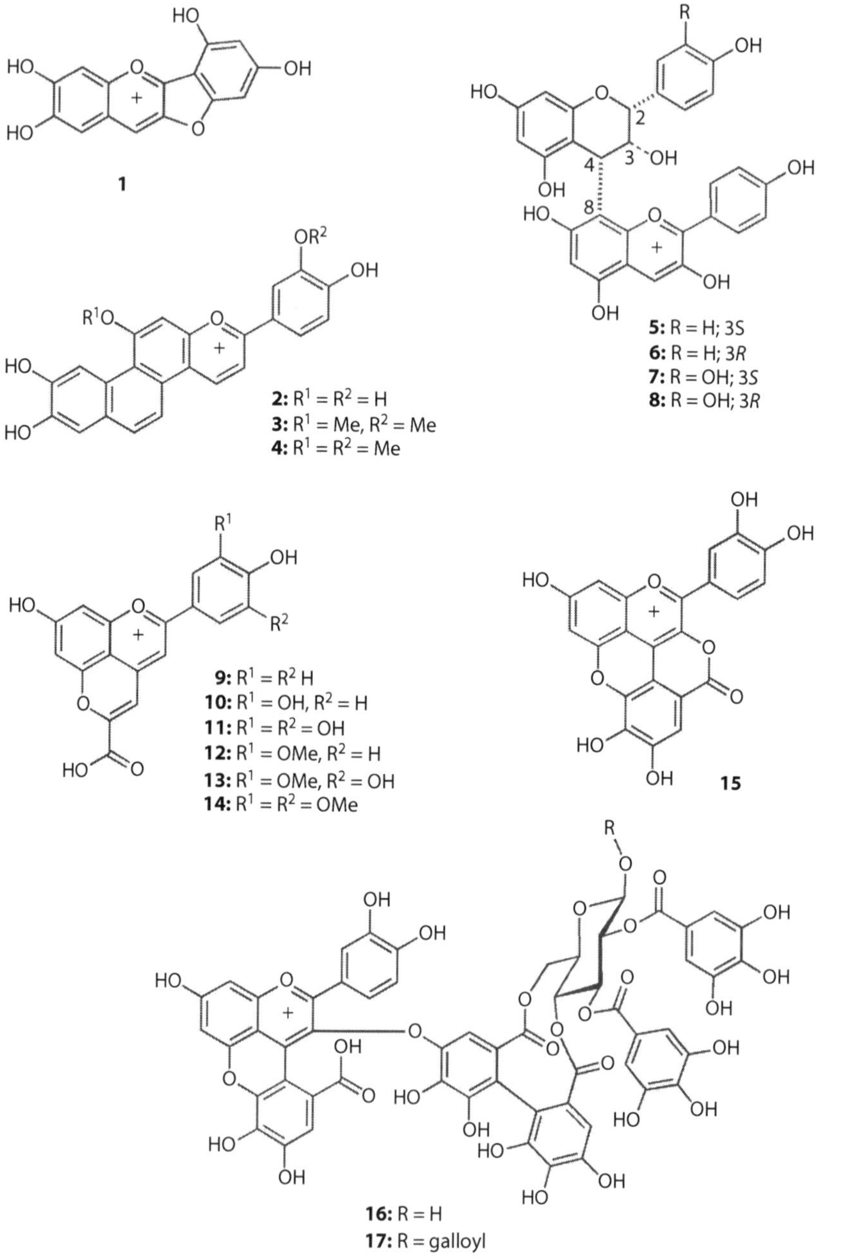

FIGURE 2.1 Structures of rare anthocyanidins with extended C15 skeletons isolated from plants. Riccionidin A (**1**), sphagnorubins a–c (**2–4**), flavanol-anthocyanidin dimers (**5–8**), pyranoanthocyanidins (**9–14**), rosacyanin B (**15**), rosacyanin A1 (**16**), and A2 (**17**). The anthocyanidins with normal C15 skeletons are presented in Table 2.1.

TABLE 2.2

^{1}H and ^{13}C NMR Spectral Data for the Flavylium Cation (AH$^+$), Hemiketal (B$_a$-Major, B$_b$-Minor), and Quinonoidal (A$_7$) Forms and the the Anomeric Sugar Signal of Malvidin 3-Glucoside, Recorded in CF$_3$CO$_2$D–CD$_3$OD (5:95, v/v), Pure CD$_3$OD, and 2,2,6,6-Tetramethylpiperidine-CD$_3$OD (2:98, v/v) at 25°C, Respectively

	^{1}H δ (ppm), J (Hz)				^{13}C δ (ppm)			
	AH$^+$	B$_a$	B$_b$	A$_7$	AH$^+$	B$_a$	B$_b$	A$_7$
2					163.7	103.1	103.6	155.0
3					145.7	145.4	145.3	139.2
4	9.13 d 0.7	6.58 d 0.8	6.65 d 0.8	8.32 d 0.8	136.9	98.6	99.8	133.0
5					159.2	154.6	153.2	170.3
6	6.76 d 2.0	6.07 d 2.2	6.06 d 2.2	5.99 d 1.8	103.2	97.2	97.1	106.9
7					170.5	158.4	158.4	184.3
8	7.05 dd 0.7, 1.9	6.05 dd 0.7, 2.2	6.04 dd 0.7, 2.2	6.09 d 0.7	95.3	95.1	95.3	96.3
9					157.7	153.0	154.1	158.6
10					113.1	101.8	101.9	118.2
1'					119.7	132.2	132.6	114.4
2'	8.08 s	6.97 s	6.94 s	7.73 s	110.5	105.9	103.6	108.4
3'	4.10 s	3.91 s	3.91 s	3.97 s	149.5	148.5	148.8	154.3
3' (OCH$_3$)					57.3	56.7	56.8	56.5
4'					146.0	136.8	135.5	151.5
5'					149.5	148.5	145.6	154.3
5' (OCH$_3$)	4.10 s	3.91 s	3.91 s	3.97 s	57.3	56.7	56.8	56.5
6'	8.08 s	6.97 s	6.94 s	7.73 s	110.5	105.9	103.6	108.4
1''	5.44 d 7.7	4.97 d 7.8	4.83 d 7.8	5.09 d 7.7	103.7	102.1	102.6	103.7

Note: s = singlet, d = doublet, dd = double doublet.

anthocyanins of fruits and vegetables reported in Tables 2.3 and 2.4 are based on the common anthocyanidins—in most species just one anthocyanidin with dominance of Cy. However, some fruits and vegetables such as black currants, plums, and sweet potatoes contain anthocyanins, which are based on two types of anthocyanidins, while blueberries, beans, grapes, and potatoes on the other hand contain anthocyanins grounded on several anthocyanidins, including some, which have methoxyl-groups on their B-rings (see Section 2.3). The common anthocyanidins (Table 2.1) are now and then reported to occur *in vivo* as free aglycones; however, these reports are in most cases artifacts formed from anthocyanins during the extraction and isolation stages. Nevertheless, the natural presence of Cy, Pn, and Pg in extracts of

TABLE 2.3

Qualitative and Quantitative Anthocyanin Content of Selected Fruits Used in the Human Diet

Samples	Major[a] Anthocyanin[b]	Content (mg/100 g)		References
		FW	DW	
		Adoxaceae		
Sambucus canadensis (American elderberry)	Cy3-[2-(xyl)glc]-5-glc, Cy3-[2-(xyl)-6-(-E-cum)glc]-5-glc	104–444		Johansen et al. (1991), Nakatani et al. (1995), Lee and Finn (2007)
Sambucus nigra (European elderberry)	Cy3-glc, Cy3-[2-(xyl)glc]	170–1374		Andersen et al. (1991), Wu et al. (2004a), Määttä-Riihinen et al. (2004a), Lee and Finn (2007)
		Anacardiaceae		
Mangifera indica (Mango, peel)	7-MeCy3-gal		0.02–0.40	Berardini et al. (2005a,b)
Pistacia vera (Pistachio nut[c])	Cy3-gal, Cy3-glc		2–47	Miniati (1981), Wu et al. (2005, 2006), Bellomo and Fallico (2007)
		Arecaceae		
Euterpe edulis (Jussara)	Cy3-glc, Cy3-[6-(rha)glc]		2956	Sousa de Brito et al. (2007)
Euterpe oleracea (Acai)	Cy3-glc, Cy3-[6-(rha)glc]	30–303	3410	Neida and Elba (2007), Mertens-Talcott et al. (2008), Vera de Rosso et al. (2008), Agawa et al. (2011)
Phoenix dactylifera (Dates)		0.2–1.5		Al-Farsi et al. (2005)
		Asteraceae		
Helianthus annuus (Sunflower seeds, purple)	Cy3-glc, Cy3-[(mal)glc], Cy3-xyl, Cy3-[(mal) xyl]		5200	Mazza and Gao (1994), Wiesenborn et al. (1995)

continued

Species (Common name)	Family	Value	Anthocyanins	Reference
	Chrysobalanaceae			
Chrysobalanus icaco (Guajiru/paradise plum)		104	Pt3-[(ace)glc], Pt3-[(suc)rha]*	Sousa de Brito et al. (2007)
	Davidsoniaceae			
Davidsonia jerseyana (Davidson's plum)		57	Cy3-[2-(xyl)glc], Dp3-[2-(xyl)glc], Pn3-[2-(xyl)glc]	Netzel et al. (2006)
	Elaeagnaceae			
Hippophae sp. (Sea-buckthorn)		0.84	Cy3-gal	Hosseinian and Beta (2007)
	Elaeaocarpaceae			
Aristotelia chilensis (Maqui)		138	Dp3-[2-(xyl)glc]-5-glc	Escribano-Bailon et al. (2006)
	Ericaceae			
Empetrum nigrum (Crowberry)		398–768	Cy3-gal, Cy3-ara, Dp3-gal, Dp3-ara, Mv3-gal, Mv3-ara, Pn3-gal, Pn3-ara, Pt3-gal, Pt3-ara	Kähkönen et al. (2001), Ogawa et al. (2008), Koskela et al. (2010), Määttä-Riihinen et al. (2004a)
Vaccinium angustifolium (Blueberry, lowbush)		91–234	Cy3-glc, Cy3-gal, Dp3-gal, Dp3-glc, Mv3-glc, Mv3-gal, Pn3-glc, Pn3-gal, Pt3-glc, Pn3-gal	Francis et al. (1966), Prior et al. (1998), Kalt et al. (1999), Connor et al. (2002)
Vaccinium ashei (Blueberry, rabbiteye)		87–484	Cy3-gal, Cy3-glc, Dp3-gal, Dp3-glc, Mv3-gal, Mv3-glc, Pt3-gal, Pt3-glc, Dp3-ara, Mv3-ara	Prior et al. (1998), Prior et al. (2001), Moyer et al. (2002)
Vaccinium constablaei (Blueberry, mountain highbush)		168–290		Connor et al. (2002)

TABLE 2.3 (continued)

Qualitative and Quantitative Anthocyanin Content of Selected Fruits Used in the Human Diet

Samples	Major[a] Anthocyanin[b]	Content (mg/100 g) FW	Content (mg/100 g) DW	References
Vaccinium corymbosum (Blueberry, highbush)	Dp3-gal, Dp3-glc, Dp3-ara, Mv3-gal, Mv3-glc, Mv3-ara, Pt3-gal, Pt3-glc, Pt3-ara	63–438	2762	Prior et al. (1998), Kalt (1999), Connor (2002), Moyer (2002), Määttä-Riihinen (2004a), Cho (2004), Müller (2012)
Vaccinium myrtilloides (Blueberry, Canadian)		218–298		Moyer et al. (2002), Connor et al. (2002)
Vaccinium myrtillus (Bilberry/whortleberry)	Cy3-glc, Cy3-gal, Cy3-ara, Dp3-glc, Dp3-gal, Dp3-ara, Mv3-glc, Mv3-gal, Mv3-ara, Pt3-glc, Pt3-gal, Pt3-ara	300–1017	2298–7465	Prior et al. (1998), Kähkönen et al. (2001, 2003), Määttä-Riihinen et al. (2004a), Müller et al. (2012)
Vaccinium uliginosum (Bog whortleberry/bilberry)	Cy3-glc, Cy3-gal, Cy3-ara, Dp3-glc, Dp3-gal, Dp3-ara, Mv3-glc, Mv3-gal, Mv3-ara, Pt3-glc, Pt3-gal, Pt3-ara, Pn3-glc	256–432	1425	Andersen (1987), Määttä-Riihinen et al. (2004a), Lätti et al. (2010)
Vaccinium macrocarpon (Cranberry, American)	Cy3-glc, Cy3-ara, Cy3-gal, Pn3-gal, Pn3-ara	20–360	395–480	Wang and Stretch (2001), Prior et al. (2001), Määttä-Riihinen et al. (2004a), Wu et al. (2006), Ogawa et al. (2008), Brown (2011)
Vaccinium oxycoccus (Cranberry, small)	Cy3-gal, Cy3-ara, Cy3-glc, Pn3-gal, Pn3-ara, Pn3-glc		397	Andersen (1989), Huopalahti et al. (2000), Kähkönen et al. (2001)
Vaccinium vitis-idaea (Cowberry/lingonberry)	Cy3-gal, Cy3-ara,	35–130	225–355	Andersen (1985), Kähkönen et al. (2001, 2003), Määttä-Riihinen et al. (2004a), Lee and Finn (2012)
Vaccinium membranaceum (Huckleberry, thinleaf)	Cy3-gal, Cy3-glc, Cy3-ara, Dp3-gal, Dp3-glc, Dp3-ara	116–167		Moyer et al. (2002), Lee et al. (2004)
Vaccinium ovalifolium (Huckleberry, black)		185–400		Moyer et al. (2002), Lee et al. (2004)
Vaccinium parvifolium (Huckleberry, red)		34		Moyer et al. (2002)

Fabaceae

Species	Anthocyanins	Content	References
Lens culinaris (Lentil, black)	Dp3-[2-(glc)ara]	16–68	Takeoka et al. (2005), Xu and Chang (2010)
Glycine spp. (Soybean, black)	Cy3-glc, Dp3-glc	158–2040	Yoshida et al. (1996), Choung et al. (2001)
Phaseolus vulgaris (Bean, black)	Dp3-glc, Mv3-glc, Pt3-glc	214–278	Takeoka et al. (1997), Choung et al. (2003), Wu et al. (2005, 2006)
Phaseolus vulgaris (Bean, red)	Cy3-glc, Cy3-[2-(xyl)glc], Pg3-glc, Pg3-[2-(xyl)glc]	27–74	Choung et al. (2003), Wu et al. (2005, 2006), Macz-Pop et al. (2006b)
Vigna unguiculata (Cowpeas)	Dp3-glc, Cy3-glc	88–210	Ha et al. (2010), Ojwang et al. (2012)

Grossulariaceae

Species	Anthocyanins	Content	References
Ribes alpinum (Alpine currant)	Cy3-glc, Cy3-[6-(rha)glc]	5	Jordheim et al. (2007b)
Ribes aureum (Golden currant)	Cy3-glc, Cy3-[6-(rha)glc]	170	Jordheim et al. (2007b)
Ribes nidigrolaria (Jostaberries)	Cy3-glc, Cy3-[6-(rha)glc], Dp3-glc, Dp3-[6-(rha)glc]	40–89	Moyer (2002). Jordheim et al. (2007b)
Ribes nigrum (Black currant)	Cy3-glc, Cy3-[6-(rha)glc], Dp3-glc, Dp3-[6-(rha)glc]	744–2120, 96–587	Kähkönen et al. (2001), Moyer et al. (2002), Kampuse et al. (2002), Kähkönen et al. (2003), Wu et al. (2004a), Määttä-Riihinen et al. (2004a), Ogawa et al. (2008)
Ribes odoratum (Buffalo currant)		273	Moyer et al. (2002)
Ribes rubrum (Red currant)	Cy3-[6-(rha)glc], Cy3[2(-xyl)glc], Cy3[2-(xyl)-6-(rha)glc]	1–34	Kähkönen et al. (2001), Benvenuti et al. (2004), Wu et al. (2004a), Määttä-Riihinen et al. (2004a), Ogawa et al. (2008)

continued

TABLE 2.3 (continued)
Qualitative and Quantitative Anthocyanin Content of Selected Fruits Used in the Human Diet

Samples	Major[a] Anthocyanin[b]	Content (mg/100 g)		References
		FW	DW	
Ribes uva-crispa (Red gooseberry)	Cy3-xyl, Cy3-[6-(rha)glc], Cy3-[6-(cum)glc], Cy3-[6-(caf)glc], Pn3-glc	0.1–46	81–85	Kähkönen et al. (2001), Wu et al. (2004a), Jordheim et al. (2007b), Pantelidis et al. (2007)
Lauraceae				
Persea americana (Avocado)	Cy3-glc	14–64		Cox et al. (2004)
Lythraceae				
Punica granatum (Pomegranate juice[A], peel[B])	Cy3-glc, Cy3,5-di-glc, Dp3-glc	9–935 mg/L[A]	6100–8600[B]	Alighourchi et al. (2008), Hasnaoui et al. (2011), Elfalleh et al. (2011), Varasteh et al. (2012)
Malpighiaceae				
Malpighia emarginata	Cy3-*rha*, Pg3-*rha*	4–60	261–528	Lima et al. (2003), Hanamura et al. (2005), Sousa de Brito et al. (2007), Vera de Rosso et al. (2008)
Moraceae				
Ficus carica (Fig)	Cy3-glc, Cy3-[6-(rha)glc], Cy3,5-di-glc, Pg3-[6-(rha)glc], 5-CCy3-glc	3	4.3	Dueñas et al. (2008), Rababah et al. (2011)
Morus nigra (Mulberry, red[A], black[B])	Cy3-glc, Cy3-[6-(rha)glc], Pg3-glc	11[A] / 72[B]	1610	Ogawa et al. (2008), Stefanut et al. (2011), Ercisli et al. (2010)
Myrtaceae				
Eugenia reinwardtiana (Cedar bay cherry)	Cy3-glc	43		Netzel et al. (2006)
Eugenia umbelliflora (Baguacu)	Cy3-glc, Dp3-glc, Mv3-glc, Pn3-glc and Pt3-glc	342		Kuskoski et al. (2003)

Kunzea pomifera (Muntries)	Cy3-glc, Dp3-glc	38		Netzel et al. (2006)
Myrtus communis (Myrtle berry)	Cy3-glc, Dp3-glc, Mv3-glc	171–630		Mulas et al. (2002), Messaoud and Boussaid (2011)
Syzygium cumini (Java plum, jambolão)	Dp3,5-di-glc, Mv3,5-di-glc, Pt3,5-di-glc	79	771	Sousa de Brito et al. (2007)
Oleaceae				
Olea europaea (Olive, black)	Cy3-glc, Cy3-[6-(rha)glc]	48–443		Romani et al. (1999)
Passifloraceae				
Passiflora spp. (Passion fruit)	Cy3-glc			Kidøy et al. (1997)
Poaceae				
Oryza sativa (Rice, black)	Cy3-glc, Pn3-glc	10–493		Ryu et al. (1998), Lee (2010)
Sorghum spp. (Sorghum, brown[A]/black[B])	Ap, Lt	160–390[A] 400–980[B]		Awika et al. (2004a,b)
Triticum aestivum (Wheat, purple[A]/blue[B])	Cy3-glc, Cy3-[6-(rha)glc], Dp3-glc, Dp3-[6-(rha)glc]	20[A] 50[B]	0.4–3.4	El-Sayed and Hucl (2003), Tyl and Bunzel (2012)
Zea mays (Corn/maize)	Cy3-glc, Cy3-[6-(mal)glc], Cy3-[3,6-di-(mal)glc], Pg3-glc, Pn3-glc	54–1734	1680–1878	Cevallos-Casals and Cisneros-Zevallos (2003), Moreno et al. (2005)
Rosaceae				
Amelanchier alnifolia (Saskatoon berry)	Cy3-gal, 3-glc, 3-ara	190–518	562–2490	Hosseinian and Beta (2007), Bakowska-Barczak and Kolodziejczyk (2008), Lavola et al. (2012)
Aronia melanocarpa (Chokeberry, black)	Cy3-ara, Cy3-gal	480–1480	177–1052	Strigl et al. (1995), Kähkönen et al. (2001), Wu et al. (2004a), Määttä-Riihinen et al. (2004a), Slimestad et al. (2004), Hosseinian and Beta (2007)

continued

TABLE 2.3 (continued)
Qualitative and Quantitative Anthocyanin Content of Selected Fruits Used in the Human Diet

Samples	Major[a] Anthocyanin[b]	Content (mg/100 g)		References
		FW	DW	
Fragaria × ananassa (Strawberry)	Pg3-glc	8–64	97–520	Kalt et al. (1999), García-Vigera et al. (1999), Kähkönen et al. (2001), Wu et al. (2006), Hosseinian et al. (2007), Ogawa et al. (2008), Rababah et al. (2011), Riedl et al. (2011), Aaby et al. (2012)
Malus spp. (Apple)	Cy3-gal	1–117	3–4	Kähkönen et al. (2001), Wolfe et al. (2003), Rababah et al. (2005), Hagen et al. (2006), Wu et al. (2006), Vieira et al. (2011)
Prunus spp. (Cherry)	Cy3-glc, Cy3-[6-(rha)glc]	1–297		Gao and Mazza (1995), Wu et al. (2006), Sugawara and Igarashi (2008), Rababah et al. (2011)
Prunus armeniaca (Apricot)	Cy3-glc, Cy3-[6-(rha)glc]	3		Ruiz et al. (2006), Rababah et al. (2011)
Prunus spp. (Plum)	Cy3-glc, Cy3-[6-(rha)glc], Pn3-glc, Pn3-[6-(rha)glc]	5–449	1093–1161	Los et al. (2000), Tomás-Barberán et al. (2001), Cevallos-Casals et al. (2002, 2006), Wu et al. (2006)
Prunus elattus (Plum, Illawarra)	Cy3-glc	871		Netzel et al. (2006)
Prunus persica (Peach)	Cy3-glc	4–50	168–230	Tomás-Barberán et al. (2001), Cevallos-Casals et al. (2002, 2006), Wu et al. (2006)
Prunus persica var. *nucipersica* (Nectarine)	Cy3-glc	0.1–34		Tomás-Barberán et al. (2001), Wu et al. (2006)
Prunus serotina (Capuli ssp.)	Cy3-glc, Cy3-[6-(rha)glc]	95–140		Jimenez et al. (2011)
Prunus spinosa (Blackthorn)	Cy3-glc, Cy3-[6-(rha)glc], Pn3-[6-(rha)glc]	54		Määttä-Riihinen et al. (2004a)

Species (common name)	Anthocyanins	Content	Content	References
Prunus timorense (Plum, Burderkin)	Cy3-glc	273		Netzel et al. (2006)
Pyrus spp. (Pear)	Cy3-gal	5–10		Dussi et al. (1995)
Rubus arcticus (Raspberry, arctic)	Cy3-glc, Cy3-[6-(rha)glc]	89		Määttä-Riihinen et al. (2004b)
Rubus chamaemorus (Cloudberry)	Cy3-glc, Cy3-[6-(rha)glc], Cy3-[2-(glc)glc], Cy3-[(gly)-6-(rha)glc]*	2		Määttä-Riihinen et al. (2004b)
Rubus fruticosus (Blackberry)	Cy3-glc	70–332	1010	Cho et al. (2004), Fan-Chiang and Wrolstad (2005), Wu et al. (2006), Ogawa et al. (2008), Jordheim et al. (2011a)
Rubus idaeus (Raspberry, red)	Cy3-glc, Cy3-[6-(rha)glc]	2–332	171–1030	Wang and Lin (2000), Kähkönen et al. (2001), Wu et al. (2006), Hosseinian and Beta (2007), Ogawa et al. (2008), Riedl and Murkovic (2011), Bobinaite et al. (2012)
Rubus moluccanus var. *autropacificus* (Raspberry, molucca)	Cy3-glc, Cy3-[6-(rha)glc]	115		Netzel et al. (2006)
Rubus occidentalis (Raspberry, black)	Cy3-glc, Cy3-[6-(rha)glc], Cy3-[2-(xyl)glc]	194–687	932–973	Wang and Lin (2000), Wu et al. (2006)
Sorbus aucuparia (Rowanberry, wild)	Cy3-gal, 3-ara		10–14	Kylli et al. (2010)
Rutaceae				
Citrus sinensis (Blood orange)	Cy3-glc, Cy3-[6-(mal)glc]	3		Maccarone et al. (1998), Lee (2002b), Rababah et al. (2011)
Sapindaceae				
Litchi chinensis (Litchi)	Cy3-[6-(rha)glc]	48–177		Lee and Wicker (1991), Sarni-Manchado et al. (2000)

continued

TABLE 2.3 (continued)

Qualitative and Quantitative Anthocyanin Content of Selected Fruits Used in the Human Diet

Samples	Majorᵃ Anthocyaninᵇ	Content (mg/100 g)		References
		FW	DW	
Solanaceae				
Lycium ruthenicum (Lycii)	Pt-3-[(rha)glc]-(E-p-cum)-5-glc*	450–530	220	Kosar et al. (2003), Zheng et al. (2011)
Solanum betaceum (Tamarillo)	Cy3-[6-(rha)glc], Dp3-[6-(rha)glc], Pg3-[6-(rha)glc]		7818[ARE]	Wrolstad and Heatherbell (1974), Osorio et al. (2012)
Solanum melongena (Eggplant)	Dp3-[6-(rha)glc], Dp3-[6-(rha)glc]-5-glc, Dp3-[6-(4-(E-cum)rha)glc]-5-glc	86		Ichiyanagi et al. (2005c), Wu et al. (2006)
Vitaceae				
Vitis spp. (Grape)	Cy3-glc, Dp3-glc, Mv3-glc, Mv3-[6-(ace)glc], Mv3-[6-(E-p-cum)glc], Pn3-glc, Pt3-glc,	2–791	113	Revilla et al. (1999), Cho 2004, Amico et al. (2004), Wu et al. (2006), Gómez Gallego et al. (2012)
Winteraceae				
Tasmanian lanceolata (Tasmanian pepper)	Cy3-glc, Cy3-[6-(mal)glc]	949		Netzel et al. (2006)

Note: FW = fresh weight, DW = dry weight, ARE = anthocyanin-rich extract, *tentative structure.

ᵃ Major: Compounds estimated to occur in relative anthocyanin amounts higher than 10%. In a few papers there exist no discrimination between major and minor compounds, and all reported anthocyanins have been included.

ᵇ Cy = cyanidin, Dp = delphinidin, Mv = malvidin, Pg = pelargonidin, Pn = peonidin, Pt = petunidin, 5-CCy = 5-carboxypyranocyanidin, Ap = apigeninidin, Lt = luteolinidin, ace = acetyl, caf = caffeoyl, cum = p-coumaroyl, fer = feruloyl, mal = malonyl, hba = p-OH-benzoyl, sin = sinapoyl, ara = arabinosyl, gal = galactosyl, glc = glucosyl, gly = glycosyl, rha = rhamnosyl, xyl = xylosyl. See Table 2.1 for anthocyanidin structures, and Figure 2.4 for acyl structures.

TABLE 2.4

Qualitative and Quantitative Anthocyanin Content of Selected Vegetables Used in the Human Diet

Sample	Major[a] Anthocyanin[b]	Content (mg/100 g) FW	DW	References
	Amaryllidaceae			
Allium cepa (Red onions)	Cy3-glc, Cy3-[3-(glc)glc], Cy3-[6-(mal)glc], Cy3-[6-(mal)-3-(glc)glc]	49		Terahara et al. (1994), Fossen et al. (1996), Wu et al. (2006)
	Apiaceae			
Coriandrum sativum (Coriander leaves[d])	Pn3-[(fer)glc]-5-glc		$(6-7) \times 10^{-4}$	Barros et al. (2012)
Daucus carrota (Carrot, black/purple)	Cy3-[2-(xyl)gal], Cy3-[2-(xyl)-6-(glc)gal], Cy3-[2-(xyl)-6-(6-(sin)glc) gal], Cy3-[2-(xyl)-6-(6-(fer)glc)gal], Cy3-[2-(xyl)-6-(6-(cum)glc)gal]	1–44	4–1799	Kammerer et al. (2004), Elham et al. (2006), Montilla et al. (2011)
	Asparagaceae			
Asparagus officinalis (Asparagus)	Cy3-[6-(rha)glc], Cy3-[3″-(glc)-6″-(rha)glc]		24	Sakaguchi et al. (2008), Li et al. (2012)
	Araceae			
Colocasia esculenta (Taro)	Cy3-glc, Cy3-rha, Pg3-glc	3	27	Chan et al. (1977), Champagne et al. (2011)
Xanthosoma sagittifolium (Cocoyam)		7	38	Champagne et al. (2011)
	Asteraceae			
Cichorium intybus (Chicory)	Cy3-glc, Cy3-[6-(mal)glc], Dp3-[6-(mal)glc]	126–590		Bridle et al. (1984), Mulinacci et al. (2001), Innocenti et al. (2005)
Lactuca sativa (Lettuce, red leaf)	Cy3-[6-(mal)glc]	0.7–4		Yamaguchi et al. (1996), Wu et al. (2005), Wu et al. (2006)

continued

TABLE 2.4 (continued)

Qualitative and Quantitative Anthocyanin Content of Selected Vegetables Used in the Human Diet

Sample	Major[a] Anthocyanin[b]	Content (mg/100 g) FW	Content (mg/100 g) DW	References
	Brassicaceae			
Brassica oleracea (Broccoli sprouts)	Cy3-[(p-cum)(sin)glcglc]-5-glc, Cy3-[(sin)(fer)glcglc]-5-glc, Cy3-[(sin)(sin)glcglc]-5-glc, Cy3-[(sin)(sin)glcglc]-5-[(mal)glc]*	0.2–0.6		Moreno et al. (2010)
Brassica oleracea L. var. *botrytis* (Cauliflower, violet)	Cy3-[2-(glc)glc]-5-glc, Cy3-[2-(6-(p-cum)glc)glc]-5-glc, Cy3-[2-(6-(fer)glc)glc]-5-glc, Cy3-[2-(6-(sin)glc)glc]-5-glc, Cy3-[2-(6-(p-cum)glc)glc]-5-[6-(sin)glc], Cy3-[2-(6-(fer)glc)glc]-5-[6-(sin)glc]	4–40	200	Lo Scalzo et al. (2008), Viscardi et al. (2009), Li et al. (2012)
Brassica oleracea L. var. *capitata* (Cabbage, red)	Cy3,5-di-glc, Cy3-[2-(glc)glc]-5-glc, Cy3-[2-(2-(sin)glc)glc]-5-glc, Cy3-[6-(sin)-2-(2-(sin)glc)glc]-5-glc	75–363	199	Hrazdina (1977), Wu et al. (2006), Lo Scalzo et al. (2008), Li et al. (2012)
Raphanus sativus (Radish, red)	Pg3-[2-(glc)-6-(cum)glc]-5-[6-(mal)glc], Pg3-[2-(glc)-6-(fer)glc]-5-[6-(mal)glc]	70–130		Giusti et al. (1998), Wu et al. (2006), Wu et al. (2005)
	Convolvulaceae			
Ipomoea batatas (Sweet potato)	Pn3-[2-(6-(hba)glc)-6-(caf)glc]-5-glc, Pn3-[2-(glc)-6-(hba)glc]-5-glc, Pn3-[2-(6-(fer)glc)-6-(caf)glc]-5-glc, Cy3-[2-(6-(hba)glc)-6-(caf)glc]-5-glc, Pn3-[2-(glc)-6-(caf)glc]-5-glc, Cy3-[2-(glc)-6-(hba)glc]-5-glc	180–1407	611–625	Cevallos-Casals and Cisneros-Zevallos (2003), Kim et al. (2012)
	Dioscoreaceae			
Dioscorea alata (purple yam)	Cy3-[6-(6-(sin)-3-(glc)glc)glc]-3'-[glc]-7-[6-(sin)glc], Cy3-[6-(6-(sin)-3-(glc)glc)glc]-3'-[6-(sin)glc], Cy3-[6-(6-(sin)glc)glc]	27	93	Yoshida et al. (1991), Champagne et al. (2011)

Euphorbiaceae

Species	Anthocyanins	Amount	References
Manihot esculenta (Cassava)	Cy3-[6-(rha)glc], Dp3-[6-(rha)glc]	124	Byanukama et al. (2009)
Lamiaceae			
Ocinum basilicum (Purple basil)	6 Cy-(cum)*, Cy-(cum)(mal)*	8–19	Phippen and Simon (1998), Kwee and Niemeyer (2011)
		33–1337	
Oxalidaceae			
Oxalis triangularis (Shamrock, purple)	Mv3-[6-(rha)glc]-5-glc, Mv3-[6-(4-(mal)rha)glc]-5-glc	195	Pazmino-Duran et al. (2001), Fossen et al. (2005)
Polygonaceae			
Rheum rhabarbarum (Rhubarb)	Cy3-glc, Cy3-[6-(rha)glc]	4[c]	Koponen et al. (2007), Wrolstad and Heatherbell (1968)
Solanaceae			
Solanum spp. (Potato, red/purple)[e]	Cy3-[6-(4-(cum)rha)glc]-5-glc, Dp3-[6-(4-(cum)rha)glc]-5-glc, Mv3-[6-(4-(cum)rha)glc]-5-glc, Pg3-[6-(4-(cum)rha)glc]-5-glc, Pn3-[6-(4-(cum)rha)glc]-5-glc, Pt3-[6-(4-(cum)rha)glc]-5-glc, Pn3-[6-(4-(caf)rha)glc]-5-glc, Pt3-[6-(4-(caf)rha)glc]-5-glc, Pg3-[6-(4-(fer)rha)glc]-5-glc, Pn3-[6-(4-(fer)rha)glc]-5-glc	3–1382	Rodriguez–Saona et al. (1998), Lewis et al. (1998a,b), Fossen and Andersen (2000), Fossen and Andersen (2003b)

Note: FW = fresh weight, DW = dry weight, *tentative structure.

a   Major: Compounds estimated to occur in relative anthocyanin amounts higher than 10%. In a few papers, there exists no discrimination between major and minor compounds, and all reported anthocyanins have been included.

b   See Table 2.3 for abbreviations.

c   Anthocyanidin mg/100 g.

d   *In vitro* grown plants with purple pigmentation.

e   For more details on different *Solanum* species and cultivars, see Lewis et al. (1998a,b) and Rodriguez-Saona et al. (1998).

beans has been suggested after careful consideration of the process of extraction and purification followed by application of LC-MS for identification purposes (Macz-Pop et al. 2006b).

Among the rare anthocyanins with the normal C_{15}-skeleton isolated from plants, 19 are found to be based on nine different A-ring methoxylated anthocyanidins, while 10 anthocyanins have an extra 6-hydroxyl group on their A-rings (Table 2.1). Among the former, the structures of pulchellidin, europinidin, and capensinidin have recently been revised to be 5,7-dimethoxy-3,3',4',5'-tetrahydroxyflavylium, 5,7,3'-trimethoxy-3,4',5'-trihydroxyflavylium, and 5,7,3',5'-tetramethoxy-3,4'-dihydroxyflavylium cations, respectively (Skaar et al. 2012), while several 7-O-methylated anthocyanidin glycosides have been isolated from flowers of *Catharanthus roseus* (Toki et al. 2008, Toki, et al. 2011). Among fruits and vegetables only the major pigment, 7-O-methylCy3-galactoside (**18**) (Figure 2.3) of mango (cultivar "Tommy Atkins"), contains this type of anthocyanidin (Berardini et al. 2005b). This pigment has previously most probably been wrongly reported to occur in mango as Pn3-galactoside (Proctor 1969).

Among the 22 3-deoxyanthocyanins, which are lacking a 3-hydroxyl or 3-O-glycosyl unit, 14 of them are without any glycosidic moiety (and should more strictly be named as 3-deoxyanthocyanidins) (Table 2.1). The 3-deoxyanthocyanins have a limited distribution in angiosperms, however, apigeninidin and luteolininidin and their glycosides have been found in significant levels in grains of sorghum (*Sorghum* sp.), especially in black varieties (Lo et al. 1996, Pale et al. 1997, Awika et al. 2004b). Riccionidin A (**1**), which has been isolated from liverworts (Kunz et al. 1994) and adventitious root cultures of *Rhus javanica* (Anacardiaceae) (Taniguchi et al. 2000), is a special variant of 3-deoxyanthocyanidins. This pigment is accompanied by riccionidin B, which most probably is based on two molecules of riccionidin A linked via the 3'- or 5'-positions. Another special variants, the sphagnorubins A–C (**2–4**), have been isolated from peat moss (*Sphagnum* spp.) (Vowinkel 1975). The 3-deoxyanthocyanidins are relative stable toward pH changes (Sweeny et al. 1981, Mazza and Brouillard, 1987, Bjorøy et al. 2009). Some 3-deoxyanthocyanidins have been demonstrated to be more cytotoxic to cancer cells than their anthocyanidin analogs (Shih et al. 2007).

With respect to anthocyanin compounds formed during storage and processing in plant-derived foods, the group of pyranoanthocyanidins has gained much attention mostly because of their influence on color evolution in wine during maturation (Cheynier 2006, de Freitas and Mateus 2011). From intact plant materials, the rosacyanins (**15–17**) have been isolated from *Rosa hybrida* petals (Fukui et al. 2006; Fukui and Tanaka, 2010), 5-carboxypyranopelargonidin 3-glucoside (**9**) from strawberry fruits (Andersen et al. 2004) and 5-carboxypyranocyanidin 3-glucosides from edible scales as well as from the dry outer scales of red onion (*Allium cepa*) (Fossen and Andersen, 2003b).

Most anthocyanins are monomeric in nature. However, anthocyanin oligomers, anthocyanin–flavanol polymers, linked together or through a methylmethine bridge, are formed during storage and processing. A heterodimer type of anthocyanins consisting of anthocyanidin moiety(ies) covalently linked to flavanol aglycone(s) can be the typical representative here. Flavanol–anthocyanidin heterodimers seem to occur, although in small quantities, also in extracts of unprocessed plants. In extracts of fresh strawberries, four purple-colored anthocyanins were characterized as catechin(4α → 8)Pg3-glc, epicatechin(4α → 8)Pg3-glc, afzelechin(4α → 8)

Pg3-glc, and epiafzelechin($4\alpha \rightarrow 8$)Pg3-glc (Fossen et al. 2004) (**5–8**). The same four heterodimers together with afzelechin($4 \rightarrow 8$) Pg3-[6-(rha)glc] were tentatively identified in extracts of the strawberry cultivar "Camarosa" (Gonzalez-Paramas et al. 2006). Similarly, (epi)catechin-Cy3,5-diglc has been identified in the extract of purple corn, (epi)catechin-Pn3-glc and (epi)catechin-Mv3-glc in extract of grape skin, while (epi)catechin-Cy3-glc, (epi)gallocatechin-Dp, and (epi)catechin linked to Cy, Pt, and Pn have been reported to occur in extracts of various beans (*Phaseolus coccineus* and *P. vulgaris*) (Macz-Pop et al. 2006a,b, Gonzalez-Paramas et al. 2006). Putative flavanol–anthocyanin condensation products have also been detected in a concentrate from black currant (*Ribes nigrum*) fruits and in extracts of the fig (*Ficus carica*) (McDougall et al. 2005, Dueñas et al. 2008).

2.2.3 SECONDARY STRUCTURES OF ANTHOCYANIDINS: EQUILIBRIUM FORMS

The basic unit of anthocyanidins is often named by chemists as 2-phenyl-benzo-pyrylium or flavylium cation. Since anthocyanins are fairly stable under relatively strong acidic conditions, the procedures for their extraction, isolation, and structure elucidation are normally performed with acidic solvents in all steps. The flavylium cations with their reddish nuances constitute the predominant secondary anthocy-anidin form under relative strong acidic condition. As a consequence, the structures of nearly all anthocyanins have been elucidated when their anthocyanidins occur on the flavylium cation form. Nevertheless, in the transport of anthocyanins from the food into and through the human body, the anthocyanins experience similar strong acidic conditions only in the stomach. Anthocyanins are outstanding compounds by the way their anthocyanidins under *in vitro* conditions are shown to be involved in a series of equilibriums giving rise to several secondary structures dependent espe-cially on pH (Figure 2.2). Although there exists very inadequate information about these forms under *in vivo* conditions, it is recognized that they have different proper-ties (stability/reactivity and colors). Here, follows some details regarding the anthocy-anidin secondary structures of anthocyanins under *in vitro* conditions.

On the basis of the observations of some simple anthocyanidin 3-mono- or digly-cosides in aqueous buffer solutions, the following scheme (Figure 2.2) is generally accepted for the major anthocyanins in for instance blueberries, grapes, cherries, strawberries, and black currants: At a pH of approx. 3 or lower, the flavylium cation (**AH⁺**) with reddish nuances is the predominant form (Brouillard and Dangles, 1994; Pina et al. 2012). While the pH is raised, thermodynamic and kinetic competition occur between (a) the hydration reaction on position 2 of **AH⁺** followed by proton loss giving colorless hemiketal forms (**B**, also called hemiacetal or carbinol pseudo-bases), which can further undergo ring opening by a tautomeric process giving the yellow *cis*-chalcone **C$_E$**, and (b) deprotonation of the acidic hydroxyl groups at **AH⁺** giving violet quinonoidal bases (**A**). Further deprotonation of the quinonoidal bases can take place at pH 6–7 with the formation of more bluish resonance-stabilized qui-nonoidal anions (**A⁻**). The *trans*-chalcone **C$_Z$** is thereafter formed by the isomeriza-tion of **C$_E$**. The deprotonation step (b) is faster than the hydration step (a). However, at equilibrium **B**, **C$_E$** and **C$_Z$** might be more stable than **A**. This can cause **A** first to be formed as a kinetic product, which later totally or partially disappears, while the

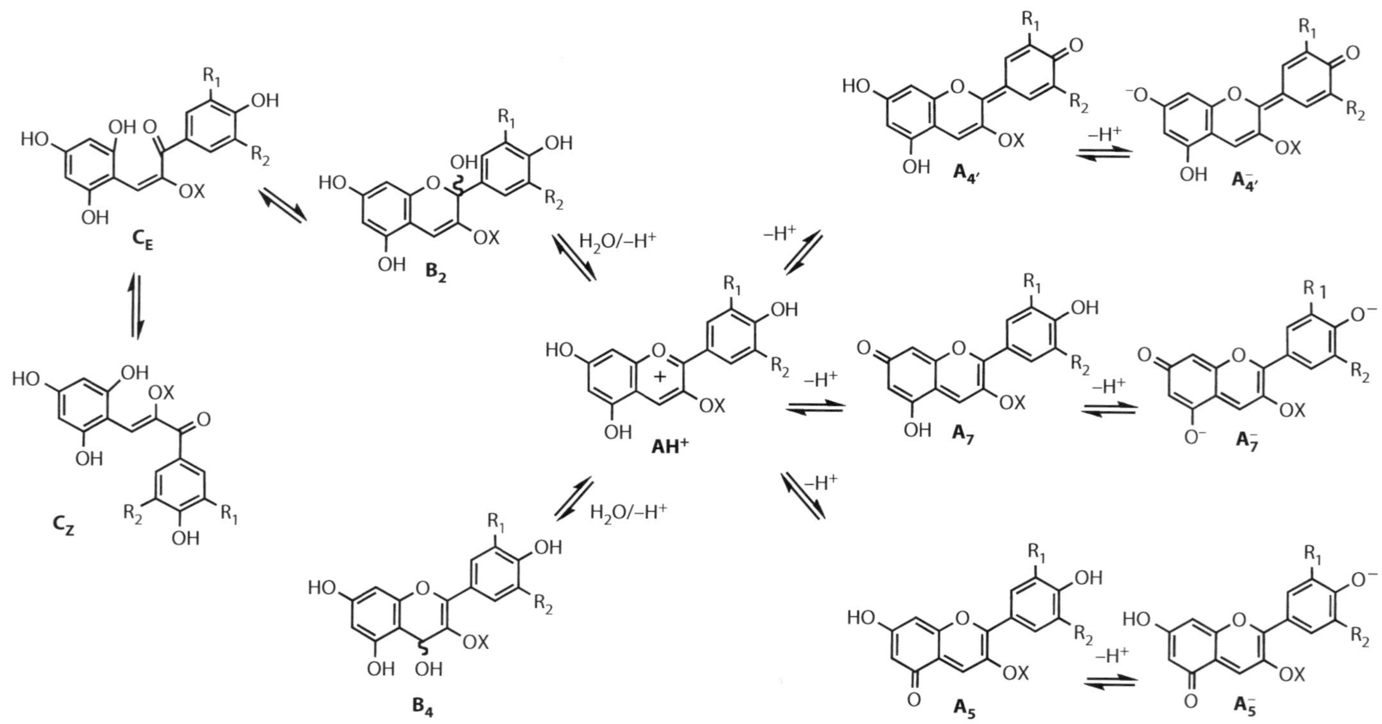

FIGURE 2.2 General scheme for anthocyanidin equilibrium forms: Flavylium cation (AH^+), quinonoidal bases = anhydrobases (A_4, A_7, and A_5), ionized quinonoidal bases (A_4^-, A_7^- and A_5^-), hemiketals = carbinol pseudobases (B_2 and B_4), cis-chalcone C_E and trans-chalcone C_Z (more correct: retrochalcones). X = glycoside, R_1 and R_2 can be hydroxyl and/or methoxyl groups, depending on the type of aglycone.

yield of hemiketal and/or chalcone forms increase toward the thermodynamic equilibrium. For malvidin 3,5-diglucoside in weakly acidic aqueous solutions at room temperature the ratio at equilibrium between the **B** and C_E is 4:1 (Brouillard and Lang, 1990). In a recent comprehensive review by Pina et al. (2012), the historical advancements in the understanding of the equilibriums of the chemical reaction networks of anthocyanidins, and new thermodynamic and kinetics properties of these equilibriums, mainly based on synthetic analogs, have been described.

Complete ^1H and ^{13}C structural assignments for the various anthocyanidin secondary structures, but the flavylium cation form, have been incomplete for all anthocyanins. One- and two-dimensional NMR have been used to assign the H-atoms of the flavylium cation, two hemiketals (**B**) and the C_Z- and C_E-chalcones of Mv3,5-diglc in aqueous solution in the pH range 0.3–4.5, and also to determine their molar fractions as a function of pH (Santos et al. 1993). All ^1H and ^{13}C atoms of the two epimeric 2-hydroxy-hemiketals (**B₂**) of the 3-glucosides of Dp, Pt, and Mv in addition to Cy3-gal dissolved in deuterated methanol have been completely assigned (Jordheim et al. 2006). For each of these anthocyanins dissolved in deuterated methanol, the equilibriums between the two hemiketals and the corresponding flavylium cation were confirmed. The molar proportions of the flavylium cation and the two hemiketals of the four anthocyanins in deuterated methanol were very similar (70:30), even during storage for weeks. Even though 4-hydroxy-hemiketal forms (**B₄**) often are presented in the general equilibrium scheme related to secondary structures of anthocyanidins (Figure 2.2), no 4-hydroxy-hemiketal form was detected experimentally for any of the pigments (Jordheim et al. 2006). In Table 2.2, we have presented the first complete ^1H and ^{13}C NMR assignments for any anthocyanidin quinonidal form, here demonstrated for Mv3-glc. These data were in accordance with only one quinonoidal form of Mv3-glc (**A₇**).

The pH changes considerably in the human body from the stomach (1.5), intestine microsurface (5.3), duodenum (5.5), urine (5.75), saliva (6.4), liver (7.0), feces (7.15) to the blood (7.40) (Newton 1978). The various anthocyanins will thus most probably under physiological conditions occur on different anthocyanidin forms similar to what has been described for anthocyanins in model systems. The distribution of the various secondary structures will also under *in vivo* conditions depend on anthocyanin structure, most probably with consequences for the nutraceutical impact of the diet. However, the reader should be aware that more complex anthocyanins, like those reported to be in many vegetables, have a different distribution of their secondary structures than that suggested for simple anthocyanins in Figure 2.2. In fact, the equilibrium schemes of the more complex anthocyanins are far from being fully elucidated! For instance, when a covalently linked anthocyanin–flavone *C*-glycoside isolated from purple leaves of purple shamrock *Oxalis triangularis* dissolved in deuterated methanol and trifluoroacetic acid (95:5) was observed by NMR 45 min after sample preparation, the pigment occurred mainly as flavylium cation (38%) and two equilibrium forms assigned to be quinonoidal bases (54%) (Fossen et al. 2007). More simple anthocyanins are normally considered to be on their flavylium cation form in this solvent (Jordheim et al. 2006). The NMR results indicated the presence of vertical π–π stacking between the B-ring of the flavone unit and the A-ring of each of the two quinonoidal bases (Fossen et al. 2007). It was not possible to discriminate between inter- or intramolecular association mechanisms. Only minor amounts of the two hemiketal forms were present. After 5 days of storage at 27°C, the hemiketals (39%) and flavylium cation (38%) constituted the main forms of the pigment.

2.2.4 Glycosyl Units of Anthocyanins, Including Acyl Moieties

The glycosyl moieties of anthocyanins help to increase the water solubility of the pigments and to improve anthocyanidin stability. They are normally connected to the anthocyanidins with *O*-linkages, however, two Cy3-*O*-glycoside-8-*C*-glycosides have been isolated from the purple flowers of toad lily (*Tricyrtis formosana*) cultivar Fujimusume (Liliaceae) (Saito et al. 2003). It is possible to make 3-deoxyanthocyanidin *C*-glycosides from their respective flavone 6-*C*-glycosides (Bjorøy et al. 2009). Two 3-deoxyanthocyanidin 6,8-di-*C*-glucosides with *C–C* linkages between the sugar moieties and the aglycone were found to be far more stable toward acid hydrolysis than pelargonidin 3-*O*-glucoside, which has the common anthocyanidin *C–O* linkage between the aglycone and the sugar (Bjorøy et al. 2009). With the exception of the 3-deoxyanthocyanins, nearly all anthocyanins have a glycosyl moiety located at the 3-position, while additional moieties might be connected through their 5-, 7-, 3′-, 4′-, or 5′-hydroxyl groups. Among the food plants, red onion is extraordinary containing four Cy4′-glucosides (**21**) in their pigmented scales (Fossen et al. 2003a), while the alatanins isolated from edible purple yam (*Dioscorea alata*) (Yoshida et al. 1991) and one of the pigments (**19**) in tamarillo (*Solanum betaceum*) (Osorio et al. 2012) contain 3′-glucoside units (Figure 2.3).

The various mono-, di-, and trisaccharides found in anthocyanins are formed by glucosyl, galactosyl, rhamnosyl, arabinosyl, xylosyl, glucuronosyl, and apiosyl units. One or more glucosyl units have been identified in more than 90% of the various anthocyanins, while the most unusual glycosyl unit of anthocyanins, apiosyl, is hitherto found limited to five anthocyanins in the African milk bush (*Synadenium grantii*) (Andersen and Jordheim 2010). Cassava (*Manihot esculenta*) and several fruits like black olive (*Olea europaea*), elderberry (*Sambucus nigra*), and species in the genera *Ribes*, *Rubus*, and *Prunus* are rich in common anthocyanidin 3-disaccharides based on rutinosyl or sambubiosyl (Table 2.3). However, the disaccharide 2-(glucosyl)arabinoside has only been detected in a Dp-glycoside (**20**) from black lentil (*Lens culinaris*), and the disaccharide laminariobiose has a vey restricted occurrence in anthocyanins being identified mainly in anthocyanins (**21**) isolated from the genus *Allium* (Fossen et al. 2003a) and as part of a Cy3-trisaccharide (**22**) from purple asparagus (*Asparagus officinalis*) (Sakaguchi et al. 2008) (Figure 2.3).

More than 69% of the reported anthocyanins contain one or more acyl groups (Figure 2.4) linked to their sugar moieties, and anthocyanin properties are highly affected by the nature, number, and linkage positions of these acyl groups. Malonyl, which have been identified in 27% of the various anthocyanins, is the most frequently occurring acyl moiety of anthocyanins. In vegetables, this acyl group is part of some anthocyanins isolated from the family Brassicaceae (radish and red cabbage), genus *Allium* (red onion, chive, alpine leek, garlic), red leaf lettuce (*Lactuca sativa*), purple shamrock (*Oxalis triangularis*), and chicory (*Cichorium intybus*) (Table 2.4). In anthocyanins from fruits in our diet, it has been detected in blood orange (*Citrus*

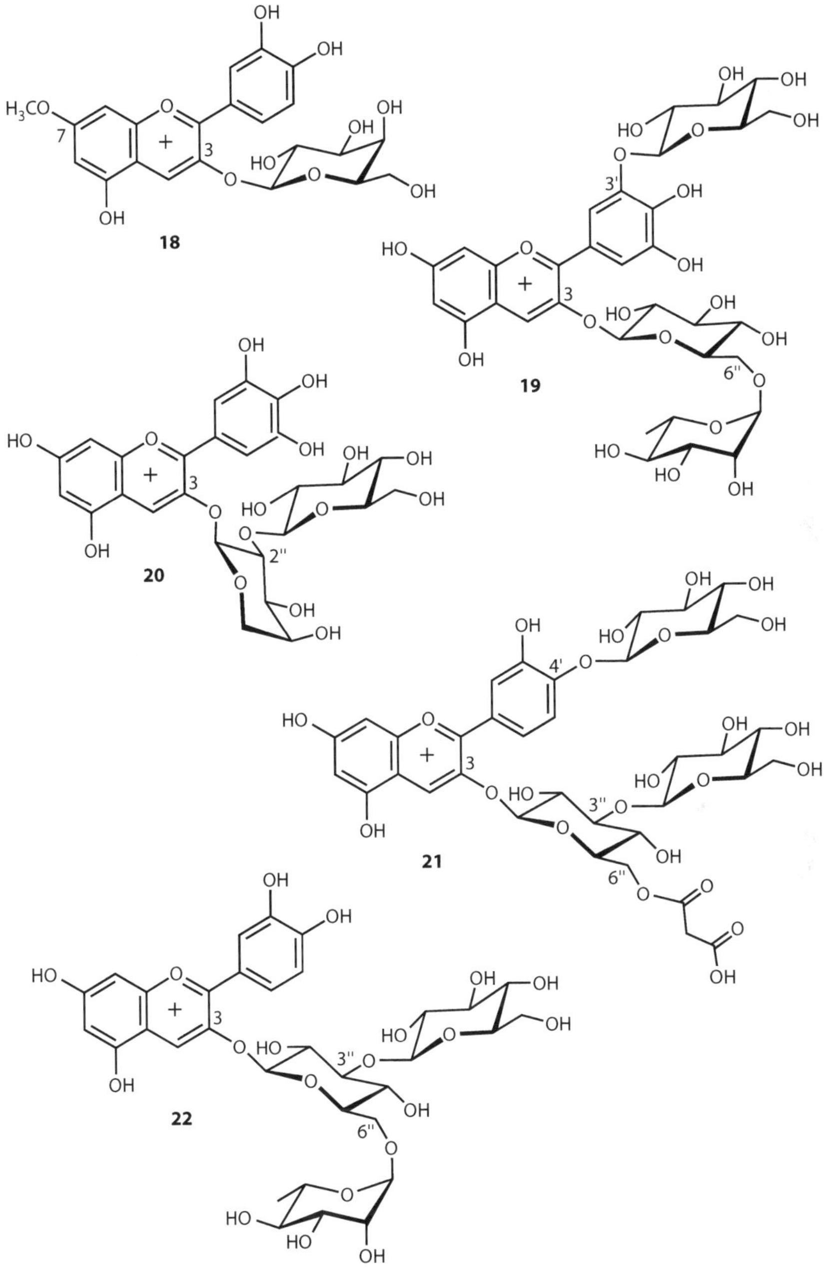

FIGURE 2.3 Structures of some rare anthocyanins, which have been found only in food plants. 7-*O*-methylCy3-galactoside (**18**) from mango, Dp3-[6-(rhamnosyl)glucoside]-3′-glucoside (**19**) isolated from tamarillo, Dp3-[2-(glucosyl)arabinoside] (**20**) from black lentil, Cy3-[3-(glucosyl)-6-(malonyl)glucoside]-4′-glucoside (**21**) from red onion, and Cy3-[3-(glucosyl)-6-(rhamnosyl) glucoside] (**22**) from purple asparagus.

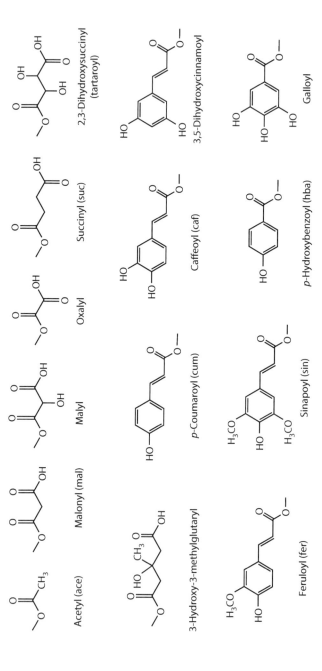

FIGURE 2.4 Structures of the aromatic and alipahtic acyl units, which have been found connected to a glycosyl moiety of acylated anthocyanins.

sinensis), sunflower (*Helianthus annuus*), passion fruit (*Passiflora suberosa*), and certain types of corn (*Z. mays*). Among the other aliphatic acyl groups of anthocyanins (Figure 2.4) detected in food plants, acetyl has been identified in anthocyanins isolated from grape (*Vitis* spp.), blood orange (*Citrus sinensis*), and litchi (*Litchi chinensis*), while 3-hydroxy-3-methylglutaryl has recently been identified in Cy3-[3-(hydroxy-3-methylglutaroyl)glucoside] from blackberries (*Rubus fruticosus*) (Jordheim et al. 2011a) and blood orange (Jordheim et al. 2011b). This anthocyanin has most probably previously been identified as Cy3-dioxalylglucoside in various blackberry samples and blood orange.

Around 380 different anthocyanins from plants have hitherto been reported to have aromatic acylation, which include five hydroxycinnamic and two hydroxybenzoyls acyl groups (Figure 2.4). One or more of the hydroxycinnamic acyl groups, *p*-coumaroyl, caffeoyl, feruloyl, and sinapoyl occur in most anthocyanins found in vegetables like red cabbage, red and purple raddish, eggplant, blue potatoes, and sweet potatoes (see Table 2.4). In anthocyanins from fruits, *p*-coumaroyl is nearly the only aromatic acyl group, isolated from extracts of grape (Mercadante and Bobbio 2008), the eggplant (Ichiyanagi et al. 2005c), and some gooseberry cultivars (Jordheim et al. 2007a). In these fruits, this type of anthocyanins occurs in relative small amounts.

Altogether 153 anthocyanins are reported to be acylated with both aromatic and aliphatic acyl groups. The largest monomeric anthocyanin recorded to date, ternatin A1 (from butterfly pea, *Clitoria ternatea*) (Terahara et al. 1990), is composed of Dp with seven molecules of glucosyl, four of coumaroyl, and one of malonyl (molecular mass: 2107 g/mol). In addition to aromatic and aliphatic acyl groups, two anthocyanins conjugated with sulfate, malvidin 3-glucoside-5-[2-(sulfato)glucoside] and malvidin 3-glucoside-5-[2-(sulfato)-6-(malonyl)glucoside], have been isolated from violet flowers of *Babiana stricta* (Iridaceae) (Toki et al. 1994). The acyl groups of anthocyanins are mostly linked to the sugar 6-position (86%), although 2-, 3-, and 4-acylations have been reported.

2.2.5 ANTHOCYANIN REACTIVITY AND STABILITY

Independent of the secondary form in which they occur (Section 2.2.3), the anthocyanins possess hydrophilic properties through the hydroxyl groups on their sugar moieties, anthocyanidin part and possible acyl groups, and have thus high hydrogen-bonding capacity both as hydrogen-bonding donor and as acceptor. As a consequence, the anthocyanins often are classified as water-soluble compounds. Around 29% of the anthocyanins also contain dicarboxylic aliphatic acyl groups, which render these anthocyanins as zwitterions, at least under acidic conditions, with a cation flavylium form and an anionic carboxylic group, which increase their polar nature further. However, the water solubility of anthocyanins depends furthermore on their aromatic rings, which to some degree give them a hydrophobic character. This effect is strengthened for those anthocyanins bearing aromatic or aliphatic acyl groups or both (69%), which provide these anthocyanins with a nearly amphiphilic (both hydrophilic and lipophilic) nature. The planar form of the aromatic rings allows intra- and intermolecular associations giving increased anthocyanin stability, which varies to some extent depending on the structural parts involved in the associations.

With focus on the reactivity of the anthocyanidin part of the anthocyanins, the chemistry is dominated by the aromatic rings with their delocalized π-electron system, which are influenced by the electron-donating oxygen functions of the various rings. This generates differences in the electron density at particular sites in the anthocyanidins. Specific positions will thus be more susceptible to react as nucleophiles or electrophiles. As shown in Figure 2.5, we can thus describe the major reactions of anthocyanidins according to different assets of the three rings (A–C). In addition to the specific reactions types indicated in Figure 2.5, the phenol groups (or more probably their phenolate anions) of the anthocyanidins can most probably be oxidized to generate radicals, which can be delocalized through resonance, participating in carbon–carbon and carbon–oxygen bond-forming couplings in oligo/polymeric compounds.

2.2.5.1 Reactions on the A-Ring: Formation of Heterodimer Anthocyanins

According to the biosynthetic route to anthocyanins, the oxygen-containing groups in the 5-, 7-, and 9-positions of the anthocyanidin A-ring are in *meta*-position to each other. Owing to mesomeric effects of these electron-donating groups, the carbon atoms in the 6- and 8-positions receive increased electron density. The chemistry of ring A is thus dominated by the nucleophilic nature of the carbon atoms in positions 6 and 8 (Figure 2.5), which facilitates electrophilic aromatic substitution in these positions. As a consequence, the hydrogens at the 6- and 8-positions are relatively

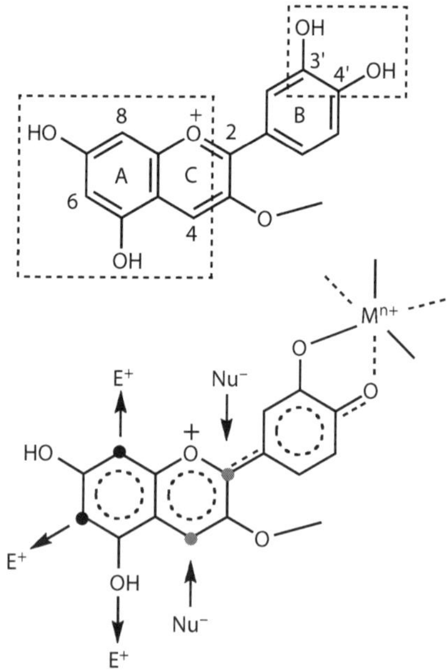

FIGURE 2.5 The major type of reactions for the three anthocyanidin rings (A–C), here presented in their flavylium cation form. E^+ and Nu^- are electrophiles and nucleophiles, respectively, which do not have to be formal charged molecules. The two ortho-positioned hydroxyl groups also facilitate *O*-methylation in formation of phase II metabolites, as well as involvement in (anti)oxidation reactions.

acidic and involved, for instance, in exchange with deuterium in deuterated solvents—in contrast to the hydrogens of the B- and C-rings, which are directly linked to the carbon skeleton (Jordheim et al. 2007c). The increased nucleophilic reactivity of the carbon 8-position, in particular, can be detected in aqueous media (juices, wines, etc.), where anthocyanins readily react with other phenolic compounds such as flavanols (catechins), giving rise to flavanol($4\alpha \rightarrow 8$)anthocyanins. This type of heterodimer compounds resulting from direct condensation between an anthocyanidin unit and a flavanol are usually assumed to be formed exclusively during storage and processing in plant-derived foods, including wines (Remy-Tanneau et al. 2003, He et al. 2006b). These heterodimers can also be detected in small quantities in extracts of unprocessed plants like strawberries (**5–8**), black currants, and various beans (see Section 2.2.2).

2.2.5.2 Reactions on the C-Ring: Hemiketals and Sulfites

The reactions of ring C are dominated by nucleophilic attack on the electrophilic carbon atoms at positions 2 and 4. In fact, the most important reaction of the anthocyanins is the water attack on C-2 of the flavylium cation forming the hemiketal forms (see Subsection 2.2.5.5 *Stability*, and Section 2.2.3). The electrophilic character of C-4 is mostly reported in connection with the use of SO_2 as preservative in the food industry. SO_2 (as bisulfite, HSO_3^-) will attack C-4 and form colorless adducts (Berké et al. 1998) in a reversible reaction. The ability of bisulfite to form colorless 4-adducts with free anthocyanins on their flavylium cation form, and not with polymeric anthocyanins (e.g., pyranoanthocyanins) having their 4-positions occupied, has been used to calculating the content of polymeric compounds in aqueous extracts, juices, and wines (Lee et al. 2002a, Giusti and Wrolstad 2005, Brownmiller et al. 2008).

2.2.5.3 Reactions Involving Both the A- and the C-Rings: Formation of Pyranoanthocyanins

Small organic molecules such as cinnamic acids, acetaldehyde, pyruvic acid, acetoacetic acid, vinyl-phenol, and so on may also react with anthocyanins. This happens in cycloaddition reactions involving the nucleophilic 5-OH group on the A-ring, the organic compound, and the electrophilic carbon atom at position 4 on the C-ring, which are followed by dehydratization, oxidation, and rearomatization steps creating different types of pyranoanthocyanins (Figure 2.6). This newly formed D-ring will thus have different types of substituents according to the "small organic molecule." The great variety of pyranoanthocyanin structures, namely carboxypyranoanthocyanins, methylpyranoanthocyanins, pyranoanthocyanin-flavanols, pyranoanthocyanin-phenols, portisins, oxovitisins, and pyranoanthocyanin dimers in red wines have recently been reviewed by de Freitas and Mateus (2011). Pyranoanthocyanins have also been identified in small amounts in extracts of strawberries (Andersen and Francis 2004) and red onions (Fossen et al. 2003b). Four reported methylpyranoanthocyanins isolated from black currant seeds were later shown to be the oxidative cycloaddition products of the extraction solvent (acetone) and the natural anthocyanins (Lu and Foo 2001).

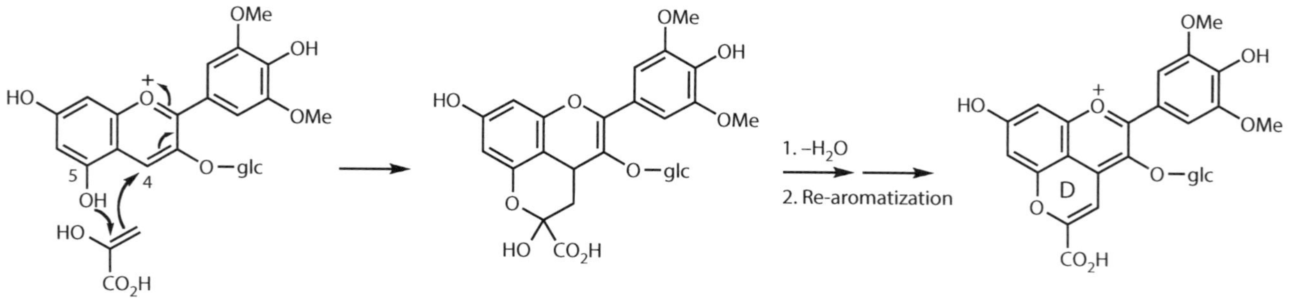

FIGURE 2.6 Postulated mechanism for the reaction between malvidin 3-glucoside and pyruvic acid forming carboxypyranomalvidin 3-glucoside (vitisin A). (Modified after Fulcrand, H. et al. 1998. *Phytochemistry* 47: 1401–7.) This is a typical example of a reaction going on between simple anthocyanins and small organic molecules in processed foods and wine during processing and storage.

2.2.5.4 Reactions on the B-Ring: Metal–Complexation, O-Methylation, and Antioxidant Activity

Owing to the biosynthetic route, anthocyanidins with more than one hydroxyl group on their B-ring (like Cy) have the hydroxyl groups in *ortho*-position to each other, which give no similar keto–enol tautomerism influencing the electron density of the carbon atoms of ring B, as experienced for ring A. Thus, the reactions of ring B depend to a large extent on the number and type of oxygen functions on this ring. The presence of at least two adjacent hydroxyl groups on the B-ring (like in Cy, Dp, or Pt) opens the door for metal chelation (Figure 2.5). However, it is just in <10 cases related to blue flower color that anthocyanins together with flavones/flavonols have been reported to be in complexation with metal ions as supramolecules in intact plants (Yoshida et al. 2009, Andersen and Jordheim 2010). A more discussed function of the *ortho*-positioned hydroxyl groups found in the B-ring of delphinidin (pyrogallol) and in cyanidin and petunidin is their catechol (more precisely phenol groups) of the B-ring, and their ability to act as antioxidants (Bueno et al. 2012). By donating a hydrogen atom to the free radical from one of the hydroxyl groups, the B-ring is made into a phenoxy radical. The purpose of the *ortho*-positioned hydroxyl group is to stabilize the resulting phenoxy radical. It is also interesting to note that a prerequisite for *O*-methylation in formation of phase II metabolites seems to be two hydroxyl groups in *ortho*-position to each other on the anthocyanidin B-ring (see Section 2.5.2).

2.2.5.5 Stability

Stability is a very important (and challenging) element when considering the impact of dietary anthocyanins on the human metabolism. The common simple anthocyanins show instability toward a variety of chemical and physical parameters, including oxygen, high temperatures, light, and most pH values. They react with sulfur dioxide, some cations and proteins, and will undergo transformations under storage, for instance during maturation of wine. Decolorization of anthocyanin solutions caused by enzymatic activity of glycosidases and polyphenoloxidases have been shown to vary with purity, product, and plant involved. On the other hand, under *in vivo* conditions the arrangements of the anthocyanins, often in supramolecules, and their surrounding matrixes will stabilize these compounds under conditions in which they normally would degrade as isolated compounds, as typical seen in flowers!

The stability of the anthocyanin molecules is highly linked to the distribution of the anthocyanins on their various secondary anthocyanidin forms (Brouillard and Dangles, 1994, Cabrita et al. 2000, Torskangerpoll and Andersen 2005). In this context, we suggest the extent of formation of hemiketal forms as the most critical component for experiencing anthocyanin instability. Under acidic conditions in room temperature, when the flavylium cation form predominates, simple anthocyanins seem to be rather stable for weeks. However, under near neutral conditions a simple anthocyanin like malvidin 3-glucoside dissolved in aqueous solutions with pH 6 and 6.5 showed only some faint color after 1 h and no color at all after 1 day of storage (Cabrita et al. 2000). This anthocyanin will partly exist on its hemiketal or chalcone forms under these conditions. The opening of the C-ring of the flavylium cation by hydration at the 2-position making hemiketal forms followed by chalcone formation (Figure 2.2) is postulated as the first degradation step of anthocyanins. In connection with this, the following factors are highlighted:

1. The common anthocyanidins with free 3-hydroxyl groups are degraded much faster than their corresponding glycosides (Iacobucci and Sweeny 1983, Furtado et al. 1993). All the anthocyanins found in the diet described in Section 2.3 have at least one O-glycosyl unit connected to the anthocyanidin 3-position, which stabilize the anthocyanins versus the anthocyanidins. Strong acidic conditions and enzymes will hydrolyze the glycosidic bonds in anthocyanins and anthocyanin derivatives, both in dietary products and in the human body. It is interesting to note that cyanidin 3-(2″-glucosylglucoside)-5-glucoside has been reported to be more unstable than cyanidin 3-glucoside at most pH values (Torskangerpoll and Andersen 2005).

2. Increased temperatures seem to increase degradation of simple anthocyanins since the equilibriums in Figure 2.2 are displaced toward the *trans*-chalcone C_Z form (Brouillard and Lang 1990). This effect is observed as paler anthocyanin solution after heating. Higher temperatures will also imply higher hydrolysis rates of the O-glycosyl moieties forming unstable anthocyanidins.

3. Anthocyanins with aromatic acyl groups (especially polyacylated anthocyanins) are relative stable in neutral or weakly acidic aqueous solutions. The increased stability is achieved by intramolecular association between the anthocyanidin and the aromatic acyl group(s), which makes it difficult for a water molecule to attack the anthocyanidin 2-position of the C-ring keeping the anthocyanidins in their stable forms (flavylium cation or quinonidal forms) (Nerdal and Andersen 1992, Dangles et al. 1993, Brouillard and Dangles 1994, Gakh et al. 1998, Noda et al. 2000, Matsui et al. 2001, Yoshida et al. 2009, Andersen and Jordheim 2010).

4. In principle, the same effect as described in (3) rule when anthocyanin stability increases in more concentrated anthocyanin solutions, and when co-pigmentation is observed. In these cases, inter-molecular association between anthocyanins (self-association effects), and between anthocyanins and other molecules protects against water attack on the anthocyanidin 2-position, and thereby the formation of hemiketal forms (Brouillard and Dangles 1994, Andersen and Jordheim 2010).

5. Anthocyanin derivatives with substituents connected to the C-ring, for instance pyranoanthocyanins, have different reactivity than the original anthocyanins, and are seen to be more stable in a wider pH range. This is because the substitution at C-4 of the flavylium cation affects the distribution of the charge throughout the C-ring. As a result, the C-2 and C-4 positions at the flavylium cation become less reactive toward nucleophilic attack (hydration), which increases the stability of this type of anthocyanins in weakly acidic and neutral aqueous solutions. It has thus been shown that C-4-substituted anthocyanins exhibit higher resistance to bleaching by

sulfur dioxide, higher color intensities, and restricted formation of hemik-etals under weakly acidic to neutral solvent conditions as compared to analogous anthocyanidin 3-glucosides (Bakker and Timberlake 1997, Mateus and de Freitas 2001, Andersen et al. 2004).

2.3 ANTHOCYANIN CONTENT IN FRUITS, VEGETABLES, AND PROCESSED FOODS

Most scientific literature written in English reporting major anthocyanins and the total anthocyanin content of fruits and vegetables used in the human diet have been tabulated in Tables 2.3 and 2.4. The division of the two tables is based on botanical and not culinary terms. For instance, avocado, eggplant, grains/cereals, peas, beans, lentils, and corn/maize have been listed as fruits, while rhubarb has been placed among the vegetables. Regarding the qualitative anthocyanin content of the various sources, those anthocyanins, which are estimated to occur in relative anthocyanin amounts higher than approximately 10%, have been included. However, a few of the sources have also been represented with minor compounds, since the relative quantitative content of individual anthocyanins in these sources has not been reported. To be able to present as much quantitative information as possible about each of the various sources, we have included publications reporting the quantitative content obtained by HPLC analysis as well as by spectrophotometric measurements made alone. Those publications, which are reporting the quantitative content based on anthocyanidins (and not anthocyanins) have, with exception of blood orange and rhubarb, been excluded from Tables 2.3 and 2.4. With focus on the anthocyanins in our diet, the qualitative content of fruits and vegetables is very different (Section 2.3.1).

With respect to the regulations on the use of anthocyanins as food colorants, in the United States the Food and Drug Administration (FDA) has a list of "natural" colors that do not require certification (Colorants Exempt from Certification). Anthocyanins can according to this list be obtained either from "grape color extract," "grape skin extract," or "fruit or vegetable juice." According to the list of colorants permitted in food in the European Union's directive, anthocyanins (as grape color extract, grape skin extract, concentrated juice of black currant, cherry, cranberry, elderberry, strawberry, red cabbage, etc.) have been given just one common code (E163), in contrast to carotenoids and carotenoid containing sources, which have been given several (sub) codes. This may indicate that anthocyanins, at least from a regulatory point of view, are looked upon as one homogeneous group of harmless compounds.

2.3.1 ANTHOCYANINS IN FRUITS AND VEGETABLES

In Tables 2.3 and 2.4, overall 101 different nonprocessed anthocyanin dietary plant sources have been surveyed, based on 154 different publications. Among the fruits, 81 species of 46 genera belonging to 25 families have been included, while 17 species in 13 families have been covered as vegetables. There is no reference to the anthocyanin content of different cultivars in these tables. With focus on structure diversity of the major anthocyanins of fruits and vegetables in the human diet, altogether 96 different anthocyanins have been reported. This constitutes 14% of the 702 different anthocyanins, which have been reported to occur in plants (Andersen 2012). Their molecular masses stretch from 433.1 Da for Pg3-glc, which is the major anthocyanin in strawberry, to 1509.4 Da for Alatanin A isolated from purple yam (*D. alata*). This latter anthocyanin has five glucosyl and two sinapoyl units connected to Cy.

When comparing the content of the vegetables in Table 2.4 and fruits in Table 2.3, it is very striking that among the different anthocyanins found in vegetables (51) and fruits (53), only 8 of them (Pg3-glc, Cy3-glc, Cy3-rha, Cy3,5-diglc, Cy3-[(6-(rha)glc], Dp3-[(6-(rha)glc], Cy3-[(6-(mal)glc], and Pt3-[6-(4-(cum)rha)glc]-5-glc) occur in both tables. It is also obvious that the anthocyanins in the vegetables are considerably more complex than those of the fruits: The proportion of simple anthocyanins without acyl groups and just one or two monosaccharide units is 16%. The corresponding number in the fruits of Table 2.3 is 74%. Above 77% of the anthocyanins in the vegetables have one or more acyl groups, and 70% contains one or more aromatic acyl groups (Figure 2.7, bottom). Among the anthocyanins of the fruits, 23% contains acyl groups and only 11% has one aromatic acyl groups. When seen together, an "average anthocyanin" representing all the anthocyanins of the fruits in Table 2.3 has a molecular weight of 556.3 Da, and consists of the anthocyanidin connected to

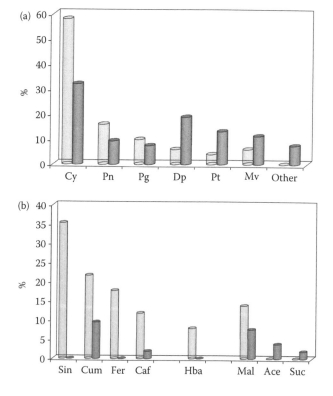

FIGURE 2.7 Comparison of relative occurrences of the various anthocyanidins (top) and acyl groups (bottom) in anthocyanins from vegetables (light gray bars) and fruits (dark gray bars) in human diet (from Tables 2.4 and 2.3, respectively). See Table 2.3 for abbreviations.

1.4 monosaccharide units and 0.1 aromatic acyl group. Likewise an "average antho-cyanin" of the vegetables (Table 2.4) has a molecular weight of 887.2 Da and consists of the anthocyanidin, 2.7 monosaccharide units and 0.9 aromatic acyl group. Most of the anthocyanins in the vegetables (84%) are based on anthocyanidins with just two or one oxygen function(s) on their B-rings (Cy, Pn, or Pg), while fruits contain relative more (43% versus 16% in vegetables) of anthocyanidins with three oxygen functions on their B-rings (Dp, Pt, and Mv) (Table 2.7, top).

Among the vegetables, the families Brassicaceae (red cabbage, radish), Apiaceae (purple carrot), Solanaceae (purple potato), and Convolvulaceae (purple sweet potato) are especially rich in anthocyanins with aromatic acylation. Aromatic acyl-ated anthocyanins based on Cy are typically found in red cabbage and purple car-rot, while aromatic acylated Pg-, Pn-, and Pt-glycosides prevail in radish, purple potato, and sweet potato, respectively. The anthocyanins of purple yam (*D. alata*) consist of extremely complex Cy-glycosides having up to five monosaccharides and two sinapoyl moieties. Anthocyanins with aliphatic acylation are common in Amaryllidaceae (red onion), Asteraceae (chicory, red leaf lettuce), and Oxalidaceae (purple shamrock). These anthocyanins are based on Cy in red onion, chicory, and red leaf lettuce, and Mv in purple shamrock.

The most commonly eaten anthocyanin sources belong to the fruits in family Rosaceae (blackberries, raspberries, strawberry, cherries, plums, apples) and the gen-era *Vaccinium* (blue- and bilberries, cranberries, etc.) and *Ribes* (black and red cur-rants, etc.), represented by 21, 13, and 7 species in Table 2.3, respectively. Most of the species in Rosaceae contain only Cy3-glc, or Cy-3-gal, or both 3-glc and 3-[6-(rha) glc] of Cy, as their major anthocyanins. Strawberry contains mainly Pg3-glc, while the 3-glc and 3-[6-(rha)glc] of both Cy and Pn occur in some plums. The anthocyanin con-tent of *Ribes* species is characterized with 3-glc and 3-[6-(rha)glc] of Cy. Black currant contains in addition the 3-glc and 3-[6-(rha)glc] of Dp, some cultivars of red currant contain as well Cy3-triglycosides, while some cultivars of gooseberries contain addi-tionally (and extraordinary), a relative high content of aromatic acylated Cy-glycosides. In *Vaccinium*, blue- and bilberries are characterized with mixtures of the 3-glc, 3-gal, and 3-ara of all the common anthocyanidins, except Pg. The red fruits of the *Vaccinium* species contain the same three simple 3-monoglycosides based on Cy (cowberry), or the same 3-monoglycosides based on both Cy and Pn (cranberries). Red grapes and their derived wine have for many people a unique position in the diet. The various *Vitis* species and cultivars contain different anthocyanins (Mercadante and Bobbio 2008), however, a typical representative like "Cabernet Sauvignon" contains the 3-glc of Dp, Pt, and Mv, and 3-[6-(ace)glc] and 3-[6-(cum)glc] of Mv as major anthocyanins.

The total anthocyanin content reported to occur in fruits varied from 2.5 mg/100 g (fresh weight, FW) in figs (*Ficus carica*) to 1734 mg/100 g in purple corn (*Z. mays*) (Table 2.3), and in vegetables from 0.2 mg/100 g in broccoli sprouts (*Brassica olera-cea*) to 1407 mg/100 g in sweet potato (*Ipomoea batatas*) (Table 2.4). Commonly consumed berries such as black currant, blue-/bilberries, and chokeberry (*Aronia melanocarpa*) have in general a very high anthocyanin content with reported maxi-mum values of 587, 1017, and 1480 mg/100 g (FW), respectively. Red and purple potatoes (*Solanum tuberosum*) and purple sweet potato (*I. batatas*) are the veg-etables with highest reported anthocyanin content, with maximum values of 550,

1383, and 1407 mg/100 g (FW), respectively. Anthocyanins, especially from color-ful potatoes and red cabbage, have been suggested as potential sources for food colorants (Opheim and Andersen 1992, Rodriquez-Saona et al. 1998), mainly because of their high content of relative stable anthocyanins with aromatic acylation. However, as seen from the data in Tables 2.3 and 2.4, the quantitative content of anthocyanins seem to vary a lot, even for the same species; The amount of anthocyanins in purple carrots (*D. carrota*) have been reported to be 4–1799 mg/100 g (DW), while the amounts in plums (*Prunus domestica*) and apples (*Malus sylvestris*) are 5–449, and 1–117 mg/100 g (FW), respectively.

Some of the factors, which influence the anthocyanin content in fruits and vegetables, and the reliability and reproducibility in determination of the anthocyanin content are discussed in Sections 2.4 and 2.6, respectively. It is complicated, if not impossible, to quantify all effects affecting the real anthocyanin content of the fruits and vegetables in our diet. It is thus challenging to operate with databases on anthocyanin content for estimation of daily intake, and so on. In this context, it is likely to improve the estimates by using averaged values obtained from a wide range of sources.

2.3.2 Anthocyanins in Beverages and Some Processed Foods

Color is one of the most important sensory quality attributes of food, and the major focus on anthocyanins in processed foods has been related to their use as natural food colorants. However, many people get a substantial part of their anthocyanin intake through consumption of processed plant products, with possible nutraceutical effects also from the derivatives and degradation products made from the anthocyanins in the products.

The qualitative and quantitative anthocyanin content in processed products derived from fruits and vegetables have, except for wine, been much less studied than the anthocyanin content of intact plant sources. In Table 2.5, the qualitative and quantitative anthocyanin content in various jams, juices, teas, wines, and some selected food products have been collected. These products are mainly based on fruits, except for the tea varieties of *Camellia sinensis* leaves, dried roselle calyxes from *H. sabdariffa*, and some purple-fleshed sweet potato (*I. batatas*) products available in Japan. Anthocyanins are to some extent also found in some dairy products, candies and as powder or mixtures in nutritional supplements/herbal remedies. Certain berries such as blueberries, blackberries, and black currants have very high anthocyanin content (Table 2.4), desirable flavors, and are consumed both in unprocessed and processed plant products, and most studies on anthocyanin content of plant products are related to those berries. Anthocyanin containing products derived from vegetables have minor occurrence on the commercial marked mainly due to less favorable aroma and flavor characteristics. However, some vegetable sources rich in anthocyanins such as purple sweet potato (*I. batatas*), black carrot (*D. carota*), and red cabbage (*B. oleracea*) have received high attention as coloring agents, mainly due to their content of relative stable anthocyanins.

By comparing Tables 2.3 and 2.5, it is seen that elevated anthocyanin contents of intact plant sources in most cases are, naturally enough, correlated with elevated anthocyanin contents in their derived products. Strawberry jam, for instance, has a

TABLE 2.5
Anthocyanin Content of Beverages and Processed Foods

Product	Content		References
	FW	DW	
	Fruits		
Jams and Marmalade			
Apricot[a]			Rababah et al. (2011)
a. Fruit		a. 2.5 mg/100 g	
b. 0 days storage, jam		b. 0.7 mg/100 g	
c. 5 months storage 25°C		c. 0.9 mg/100 g	
Cherry jam	10–30 mg/100 g[b]		García-Vigera et al. (1997)
	5–10 mg/100 g[a]		Kim and Padilla-Zakour (2004)
Cherry[a]			Rababah et al. (2011)
a. Fruit		a. 22 mg/100 g	
b. 0 days storage, jam		b. 7 mg/100 g	
c. 5 months storage 25°C		c. 5.6 mg/100 g	
Plum jam[a]	7–9 mg/100 g		Kim and Padilla-Zakour (2004)
Fig[a]			Rababah et al. (2011)
a. Fruit		a. 4.2 mg/100 g	
b. 0 days storage, jam		b. 1.7 mg/100 g	
c. 5 months storage 25°C		c. 1 mg/100 g	
Blackberry jam[b]	0.2–27 mg/100 g		Garcia-Vigera et al. (1997)
Black currant jam[b]	0.4–49 mg/100 g		Garcia-Vigera et al. (1997), Hollands et al. (2008)
Blood orange[a]			Rababah et al. (2011)
a. Fruit		a. 2.8 mg/100 g	
b. 0 days storage, jam		b. 0.8 mg/100 g	
c. 5 months storage 25°C		c. 0.7 mg/100 g	
Blood orange marmalade[a]			Licciardello and Muratore (2011)
a. 0 days of storage 20°C	a. 12 mg/100 g		
b. 60 days of storage 20°C	b. 2.4 mg/100 g		
c. 31 days of storage 35°C	c. 1.7 mg/100 g		
Blueberry jam[b]	18–22 mg/100 g		Garcia-Vigera et al. (1997)
Blueberry[a]			Howard et al. (2010)
a. 1 day fresh fruit	a. 601 mg/100 g		
b. 1 day after processing, jam	b. 471 mg/100 g		
c. 6 months storage 25°C, jam	c. 226 mg/100 g		
d. 6 months storage 4°C, jam	d. 313 mg/100 g		
Blueberry (without sugar)[a]			Howard et al. (2010)
a. 1 day fresh fruit	a. 601 mg/100 g		
b. 1 day after processing, jam	b. 472 mg/100 g		
b. 6 months storage 25°C, jam	c. 304 mg/100 g		
c. 6 months storage 4°C, jam	d. 386 mg/100 g		

continued

TABLE 2.5 (continued)
Anthocyanin Content of Beverages and Processed Foods

Product	Content		References
	FW	DW	
Raspberry jam	0.03–6 mg/100 g[b]		Garcia-Vigera et al. (1997),
	30 mg/100 g[a]		Kim et al. (2004)
Strawberry jam[b]	0.4–3 mg/100 g		Garcia-Vigera et al. (1997)
Strawberry[a]			Rababah et al. (2011)
a. Fruit		a. 232 mg/100 g	
b. 0 days storage 25°C, jam		b. 8 mg/100 g	
c. 5 months storage 25°C, jam		c. 7 mg/100 g	
Strawberry[a]			García-Vigera et al. (1999)
a. Fruit	a. 11–19 mg/100 g		
b. 0 days storage 20°C, jam	b. 6–12 mg/100 g		
c. 200 days storage 20°C, jam	c. 0.2–0.4 mg/100 g		
Juices, Health Drinks			
Black currant, 100% juice[b,c]	176–1298 mg/L		Weidel et al. (2011)
Black currant fruit drinks, 25–35% fruit content[b]	15–723 mg/L		Weidel et al. (2011)
Black currant juice[b]			Mattila et al. (2011)
German product	152 mg/L		
Polish product	128 mg/L		
Finnish product	48 mg/L		
British product	19 mg/L		
Black currant juices and concentrates[b]	1.4–492 mg/L		Nielsen et al. (2003)
Bilberry "nectar" (*V. myrtillus*)[b,c]			Müller et al. (2012)
65%	1529 mg/L		
70%	656 mg/L		
Bilberry juice (*V. myrtillus*)[b,c]			Müller et al. (2012)
100%	1610–5963 mg/L		
Blueberry "nectar" (*V. corymbosum*)[b,c]			Müller et al. (2012)
70%	258–386 mg/L		
Blueberry juice (*V. corymbosum*)[a,c]			Müller et al. (2012)
100%	417 mg/L		
Blueberry juice[b]	907 mg/L		Fanali et al. (2011)
Cherry juice[b]	54 mg/L		Fanali et al. (2011)
Cranberry juice[b]	2.2 mg/L		Prior et al. (2001)
Pomegranate juice[b]	43 mg/L		Fanali et al. (2011)
Black currant juice concentrate[b]	7800 mg/L		Bermúdez-Soto and Tomás-Barberán (2004)
Cherry juice concentrate[b]	200 mg/L		Bermúdez-Soto and Tomás-Barberán (2004)
Chokeberry juice concentrate[b]	8000 mg/L		Bermúdez-Soto and Tomás-Barberán (2004)

TABLE 2.5 (continued)
Anthocyanin Content of Beverages and Processed Foods

Product	Content		References
	FW	DW	
Elderberry juice concentrate[b]	11,300 mg/L		Bermúdez-Soto and Tomás-Barberán (2004)
Plum juice concentrate[b]	20 mg/L		Bermúdez-Soto and Tomás-Barberán (2004)
Raspberry juice concentrate[b]	400 mg/L		Bermúdez-Soto and Tomás-Barberán (2004)
Red currant juice concentrate[b]	200 mg/L		Bermúdez-Soto and Tomás-Barberán (2004)
Red grape juice concentrate[b]	1400 mg/L		Bermúdez-Soto and Tomás-Barberán (2004)
Strawberry juice concentrate[b]	80 mg/L		Bermúdez-Soto and Tomás-Barberán (2004)
Wine			
Red wine, vintage (1993–1998)			Sanchez-Moreno et al. (2003)
Cabernet Sauvignon	60 mg/L		
Chambourcin	112–170 mg/L		
Merlot	53 mg/L		
Montepulciano Sangiovese	95 mg/L		
Tempranillo	72 mg/L		
Villard Noir	59 mg/L		
Port wine			
Touriga Francesa (1997–1999)	40–120 mg/L (LA) 117–361 mg/L (HA)		Mateus et al. (2002)
Touriga Nacional (1997–1999)	187–483 mg/L (LA) 328–511 mg/L (HA)		
Port wine, mixtures of var. Touriga Nacional, Touriga Francesa, Tinta Roriz, Tinta Barroca			Pinho et al. (2012)
Ruby	92–156 mg/L		
Late Bottled Vintage	32–78 mg/L		
Tawny Reserve	0–41 mg/L		
Tawny 10 years	n.d.		
Vegetables			
Purple-fleshed sweet potato (*Ipomoea batatas*)[a,d]			
10 primary processed foods (dried powders, dried flakes, and dried sweet potatoes)		53–936 mg/100 g	Oki et al. (2010)

continued

TABLE 2.5 (continued)
Anthocyanin Content of Beverages and Processed Foods

Product	Content		References
	FW	DW	
13 beverages (including soft drinks and fermented vinegars)		1–132 mg/100 mL	Oki et al. (2010)
11 deep-fried foods (including chips and fried dough cookies)		33–376 mg/100 g	Oki et al. (2010)
27 secondary processed foods (including Western and Japanese sweets, and jams)		1–70 mg/100 g	Oki et al. (2010)
Tea, Dried Plant Material			
Leaf tea "Sunrouge" (*Camellia taliensis* × *C. sinensis*)		164–258 mg/ 100 g	Saito et al. (2011b)
Hibiscus, lixiviated roselle calyxes (*Hibiscus sabdariffa*)		181–236 mg/ 100 g	Daniel et al. (2012)

Notes: FW = fresh weight, DW = dry weight, LA = grown at low altitude, HA = grown at high altitude, n.d. = not detected.

[a] Processing and storage by scientific group.

[b] Processing and storage commercially.

[c] Pharmacies and health food shops.

[d] Dry weight (or fresh weight?).

reported mean value of approximately 1.7 mg anthocyanins/100 g compared to blueberry jam with 20 mg anthocyanins/100 g (Garcia-Vigera et al. 1997). Also expectedly, the percentage of anthocyanin containing fruit in the processed product is reflected in the anthocyanin content of the product. Weidel et al. (2011) have reported that black currant juice based on 100% fruit contained 176–1298 mg anthocyanins/L juice, while juice based on 25–35% fruit contained 15–723 mg/L. The same pattern has also been described for bilberry "nectars" and juices (Müller et al. 2012). However, the wide range in anthocyanin content of products based on the same type of fruits as seen in Table 2.5 is not always obvious for the consumer!

Mattila et al. (2011) used HPLC methods to study the polyphenol (including anthocyanin) and vitamin C content in commercial black currant juices purchased from various European countries (Finland, Poland, Germany, and the United Kingdom). With respect to the anthocyanin content in the 12 analyzed samples, there was a 14-fold variation from 17.2 in an UK sample to 232 mg anthocyanins/1 L juice in a German sample. The UK and Finnish products contained on average lower levels of anthocyanins than the German and Polish products (Table 2.5). Generally, the juice samples near to their sell-by dates showed much lower anthocyanin levels than fresh samples. The authors concluded that the variation within each juice brand was probably more influenced by different storage stages than different batches. Similarly, 13 commercially black currant juices and juice concentrates purchased from Danish grocery stores showed high variation of their total anthocyanin content ranging from 1.4 to 492 mg/L (Nielsen et al. 2003). It was concluded in this study that the large variation indicated that habitual dietary intakes of black currant anthocyanins cannot merely be calculated from the intake of black currant juices in general, but must be determined by intake data of specific beverages or by the use of a biomarker estimating the total intake of anthocyanins. Similar variation in anthocyanin content has also been reported for other processed foods like jams (Table 2.5), reflecting how product specification, methods of processing, storage time, and so on considerably influence the real anthocyanin content. For instance, it has been reported that anthocyanins were better retained in canned berries (66% in water, 72% in syrup) and puree (52%) than in clarified juice (41%) in processing of blueberries (Brownmiller et al. 2008). With respect to storage effects, it has been shown that at 20°C as much as 94% of the anthocyanins in strawberry jam was degraded after 200 days, 80% in blood orange marmalade after 60 days, while 52% of the anthocyanins in blueberry jam was gone after 60 days (Table 2.5).

What is happening with the anthocyanins during processing and storage? As described above, a considerable part of the anthocyanin content may degrade. Some details about the degradation pathways of anthocyanins in model systems influenced by heating and pH are described in Section 2.5.1. On the other hand, anthocyanins may also be transformed into other forms in processed foods. In red wines, the intact anthocyanins (mainly Mv3-glc and its derivatives) are present in considerable amounts in young wines. However, during wine aging, they are gradually replaced by other relative stable anthocyanin-derived compounds. The effects of processing and storage on anthocyanins have been studied thoroughly in relation to the chemical transformations going on in wine and their impact on the wine colors. During the last two decades, a variety of compounds have been suggested to contribute to red wine color, including anthocyanin oligomers, anthocyanin–flavanol polymers, linked together or through a methylmethine bridge, xanthylium salts, pyranoanthocyanins, and many others that remain to be described (Cheynier 2006, de Freitas and Mateus 2011). Some details about the mechanism involved in forming these anthocyanin derivatives are covered in Section 2.2.5. The effects of processing and aging can be exemplified by the reported content of anthocyanins and anthocyanin derivatives (mainly pyruvic acid adducts) in different port wines. Anthocyanins expressed as Mv3-glc equivalents were nearly absent in 10-year-old Tawny, while Ruby port wine contained 92–156 mg/L (Pinho et al. 2012). The lower amounts of anthocyanic compounds in old Tawny seemed to be caused by "oxidative" aging in wooden barrels. A basic mechanism for formation of Mv3-glc pyruvic acid adducts like vitisin A is shown in Figure 2.6.

The development of anthocyanin derivatives formed during wine fermentation and storage for years may look different from what is happening in processing and storage of jams and juices. But as in wine a rather complex heterogeneous class of polymeric compounds is formed in reactions between anthocyanins and other constituents in the food matrix, such as other flavonoids/phenolic compounds or molecules with relative low molecular weights, following the same basic principles as described in Section 2.2.5. To illustrate what is happening and also to indicate the positive potential of this type of reactions: Some vinylphenolic pyranoanthocyanins were identified after ferulic or sinapic acid was added to strawberry and raspberry juices (Rein et al. 2005). These anthocyanin derivatives, which were formed according to the same mechanism as described in Figure 2.6, both enhanced and stabilized the colors of the juices.

In a recent review on processing and storage effects on berry polyphenols, it has been concluded that formation of polymeric pigments accounts for some of the losses of anthocyanins and procyanidins during processing and storage, but the complete fate of anthocyanins remains unclear (Howard et al. 2012). The same group has therefore reported that new improved methods are required to stabilize the anthocyanins during heating and storage (Brownmiller et al. 2008). Thus, both degradation and derivatization of anthocyanins in processed foods have certain impact on the anthocyanin content, and most probably the nutritional properties of the foods (Pappas and Schaich 2009, Patras et al. 2010, Howard et al. 2012, Soto-Vaca et al. 2012).

2.3.3 Dietary Intake of Anthocyanins

Flavonoids as a group are the most abundant polyphenols in our diet, and anthocyanins, flavanols (catechins plus proanthocyanidins), and their derivatives constitute the major contributors. In the classical study by Kühnau in 1976, the average daily intake of anthocyanins in the United States was estimated to be 185 mg/person during the winter and 215 mg during the summer. In a more recent study where over 100 common foods in the United States were screened for anthocyanin, the daily intake was assessed to be considerably lower (12.5 mg/person) (Wu et al. 2006). Similar numbers (5–9 mg/day) have been estimated as the mean anthocyanin intake in UK adults obtained by diet diaries and a food composition database produced at the University of Surrey (Gosnay et al. 2002, Woods et al. 2003, Clifford and Brown 2006). However, the anthocyanidin (aglycone) consumption in Finnish adults has been estimated to be as high as 47 mg/day (Koponen et al. 2007). This outstanding intake in the Finnish population has been explained by relatively high berry consumption (52 g/day) among Finnish adults (Ovaskainen et al. 2008). In recent years, several groups have also used the USDA Database on Flavonoids for estimation of daily intake in different countries (Chun et al. 2007, Mullie et al. 2008, Zamora-Ros et al. 2010, Drossard et al. 2011). In this database, USDA scientists have converted all anthocyanidin–glycoside values into aglycone (anthocyanidin) values using conversion factors based on the molecular weight of the specific compound (USDA Database on Flavonoids, Release 3, Bhagwat et al. 2011). The estimates vary from 3.1 mg/day for US adults (Chun et al. 2007) to 18.9 mg/day in a Spanish adult population (Zamora-Ros et al. 2010). A higher consumption of anthocyanins in the Mediterranean countries has been explained by a higher intake of red wine, which was estimated to be 46% of the total anthocyanidin intake (Zamora-Ros et al. 2010, Drossard et al. 2011).

It is difficult to estimate anthocyanin consumption homogenously and properly for several reasons. The estimated intakes are often based on different techniques for assessing dietary intake (Chun et al. 2007, Mullie et al. 2008, Zamora-Ros et al. 2010, Drossard et al. 2011), and structural diversity of more or less fragile anthocyanins (and their continuously formed derivatives in processed food) complicates identification and quantification. As seen from Tables 2.3 through 2.5, it is indeed experienced that the qualitative and especially the quantitative anthocyanin content within a certain type of foods can vary tremendously. Despite these considerations, the more recent estimates presented above indicates that the average adult intake of anthocyanins in Europe and the United States is in the scale of 10 mg/day.

Finally, our opinion is that anthocyanin databases used for estimates of anthocyanin intake should differentiate more on structural differences between anthocyanins occurring in the various parts of our diet. As seen from Tables 2.3 and 2.4, the anthocyanins identified in fruits and vegetables are remarkably different. Fruits contain in general unacylated simple anthocyanins, while vegetables have been reported to have high amounts of aromatic acylated anthocyanins. Aromatic acylated anthocyanins are, as discussed in Section 2.2.5, considered to be more stable than nonacylated anthocyanins, and they seem to have different properties, including different hydrolysis rates, phase II derivative formation and bioavailability, than more simple anthocyanins. A more differentiated estimate of specific types of anthocyanins in the diet will most probably improve the value of epidemiological analysis including these compounds. Such epidemiological analysis might provide a better understanding of the relationship between intake of specific anthocyanins and possible development/prevention of various diseases.

2.4 FACTORS INFLUENCING THE ANTHOCYANIN CONTENT IN FRUITS AND VEGETABLES

The reported anthocyanin content in plants depends upon an assortment of factors (Figure 2.8), as experienced by the variation in results presented for each species in Tables 2.3 and 2.4. Both the qualitative and quantitative anthocyanin contents are obviously under control of complex interactions between genotype, environmental factors, and agricultural practice. Many studies have focused on elucidating these interactions with the purpose of obtaining increased yield of anthocyanins. The variation in anthocyanin content caused by the influence of temperature, sun radiation, shading, water deficiency, mineral nutrition, plant hormones, and so on, has in particular been examined systematically in fruits of the family Rosaceae (apples, pears, peaches, cherries, plums, strawberries) and in grapes (Vitaceae) (Awad and Jager 2002, Mozetic et al. 2004, Steyn et al. 2009, Crespo et al. 2010, He et al. 2010a, Telias et al. 2011, Vega et al. 2011, Aaby et al. 2012). In a recent review, the current progress on the development of cell cultures for the production of anthocyanins has been reported (Deroles et al. 2009). In this review, the processes that enhance anthocyanin production as well as factors that place limits on the final yield were detailed. In addition to the real variation in anthocyanin content caused by genotype diversity and phenotype plasticity occurring in plants belonging to either the same species, cultivar or population, variations due to different ripeness stages (Bureau et al. 2009), and the impact of sample storage and processing (see Section 2.3.2) have also to be considered.

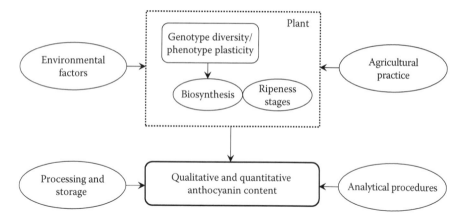

FIGURE 2.8 General scheme indicating factors influencing the anthocyanin content in fruits and vegetables.

2.4.1 GENETIC DIVERSITY

Anthocyanins are end-products in the flavonoid pathway, and they are found to be synthesized under complex regulation of multiple genes at the transcriptional level (Springob et al. 2003). Floral pigmentation caused by anthocyanins has been used to help elucidation of fundamental genetic principles since the days of Mendel, and knowledge acquired through understanding of the various steps in flavonoid biosynthesis is today used in genetic engineering to expand the horticultural gene pool.

The genetic relationships between plant taxa are in many cases expressed by their qualitative anthocyanin content, even on the family level. For instance, among the 98 anthocyanins, which altogether have been identified from species in the family Brassicaceae (formerly Cruciferae), all but four anthocyanins are acylated, and all but one can be classified as either an anthocyanidin 3-sophoroside-5-glucoside or an anthocyanidin 3-sambubioside-5-glucoside (Andersen and Jordheim 2006, Tatsuzawa et al. 2012). The first glycosidic pattern (3-sophoroside=3-[2-(glc)glc]) is found in species belonging to genera *Raphanus* (radish), *Brassica* (cabbage), *Iberis, Sinapis,* and *Moricandia,* while the latter (3-sambubioside=3-[2-(xyl)glc]) is found in *Arabidopsis, Cheiranthus, Heliophila, Lobularia, Lunaria, Malcolmia, Matthiola,* and *Orychophragonus.* When comparing just red cabbage and radish, it is clear that the anthocyanins in red cabbage are mainly Cy-glycosides, while the radish cultivars contain very similar Pg-glycosides. Recently, the juice of *Raphanus sativus* cv. Sango sprouts was, however, shown to contain Cy-glycosides (Matera et al. 2012)!

The qualitative anthocyanin content reported for the various *Vaccinium* species (blueberries, bilberries, cranberries, cowberries) indicates anthocyanin-type variation on the genus level. All of the *Vaccinium* species in Table 2.3 have been collected in Europe or in North America. They contain combinations of anthocyanins based on three 3-monosaccharides (3-glc, 3-gal, and 3-ara) linked to the anthocyanidins, Dp, Cy, Pt, Pn, and Mv (see Table 2.1 for abbreviations). However, berries from *V. japonicum,* which is native to Eastern Asia, contain the 3-arabinosides of Pg and Cy as major anthocyanins (Andersen 1987). Pg-glycosides are relatively infrequent in the diet (except for strawberry and radish), and the findings in *V. japonicum* might be beneficial for future breeding within *Vaccinium*. From a chemotaxonomic point of view, it is also interesting to note the presence of three anthocyanidin 3-triglycosides, six anthocyanidin 3-diglycosides, and Dp3-rha in berries of *V. padifolium*, in addition to the "normal" 15 anthocyanidin 3-monosaccharides occurring in the *Vaccinium* species mentioned in Table 2.3 (Cabrita et al. 2000). *V. padifolium* is endemic to Madeira island.

In *V. vinifera* grapes, the anthocyanins are 3-monoglucosides, since they are lacking the dominant allele involved in the production of anthocyanidin 3,5-diglucosides, while non-*V. vinifera* species are able to produce anthocyanidin 3,5-diglycosides (Jánváry et al. 2009). As a consequence, most red French–American grape hybrids synthesize anthocyanidin 3,5-diglycosides. The identification of these anthocyanins can thus be used to detect the usage of interspecific hybrid grapes in Appellation-controlled red wines. When comparing grapes of different *V. vinifera* cultivars, some cultivars like "Pinot Noir," red "Chardonnay," and pink "Sultana" are not able to synthesize acylated anthocyanins, opposite to the majority of red *V. vinifera* cultivars (Boss et al. 1996, Mazza et al. 1999).

The anthocyanin content in *V. myrtillus* has been subjected to analysis on the population level including berries from 179 individual plants in 20 populations growing on a 1000 km south–north axis in Finland (Lätti et al. 2008). Extensive variation in the anthocyanin content within and between the populations was observed; a significantly lower total anthocyanin content was observed in the berries growing in the southern region compared to those growing in the central and northern regions. The Dp3-glycosides dominated in berries of the northern populations, whereas the Cy3-glycosides were most common in southern populations. Comparable studies on bog bilberries (*V. uliginosum*) revealed similar results showing the highest total anthocyanin content and proportions of Dp3-glycosides in populations growing in northern Finland (Lätti et al. 2010). The authors of these latter studies concluded that the northern climate appeared to favor biosynthesis of more hydroxylated anthocyanidins like Dp. If the results are a consequence of the northern climate conditions or genetic adaptation on the growth environment is not clear (Jaakola and Hohtola 2010).

In connection with variation of the quantitative anthocyanin content, Cho et al. (2004) have demonstrated the success of achieving increased anthocyanin content by advanced breeding selection. The anthocyanin content of fully mature fruits from five genotypes of blueberries (*Vaccinium*) and red wine grapes (*Vitis*) and six genotypes of blackberries (*Rubus*) have been analyzed. The different genotypes with advanced breeding had in general higher total anthocyanin content than those without. Several papers have reported a higher anthocyanin content in some wild species (for instance, raspberries and blackberries) compared to cultivated ones (Cekic and Oezgen 2010, Cuevas-Rodriguez et al. 2010), and further exploration of the wild genetic resources for improving fruit and processing quality might be beneficial. Liang et al. (2012) have characterized the polyphenolic composition and content in ripe berries of 147 grape accessions from 16 *Vitis* species for two consecutive years. They observed tremendous variation for all examined compounds, and documented much higher content of many of these compounds in the wild species compared to those reported for the most widely cultivated species of *V. vinifera.*

2.4.2 ENVIRONMENTAL FACTORS AND AGRICULTURAL PRACTICE

Environmental factors and agricultural practice, including light/UV irradiation, temperature, water availability/irrigation, and mineral nutrition/fertilization, have for a long time been known to influence the anthocyanin biosynthesis and accumulation (McClure 1975, Steyn 2009b), mainly with respect to the quantitative anthocyanin content in fruits and vegetables. Light, for instance, is assumed to be one of the most important stimulators for anthocyanin production. During red coloration of lettuce leaves after UV-B irradiation, some genes were shown to be upregulated (Park et al. 2007). Low and high temperatures similarly have been reckoned to either increase or decrease anthocyanin accumulation, respectively (Piero et al. 2005, Ubi et al. 2006, Mori et al. 2007, Crifó et al. 2011). Several endogenous and exogenous chemical factors have also been found to regulate anthocyanin biosynthesis (Deroles 2009, He et al. 2010a). It has for instance been reported that sugars can enhance the expression of the flavanone 3-hydroxylase gene, and thus induce accumulation of anthocyanins (Zheng et al. 2009), while low-level ethanol can trigger grape gene expression leading to anthocyanin accumulation during ripening (El Kereamy et al. 2002). Similarly, phytohormones, such as abscisic acid, jasmonic acid, and ethylene, have been found to induce enhancement in the anthocyanin content (Jeong et al. 2004, Moreno et al. 2010). Low concentrations of 2,4-dichlorophenoxyacetic acid in the medium used for strawberry suspension cultures have, for instance, limited cell growth and enhanced both anthocyanin production and anthocyanin *O*-methylation (Nakamura et al. 1998). In the strawberry suspension cells, mainly Cy3-glc and Pn3-glc were produced. The ratio of Pn3-glc to the total anthocyanin content increased significantly under the influence of 2,4-dichlorophenoxyacetic acid, most probably because the activity of anthocyanin methyltransferase was enhanced (Nakamura et al. 1998). It is interesting to note that a *O*-methylated anthocyanin like Pn3-glc is not found in intact strawberries, which contains Pg3-glc as a major pigment.

Long-term studies that focus on the gene–environment interaction in connection with flavonoid biosynthesis have been scarcely reported (Jaakola and Hohtola 2010), but during the last years more publications and reviews have been aimed at explaining the complex interactions between anthocyanin biosynthesis, anthocyanin biosynthesis genes, and environmental factors in the regulation of anthocyanin metabolism (He et al. 2010a, Dai et al. 2011, Vega et al. 2011). Recent genetic and molecular studies have characterized a diversity of biosynthetic genes involved in anthocyanin production and transportation. The sequencing of the grapevine genome has for instance now been completed (Jaillon et al. 2007, Velasco et al. 2007, Fournier-Level et al. 2009), and new regulatory genes shown to be sensitive to environmental conditions, leading to changes in the anthocyanin accumulation, have been identified in various fruits (Hugueney et al. 2009, Matus et al. 2009, Feng et al. 2010, Dai et al. 2011). The challenge today is to fully decipher the molecular bases for quantitative genetic variation in highly diversified resources, and to integrate the existing phenotypic variation extensively (Nordborg and Weigel 2008, Castellarin and Di Gaspero

2007). Ongoing climate changes will further increase the need for evaluating genetic diversity and phenotypic plasticity.

2.5 ANTHOCYANIN DERIVATIVES AND DEGRADATION PRODUCTS

The anthocyanins in natural systems are highly influenced by their physiological conditions, including pH and other compounds and enzymes, which might participate in the condensation, oxidation, and hydrolysis reactions of the various anthocyanins (see Sections 2.2.3, 2.2.5, and 2.3.2). For understanding of the exact impact of dietary anthocyanins on various health aspects, it is vital to know the exact nature and amount of the circulating anthocyanin metabolites coming from human consumption of anthocyanins, including derivatives and degradation products. At present, this type of knowledge is lacking to a large extent due to the complex nature of anthocyanin chemistry. Even when considering turnover and degradation of anthocyanin in plants, very little is known about their catabolism (Oren-Shamir 2009). Therefore, several of the studies related to the field have evaluated just one specific parameter, for instance, the effect of temperature on anthocyanin degradation, often performed in simplified model systems, sometimes based on pure anthocyanins in aqueous buffer solutions. However, the most detailed studies on anthocyanin stability and degradation have been carried out with fruit extracts. The results from even this type of studies are still far from being complete (e.g., Section 2.5.1). Nevertheless, they give a good background for understanding the formation of possible derivatives and degradation products of anthocyanins in intact plants, processed plant products, or human metabolism. The procedures for preparation and analysis of flavonoid phase II metabolites of mainly quercetin and catechins have recently been reviewed by Santos-Buelga et al. (2012).

2.5.1 ANTHOCYANIN DEGRADATION PRODUCTS IN FOOD PRODUCTS AND MODEL SYSTEMS

The anthocyanins in food products are influenced by various factors, such as heat, pH, light, and H_2O_2. They react with sulfur dioxide, some cations and proteins, other phenolic compounds (see Section 2.2.5), and so forth, and are highly influenced by enzymatic activity of glycosidases and polyphenoloxidases. Among the factors affecting anthocyanin stability both in foods and in model solutions, the effects of pH and heating have been studied thoroughly. Most of these studies have been concentrating on whether pH or heating separately influence anthocyanin degradation. However, when the various proposals for thermal degradation pathways in the literature were summarized, the strong influence of pH conditions for shaping the different pathways was for us more obvious than previously reported. With this in mind, we have proposed three different thermal degradation pathways of anthocyanins (Figure 2.9 a–c) modified after results reported in this field.

Hrazdina (1971) has reported that common anthocyanidin 3,5-diglucosides during heating in solutions with a pH around 7 formed 3,5-diglucosyl-7-hydroxy-coumarin with loss of the anthocyanidin B-ring (Figure 2.9a). It is important to note that the

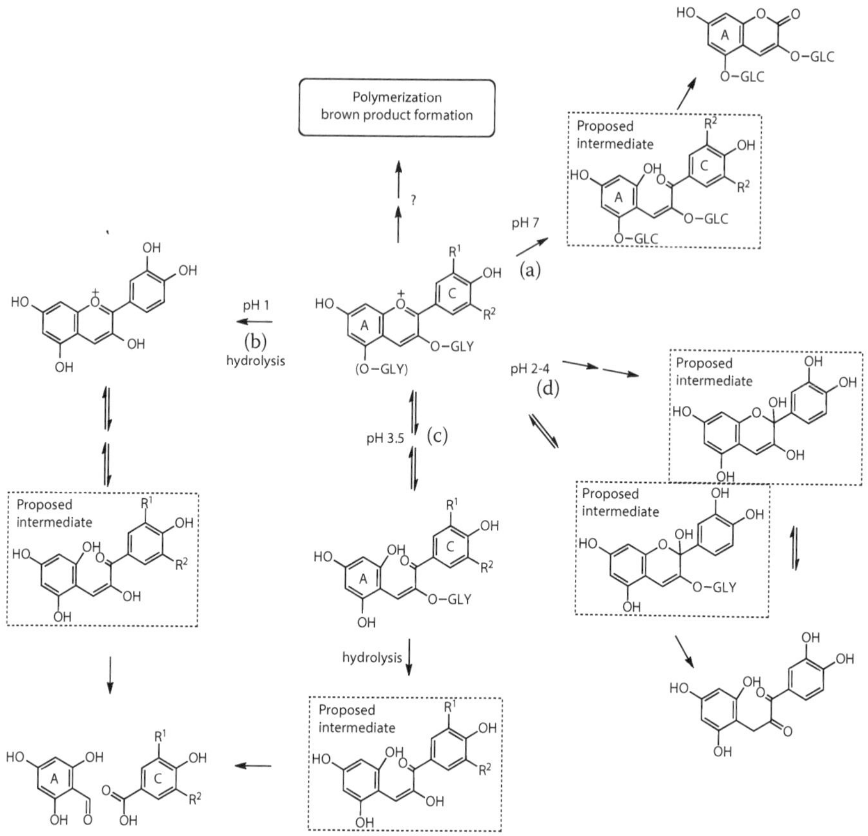

FIGURE 2.9 Thermal degradation pathways of simple anthocyanins influenced by pH. The various pathways (a–d) are modified from literature (see text). GLY = glycosyl, GLU = glucuronosyl.

degradation product (coumarin-glucoside) contained the two glucosyl units, which had not been hydrolyzed from the anthocyanidin fragment at this pH. It was also assumed that the formation of the coumarin-glucoside required the presence of the anhydrobase (quinonoidal form), since formation of the coumarin-glucoside seemed to be directly proportional to the decomposition of the anhydrobase (Hrazdina 1971). However, this can be a misinterpretation of observations since the anhydrobase of these anthocyanins at pH 7 most probably is formed as relative unstable kinetic products (see Section 2.2.3). Hrazdina did as well state that the anhydrobase faded and finally became yellow as a result of the transformation of the anhydrobase to a chalcone (even though this was not a part of the mechanism, which was proposed (Hrazdina 1971)). Anyway, we have proposed the chalcone form (instead of the quinonoidal base) as intermediate in the degradation pathway represented by Figure 2.9a.

At pH 1, the situation seems to be completely different. When acylated and non-acylated anthocyanins in purified strawberry, elderberry, and black carrot samples were heated at 95°C at pH 1, they were cleaved by successive loss of the sugar moieties (Sadilova et al. 2006). The aglycones (Pg and Cy) were further degraded by scission into 2,4,6-trihydroxybenzaldehyde (from the A-rings of Cy, Pg), 4-hydroxybenzoic acid (B-ring of Pg), and 3,4-dihydroxybenzoic acid (protocatechuic) (B-ring of Cy), respectively (Figure 2.9b). It was also reported that intermediate degradation products such as chalcones could not be detected because of the low pH value favoring the flavylium cation, and the flavylium form was suggested cut into the final degradation products (Sadilova et al. 2006). However, in a thermal degradation study of four common anthocyanidins (Pg, Cy, Dp, and Mv) in aqueous acidic media, it has been reported that all the anthocyanidins showed a similar thermal degradation pattern, and that chalcone was detected as the intermediary product undergoing cleavage yielding the final decomposition products, even at low pH values (Furtado et al. 1993). We suggest that the degradation products of anthocyanins following route (b) similarly are formed by scission of the chalcone (or flavylium cation?) and have proposed this in Figure 2.9b, even at pH 1. The reported lack of detection of the chalcone by Sadilova et al. (2006) can be explained by Le Chatelier's principle; the flavylium cation can be in equilibrium with small amounts of the hemiketal form, which immediately is converted into the labile chalcone.

Sadilova et al. (2007) have reported that the thermal degradation pathway of anthocyanins is pH-dependent. When they investigated the structural changes of the same anthocyanins as described above (Sadilova et al. 2006), they observed that the first step in thermal degradation at pH 3.5 was the opening of the pyrylium ring giving chalcone-glycosides (Figure 2.9c). The chalcone-glycosides would then be cleaved to yield the corresponding chalcones, which degraded to the same final products as observed in the experiments performed at pH 1 (Sadilova et al. 2006). We have also included in Figure 2.9a,d route, which is in accordance with the breakdown of several Cy3-glycosides at 100°C in the pH range 2–4 (Adams 1973). It was here reported that the essential feature was the lability of the aglycone-sugar bond to hydrolysis prior to the formation of a α-diketone of Cy. Hemiketal forms were suggested as intermediates in the formation of this α-diketone, however, the hydrolysis step was indicated both from the flavylium cation and from the hemiketal form. We suggest that this α-diketone might be looked upon as the keto-tautomer of the chalcone (the enol-form). Anyway, this compound was stable under anaerobic conditions but it easily degraded in atmospheres of oxygen (Adams 1973).

Therefore, the final thermal degradation products of anthocyanins seem to be 2,4,6-trihydroxybenzaldehyde originating from ring A, and the 4-hydroxybenzoic acid derivatives originating from ring B (e.g., 3,4-dihydroxybenzoic acid for Cy and 4-hydroxy-3,5-dimethoxybenzoic acid for Mv) at pH values below 7. In some cases, phloroglucinol (originating from ring B) or 2,4,6-trihydroxyphenylacetaldehyde (originating from ring B and partly from ring C) has been suggested. However, some papers have mentioned explicit that these latter compounds have not been detected in their analysis. At pH 7, nonhydrolyzed coumarins have been detected as thermal degradation products. In a study where several Cy3-glycosides were degraded at 100°C in weakly acidic solutions, the anthocyanidin-sugar bond was at pH 2–4 found to be most labile, while at pH 1.0 all the glycosidic bonds within the glycosyl moieties also were susceptible to hydrolysis (Adams 1973). When reflecting what type of secondary anthocyanidin form of the various anthocyanins, which really is degraded, the flavylium cation, hemiketal, and chalcone forms have been suggested in various papers examining this issue at various pH levels. Because of the equilibriums between the flavylium cation, the hemiketals and the labile chalcone, we suggest in Figure 2.9 that the C15 skeletons of the anthocyanins are degraded at elevated temperatures at all pH values between 1 and 7 from their chalcone forms. However, the situation might be different when the anthocyanidin (not anthocyanin) is the precursor of the degradation products according to route b (Figure 2.9), since the well-known equilibrium scheme reported for simple anthocyanidin 3-glycosides (Figure 2.2) has similarly not been established for anthocyanidins.

2.5.2 Intact Anthocyanins and Their Phase II Derivatives in Humans and Animals

Despite considerable efforts, the understanding of what is happening with anthocyanins from the diet in humans and experimental animals still remains very vague. Many studies have indeed demonstrated that anthocyanins from the diet occur as intact anthocyanidin glycosides in urine and blood plasma (see, for instance, Matsumoto et al. 2001, Murkovic et al. 2001, Wu et al. 2002, McGhie et al. 2003, Netzel et al. 2005, Manach et al. 2005). Several studies have also demonstrated that anthocyanins from the diet can be metabolized (phase II reactions) into O-methylated and/ or glucuronidated and possibly sulfonated derivatives (Wu et al. 2004b, Kay et al. 2005, El Mohsen et al. 2006, Felgines et al. 2007). The proportion of the absorbed anthocyanins including their metabolized derivatives has, however, been found to be very low. The typical urinary excretion of the total amount of anthocyanins and their derivatives has in most studies been far <1% of the ingested amount. Most of the anthocyanins are thus assumed by several groups to be degraded into various phenolic acids and their derivatives by intestinal conditions/bacteria (see Section 2.5.3), or they may interact with the cells of the gastrointestinal tract in their intact form, even without being absorbed (He et al. 2005).

2.5.2.1 Effects of Anthocyanin Structure on Absorption

There are several attributes of the various anthocyanin structures, which seem to influence the rate and extent of absorption, and the degree and type of derivatization. Several studies mainly on simple anthocyanidin 3-monoglycosides or 3-disaccharides from common fruits of our diet have indicated different impact of the various B-rings of the anthocyanidins (e.g., Wu et al. 2004b, 2005, Felgines et al. 2007, Carkeet et al. 2008). After a bilberry extract has orally or intravenously been administered to rats, the half disappearance time of anthocyanins in blood plasma was found to be in the following order Dp > Cy > Pt = Pn > Mv, when anthocyanins carrying the same sugar moiety on different anthocyanidins were compared (Ichiyanagi et al. 2006a). However, in the same study, it was summarized that gastrointestinal uptake of anthocyanins to a larger extent was governed by the 3-glycosidic moiety; the anthocyanidin 3-galactosides were absorbed more effectively than anthocyanidin 3-arabinosides. In this context, it is also interesting to note that after marionberry, elderberry, or black currant were consumed by pigs, most of the anthocyanidin

3-monoglucosides were excreted in the form of anthocyanin derivatives, whereas over 80% of the anthocyanidin 3-disaccharides containing rutinosyl or sambubiosyl were excreted as intact molecules (Wu et al. 2004b, Wu and Prior 2005). When urine and plasma samples were collected from 32 rats receiving chokeberry-, bilberry-, and grape-enriched diet for 14 weeks, all the anthocyanidin 3,5-diglucosides from grape were consistently excreted in the urine in higher proportion than their corresponding 3-monoglucosides (He et al. 2006a). Thus, it seems that more than one monosaccharide unit as part of the anthocyanin can prevent or restrict their conversion into anthocyanidin-glucuronides.

Vegetables are rich in anthocyanins containing aromatic acyl groups (Table 2.4). Acylated flavonoids have generally been recognized as nonabsorbable in the small intestine owing to their larger molecular size and lack of a free sugar moiety for transporter binding (He and Giusti, 2010b). Several studies have, however, reported that Cy-glycosides with aromatic acyl groups (like those from red cabbage, purple carrots) (Kurilich et al. 2005, He et al. 2006a, Novotny et al. 2012) and Pt3-[6-(cum)glc] from grapes (He et al. 2006a) have been absorbed as intact anthocyanins, although they showed substantial lower bioavailability than nonacylated anthocyanidin-glycosides.

2.5.2.2 Various Anthocyanin Phase II Metabolites

When considering derivatization of anthocyanins after ingestion (phase II reactions), *O*-methylation of hydroxyl groups of the anthocyanidin B-ring by catechol-*O*-methyl transferase (Figure 2.10) seems to be a major metabolic pathway for some nonacylated anthocyanins (Miyazawa et al. 1999; Tsuda et al. 1999, Wu et al. 2002). However, a prerequisite for this *O*-methylation seems to be two hydroxyl groups in *ortho*-position to each other on the anthocyanidin B-ring, a catechol or pyrogallol unit as found in Cy-/Pt- and Dp-glycosides, respectively. According to this, as far as we know, no *O*-methylated Pg-derivative has been reported formed after intake of Pg-glycosides, since Pg is lacking this catechol B-ring. On Cy-glycosides *O*-methylation seems to occur on either the 3'- or 4'-position of the anthocyanidin B-ring, with the 3'-position as the preferred position for *O*-methylation (Ichiyanagi et al. 2005a). For Pt-glycosides, it has been shown that catechol-*O*-methyl transferase has catalyzed the B-ring methylation of Pt3-glc into Mv3-glc and most probably 4'-*O*-methyl-Pt3-glc (Zimman and Waterhouse 2002), while both 4'-*O*-methyl-Pt3-glc and 4'-*O*-methyl-Pt3-gal have been reported among the metabolites in liver and kidney tissue after consumption of bilberry extracts by rats (Ichiyanagi et al. 2006a). Dp3-glc carrying the pyrogallol structure has, however, been reported metabolized exclusively to 4'-*O*-methyl-Dp3-glc (Ichiyanagi et al. 2004), and not into 3'-*O*-methyl-Dp3-glc. Contrary to what has been reported for the simple anthocyanins described above, the aromatic acylated anthocyanin nasunin (Dp3-(6-(4-*p*-coumaroylrhamnosyl) glucoside)-5-glucoside) seems not to be modified by catechol-*O*-methyl transferase into its methylated derivative (Ichiyanagi et al. 2006b).

Glucuronidated anthocyanin metabolites, such as Pn-monoglucuronide and Cy3-glc monoglucuronide, were first detected by Wu et al. (2002) in elderly women after consumption of elderberry. In later studies, Cy-monoglucuronides have been reported to be the main (or among the main) derivatives detected in urine after anthocyanin

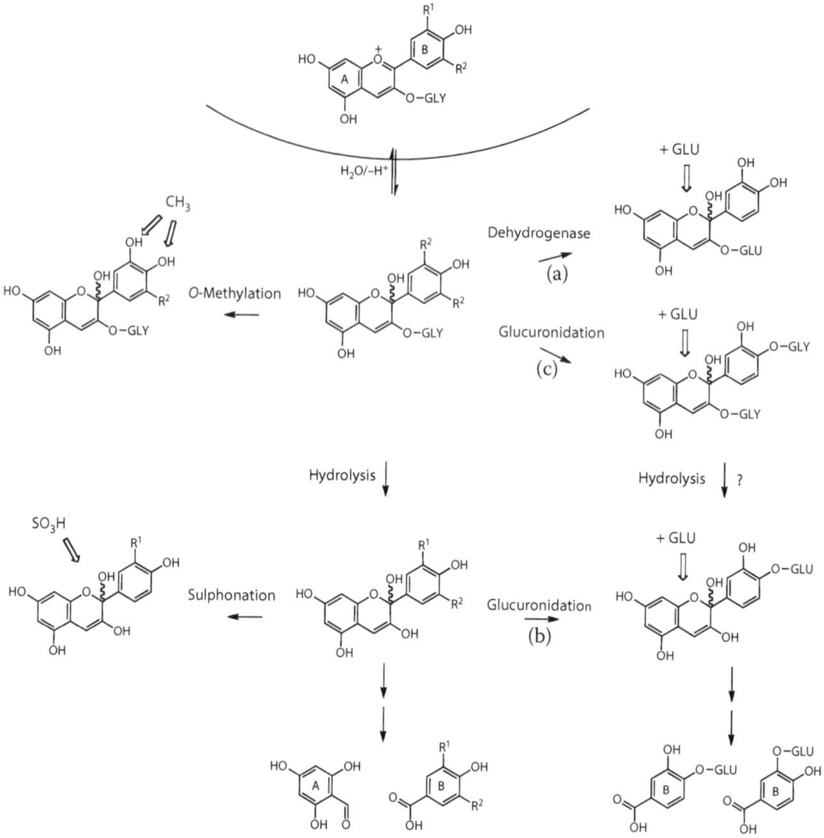

FIGURE 2.10 A summary of reported phase II derivatives and degradation products of anthocyanins in metabolism. The flavylium cation form represents the anthocyanins in the diet. The hemiketal form represents the anthocyanins under physiological conditions, however, this will certainly change with pH, and so on in the various organs. The glucuronidated derivatives may contain several glucuronosyl units in various positions. They may also be O-methylated. a–c show the different mechanisms suggested for glucuronidation. GLY = glycosyl, GLU = glucuronosyl. The substitution patterns on the various anthocyanidin B-rings indicate the type of anthocyanidins, which have been subjected.

intake (Felgines et al. 2005, Kay et al. 2005, Prior and Wu 2006, Tian et al. 2006). It has also been shown that Pg3-glc had a much higher apparent conversion into anthocyanidin-glucuronides than similar Cy-based anthocyanins, as indicated by analysis of urinary excretion in pigs (Wu and Prior 2005). Two pathways have been proposed explaining the formation of Cy-monoglucuronides (Prior and Wu 2006). The first possibility is that the anthocyanidin-glucuronide is formed directly from Cy3-glc by a dehydrogenase, which oxidizes the –CH$_2$OH moiety of glucose into an –CO$_2$H moiety (Figure 2.10a). The second pathway (Figure 2.10b) requires first the hydrolysis of Cy-glycosides to cyanidin followed by glucuronidation of the anthocyanidin. This second possibility is similar to what has been suggested as the pathway used for glucuronidation of other flavonoid groups. However, Ichiyanagi et al. (2005b) have reported a third alternative (Figure 2.10c), which they call extended glucuronidation, since the glucuronosyl moiety is connected to intact anthocyanins and *O*-methyl derivatives like Cy3-glc and *O*-methyl-Cy3-glc. This pathway is claimed to be a specific phase II metabolic route for anthocyanins.

According to what is reported in the literature, anthocyanins seem only to a small extent to be converted into anthocyanidin sulfoconjugates. Among the very few reports of this type of metabolic anthocyanin-derivatives, Pg-sulfate together with other Pg derivatives has been recovered in human urine after strawberry consumption (Felgines et al. 2003, Mullen et al. 2008), and very small amounts of Cy-sulfate after blackberry consumption (Felgines et al. 2005). Anthocyanidin sulfate formation requires hydrolysis of the anthocyanin into the anthocyanidin followed by sulfoconjugation of the aglycon (Runge-Morris 1997).

2.5.3 Anthocyanin Degradation Products in Humans and Animals

As described above, only a small proportion of the dietary anthocyanins including their metabolized derivatives are absorbed in their intact forms. Kay et al. (2009) have even given the following title on a recent paper: *The bioactivity of dietary anthocyanins is likely to be mediated by their degradation products*. By using an *in vitro* model, Keppler and Humpf (2005) investigated the microbial deglycosylation and degradation of six anthocyanins based on three different anthocyanidins with mono- or diglycosidic bonds. They found that all the anthocyanidin-glycosides were hydrolyzed by the microflora within 20 min and 2 h of incubation depending on the sugar moiety. Due to the high instability of the liberated anthocyanidins at neutral pH, primary phenolic degradation products were detected already after 20 min of incubation. Further metabolism of these phenolic acids was accompanied by *O*-demethylation. This microbial pathway of *O*-demethylation is contrary to the enzymatic activity of intestinal and hepatic tissues, which exhibited catechol-*O*-methyltransferase activity giving *O*-methylation, as described under Section 2.5.2. Similarly, when Mv3-glc, Pn3-glc, and Cy3-glc were incubated with human fecal flora, more than 90% of the anthocyanins were degraded after 2 h (Fleschhut et al. 2006). In control incubations with heat-inactivated fecal suspension, only a slight decrease in anthocyanin content could be observed, suggesting that the microflora is primarily responsible for degradation of the anthocyanins. To determine the (in) stability of the common anthocyanidins in more detail, Mv, Cy, Pg, Dp, and Pn have been incubated at 37°C in cell culture medium with pH 7.4 (Fleschhut et al. 2006). After 60 min, Mv, Cy, Pn, and Dp had almost completely disappeared, while only about 20% of Pg had disappeared during a 30 min incubation period. Pg thus seemed to be unstable, however, relatively more stable than the other common anthocyanidins at this pH. Opposite to this, glycosides of Dp and Pt have recently been found to be more susceptible to degradation than those of Cy, Pg, Pn, and Mv in both intact and artificial saliva, suggested to be mediated by oral microbiota (Kamonpatana 2012).

To clarify which metabolic products are formed from anthocyanins, rats have been orally given Cy3-glc (Tsuda et al. 1999). Although the intact aglycone, Cy, was found to be present in the jejunum in this study, 3,4-dihydroxybenzoic acid (protocatechuic acid, PCA) corresponding to the B-ring of Cy (and not Cy intact) was detected in high amounts in plasma. More recently, Vitaglione et al. (2007) have reported that PCA accounted for *ca* 44% of the ingested Cy-glycosides from blood orange juice in 6 h post-consumption plasma, and almost 73% excretion of the ingested Cy-glycosides. In another study, incubation of Mv3-glc with fecal bacteria resulted mainly in the formation of 4-hydroxy-3,5-dimethoxybenzoic acid (syringic acid), while similar incubation of a mixture of 3-glucosides of Mv, Dp, Pn, Pt, and Cy resulted in the formation of 3,4,5-trihydroxybenzoic acid (gallic), syringic, and *p*-coumaric acids (Hidalgo et al. 2012). However, in contrast to the high proportions of phenolic acids, which are reported above, Ichiyanagi et al. (2007) detected after oral or intravenous administration of Cy3-glc to rats no PCA at any time in the blood plasma. It was here concluded that PCA was neither absorbed as the major intestinal metabolite of Cy3-glc after oral administration nor formed as the metabolite in liver, and thus PCA does not exhibit high contribution to the functionality of Cy3-glc *in vivo*.

The major stable products of anthocyanin degradation in humans seem thus in most studies to be the corresponding phenolic acids derived mainly from the B-ring of the anthocyanin skeleton (Aura et al. 2005, El Mohsen et al. 2006, Ávila et al. 2009, Kay et al. 2009) as well as their glucuronide conjugates (Woodward et al. 2011) (Figure 2.10). In some cases, 2,4,6-trihydroxybenzaldehyde (phloroglucinol aldehyde) corresponding to the anthocyanidin A-ring (Keppler and Humpf 2005) and *p*-coumaric acid (Hidalgo et al. 2012) have been identified.

2.6 RELIABILITY AND REPRODUCIBILITY IN ANTHOCYANIN ANALYSIS

For determination of expected anthocyanin content, a routine analysis may consist of just one HPLC (or LC-MS) injection (5 µL) of a crude anthocyanin extract based on a couple of berries, interpreted toward a known reference compound or standard curve. On the contrary, a typical complete structure determination should include running two-dimensional NMR experiments on the pure, isolated anthocyanin. It will then be possible to achieve proper assignments of proton and carbon shift values, which are necessary for elucidation of linkage points between the structure moieties. This latter procedure might depend on extraction of at least 100 g plant material with acidified alcoholic solvent, followed by purification and separation using various chromatographic techniques before structure elucidation by NMR spectroscopy, high-resolution mass spectrometry, and sometimes chemical degradation.

However, experimental measurements including anthocyanins are almost always prone to some sorts of errors. As noted previously in this chapter, anthocyanins will to various degrees show instability toward several chemical and physical parameters, including high temperatures, oxygen, and most pH values. This instability makes it especially difficult to analyze anthocyanin samples under physiological conditions (aqueous solutions with slightly acidic toward neutral pH values). Many anthocyanins have similar structures, and several sources used in the human diet, especially vegetables, contain rather complex anthocyanin mixtures, which are rather difficult to separate and isolate in pure state. The major aim of this section is to indicate factors, which should be considered during anthocyanin analysis for obtaining reliable and reproducible results in different labs. These considerations should also be kept in mind when comparing the qualitative, and especially the quantitative anthocyanin content reported in Tables 2.3 through 2.5.

We will in this chapter make no attempt to cover methods used for isolation and identification of anthocyanins in plants and derived products, which have been treated thoroughly in several reviews (Rivas-Gonzalo 2003, Andersen and Francis 2004, Andersen and Fossen 2005, Rodriguez-Saona and Wrolstad 2005, Giusti and Jing 2008). Methodology for assessment of anthocyanins in the blood, plasma, urine, and tissues is treated in Chapter 4.

2.6.1 EXTRACTION

Anthocyanins in plants and plant-derived products are normally extracted from fresh, frozen, or freeze-dried material, which prior to extraction often have been macerated in a blender or cut into pieces by a knife. The most common solvents include acidified ethanol, methanol, and acetone; however, anthocyanins might react with acetone and give rise to pyranoanthocyanins (Lu and Foo 2001). The acids used as part of the solvents are for instance trifluoroacetic acid (TFA) or acetic acid, since mineral acids like aqueous hydrochloric acid (HCl) lead to hydrolysis of the ester linkage of potential aliphatic acyl groups attached to the glycosyl moieties. When nearly no anthocyanins were reported to be acylated with aliphatic acids two decades ago due to the use of HCl during extraction and isolation, seven different aliphatic acyl groups (Figure 2.4) have hitherto been found in altogether 239 anthocyanins. Malonyl groups have now been found in more than 27% of the different anthocyanins, which makes this acyl group to be the most common acylation unit of anthocyanins. Here, we just want to add that methyl esterification of the terminal carboxyl group of dixcarboxylic aliphatic acyl groups of anthocyanins can easily happen in acidified methanolic solvents during extraction and in the isolation and identification processes.

The various extraction procedures will normally achieve the same qualitative anthocyanin profile. However, they are nonselective and give often anthocyanin solutions with large amounts of by-products, which might complicate interpretation of results. Differences with respect to solvents, extraction times, number of extractions, and so on will certainly influence the total amount of extracted anthocyanins. Other factors affecting the "true picture" of quantitative anthocyanin content are difficulties with assessing the anthocyanin content, which is left in the residue (press-cake) after extraction, as well as the water content in purified samples.

2.6.2 SPECTROPHOTOMETRIC DETECTION

A common method used for qualitative and quantitative anthocyanin analysis combines HPLC separation with UV–vis spectrophotometric detection (often with diode-array detector). Anthocyanins are individually separated on the LC column before detection, and some information about identities and amounts of individual anthocyanins will be available afterwards. When several anthocyanins are present in mixtures, it might be difficult to obtain proper resolution of individual compounds in the analytical system. Several laboratories have therefore chosen to analyze only the anthocyanidin part of the anthocyanin content after acid hydrolysis. Although this method simplifies identification of the qualitative anthocyanin (or rather the anthocyanidin) content, and also provides important information concerning total amounts of individual anthocyanin aglycones, information about the number and nature of intact anthocyanins with their sugar and possible acyl moieties is lost.

Several methods used to quantify the total anthocyanin content are based solely on UV–vis spectrophotometric instrumentation (Rapisarda et al. 2000, Lee et al. 2008, Bordonaba et al. 2011, Gil-Munoz et al. 2011). These methods provide very little qualitative information about the anthocyanin content including number and nature of individual compounds. Using the pH differential method (Fuleki and Francis 1968a,b, Jackman and Smith 1996, Giusti and Wrolstad 2005), the anthocyanin amount present in a sample is determined by measuring the change in absorbance at two different pH values based upon expected anthocyanin structural differences at pH 1.0 (colored flavylium cation form) and pH 4.5 (colorless hemiketal forms) (Lee et al. 2005). The measured difference in $\lambda_{vis\text{-}max}$ absorbance (~520 nm) at the two pH values is in theory proportional to the anthocyanin concentration. The method expects similar structural changes for all monomeric anthocyanins, which is likely for simple nonacylated anthocyanins found in most fruits. However, the situation is different for most anthocyanins found in vegetables (see Section 2.3.1). These anthocyanins contain aromatic acyl groups, which prevent the same extent of formation of colorless anthocyanin forms at pH 4.5, as found for simple nonacylated anthocyanins. They have thus a relative high absorption even at pH 4.5, which might influence considerably the interpretation of the quantitative results obtained by this method.

2.6.3 REFERENCE COMPOUNDS, MOLAR ABSORPTIVITY, AND ANTHOCYANIN PURITY

In quantitative determinations of anthocyanin content, both HPLC with spectrophotometric detection and UV–vis spectrophotometric methods are based upon the use of Lambert–Beer's law ($A = \varepsilon cl$). In some cases, the difficult point for obtaining correct values is the use of the right molar absorptivity (ε) value, which often is different for different anthocyanins. In other cases, the complex chemistry of anthocyanins (instability, etc.) combined with other experimental problems with preparing pure anthocyanins (Lee et al. 2005, Jordheim et al. 2007a) has made it difficult to determine exact masses of individual anthocyanins. This is indeed reflected in the huge variations found among the reported ε-values for these compounds (Jordheim et al. 2007a and references therein). In addition to substantial differences between ε-values given for the same anthocyanin, even in the same solvent, there exist inconsistent differences in literature between ε-values of structurally very similar anthocyanins.

Caused by the lack of "genuine" molar absorptivity values, or in many cases lack of commercially available authentic anthocyanin compounds, most laboratories use one fixed reference compound, normally Cy3-glc or Mv3-glc, for calibration purposes. In HPLC-analyses of black currant, bilberry, and cowberry, Kähkönen et al.

(2003) found an underestimation of 10–20% in anthocyanin content by using only Cy3-glc as reference, compared to the use of authentic standards. Similar studies, based upon the use of 15 authentic anthocyanin standards, showed up to 53% and 64% higher anthocyanin content in fresh bilberries and blueberries, respectively, compared to previous reports using only Cy3-glc as reference compound (Müller et al. 2012). The impact of reference structure was also demonstrated when the anthocyanin content was calculated in Cy3-glc and Mv3-glc equivalents after analyses by both pH differential and HPLC methods (Lee et al. 2008). There was high correlation between the results obtained by the two methods applied on fruit products with mainly nonacylated anthocyanins. However, all the quantitative results were highest when calculated based on Mv3-glc, regardless of the analytical method used, due to differences in the molar absorptivity values of Cy3-glc and Mv3-glc, and of course the molecular masses of the reference compounds. Similarly, it has been shown that quantification based on Cy3-glc equivalents by the pH differential method led to lower results due to underestimation of acylated compounds or anthocyanins with higher molecular weights than Cy3-glc, when they were present (Riedl and Murkovic 2011). Finally, we will in this context mention that the use of molecular weight correction factors has been suggested for making more correct calculations (Wu et al. 2006). These factors were obtained for individual anthocyanins by dividing the molecular weight of each anthocyanin to be quantified by that of the standard anthocyanin.

2.7   CONCLUSION

This chapter describes accurately the qualitative and quantitative anthocyanin content of the fruits and vegetables, which are the major anthocyanin sources typically consumed in Europe, the United States, and Australia. Among the 702 different anthocyanins from plants, which hitherto have been properly identified, altogether 96 of them occur as major compounds in one or more of the approximately 100 nonprocessed anthocyanin-containing plant sources. The molecular masses of the anthocyanins in these fruits (81 species) and vegetables (17 species) vary from 298 Da for apigeninidin (sorghum) via 433 Da for pelargonidin 3-glucoside (strawberry), to 1509 Da for Alatanin A (purple yam).

The anthocyanins of the vegetables in our diet have considerably more complex glycosyl moieties than those of the fruits. The "average anthocyanin" made from the 51 major anthocyanins occurring in vegetables has a molecular weight of 887.2 amu and consists of the anthocyanidin, 2.7 monosaccharide units and 0.9 aromatic acyl group. The corresponding "average anthocyanin" representing the 53 major anthocyanins of the fruits has a molecular weight of 556.3 Da, and consists of the anthocyanidin connected to 1.4 monosaccharide units and 0.1 aromatic acyl group. Most of the anthocyanins in the vegetables (84%) are based on anthocyanidins with just two or one oxygen-containing functional groups on their B-rings (cyanidin, peonidin, or pelargonidin), while fruits contain relatively more (43% versus 16% in vegetables) anthocyanins with three oxygen functions on their anthocyanidin B-rings (delphinidin, petunidin, and malvidin).

Among the vegetables, the families Solanaceae (purple potato), Convolvulaceae (purple sweet potato), Apiaceae (purple carrot), and Brassicaceae (red cabbage, radish) are especially rich in complex anthocyanins with at least three monosaccharide units and aromatic acylation. Purple sweet potato (*I. batatas*) and red and purple potatoes (*Solanum tuberosum*) are the vegetables with highest reported anthocyanin content, with maximum values of 1407, 1383, and 550 mg/100 g fresh weight, respectively. The most commonly eaten anthocyanin-rich sources belong to fruits in family Rosaceae (blackberries, raspberries, plums, cherries, strawberry, apples), and the genera *Vaccinium* (bilberries, blueberries, cranberries), *Ribes* (black and red currants), and *Vites* (grapes). They contain virtually only simple anthocyanidin 3-monoglycosides or 3-diglycosides without acylation. Purple corn (*Z. mays*), chokeberry (*Aronia melanocarpa*), European elderberry (*Sambucus nigra*), and bilberry (*Vaccinium myrtillus*) have been reported to have the highest anthocyanin content among the fruits, 1734, 1480, 1374, and 1017 mg/100 g fresh weight, respectively, while black grapes, black raspberry, black currant, and more rare fruits in the diet like black sorghum, black rice, Tasmanian pepper (*Tasmanian lanceolata*), crowberry (*Empetrum nigrum*), myrtle berry (*Myrtus communis*), saskatoon berry (*Amelanchier alnifolia*), and lycii (*Lycium ruthenicum*) also are rich anthocyanin sources.

This chapter has, to some extent, described how the anthocyanin content in the various diet sources is influenced by genetic diversity, phenotype plasticity, and agricultural practice. Many studies have focused on elucidating these interactions with the purpose of obtaining increased yield of anthocyanins in food products. To improve production of anthocyanins, efforts have also been devoted to advanced breeding selection with considerable success, and optimization of biosynthetic pathways in cell cultures of several fruits and vegetables, with varying success. By introducing new genes in plants encoding for novel enzyme activities, transcription factors, or inactivation of endogenous genes used in anthocyanin biosynthesis, several new plant varieties with modified colors and coloration pattern have already been created, hitherto mainly flowers. The future will tell how much genetic bioengineering approaches will influence the anthocyanin content in our diet.

Many people get a substantial part of their anthocyanin intake through consumption of processed plant products, with possible nutraceutical effects also from the derivatives and degradation products made from the anthocyanins in the products. The effects of processing and storage on anthocyanins have in particular been studied thoroughly in relation to the chemical transformations going on in wine, and a variety of new anthocyanidin derivatives have been identified during the two last decades. As in wine a rather complex heterogeneous class of polymeric compounds is formed in reactions between anthocyanins and other constituents in the food matrix, such as other flavonoids/phenolic compounds or molecules with relative low molecular weights, during processing and storage. As background for explaining the formation of these derivatives, this chapter has focused in detail on the reactivity of the various positions on the three rings of the anthocyanidin part of the anthocyanins.

Several sources used in the human diet, especially vegetables, contain rather complex anthocyanin mixtures, which are rather difficult to separate and isolate in pure state. The anthocyanins will to various degrees show instability toward several chemical and physical parameters, and they are prone to form derivatives both in processed foods and in human metabolism. Just a few simple anthocyanins are available as commercial standards. Some analytical difficulties are for instance reflected in the huge variations found among the molar absorptivity (ε) values reported for many anthocyanins. This chapter has focused on the necessity of careful considerations regarding solvents, spectrophotometric detection, reference compounds, molar absorptivity values, and anthocyanin purity for obtaining reliable and reproducible results during anthocyanin analysis.

For understanding of the exact impact of dietary anthocyanins on various health aspects, it is vital to know the exact nature and amount of the circulating anthocyanin metabolites coming from human consumption of anthocyanins, including derivatives and degradation products. At present this type of knowledge is to a large extent lacking due to the complex nature of the anthocyanins both in intact plants, processed foods, and in metabolism. One crucial point here is that each anthocyanin has their aglycone (anthocyanidin) involved in a series of equilibriums giving rise to different forms (secondary structures), which exhibit their own properties including stability and colors. There exist very inadequate information about these forms, in particular about their distribution under *in vivo* conditions. Complete ^{1}H and ^{13}C structural assignments for the various anthocyanidin secondary structures, but the flavylium cation form, have been incomplete for all reported anthocyanins. In this chapter, we have presented the first complete ^{1}H and ^{13}C NMR assignments for any anthocyanidin quinonidal form, here demonstrated for malvidin 3-glucoside. The stability of the anthocyanin molecules is highly linked to the distribution of the anthocyanins on their various secondary anthocyanidin forms. We have suggested the extent of formation of hemiketal forms as the most critical component for experiencing anthocyanin instability—the hemiketal forms are in equilibrium with the labile chalcone forms. In this chapter, we have therefore discussed various factors, especially structural features, which influence the formation of the hemiketal forms, as background for understanding the different stability of different types of anthocyanins.

Anthocyanins to various degrees show instability toward several chemical and physical parameters, including high temperatures, oxygen, most pH-values, specific enzymes, and so on. Among these factors, the influence of pH and heating have been studied most thoroughly, especially in model systems. In this chapter, three thermal degradation pathways of anthocyanins highly influenced by pH slightly and modified by literature, and the various reported anthocyanin phase II metabolites (in particular *O*-methylated and glucuronidated derivatives), have been presented. According to the existing literature, anthocyanins seem only to a small extent to be converted into anthocyanidin sulfoconjugates.

As shown in this chapter, there exist a distinct difference between the anthocyanin content in vegetables and fruits of our diet at least with respect to aromatic acylation and number of monosaccharide units. It is also shown that different anthocyanin structures (aromatic acylation versus nonacylation, number and type of glycosyl moieties, as well as anthocyanidin nature) have different stability, reactivity, and other properties, which influence their bioavailability, degradation routes, and their ability to form various phase II metabolites. The questions are accordingly not about if anthocyanin-rich vegetables give different metabolites than fruits under physiological conditions. The questions are related to the identity and quantity of these anthocyanin-metabolites under various physiological conditions, and their potential specific health effects. Thus, if anthocyanins or their derivatives have impact on our health, we have to design our diet with respect to choice of fruits or vegetables in a far more precise way than "5 A Day" to obtain optimum effects!

REFERENCES

Aaby, K.; Mazur, S.; Nes, A.; Skrede, G. 2012. Phenolic compounds in strawberry (*Fragaria × ananassa* Duch.) fruits: Composition in 27 cultivars and changes during ripening. *Food Chem.* 132: 86–97.

Adams, J.B. 1973. Thermal degradation of anthocyanins with particular reference to the 3-glycosides of cyanidin. I. In acidified solution at 100°C. *J. Sci. Food Agric.* 24: 747–62.

Agawa, S.; Sakakibara, H.; Iwata, R.; Shimoi, K.; Hergesheimer, A.; Kumazawa, S. 2011. Anthocyanins in mesocarp/epicarp and endocarp of fresh acai (*Euterpe oleracea* Mart.) and their antioxidant activities and bioavailability. *Food Sci. Technol. Res.* 17: 327–34.

Al-Farsi, M.; Alasalvar, C.; Morris, A.; Baron, M.; Shahidi, F. 2005. Comparison of antioxidant activity, anthocyanins, carotenoids, and phenolics of three native fresh and sun-dried date (*Phoenix dactylifera* L.) varieties grown in Oman. *J. Agric. Food Chem.* 53: 7592–9.

Alighourchi, H.; Barzegar, M.; Abbasi, S. 2008. Anthocyanins characterization of 15 Iranian pomegranate (*Punica granatum* L.) varieties and their variation after cold storage and pasteurization. *Eur. Food Res. Technol.* 227: 881–7.

Amico, V.; Napoli, E.M.; Renda, A.; Ruberto, G.; Spatafora, C.; Tringali, C. 2004. Constituents of grape pomace from the Sicilian cultivar 'Nerello Mascalese'. *Food Chem.* 88: 599–607.

Andersen, Ø.M. 1985. Chromatographic separation of anthocyanins in cowberry (Lingonberry), *Vaccinium vites-idaea* L. *J. Food Sci.* 50: 1230–2.

Andersen, Ø.M. 1987. Anthocyanins in fruits of *Vaccinium japonicum*. *Phytochemistry* 26: 1220–1.

Andersen, Ø.M. 1989. Anthocyanins in fruits of *Vaccinium oxycoccus* L. (small cranberry). *J. Food Chem.* 54: 383–7.

Andersen, Ø.M. 2012. Personal database on anthocyanins.

Andersen, Ø.M.; Aksnes, D.W.; Nerdal, W.; Johansen, O.P. 1991. Structure elucidation of cyanidin-3-sambubioside and assignments of the 1H and 13C NMR resonances through two-dimensional shift-correlated NMR techniques. *Phytochem. Anal.* 2: 175–83.

Andersen, Ø.M.; Fossen, T. 2005. Characterization of anthocyanins by NMR. In *Handbook of Food Analytical Chemistry: Pigments, Colorants, Flavors, Texture, and Bioactive Food Components,* eds., R. Wrolstad et al., 47–69. Hoboken: John Wiley & Sons.

Andersen, Ø.M.; Fossen, T.; Torskangerpoll, K.; Fossen, A.; Hauge, U. 2004. Anthocyanin from strawberry (*Fragaria ananassa*) with the novel aglycone, 5-carboxypyranopelargonidin. *Phytochemistry* 65: 405–10.

Andersen, Ø.M.; Francis, G.W. 2004. Techniques of pigment identification. In *Plant Pigments and their Manipulation,* ed., K. Davies, 293–341. London: Blackwell Publishing.

Andersen, Ø.M.; Jordheim, M. 2006. The anthocyanins. In *Flavonoids: Chemistry, Biochemistry and Applications*, eds., Ø. M. Andersen, K. R. Markham, 263–318. Boca Raton, FL: CRC Press.

Andersen, Ø.M.; Jordheim, M. 2010. Chemistry of flavonoid-based colors in plants. In *Comprehensive Natural Products II: Chemistry and Biology*, eds. in chief; N. L. Mander, H.-W. Liu, 3: 547–614, Oxford, UK: Elsevier.

Andersen, Ø.M.; Jordheim, M.; Byamukama, R.; Mbabazi, A.; Ogweng, G.; Skaar, I.; Kiremire, B. 2010. Anthocyanins with unusual furanose sugar (apiose) from leaves of *Synadenium grantii* (Euphorbiaceae). *Phytochemistry* 71:1558–63.

Aura, A.M. et al. 2005. *In vitro* metabolism of anthocyanins by human gut microflora. *Eur. J. Nutr.* 44: 133–42.

Ávila, M.; Hidalgo, M.; Sánchez-Moreno, C.; Pelaez, C.; Requena, T.; de Pascual-Teresa, S. 2009. Bioconversion of anthocyanin glycosides by *Bifidobacteria* and *Lactobacillus*. *Food Res. Int.* 42: 1453–61.

Awad, M.A.; Jager, de A. 2002. Formation of flavonoids, especially anthocyanin and chlorogenic acid in 'Jonagold' apple skin: Influences of growth regulators and fruit maturity. *Sci. Hort.* 93: 257–66.

Awika, J.M.; Rooney, L.W.; Waniska, R.D. 2004a. Properties of 3-deoxyanthocyanins from sorghum. *J. Agric. Food Chem.* 52: 4388–94.

Awika, J.M.; Rooney, L.W.; Waniska, R.D. 2004b. Anthocyanins from black sorghum and their antioxidant properties. *Food Chem.* 90: 293–301.

Bae, R.N.; Kim, K.W.; Kim, T.C.; Lee, S.K. 2006. Anatomical observations of anthocyanin rich cells in apple skins. *HortSci.* 41: 733–6.

Bakker, J.; Timberlake, C.F. 1997. Isolation, identification, and characterization of new color-stable anthocyanins occurring in some red wines. *J. Agric. Food Chem.* 45: 35–43.

Bakowska-Barczak, A.M.; Kolodziejczyk, P. 2008. Evaluation of saskatoon berry (*Amelanchier alnifolia* Nutt.) cultivars for their polyphenol content, antioxidant properties, and storage stability. *J. Agric. Food Chem.* 56: 9933–40.

Barros, L.; Dueñas, M.; Dias, M.I.; Sousa, M.J.; Santos-Buelga, C.; Ferreira I.C.F.R. 2012. Phenolic profiles of *in vivo* and *in vitro* grown *Coriandrum sativum* L. *Food Chem.* 132: 841–8.

Bellomo, M.G.; Fallico, B. 2007. Anthocyanins, chlorophylls and xanthophylls in pistachio nuts (*Pistacia vera*) of different geographic origin. *J. Food Comp. Anal.* 20: 352–9.

Benvenuti, S.; Pellati, F.; Melegari, M.; Bertelli, D. 2004. Polyphenols, anthocyanins, ascorbic acid, and radical scavenging activity of *Rubus*, *Ribes*, and *Aronia*. *J. Food Sci.* 69: FCT164–FCT169.

Berardini, N.; Fezer, R.; Conrad, J.; Beifuss, U.; Carle, R.; Schieber, A. 2005a. Screening of mango (*Mangifera indica* L.) cultivars for their contents of flavonol *O*- and xanthone *C*-glycosides, anthocyanins, and pectin. *J. Agric. Food Chem.* 53: 1563–70.

Berardini, N.; Schieber, I.; Klaiber, I.; Beifuss, U.; Carle, R.; Conrad, J. 2005b. 7-*O*-methylcyanidin 3-*O*-β-D-galactopyranoside, a novel anthocyanin from mango (*Mangifera indica* L. cv. 'Tommy Atkins') peels. *Z. Naturforsch., B: Chem. Sci.* 60: 801–4.

Berké, B.; Chèze, C.; Vercauteren, J.; Deffieux, G. 1998. Bisulfite addition to anthocyanins: Revisited structures of colourless adducts. *Tetrahedron Lett.* 39: 5771–4.

Bermúdez-Soto, M.J.; Tomás-Barberán, F.A. 2004. Evaluation of commercial red fruit juice concentrates as ingredients for antioxidant functional juices. *Eur. Food Res. Technol.* 219: 133–41.

Bhagwat, S.; Haytowitz, D.B.; Holden, J.M. 2011. USDA database for the flavonoid content of selected foods. U.S. Department of Agriculture. http://www.ars.usda.gov/nutrient data

Bjorøy, Ø.; Rayyan, S.; Fossen, T.; Kalberg K.; Andersen, Ø.M. 2009. *C*-glycosylanthocyanidins synthesized from *C*-glycosylflavones. *Phytochemistry* 70: 278–87.

Bobinaite, R.; Viškelis, P.; Venskutonis, P.R. 2012. Variation of total phenolics, anthocyanins, ellagic acid and radical scavenging capacity in various raspberry (*Rubus* spp.) cultivars. *Food Chem.* 132: 1495–1501.

Bordonaba, G.J.; Crespo, P.; Terry, L.A. 2011. A new acetonitrile-free mobile phase for HPLC-DAD determination of individual anthocyanins in black currant and strawberry fruits: A comparison and validation study. *Food Chem.* 129: 1265–73.

Boss, P.K.; Davies, C.; Robinson, S.P. 1996. Anthocyanin composition and anthocyanin pathway gene expression in grapevine sports differing in berry skin colour. *Aust. J. Grape Wine Res.* 2: 163–170.

Bridle, P.; Loeffler, R.S.; Timberlake, T.; Colin, F.; Self, R. 1984. Cyanidin 3-malonylglucoside in *Cichorium intybus*. *Phytochemistry* 23: 2968–9.

Brouillard, R.; Dangles, O. 1994. Flavonoids and flower colour. In *The Flavonoids: Advances in Research since 1986*, ed., J. B. Harborne, 565–88. London: Chapman & Hall.

Brouillard, R.; Lang, J. 1990. The hemiacetal-*cis*-chalcone equilibrium of malvin, a natural anthocyanin. *Canad. J. Chem.* 68: 755–61.

Brown, P.N. 2011. Determination of anthocyanins in cranberry fruit and cranberry fruit products by high-performance liquid chromatography with ultraviolet detection: Single-laboratory validation. *J. AOAC Int.* 94: 459–66.

Brownmiller, C.; Howard, L.R.; Prior, R.L. 2008. Processing and storage effects on monomeric anthocyanins, percent polymeric color, and antioxidant capacity of processed blueberry products. *J. Food Sci.* 73: 72–9.

Bueno, J.M.; Saez-Plaza, P.; Ramos-Escudero, F.; Jimenez, A.M.; Fett, R.; Asuero, A.G. 2012. Analysis and antioxidant capacity of anthocyanin pigments. Part II: Chemical structure, color, and intake of anthocyanins. *Crit. Rev. Anal. Chem.* 42: 126–51.

Bureau, S.; Renard, C.M.G.C.; Reich, M.; Ginies, C.; Audergon, J.-M. 2009. Change in anthocyanin concentrations in red apricot fruits during ripening. *Food Sci. Technol.* 42: 372–7.

Byamukama, R.; Namukobe, J.; Kiremire, B. 2009. Anthocyanins from leaf stalks of cassava (*Manihot esculenta* Crantz). *Afr. J. Pure Appl. Chem.* 3: 020–5.

Cabrita, L.; Fossen, T.; Andersen, Ø.M. 2000. Colour and stability of the six common anthocyanidin 3-glucosides in aqueous solutions. *Food Chem.* 68: 101–7.

Carkeet, C.; Clevidence, B.A.; Novotny, J.A. 2008. Anthocyanin excretion by humans increases linearly with increasing strawberry dose. *J. Nutr.* 138: 897–902.

Castellarin, S.D.; Di Gaspero, G. 2007. Transcriptional control of anthocyanin biosynthetic genes in extreme phenotypes for berry pigmentation of naturally occurring grapevines. *BMC Plant Biol.* 7: 46.

Cekic, C.; Oezgen, M. 2010. Comparison of antioxidant capacity and phytochemical properties of wild and cultivated red raspberries (*Rubus idaeus* L.). *J. Food Comp. Anal.* 23: 540–44.

Cevallos-Casals, B.A.; Byrne, D.H.; Cisneros-Zevallos, L.; Okie, W.R. 2002. Total phenolic and anthocyanin content in red-fleshed peaches and plums. *Acta Hort.* 2: 589–92.

Cevallos-Casals, B.A.; Byrne, D.; Okie, W.R.; Cisneros-Zevallos, L. 2006. Selecting new peach and plum genotypes rich in phenolic compounds and enhanced functional properties. *Food Chem.* 96: 273–80.

Cevallos-Casals, B.A.; Cisneros-Zevallos, L. 2003. Stoichiometric and kinetic studies of phenolic antioxidants from Andean purple corn and red-fleshed sweetpotato. *J. Agric. Food Chem.* 51: 3313–9.

Champagne, A.; Hilbert, G.; Legendre, L.; Lebot, V. 2011. Diversity of anthocyanins and other phenolic compounds among tropical root crops from Vanuatu, South Pacific. *J. Food Comp. Anal.* 24: 315–25.

Chan, H.T., Jr.; Kao-Jao, T.H.C.; Nakayama, T. O.M. 1977. Anthocyanin composition of taro. *J. Food Sci.* 42: 19–21.

Cheynier, V. 2006. Flavonoids in wine. In *Flavonoids: Chemistry, Biochemistry and Applications*, eds., Ø. M. Andersen, K. R. Markham, 263–318. Boca Raton, FL: CRC Press.

Cho, M.J.; Howard, L.R.; Prior, R.L.; Clark, J.R. 2004. Flavonoid glycosides and antioxidant capacity of various blackberry, blueberry and red grape genotypes determined by high-performance liquid chromatography/mass spectrometry. *J. Sci. Food Agric.* 84: 1771–82.

Choung, M.-G. et al. 2001. Isolation and determination of anthocyanins in seed coats of black soybean *Glycine max* (L.) (Merr.). *J. Agric. Food Chem.* 49: 5848–51.

Choung, M.-G.; Choi, B.-R.; An, Y.-N.; Chu, Y.-H.; Cho, Y.-S. 2003. Anthocyanin profile of Korean cultivated kidney bean (*Phaseolus vulgaris* L.). *J. Agric. Food Chem.* 51: 7040–7043.

Chun, O.K.; Chung, S.J.; Song, W.O. 2007. Estimated dietary flavonoid intake and major food sources of U.S. adults. *J. Nutr.* 137: 1244–52.

Clifford, M.N.; Brown, J.E. 2006. Dietary flavonoids and health—Broadening the perspective. In *Flavonoids. Chemistry, Biochemistry and Applications*, eds., Ø. M. Andersen, K. R. Markham, 319–370. Boca Raton, FL: CRC Press.

Conn, S.; Franco, C.M.; Zhang, W. 2010. Characterization of anthocyanic vacuolar inclusions in *Vitis vinifera* L. cell suspension cultures. Planta 231: 1343–60.

Conn, S.; Zhang, W.; Franco, C. 2003. Anthocyanic vacuolar inclusions (AVIs) selectively bind acylated anthocyanins in *Vitis vinifera* L. (grapevine) suspension culture. *Biotechnol Lett.* 25: 835–9.

Connor, A.M.; Luby, J.J.; Hancock, J.F.; Berkheimer, S.; Hanson, E.J. 2002. Changes in fruit antioxidant activity among blueberry cultivars during cold-temperature storage. *J. Agric. Food Chem.* 50: 893–98.

Cormier, F.; Couture, R.; Do, C.B.; Pham, T.Q.; Tong, V.H. 1997. Properties of anthocyanins from grape cell culture. *J. Food Sci.* 62: 246–8.

Cox, K.A.; McGhie, T.K.; White, A.; Woolf, A.B. 2004. Skin colour and pigment changes during ripening of 'Hass' avocado fruit. *Posthar. Biol. Technol.* 31: 287–94.

Crespo, P.; Bordonaba, J.G.; Therry, L.A.; Carlen, C. 2010. Characterisation of major taste and health-related compounds of four strawberry genotypes grown at different Swiss production sites. *Food Chem.* 122: 16–24.

Crifó, T.; Puglisi, I.; Petrone, G.; Recupero, R.; Piero, A. R.L. 2011. Expression analysis in response to low temperature stress in blood oranges: Implication of the flavonoid biosynthetic pathway. *Gene* 476: 1–9.

Cuevas-Rodriguez, E.O.; Yousef, G.G.; Garcia-Saucedo, P.A.; Lopez-Medina, J.; Paredes-Lopez, O.; Lila, M.A. 2010. Characterization of anthocyanins and proanthocyanidins in wild and domesticated Mexican blackberries (*Rubus* spp.). *J. Agric. Food Chem.* 58: 7458–64.

Dai, Z.W. et al. 2011. Ecophysiological, genetic, and molecular causes of variation in grape berry weight and composition: A review. *Am. J. Enol. Viticult.* 62: 413–25.

Dangles, O.; Saito, N.; Brouillard, R. 1993. Anthocyanin intramolecular copigment effect. *Phytochemistry* 34: 119–24.

Daniel, D.L.; Huerta, B.E.B.; Sosa, I.A.; Mendoza, M. G.V. 2012. Effect of fixed bed drying on the retention of phenolic compounds, anthocyanins and antioxidant activity of roselle (*Hibiscus sabdariffa* L.). *Ind. Crops Prod.* 40: 268–276.

De Freitas, V.; Mateus, N. 2011. Formation of pyranoanthocyanins in red wines: A new and diverse class of anthocyanin derivatives. *Anal. Bioanal. Chem.* 401: 1463–73.

Deroles, S. 2009. Anthocyanin biosynthesis in plant cell cultures: A potential source of natural colourants. In *Anthocyanins*. eds., C. Winefiled, K. Davies, K. Gould, 108–67. New York: Springer.

Drossard, C.; Alexy, U.; Bolzenius, K.; Kunz, C.S.; Kersting, M. 2011. Anthocyanins in the diet of infants and toddlers: Intake, sources and trends. *Eur. J. Nutr.* 50: 705–11.

Dueñas, M.; Perez-Alonso, J.J.; Santos-Buelga, C.; Escribano-Bailon, T. 2008. Anthocyanin composition in fig (*Ficus carica* L.). *J. Food Comp. Anal.* 21: 107–15

Dussi, M.C.; Sugar, D.; Wrolstad, R.E. 1995. Characterizing and quantifying anthocyanins in red pears and the effect of light quality on fruit color. *J. Am. Soc. Hort. Sci.* 120: 785–89.

El Kereamy, A. et al. 2002. Ethanol triggers grape gene expression leading to anthocyanin accumulation during berry ripening. *Plant Sci.* 163: 449–54.

El Mohsen, M.A. et al. 2006. Absorption, tissue distribution and excretion of pelargonidin and its metabolites following oral administration to rats. *Br. J. Nutr.* 95: 51–8.

Elfalleh, W. et al. 2011. Antioxidant capacities of phenolic compounds and tocopherols from Tunisian pomegranate (*Punica granatum*) fruits. *J. Food Sci.* 76: C707–C713.

Elham, G.; Reza, H.; Jabbar, K.; Parisa, S.; Rashid, J. 2006. Isolation and structure characterisation of anthocyanin pigments in black carrot (*Daucus carota* L.). *Pak. J. Biol. Sci.* 9: 2905–8.

El-Sayed, M.A.A.; Hucl, P. 2003. Composition and stability of anthocyanins in blue-grained wheat. *J. Agric. Food Chem.* 51: 2174–80.

Ercisli, S.; Tosun, M.; Duralija, B.; Voca, S.; Sengul, M.; Turan, M. 2010. Phytochemical content of some black (*Morus nigra* L.) and purple (*Morus rubra* L.) mulberry genotypes. *Food Technol. Biotechnol.* 48: 102–6.

Escribano-Bailon, M.T.; Alcalde-Eon, C.; Munoz, O.; Rivas-Gonzalo, J.C.; Santos-Buelga, C. 2006. Anthocyanins in berries of Maqui (*Aristotelia chilensis* (Mol.) Stuntz). *Phytochem. Anal.* 17: 8–14.

Fanali, C.; Dugo, L.; D'Orazio, G.; Lirangi, M.; Dachà, M.; Dugo, P.; Mondello, L. 2011. Analysis of anthocyanins in commercial fruit juices by using nano-liquid chromatography-electrospray-mass spectrometry and high-performance liquid chromatography with UV-vis detector. *J. Sep. Sci.* 34: 150–59.

Fan-Chiang, H.-J.; Wrolstad, R.E. 2005. Anthocyanin pigment composition of blackberries. *J. Food Sci.* 70: C198–C202.

Felgines, C. et al. 2003. Strawberry anthocyanins are recovered in urine as glucuro- and sulfo-conjugates in humans. *J. Nutr.* 133: 1296–301.

Felgines, C.; Talavéra, S.; Texier, O.; Gil-Izquierdo, A.; Lamaison, J.L.; Rémésy, C. 2005. Blackberry anthocyanins are mainly recovered from urine as methylated and glucuronidated conjugates in humans. *J. Agric. Food Chem.* 53: 7721–7.

Felgines, C.; Texier, O.; Besson, C.; Lyan, B.; Lamaison, J.L.; Scalbert, A. 2007. Strawberry pelargonidin glycosides are excreted in urine as intact glycosides and glucuronidated pelargonidin derivatives in rats. *Br. J. Nutr.* 98: 1126–31.

Feng, S.Q.; Wang, Y.L.; Yang, S.; Xu, Y.T.; Chen, X.S. 2010. Anthocyanin biosynthesis in pears is regulated by a R2R3-MYB transcription factor PyMYB10. *Planta* 232: 245–55.

Fleschhut, J.; Kratzer, F.; Rechkemmer, G.; Kulling, S.E. 2006. Stability and biotransformation of various dietary anthocyanins in vitro. *Eur. J. Nutr.* 45: 7–18.

Fossen, T.; Andersen, Ø.M. 2000. Anthocyanins from tubers and shoots of the purple potato, *Solanum tuberosum. J. Hort. Sci. Biotechnol.* 75: 360–63.

Fossen, T.; Andersen, Ø.M. 2003b. Anthocyanins from red onion, *Allium cepa*, with novel aglycone. *Phytochemistry* 62, 1217–20.

Fossen, T.; Andersen, Ø.M.; Øvstedal, D.O.; Pedersen, A.T.; Raknes, A. 1996. Characteristic anthocyanin pattern from onions and other *Allium* spp. *J. Food Sci.* 64: 703–6.

Fossen, T.; Rayyan, S.; Andersen, Ø.M. 2004. Dimeric anthocyanins from strawberry (*Fragaria ananassa*) consisting of pelargonidin 3-glucoside covalently linked to four flavan-3-ols. *Phytochemistry* 65: 1421–8.

Fossen, T.; Rayyan, S.; Holmberg, M.H.; Nateland, H.S.; Andersen, Ø.M. 2005. Acylated anthocyanins from leaves of *Oxalis triangularis. Phytochemistry* 66: 1133–40.

Fossen, T.; Rayyan, S.; Holmberg, M.H.; Nimtz, M.; Andersen, Ø.M. 2007. Covalent anthocyanin–flavone dimer from leaves of *Oxalis triangularis. Phytochemistry* 68: 652–62.

Fossen, T.; Slimestad, R.; Andersen, Ø.M. 2003a. Anthocyanins with 4′-glucosidation from red onion, *Allium cepa. Phytochemistry.* 64: 1367–74.

Fossen, T.; Øvstedal, D.O.; Slimestad, R.; Andersen, Ø.M. 2003b. Anthocyanins from a Norwegian potato cultivar. *Food Chem.* 81: 433–37.

Fournier-Level, A. et al. 2009. Quantitative genetic bases of anthocyanin variation in grape (*Vitis vinifera* L. ssp. *sativa*) berry: A quantitative trait locus to quantitative trait nucleotide integrated study. *Genetics* 183: 1127–39.

Francis, F.J.; Harborne, J.B.; Barker, W.G. 1966. Anthocyanins in the lowbush blueberry, *Vaccinium angustifolium. J. Food Sci.* 31: 583–7.

Fukui, Y.; Nomoto, K.; Iwashita, T.; Masuda, K.; Tanaka, Y.; Kusumi, T. 2006. Two novel blue pigments with ellagitannin moiety, rosacyanins A1 and A2, isolated from the petals of *Rosa hybrida. Tetrahedron* 62: 9661–70.

Fukui, Y.; Tanaka, Y. 2010. Novel compound contained in blue rose. PCT Int. Appl. Patent WO 2010-JP55262.

Fulcrand, H.; Benabdeljalil, C.; Rigaud, J.; Cheynier, V.; Moutounet, M. 1998. A new class of wine pigments generated by reaction between pyruvic acid and grape anthocyanins. *Phytochemistry* 47: 1401–7.

Fuleki, T.; Francis, F.J. 1968a. Qualitative methods for anthocyanins. 1. Extraction and determination of total anthocyanin in cranberries. *J. Food Sci.* 33: 72–77.

Fuleki, T.; Francis, F.J. 1968b. Qualitative methods for anthocyanins. 2. Determination of total anthocyanin and degradation index for cranberry juice. *J. Food Sci.* 33: 78–83.

Furtado, P.; Figueiredo, P.; Chaves das Neves, H.; Pina, F. 1993. Photochemical and thermal degradation of anthocyanidins. *J. Photoche. Photobiol. A. Chem.* 75: 113–8.

Gakh, E.G.; Dougall, D.K.; Baker, D.C. 1998. Proton nuclear magnetic resonance studies of monoacylated anthocyanins from the wild carrot: Part 1. Inter- and intra-molecular interactions in solution. *Phytochem. Anal.* 9: 28–34.

Gao, L.; Mazza, G. 1995. Characterization, quantitation, and distribution of anthocyanins and colorless phenolics in sweet cherries. *J. Agric. Food Chem.* 43: 343–6.

Garcia-Vigera, C.; Zafrilla, P.; Tomas-Barberan, F. 1997. Determination of authenticity of fruit jams by HPLC analysis of anthocyanins. *J. Sci. Food Agric.* 73: 207–13.

García-Viguera, C.; Romero, F.; Abellán, P.; Artés, F.; Tomás-Barberán, F.A. 1999. Color stability of strawberry jam as affected by cultivar and storage temperature. *J. Food Sci.* 64: 243–7.

Gil-Munoz, R.; Moreno-Perez, A.; Vila-Lopez, R.; Fernandez-Fernandez, J.I.; Martinez-Cutillas, A. 2011. Determination of anthocyanin content in C.V Monastrell grapes during ripening period using several procedures. *Int. J. Food Sci. Technol.* 46: 1986–92.

Giusti, M.M.; Ghanadan, H.; Wrolstad, R.E. 1998. Elucidation of the structure and conformation of red radish (*Raphanus sativus*) anthocyanins using one- and two-dimensional nuclear magnetic resonance techniques. *J. Agric. Food Chem.* 46: 4858–63.

Giusti, M.M.; Jing, P. 2008. Analysis of anthocyanins. In *Food Colorants—Chemical and Functional Properties*, ed., C. Socaciu, 479–506. Boca Raton, FL: CRC Press.

Giusti, M.M.; Wrolstad, R.E. 2005. Characterization and measurement of anthocyanins by UV–visible spectroscopy. In *Handbook of Food Analytical Chemistry: Pigments, Colorants, Flavors, Texture, and Bioactive Food Components*, eds., R. Wrolstad et al., 19–31. Hoboken: John Wiley & Sons.

Gómez Gallego, M.A.; Gómez Garcia-Carpintero, E.; Sanchez-Palomo, E.; Hermosin-Gutierrez, I.; Gonzalez Vinas, M.A. 2012. Study of phenolic composition and sensory properties of red grape varieties in danger of extinction from the Spanish region of Castilla-La Mancha. *Eur. Food Res. Technol.* 234: 295–303.

Gonzalez-Paramas, A.M. et al. 2006. Flavanol-anthocyanin condensed pigments in plant extracts. *Food Chem.* 94: 428–36.

Gosnay, S.L.; Bishop, J.A.; New, S.A.; Catterick, J.; Clifford, M.N. 2002. Estimation of the mean intakes of fourteen classes of dietary phenolics in a population of young British women aged 20–30 years. *Proc. Nutr. Soc.* 61: 125A.

Ha, T.J.; Lee, M.-H.; Jeong, Y.N.; Lee, J.H.; Han, S.-I.; Park, C.-H.; Pae, S.-B.; Hwang, C.-D.; Baek, I.-Y.; Park, K.-Y. 2010. Anthocyanins in cowpea [*Vigna unguiculata* (L.) Walp. ssp. *unguiculata*]. *Food Sci. Biotechnol.* 19: 821–6.

Hagen, S.F.; Solhaug, K.A.; Bengtsson, G.B.; Borge, G.I.A.; Bilger, W. 2006. Chlorophyll fluorescence as a tool for non-destructive estimation of anthocyanins and total flavonoids in apples. *Postharvest Biol. Technol.* 41: 156–63.

Hanamura, T.; Hagiwara, T.; Kawagishi, H. 2005. Structural and functional characterization of polyphenols isolated from acerola (*Malpighia emarginata* DC.) fruit. *Biosci. Biotechnol. Biochem.* 69: 280–6.

Hasnaoui, N.; Jbir, R.; Mars, M.; Trifi, M.; Kamal-Eldin, A.; Melgarejo, P.; Hernandez F. 2011. Organic acids, sugars and anthocyanins contents in juices of Tunisian pomegranate fruits. *Int. J. Food* 14: 741–57.

He, F.; Mu, L.; Yan, G.-L.; Liang, N.-N.; Pan, Q.-H.; Wang, J.; Reeves, M.J.; Duan, C.-Q. 2010a. Biosynthesis of anthocyanins and the irregulation in colored grapes. *Molecules* 15: 9057–91.

He, J.; Giusti, M.M. 2010b. Anthocyanins: Natural colorants with health-promoting properties. *Annu. Rev. Food Sci. Technol.* 1: 163–87.

He, J.; Magnuson, B.A.; Giusti, M.M. 2005. Analysis of anthocyanins in rat intestinal contents—Impact of anthocyanin chemical structure on fecal excretion. *J. Agric. Food Chem.* 53: 2859–66.

He, J.; Magnuson, B.A.; Lala, G.; Tian, Q.; Schwartz, S.J.; Giusti, M.M. 2006a. Intact anthocyanins and metabolites in rat urine and plasma after 3 months of anthocyanin supplementation. *Nutr. Cancer* 54: 3–12.

He, J.; Santos-Buelga, C.; Mateus, N.; de Freitas, V. 2006b. Isolation and quantification of oligomeric pyranoanthocyanin-flavanol pigments from red wines by combination of column chromatographic techniques. *J. Chromatogr. A*, 1134: 215–25.

Hidalgo, M. et al. 2012. Metabolism of anthocyanins by human gut microflora and their influence on gut bacterial growth. *J. Agric. Food Chem.* 60: 3882–90.

Hollands, W. et al. 2008. Processing black currants dramatically reduces the content and does not enhance the urinary yield of anthocyanins in human subjects. *Food Chem.* 108: 869–78.

Hosseinian, F.S.; Beta, T. 2007. Saskatoon and wild blueberries have higher anthocyanin contents than other Manitoba berries. *J. Agric. Food Chem.* 55: 10832–38.

Howard, L.R.; Castrodale, C.; Brownmiller, C.; Mauromoustakos, A. 2010. Jam processing and storage effects on blueberry polyphenolics and antioxidant capacity. *J. Agric. Food Chem.* 58: 4022–9.

Howard, L.R.; Prior, R.L.; Liyanage, R.; Lay, J.O. 2012. Processing and storage effect on berry polyphenols: Challenges and implications for bioactive properties. *J. Agric. Food Chem.* 60: 6678–93.

Hrazdina, G. 1971. Reactions of the anthocyanidin-3,5-diglucosides: Formation of 3,5-di-(*O*-β-D-glucosyl)-hydroxyl-7-hydroxy coumarin. *Phytochemistry* 10: 1125 – 30.

Hrazdina, G.; Iredale, H.; Mattick, L.R. 1977. Anthocyanin composition of *Brassica oleracea* cv. Red Danish. *Phytochemistry* 16: 297–9.

Hugueney, P.; Provenzano, S.; Verrie`s, C.; Ferrandino, A.; Meudec, E.; Batelli, G.; Merdinoglu, D.; Cheynier, V.; Schubert, A.; Ageorges, A. 2009. A novel cation-dependent *O*-methyltransferase involved in anthocyanin methylation in grapevine. *Plant Physiol.* 150: 2057–70.

Huopalahti, R.; Jarvenpaa, E.P.; Katina, K. 2000. A novel solid-phase extraction-HPLC method for the analysis of anthocyanin and organic acid composition of Finnish cranberry. *J. Liq. Chrom. Rel. Technol.* 23: 2695–701.

Iacobucci, G.A.; Sweeny, J.G. 1983. The chemistry of anthocyanins, anthocyanidins and related flavylium salts. *Tetrahedron* 39: 3005–38.

Ichiyanagi, T.; Kashiwada, Y.; Shida, Y.; Ikeshiro, Y.; Kaneyuki, T.; Konishi, T. 2005c. Nasunin from eggplant consists of *cis–trans* isomers of delphinidin 3-[4-(*p*-coumaroyl)-L-rhamnosyl (1 → 6)glucopyranoside]-5-glucopyranoside. *J. Agric. Food Chem.* 53: 9472–7.

Ichiyanagi, T.; Rahman, M.M.; Hatano, Y.; Konishi, T.; Ikeshiro, Y. 2007. Protocatechuic acid is not the major metabolite in rat blood plasma after oral administration of cyanidin 3-*O*-β-D-glucopyranoside. *Food Chem.* 105: 1032–9.

Ichiyanagi, T. et al. 2004. Absorption and metabolism of delphinidin 3-*O*-β-D-glucopyranoside in rats. *Free Radical Biol. Med.* 36: 930–7.

Ichiyanagi, T.; Shida, Y.; Rahman, M.M.; Hatano, Y.; Konishi, T. 2005b. Extended glucuronidation is another major path of cyanidin 3-O-β-D-glucopyranoside metabolism in rats. *J. Agric. Food Chem.* 53: 7312–9.

Ichiyanagi, T.; Shida, Y.; Rahman, M.M.; Hatano, Y.; Konishi, T. 2006a. Bioavailability and tissue distribution of anthocyanins in bilberry (*Vaccinium myrtillus* L.) extract in rats. *J. Agric. Food Chem.* 54: 6578–87.

Ichiyanagi, T. et al. 2005a. Metabolic pathway of cyanidin 3-*O*-β-D-glucopyranoside in rats. *J. Agric. Food Chem.* 53: 145–50.

Ichiyanagi, T.; Terahara, N.; Rahman, M.M.; Konishi, T. 2006b. Gastrointestinal uptake of nasunin, acylated anthocyanin in eggplant. *J. Agric. Food Chem.* 54: 5306–12.

Innocenti, M.; Gallori, S.; Giaccherini, C.; Ieri, F.; Vincieri, F.F.; Mulinacci, N. 2005. Evaluation of the phenolic content in the aerial parts of different varieties of *Cichorium intybus* L. *J. Agric. Food Chem.* 53: 6497–502.

Irani, N.G.; Grotewold, E. 2005. Light-induced morphological alteration in anthocyanin-accumulating vacuoles of maize cells. *BMC Plant Biol.* 5: 1–15.

Jaakola, L.; Hohtola, A. 2010. Effect of latitude on flavonoid biosynthesis in plants. *Plant, Cell Environ.* 3: 1239–47.

Jackman, R.L.; Smith, J.L. 1996. Anthocyanins and betalains. In: *Natural Food Colourants*, 2nd ed. eds., G.F. Hendry, J.D. Houghton, Ch. 8, 244–309. London: Blackie & Son.

Jaillon, O.; Aury, J.-M.; Noel, B.; et al. 2007. The grapevine genome sequence suggests ancestral hexaploidization in major angiosperm phyla. *Nature* 449: 463–67.

Jánváry, L.; Hoffmann, T.; Pfeiffer, J.; Hausmann, L.; Töpfer, R.; Fischer, T.C.; Schwab, W. 2009. A double mutation in the anthocyanin 5-*O*-glucosyltransferase gene disrupts enzymatic activity in *Vitis vinifera* L. *J. Agr. Food Chem.* 57: 3512–8.

Jeong, S.T.; Goto-Yamamoto, N.; Kobayashi, S.; Esaka, M. 2004. Effects of plant hormones and shading on the accumulation of anthocyanins and the expression of anthocyanin biosynthetic genes in grape berry skins. *Plant Sci.* 167: 247–52.

Jimenez, M.; Castillo, I.; Azuara, E.; Beristain, C.I. 2011. Antioxidant and antimicrobial activity of capulin (*Prunus serotina* subsp *capuli*) extracts. *Rev. Mex. Ing. Quim.* 10: 29–37.

Johansen, O.P.; Andersen, Ø.M.; Nerdal, W.; Aksnes, D.W. 1991. Cyanidin 3-[6-(*p*-coumaroyl)-2-(xylosyl)-glucoside]-5-glucoside and other anthocyanins from fruits of *Sambucus canadensis*. *Phytochemistry* 30: 4137–41.

Jordheim, M.; Enerstvedt, K.H.; Andersen, Ø.M. 2011a. Identification of cyanidin 3-*O*-β-(6″-(3-hydroxy-3-methylglutaroyl)glucoside) and other anthocyanins from wild and cultivated blackberries. *J. Agric. Food Chem.* 59: 7436–40.

Jordheim, M.; Fossen, T.; Andersen, Ø.M. 2006. Characterization of hemiacetal forms of anthocyanidin 3-*O*-β-glycopyranosides. *J. Agric. Food Chem.* 54: 9340–6.

Jordheim, M.; Fossen, T.; Songstad, J.; Andersen, Ø.M. 2007c. Reactivity of anthocyanins and pyranoanthocyanins; studies on aromatic hydrogen-deuterium exchange reactions in methanol. *J. Agric. Food Chem.* 55: 8261–8.

Jordheim, M.; Måge, F.; Andersen, Ø.M. 2007b. Anthocyanins in berries of *Ribes* including gooseberry cultivars with a high content of acylated pigments. *J. Agric. Food Chem.* 55: 5529–35.

Jordheim, M.; Skare, E.; Andersen, Ø.M. 2011b. Anthocyanins in blood orange (*Citrus sinensis* L. Osbeck). Unpublished.

Jordheim, M.; Aaby, K.; Fossen, T.; Skrede, G.; Andersen, Ø.M. 2007a. Molar absorptivities and reducing capacity of pyranoanthocyanins and other anthocyanins. *J. Agric. Food Chem.* 55: 10591–8.

Kähkönen, M.P.; Heinaemaeki, J.; Ollilainen, V.; Heinonen, M. 2003. Berry anthocyanins: Isolation, identification and antioxidant activities. *J. Sci. Food Agric.* 83: 1403–11.

Kähkönen, M.P.; Hopia, A.I.; Heinonen, M. 2001. Berry phenolics and their antioxidant activity. *J. Agric. Food Chem.* 49: 4076–82.

Kalt, W.; Forney, C.F.; Martin, A.; Prior, R.L. 1999. Antioxidant capacity, vitamin C, phenolics, and anthocyanins after fresh storage of small fruits. *J. Agric. Food Chem.* 47: 4638–44.

Kammerer, D.; Carle, R.; Schieber, A. 2004. Quantification of anthocyanins in black carrot extracts (*Daucus carota* ssp. *sativus* var. *atrorubens* Alef.) and evaluation of their color properties. *Eur. Food Res. Technol.* 219: 479–86.

Kamonpatana, K.; Giusti, M. M.; Chitchumroonchokchai, C. et al. 2012. Susceptibility of anthocyanins to ex vivo degradation in human saliva. *Food Chemistry* 135: 738–47.

Kampuse, S.; Kampuss, K.; Pizika, L. 2002. Stability of anthocyanins and ascorbic acid in raspberry and black currant cultivars during frozen storage. *Acta Hort.* 85: 507–10.

Kay, C.D.; Kroon, P.A.; Cassidy, A. 2009. The bioactivity of dietary anthocyanins is likely to be mediated by their degradation products. *Mol. Nutr. Food Res.* 53: S92–S101.

Kay, C.D.; Mazza, G.J.; Holub, B.J. 2005. Anthocyanins exist in the circulation primarily as metabolites in adult men. *J. Nutr.* 135: 2582–8.

Keppler, K.; Humpf, H.U. 2005. Metabolism of anthocyanins and their phenolic degradation products by the intestinal microflora. *Bioorg. Med. Chem.* 13: 5195–205.

Kidøy, L.; Nygaard, A.M.; Andersen, Ø.M.; Pedersen, A.T.; Aksnes, D.W.; Kiremire, B.T. 1997. Anthocyanins in fruits of *Passiflora edulis* and *P. suberosa*. *J. Food Comp. Anal.* 10: 49–54.

Kim, D.O.; Padilla-Zakour, O.I. 2004. Jam processing effect on phenolics and antioxidant capacity in anthocyanin-rich fruits: Cherry, plum, and raspberry. *J. Food Sci.* 69: S395–S400.

Kim, H.-W. et al. 2012. Anthocyanin changes in the Korean purple-fleshed sweet potato, Shinzami, as affected by steaming and baking. *Food Chem.* 130: 966–72.

Koponen, J.M.; Happonen, A.M.; Mattila, P.H.; Toerroenen, A.R. 2007. Contents of anthocyanins and ellagitannins in selected foods consumed in Finland. *J. Agric. Food Chem.* 55: 1612–19.

Kosar, M.; Altintas, A.; Kirimer, N.; Baser, K. H.C. 2003. Determination of the free radical scavenging activity of Lycium extracts. *Chem. Nat. Comp.* (Translation of *Khimiya Prirodnykh Soedinenii*) 39: 531–5.

Koskela, A.K.J. et al. 2010. Variation in the anthocyanin concentration of wild populations of crowberries (*Empetrum nigrum* L. subsp. *hermaphroditum*). *J. Agric. Food Chem.* 58: 12286–91.

Kühnau, J. 1976. The flavonoids. A class of semi-essential food components: Their role in human nutrition. *World Rev. Nutr. Diet* 24: 117–191.

Kunz, S.; Burkhardt, G.; and Becker, H. 1994. Riccionidins A and B, anthocyanidins from the cell walls of the liverwort *Ricciocarpos natans*. *Phytochemistry* 35: 233–5.

Kurilich, A.C.; Clevidence, B.A.; Britz, S.J.; Simon, P.W.; Novotny, J.A. 2005. Plasma and urine responses are lower for acylated vs nonacylated anthocyanins from raw and cooked purple carrots. *J. Agric. Food Chem.* 53: 6537–42.

Kuskoski, E.M.; Vega, J.M.; Rios, J.J.; Fett, R.; Troncoso, A.M.; Asuero, A.G. 2003. Characterization of anthocyanins from the fruits of Baguaçu (*Eugenia umbelliflora* Berg). *J. Agric. Food Chem.* 51: 5450–4.

Kwee, E.M.; Niemeyer, E.D. 2011. Variations in phenolic composition and antioxidant properties among 15 basil (*Ocimum basilicum* L.) cultivars. *Food Chem.* 128: 1044–50.

Kylli, P. et al. 2010. Rowanberry phenolics: Compositional analysis and bioactivities. *J. Agric. Food Chem.* 58: 11985–92.

Lätti, A.K.; Jaakola, L.; Riihinen, K.R.; Kainulainen, P.S. 2010. Anthocyanins and flavonols in the berries of bog bilberry (*Vaccinium uliginosum* L.) from fifteen wild populations in Finland. *J. Agric. Food Chem.* 58: 427–433.

Lätti, A.K.; Riihinen, K.R.; Kainulainen, P.S. 2008. Analysis of anthocyanin variation in wild populations of bilberry (*Vaccinium myrtillus* L.) in Finland. *J. Agric. Food Chem.* 56: 190–6.

Lavola, A.; Karjalainen, R.; Julkunen-Tiitto, R. 2012. Bioactive polyphenols in leaves, stems, and berries of Saskatoon (*Amelanchier alnifolia* Nutt.) cultivars. *J. Agric. Food Chem.* 60: 1020–7.

Lee, H.S. 2002b. Characterization of major anthocyanins and the color of red-fleshed budd blood orange (*Citrus sinensis*). *J. Agric. Food Chem.* 50: 1243–6.

Lee, H.S.; Wicker, L. 1991. Quantitative changes in anthocyanin pigments of Lychee fruit during refrigerated storage. *Food Chem.* 40: 263–70.

Lee, J.H. 2010. Identification and quantification of anthocyanins from the grains of black rice (*Oryza sativa* L.) varieties. *Food Sci. Biotechnol.* 19: 391–7.

Lee, J.; Durst, R.W.; Wrolstad, R.E. 2002a. Impact of juice processing on blueberry anthocyanins and polyphenolics: Comparison of two pretreatments. *J. Food Sci.* 67: 1660–7.

Lee, J.; Durst, R.W.; Wrolstad, R.E. 2005. Determination of total monomeric anthocyanin pigment content of fruit juices, beverages, natural colorants, and wine by the pH differential method: Collaborative study. *J. AOAC Int.* 88: 1269–78.

Lee, J.; Finn, C.E. 2007. Anthocyanins and other polyphenolics in American elderberry (*Sambucus canadensis*) and European elderberry (*S. nigra*) cultivars. *J. Sci. Food Agric.* 87: 2665–75.

Lee, J.; Finn, C.E. 2012. Lingonberry (*Vaccinium vitis-idaea* L.) grown in the Pacific Northwest of North America: Anthocyanin and free amino acid composition. *J. Func. Foods* 4: 213–8.

Lee, J.; Finn, C.E.; Wrolstad, R.E. 2004. Anthocyanin pigment and total phenolic content of three *Vaccinium* species native to the Pacific Northwest of North America. *HortSci.* 39: 959–64.

Lee, J.; Rennaker C.; Wrolstad, R.E. 2008. Correlation of two anthocyanin quantification methods: HPLC and spectrophotometric methods. *Food Chem.* 110: 782–6.

Lewis, C.E.; Walkel, J.R.L.; Lancaster, J.E.; Sutton, K.H. 1998a. Determination of anthocyanins, flavonoids and phenolic acids in potatoes. I: Coloured cultivars of *Solanum tuberosum* L. *J. Sci. Food Agric.* 77:45–57.

Lewis, C.E.; Walker, J.R.L.; Lancaster, J.E.; Sutton, K.H. 1998b. Determination of anthocyanins, flavonoids and phenolic acids in potatoes. II: Wild, tuberous *Solanum* species. *J. Sci. Food Agric.* 77: 58–63.

Li, H.; Deng, Z.; Zhu, H.; Hu, C.; Liu, R.; Young, J.C.; Tsao, R. 2012. Highly pigmented vegetables: Anthocyanin compositions and their role in antioxidant activities. *Food Res. Int.* 46: 250–259.

Liang, Z.; Yang, Y.; Cheng, L.; Zhong, G.Y. 2012. Polyphenolic composition and content in the ripe berries of wild *Vitis* species. *Food Chem.* 132: 730–8.

Licciardello, F.; Muratore, G. 2011. Effect of temperature and some added compounds on the stability of blood orange marmalade. *J. Food Sci.* 76: C1094–C1100.

Lima, V.L.A.G.; Melo, E.A.; Maciel, M.I.S.; Lima, D.E.S. 2003. Avaliação do teor de antocianinas em polpa de acerola congelada proveniente de frutos de 12 diferentes aceroleiras (*Malpighia emarginata* D.C.). *Ciencia e Tecnol. de Alimentos* 23: 101–3.

Lo Scalzo, R.; Genna, A.; Branca, F.; Chedin, M.; Chassaigne, H. 2008. Anthocyanin composition of cauliflower (*Brassica oleracea* L. var. *botrytis*) and cabbage (*B. oleracea* L. var. capitata) and its stability in relation to thermal treatments. *Food Chem.* 107: 136–44.

Lo, S.-C.; Weiergang, I.; Bonham, C.; Hipskind, J.; Wood, K.; Nicholson, R.L. 1996. Phytoalexin accumulation in sorghum: Identification of a methyl ether of luteolinidin. *Physiol. Mol. Plant Pathol.* 49: 21–31.

Los, J.; Wilska-Jeszka, J.; Pawlak, M. 2000. Polyphenolic compounds of plums (*Prunus domestica*) *Pol. J. Food Nutr. Sci.* 9: 35–8.

Lu, Y.R.; Foo, L.Y. 2001. Unusual anthocyanin reaction with acetone leading to pyranoanthocyanin formation. *Tetrahedron Lett.* 42: 1371–3.

Määttä-Riihinen, K.R.; Kamal-Eldin, A.; Mattila, P.H.; Gonzalez-Paramas, A.M.; Toerroenen, A.R. 2004a. Distribution and contents of phenolic compounds in eighteen Scandinavian berry species. *J. Agric. Food Chem.* 52: 4477–86.

Määttä-Riihinen, K.R.; Kamal-Eldin, A.; Toerroenen, A.R. 2004b. Identification and quantification of phenolic compounds in berries of *Fragaria* and *Rubus* species (Familiy Rosaceae). *J. Agric. Food Chem.* 52: 6178–87.

Maccarone, E.; Rapisarda, P.; Fanella, F.; Arena, E.; Mondello, L. 1998. Cyanidin-3-(6″-malonyl)-β-glucoside. One of the major anthocyanins in blood orange juice. *Ital. J. Food Sci.* 10: 367–72.

Macz-Pop, G.A.; Gonzalez-Paramas, A.M.; Perez-Alonso, J.J.; Rivas-Gonzalo, J.C. 2006a. New flavanol-anthocyanin condensed pigments and anthocyanin composition in Guatemalan beans (*Phaseolus* spp.). *J. Agric. Food Chem.* 54: 536–42.

Macz-Pop, G.A.; Rivas-Gonzalo, J.C.; Perez-Alonso, J.J.; Gonzalez-Paramas, A.M. 2006b. Natural occurrence of free anthocyanin aglycones in beans (*Phaseolus vulgaris* L.). *Food Chem.* 94: 448–56.

Manach, C.; Williamson, G.; Morand, C.; Scalbert, A.; Rémésy, C. 2005. Bioavailability and bioefficacy of polyphenols in humans. I. Review of 97 bioavailability studies. *Am. J. Clin. Nutr.* 81: 230–42.

Markham, K.R.; Gould, K.S.; Winefield, C.S.; Mitchell, K.A.; Bloor, S.J.; Boase, M.R. 2000. Anthocyanic vacuolar inclusions—Their nature and significance in flower colouration. *Phytochemistry* 55: 327–36.

Matera, R.; Gabbanini, S.; De Nicola, G.R.; Iori, R.; Petrillo, G.; Valgimigli, L. 2012. Identification and analysis of isothiocyanates and new acylated anthocyanins in the juice of *Raphanus sativus* cv. Sango sprouts. *Food Chem.* 133: 563–72.

Mateus, N.; de Freitas, V. 2001. Evolution and stability of anthocyanin derived pigments during port wine aging. *J. Agric. Food Chem.* 49: 5217–22.

Mateus, N.; Machado, J.M.; de Freitas V. 2002. Development changes of anthocyanins in *Vitis vinifera* grapes grown in the Douro Valley and concentration in respective wines. *J. Sci. Food Agric.* 82: 1689–95.

Matsui, T.; Ueda, T.; Oki, T.; Sugita, K.; Terahara, N.; Matsumoto, K. 2001. α-Glucosidase inhibitory action of natural acylated anthocyanins. 2. α-Glucosidase inhibition by isolated acylated anthocyanins. *J. Agric. Food Chem.* 49: 1952–56.

Matsumoto, H.; Inaba, H.; Kishi, M.; Tominaga, S.; Hirayama, M.; Tsuda, T. 2001. Orally administered delphinidin 3-rutinoside and cyanidin 3-rutinoside are directly absorbed in rats and humans and appear in the blood as the intact forms. *J. Agric. Food Chem.* 49: 1546–51.

Mattila, P.H. et al. 2011. Polyphenol and vitamin C contents in European commercial black currant juice products. *Food Chem.* 127: 1216–23.

Matus. J.T. et al. 2009. Post-veraison sunlight exposure induces MYB-mediated transcriptional regulation of anthocyanin and flavonol synthesis in berry skins of *Vitis vinifera*. *J. Exp. Bot.* 60: 853–67.

Mazza, G.; Brouillard, R.J. 1987. Color stability and structural transformations of cyanidin 3,5-diglucoside and four 3-deoxyanthocyanins in aqueous solutions. *J. Agric. Food Chem.* 35: 422–6.

Mazza, G.; Fukumoto, L.; Delaquis, P.; Girard, B.; Ewert, B. 1999. Anthocyanins, phenolics, and color of Cabernet Franc, Merlot, and Pinot Noir wines from British Columbia. *J. Agr. Food Chem.* 47: 4009–17.

Mazza, G.; Gao, L. 1994. Malonylated anthocyanins in purple sunflower seeds. *Phytochemistry* 35: 237–9.

McClure, J.W. 1975. Physiology and functions of flavonoids. In *The Flavonoid*, eds., J.B. Harborne, T.J. Mabry, H. Mabry, 970–1055. London: Chapman & Hall.

McDougall, G.J.; Gordon, S.; Brennan, R.; Stewart, D. 2005. Anthocyanin-flavanol condensation products from black currant (*Ribes nigrum* L.). *J. Agric. Food Chem.* 53: 7878–85.

McGhie, T.K.; Ainge, G.D.; Barnett, L.E.; Cooney, J.M.; Jensen, D.J. 2003. Anthocyanin glycosides from berry fruit are absorbed and excreted unmetabolized by both humans and rats. *J. Agric. Food Chem.* 51: 4539–48.

Mercadante, A.Z.; Bobbio, F. 2008. Anthocyanins in foods: Occurrence and physicochemical properties. In *Food Colorants—Chemical and Functional Properties*, ed., C. Socaciu, 241–76. Boca Raton, FL: CRC Press.

Mertens-Talcott, S.U. et al. 2008. Pharmacokinetics of anthocyanins and antioxidant effects after the consumption of anthocyanin-rich Acai juice and pulp (*Euterpe oleracea* Mart.) in human healthy volunteers. *J. Agric. Food Chem.* 56: 7796–7802.

Messaoud, C.; Boussaid, M. 2011. *Myrtus communis* berry color morphs: A comparative analysis of essential oils, fatty acids, phenolic compounds, and antioxidant activities. *Chem. Biodiv.* 8: 300–10.

Miniati, E. 1981. Anthocyanin pigment in the pistachio nut. *Fitoterpia.* 52: 267–71.

Miyazawa, T.; Nakagawa, K.; Kudo, M.; Muraishi, K.; Someya, K. 1999. Direct intestinal absorption of red fruit anthocyanins, cyanidin-3-glucoside and cyanidin-3,5-diglucoside, into rats and humans. *J. Agric. Food Chem.* 47: 1083–91.

Montilla, E.C.; Arzaba, M.R.; Hillebrand, S.; Winterhalter, P. 2011. Anthocyanin composition of black carrot (*Daucus carota* ssp. *sativus* var. *atrorubens* Alef.) cultivars Antonina, Beta Sweet, Deep Purple, and Purple Haze. *J. Agric. Food Chem.* 59: 3385–90.

Moreno, D.A.; Perez-Balibrea, S.; Ferreres, F.; Gil-Izquierdo, A.; Garcia-Viguera, C. 2010. Acylated anthocyanins in broccoli sprouts. *Food Chem.* 123: 358–63.

Moreno, Y.S.; Sanchez, G.S.; Hernandez, D.R.; Lobato, N.R. 2005. Characterization of anthocyanin extracts from maize kernels. *J. Chromatogr. Sci.* 43: 483–87.

Mori, K.; Goto-Yamamoto, N.; Kitayama, M.; Hashizume, K. 2007. Loss of anthocyanins in red-wine grape under high temperature. *J. Exp. Bot.* 58: 1935–45.

Moyer, R.A.; Hummer, K.E.; Finn, C.E.; Frei, B.; Wrolstad, R.E. 2002. Anthocyanins, phenolics, and antioxidant capacity in diverse small fruits: *Vaccinium*, *Rubus*, and *Ribes*. *J. Agric. Food Chem.* 50: 519–25.

Mozetic, B.; Trebse, P.; Simcic, M.; Hribar, J. 2004. Changes of anthocyanins and hydroxycinnamic acids affecting the skin colour during maturation of sweet cherries (*Prunus avium* L.). *Lebensm.-Wissens. Technol.* 37: 123–8.

Mulas, M.; Francesconi, A.H.D.; Perinu, B.; Fadda, A. 2002. 'Barbara' and 'Daniela': Two cultivars for Myrtle berries production. *Acta Hort.* 576:169–75.

Mulinacci, N.; Innocenti, M.; Gallori, S.; Romani, A.; Vincieri, F.F. 2001. Polyphenolic content in different species of chicory. *SP–Roy. Soc. Chem.* 269: 174–8.

Mullen, W.; Edwards, C.A.; Serafini, M.; Crozier, A. 2008. Bioavailability of pelargonidin-3-*O*-glucoside and its metabolites in humans following the ingestion of strawberries with and without cream. *J. Agric. Food Chem.* 56: 713–9.

Müller, D.; Schantz, M.; Richling, E. 2012. High performance liquid chromatography analysis of anthocyanins in bilberries (*Vaccinium myrtillus* L.), blueberries (*Vaccinium corymbosum* L.), and corresponding juices. *J. Food Sci.* 77: C340–C345.

Mullie, P.; Clarys, P.; Deriemaeker, P.; Hebbelinck, M. 2008. Estimation of daily human intake of food flavonoids. *Int. J. Food Sci. Nutr.* 59: 291–8.

Murkovic, M.; Mulleder, U.; Adam, U.; Pfannhauser, W. 2001. Detection of anthocyanins from elderberry juice in human urine. *J. Sci. Food Agric.* 81: 934–7.

Nakamura, M.; Seki, M.; Furusaki, S. 1998. Enhanced anthocyanin methylation by growth limitation in strawberry suspension culture. *Enzyme Microb. Technol.* 22: 404–8.

Nakatani, N.; Kikuzaki, H.; Hikida, J.; Ohba, M.; Inami, O.; Tamura, I. 1995. Acylated anthocyanins from fruits of *Sambucus canadensis*. *Phytochemistry* 38: 755–7.

Neida, S.; Elba, S. 2007. Characterization of the acai or manaca (*Euterpe oleracea* Mart.): A fruit of the Amazon. *Arc. Latinoam. Nutr.* 57: 94–8.

Nerdal, W.; Andersen, Ø.M. 1992. Intermolecular aromatic acid association of an anthocyanin (petanin) evidenced by two-dimensional nuclear overhauser enhancement nuclear magnetic resonance experiments and distance geometry calculations. *Phytochem. Anal.* 3: 182–9.

Netzel, M.; Netzel, G.; Tian, Q.; Schwartz, S.J.; Konczak, I. 2006. Sources of antioxidant activity in Australian native fruits. Identification and quantification of anthocyanins. *J. Agric. Food Chem.* 54: 9820–6.

Netzel, M. et al. 2005. The excretion and biological antioxidant activity of elderberry antioxidants in healthy humans. *Food Res. Int.* 38: 905–10.

Newton, D. 1978. pKa values of medicinal compounds in pharmacy practice. *Drug Intell. Clin. Pharm.* 12: 546–54.

Nielsen, I.L.F.; Haren, G.R.; Magnussen, E.L.; Dragsted, L.O.; Rasmussen, S.E. 2003. Quantification of anthocyanins in commercial black currant juices by simple high-performance liquid chromatography. Investigation of their pH stability and antioxidative potency. *J. Agric. Food Chem.* 51: 5861–6.

Noda, Y.; Kaneyuki, T.; Igarashi, K.; Mori, A.; Packer, L. 2000. Antioxidant activity of nasunin, an anthocyanin in eggplant peels. *Toxicology* 148: 119–23.

Nordborg, M.; Weigel, D. 2008. Next-generation genetics in plants. *Nature* 456: 720–23.

Novotny, J.A.; Clevidence, B.A.; Kurilich, A.C. 2012. Anthocyanin kinetics are dependent on anthocyanin structure. *Br. J. Nutr.* 107: 504–9.

Nozue, M.; Yamada, K.; Nakamura, T.; Kubo, H.; Kondo, M.; Nishimura, M. 1997. Expression of a vacuolar protein (VP24) in anthocyanin-producing cells of sweet potato in suspension culture. *Plant Physiol.* 115: 1065–72.

Nozzolillo, C.; Anderson, J.; Warwick, S. 1995. Anthocyanoplasts in the Brassicaceae: Does their presence serve as a chemotaxonomic marker within the family? *Polyphenols Actu.* 12: 25–6.

Ogawa, K. et al. 2008. Anthocyanin composition and antioxidant activity of the crowberry (*Empetrum nigrum*) and other berries. *J. Agric. Food Chem.* 56: 4457–62.

Ojwang, L.O.; Dykes, L.; Awika, J.M. 2012. Ultra performance liquid chromatography-tandem quadrupole mass spectrometry profiling of anthocyanins and flavonols in cowpea (*Vigna unguiculata*) of varying genotypes. *J. Agric. Food Chem.* 60: 3735–44.

Oki, T. et al. 2010. Determination of major anthocyanins in processed foods made from purple-fleshed sweet potato. *Nippon Shokuhin Kagaku Kogaku Kaishi* 57: 128–33.

Opheim, S.; Andersen, Ø.M. 1992. Anthocyanins in the genus *Solanum*. *Phytochemistry* 11: 239–43.

Oren-Shamir, M. 2009. Does anthocyanin degradation play a significant role in determining pigment concentration in plants? *Plant Sci.* 177: 310–6.

Osorio, C.; Hurtado, N.; Dawid, C.; Hofmann, T.; Heredia-Mira, F.J.; Morales, A.L. 2012. Chemical characterisation of anthocyanins in tamarillo (*Solanum betaceum* Cav.) and Andes berry (*Rubus glaucus* Benth.) fruits. *Food Chemistry* 132: 1915–21.

Ovaskainen, M.L. et al. 2008. Dietary intake and major food sources of polyphenols in Finnish adults. *J. Nutr.* 138: 562–66.

Pale, E.; Kouda-Bonafos, M.; Nacro, M.; Vanhaelen, M.; Vanhaelen-Fastre, R.; Ottinger, R. 1997. 7-*O*-Methylapigeninidin, an anthocyanidin from *Sorghum caudatum*. *Phytochemistry* 45: 1091–2.

Pantelidis, G.E.; Vasilakakisa, M.; Manganaris, G.A.; Diamantidisa, G. 2007. Antioxidant capacity, phenol, anthocyanin and ascorbic acid contents in raspberries, blackberries, red currants, gooseberries and Cornelian cherries. *Food Chem.* 102: 777–83.

Pappas, E.; Schaich, K.M. 2009. Phytochemicals of cranberries and cranberry products: Characterization, potential health effects, and processing stability. *Crit. Rev. Food Sci. Nutr.* 49: 741–81.

Park, J.S.; Choung, M.G.; Kim, J.B.; Hahn, B.S.; Bae, S.C.; Roh, K.H.; Kim, Y.H.; Cheon, C.I.; Sung, M.K.; Cho, K.J. 2007. Genes up-regulated during red coloration in UV-B irradiated lettuce leaves. *Plant Cell Rep.* 26: 507–16.

Patras, A.; Brunton, N.P.; O'Donnell, C.; Tiwari, B.K. 2010. Effect of thermal processing on anthocyanin stability in foods; mechanisms and kinetics of degradation. *Trends Food Sci. Tech.* 21: 3–11.

Pazmino-Duran, E.A.; Giusti, M.M.; Wrolstad, R.E.; Gloria, M.B.A. 2001. Anthocyanins from *Oxalis triangularis* as potential food colorants. *Food Chem.* 75: 211–16.

Phippen, W.B.; Simon, J.E. 1998. Anthocyanins in Basil (*Ocimum basilicum* L.). *J. Agric. Food Chem.* 46: 1734–8.

Piero, A.R.L.; Puglisi, I.; Rapisarda, P.; Petrone, G. 2005. Anthocyanins accumulation and related gene expression in red orange fruit induced by low temperature storage. *J. Agr. Food Chem.* 53: 9083–8.

Pina, F.; Melo, M.J.; Laia, C.A.T.; Parola, A.J.; Lima, J.C. 2012. Chemistry and applications of flavylium compounds: A handful of colours. *Chem. Soc. Rev.* 41: 869–908.

Pinho, C.; Couto, A.I.; Valentão, P.; Andrade, P.; Ferreira, I.M.P.L.V.O. 2012. Assessing the anthocyanic composition of Port wines and musts and their free radical scavenging capacity. *Food Chem.* 131: 885–92.

Pourcel, L.; Bohórquez-Restrepo, A.; Irani, N.G.; Grotewold, E. 2012. Anthocyanin bio-synthesis, transport and regulation: New insights from model species. In *Recent Advances in Polyphenol Research*, Vol. 3, eds., V. Cheynier et al., 143–60. Chichester, Wiley-Blackwell.

Prior, R.L. et al. 1998. Antioxidant capacity as influenced by total phenolic and anthocy-anin content, maturity, and variety of *Vaccinium* species. *J. Agric. Food Chem.* 46: 2686–93.

Prior, R.L; Lazarus, S.A.; Cao, G.; Muccitelli, H.; Hammerstone, J.F. 2001. Identification of procyanidins and anthocyanins in blueberries and cranberries (*Vaccinium* spp.) using high-performance liquid chromatography/mass spectrometry. *J. Agric. Food Chem.* 49: 1270–6.

Prior, R.L.; Wu, X. 2006. Anthocyanins: Structural characteristics that result in unique meta-bolic patterns and biological activities. *Free Radic. Res.* 40: 1014–28.

Proctor, J.T.; Creasy L.L. 1969. The anthocyanin of the mango fruit. *Phytochemistry* 8: 2108.

Rababah, T.M. et al. 2011. Effect of jam processing and storage on total phenolics, antioxidant activity, and anthocyanins of different fruits. *J. Sci. Food Agric.* 91: 1096–102.

Rababah, T.M.; Ereifej, K.I.; Howard, L. 2005. Effect of ascorbic acid and dehydration on concentrations of total phenolics, antioxidant capacity, anthocyanins, and color in fruits. *J. Agric. Food Chem.* 53: 4444–7.

Rapisarda, P.; Fanella, F.; Maccarone, E. 2000. Reliability of analytical methods for determin-ing anthocyanins in blood orange juices. *J. Agric. Food Chem.* 48: 2249–52.

Rein, M.J.; Ollilainen, V.; Vahermo, M.; Yli-Kauhaluoma, J.; Heinonen, M. 2005. Identification of novel pyranoanthocyanins in berry juices. *Eur. Food Res. Technol.* 220: 239–44.

Remy-Tanneau, S.; Le Guerneve, C.; Meudec, E.; V. Cheynier, V. 2003. Characterization of a colorless anthocyanin-flavan-3-ol dimer containing both carbon-carbon and ether inter-flavanoid linkages by NMR and mass spectrometry. *J. Agric. Food Chem.* 51: 3592–7.

Revilla, I.; Perez-Magarin, S.; Gonzalez-SanJose, M.L.; Beltra, S. 1999. Identification of anthocyanin derivatives in grape skin extracts and red wines by liquid chromatography with diode array and mass spectrometric detection. *J. Chromatogr. A* 847: 83–90.

Riedl, P.; Murkovic, M. 2011. Determination and quantification of anthocyanins in fruits and berries. *Ernaehrung* 11: 464–72.

Rivas-Gonzalo, J.C. 2003. Analysis of anthocyanins. In *Methods in Polyphenol Analysis*, eds., C. Santos-Buelga; G. Williamson, 338–353. Cambridge: The Royal Society of Chemistry.

Rodriguez-Saona, L.E.; Giusti, M.M.; Wrolstad, R.E. 1998. Anthocyanin pigment composition of red-fleshed potatoes. *J. Food Sci.* 63: 458–65.

Rodriguez-Saona, L.E.; Wrolstad, R.E. 2005. Extraction, isolation, and purification of anthocyanins. In *Handbook of Food Analytical Chemistry: Pigments, Colorants, Flavors, Texture, and Bioactive Food Components,* eds., R. Wrolstad et al., 7–17. Hoboken: John Wiley & Sons.

Romani, A.; Mulinacci, N.; Pinelli, P.; Vincieri, F.F.; Cimato, A. 1999. Polyphenolic content in five tuscany cultivars of *Olea europaea* L. *J. Agric. Food Chem.* 47: 964–67.

Ruiz, D.; Egea, J.; Gill, M.I.; Tomas-Barberan, F.A. 2006. Phytonurtrient content in new apricot (*Prunus ameniaca* L.) varieties. *Acta Horticult.* 717: 363–7.

Runge-Morris, M.A. 1997. Regulation of expression of the rodent cytosolic sulfotransferases. *FASEB J.* 11: 109–17.

Ryu, S.N.; Park, S.Z.; Ho, C.-T. 1998. High performance liquid chromatographic determination of anthocyanin pigments in some varieties of blackrice. *Yaowu Shipin Fenxi* 6: 729–36.

Sadilova, E.; Carle, R.; Stintzing, F.C. 2007. Thermal degradation of anthocyanins and its impact on colour and *in vitro* antioxidant capacity. *Mol. Nutr. Food Res.* 51: 1461–71.

Sadilova, E.; Stintzing, F.C.; Carle, R. 2006. Thermal degradation of acylated and nonacylated anthocyanins. *J. Food Sci.* 71: 504–12.

Saito, N.; Tatsuzawa, F.; Miyoshi, K.; Shigihara, A.; Honda, T. 2003. The first isolation of *C*-glycosylanthocyanin from the flowers of *Tricyrtis formosana*. *Tetrahedron Lett.* 44: 6821–3.

Saito, T.; Honma, D.; Tagashira, M.; Kanda, T.; Nesumi, A.; Maeda-Yamamoto, M. Anthocyanins from new red leaf tea 'Sunrouge'. 2011b. *J. Agric. Food Chem.* 59: 4779–82.

Sakaguchi, Y. et al. 2008. Major anthocyanins from purple asparagus (*Asparagus officinalis*). *Phytochemistry* 69: 1763–66.

Sanchez-Moreno, C.; Cao, G.; Ou, B.; Prior, R.L. 2003. Anthocyanin and proanthocyanidin content in selected white and red wines. Oxygen radical absorbance capacity comparison with non-traditional wines obtained from highbush blueberry. *J. Agric. Food Chem.* 51: 4889–96.

Santos, H.; Turner, D.; Lima, J.C.; Figueiredo, P.; Pina, F.; Maçanita, A.L. 1993. Elucidation of the multiple equilibria of malvin in aqueous-solution by one-dimensional and 2-dimensional NMR. *Phytochemistry* 33: 1227–32.

Santos-Buelga, C.; Gonzalez-Manzano, S.; Dueñas, M.; Gonzalez-Paramas, A.M. 2012. Analysis and characterization of flavonoid phase II metabolites. In *Recent Advances in Polyphenol Research*, Vol. 3, eds., V. Cheynier et al., 249–86. Chichester, Wiley-Blackwell.

Sarni-Manchado, P.; Le Roux, E.; Le Guerneve, C.; Lozano, Y.; Cheynier, V. 2000. Phenolic composition of litchi fruit pericarp. *J. Agric. Food Chem.* 48: 5995–6002.

Shih, C.-H. et al. 2007. Quantitative analysis of anticancer 3-deoxyanthocyanidins in infected sorghum seedlings. *J. Agric. Food Chem.* 55: 254–9.

Skaar, I.; Jordheim, M.; Byamukama, R.; Mbabazi, A.; Wubshet, S.G.; Kiremire, B.; Andersen, Ø.M. 2012. New anthocyanidin and anthocyanin pigments from blue plumbago. *J. Agric. Food Chem.* 60: 1510–5.

Slimestad, R.; Torskangerpoll, K.; Nateland, H.S.; Johannessen, T.; Giske, N.H. 2004. Flavonoids from black chokeberries *Aronia melanocarpa*. *J. Food Comp. Anal.* 18: 61–8.

Small, C.J.; Pecket, R.C. 1982. The ultrastructure of anthocyanoplasts in red cabbage. *Planta* 154: 97–9.

Snyder, B.A.; Nicholson, R.L. 1990. Synthesis of phytoalexins in sorghum as a site-specific response to fungal ingress. *Science* 248: 1637–9.

Soto-Vaca, A.; Gutierrez, A.; Losso, J.N.; Xu, Z.; Finley, J.W. 2012. Evolution of phenolic compounds from color and flavor problems to health benefits. *J. Agric. Food Chem.* 60: 6658–77.

Sousa de Brito, E.; Pessanha de Araujo, M.C.; Alves, R.E.; Carkeet, C.; Clevidence, B.A.; Novotny, J.A. 2007. Anthocyanins present in selected tropical fruits: Acerola, jambolão, jussara, and guajiru. *J. Agric. Food Chem.* 55: 9389–94.

Springob, K.; Nakajima, J.-I.; Yamazaki, M.; Saito, K. 2003. Recent advances in the biosynthesis and accumulation of anthocyanins. *Nat. Prod. Rep.*, 20: 288–303.

Stefanut, M.N.; Cata, A.; Pop, R.; Mosoarca, C.; Zamfir, A.D. 2011. Anthocyanins HPLC-DAD and MS characterization, total phenolics, and antioxidant activity of some berries extracts. *Anal. Lett.* 44: 2843–55.

Steyn, W.J. 2009. Prevalence and functions of anthocyanins in fruits. In *Anthocyanins Biosynthesis, Functions, and Applications*, eds., K. Gould, K. Davies, C. Winefield, C., 107–168. New York: Springer Science.

Strigl, A.W.; Leitner, E.; Pfannhauser, W. 1995. Qualitative und quantitative analyse der anthocyane in schwarzen apfelbeeren (*Aronia melanocarpa* Michx. Ell.) mittels TLC, HPLC und UV/VIS-spektrometrie. *Z. Lebensm. -Untersuch. -Forsch.* 201: 266–68.

Sugawara, T.; Igarashi, K. 2008. Cultivar variation and anthocyanins and rutin content in sweet cherries (*Prunus avium* L.). *Nippon Shokuhin Kagaku Kogaku Kaishi* 55: 239–44.

Sweeny, J.G.; Wilkinson, M.M.; Iacobucci, G.A. 1981. Effect of flavonoid sulfonates on the photobleaching of anthocyanins in acid solution. *J. Agric. Food Chem.* 29: 563–7.

Takeoka, G.R. et al. 1997. Characterization of black bean (*Phaseolus vulgaris* L.) anthocyanins. *J. Agric. Food Chem.* 45: 3395–400.

Takeoka, G.R.; Dao, L.T.; Tamura, H.; Harden, Le. A. 2005. Delphinidin 3-O-(2-O-β-d-Glucopyranosyl-α-l-arabinopyranoside): A novel anthocyanin identified in Beluga black lentils. *J. Agric. Food Chem.* 53: 4932–7.

Taniguchi, S.; Yazaki, K.; Yabu-Uchi, R.; Kawakami, K.-Y.; Ito, H.; Hatano, T.; Yoshida, T. 2000. Galloylglucoses and riccionidin A in *Rhus javanica* adventitious root cultures. *Phytochemistry* 53: 357–63.

Telias, A.; Bradeen, J.M.; Luby, J.J.; Hoover, E.E.; Allan, A.C. 2011. Regulation of anthocyanin accumulation in apple peel. *Hort. Rev.* 38: 357–91.

Terahara, N.; Saito, N.; Honda, T.; Toki, K.; Osajima, Y. 1990. Structure of ternatin A1, the largest ternatin in the major blue anthocyanins from *Clitoria ternatea* flowers. *Tetrahedron Lett.* 31: 2921–4.

Terahara, N.; Yamaguchi, M.-A; Honda, T. 1994. Malonylated anthocyanins from bulbs of red onion, *Allium cepa* L. *Biosci. Biotechnol. Biochem.* 58: 1324–5.

Tian, Q.; Giusti, M.M.; Stoner, G.D.; Schwartz, S.J. 2006. Urinary excretion of black raspberry (*Rubus occidentalis*) anthocyanins and their metabolites. *J. Agric. Food Chem.* 54: 1467–72.

Tatsuzawa, F.; Ito, K.; Muraoka, H.; Namauo, T.; Kato, K.; Takahata, Y.; Ogawa, S. 2012. Triacylated peonidin 3-sophoroside-5-glucosides from the purple flowers of *Moricandia ramburii* Webb. *Phytochemistry* 76: 73–7.

Toki, K.; Saito, N.; Ueda, T.; Chibana, T.; Shigihara, A.; Honda, T. 1994. Malvidin 3-glucoside-5-glucoside sulphates from *Babina stricta*. *Phytochemistry* 37: 885–7.

Toki, K.; Saito, N.; Irie, Y.; Tatsuzawa, F.; Shigihara, A.; Honda, T. 2008. 7-O-methylated anthocyanidin glycosides from *Catharanthus roseus*. *Phytochemistry* 69: 1215–9.

Toki, K.; Saito, N.; Kitaura, M.; Tatsuzawa, F.; Shigihara, A.; Honda, T. 2011. 7-O-methylated anthocyanidin glycosides from the reddish purple flowers of *Catharanthus roseus* "Equator Lavender". *Heterocycles* 83: 2803–10.

Tomás-Barberán, F.A.; Gil, M.I.; Cremin, P.; Waterhouse, A.L.; Hess-Pierce, H.; Kader, A.A. 2001. HPLC-DAD-ESIMS analysis of phenolic compounds in nectarines, peaches, and plums. *J. Agric. Food Chem.* 49: 4748–60.

Torskangerpoll, K.; Andersen, Ø.M. 2005. Colour stability of anthocyanins in aqueous solutions at various pH values. *Food Chem.* 89: 427–40.

Tsuda, T.; Horio, F.; Osawa, T. 1999. Absorption and metabolism of cyanidin 3-*O*-β-*D*-glucoside in rats. *FEBS Lett.* 449: 179–82.

Tyl, C.E.; Bunzel, M. 2012. Antioxidant activity-guided fractionation of blue wheat (UC66049 *Triticum aestivum* L.) *J. Agric. Food Chem.* 60: 731–39.

Ubi, B.U.; Honda, C.; Bessho, H.; Kondo, S.; Wada, M.; Kobayashi, S.; Moriguchi, T. 2006. Expression analysis of anthocyanin biosynthetic genes in apple skin: Effect of UV-B and temperature. *Plant Sci.* 170: 571–8.

Varasteh, F., Arzani, K.; Barzegar, M.; Zamani, Z. 2012. Changes in anthocyanins in arils of chitosan-coated pomegranate (*Punica granatum* L. cv. Rabbab-e-Neyriz) fruit during cold storage. *Food Chem.* 130: 267–72.

Vega, A.; Gutierrez, R.A.; Pena-Neira, A.; Cramer, G.R.; Arce-Johnson, P. 2011. Compatible GLRaV-3 viral infections affect berry ripening decreasing sugar accumulation and anthocyanin biosynthesis in *Vitis vinifera*. *Plant Mol. Biol.* 77: 261–74.

Velasco, R. et al. 2007. A high quality draft consensus sequence of the genome of a heterozygous grapevine variety. *PLoS One* 2: 1326.

Vera de Rosso, V. et al. 2008. Determination of anthocyanins from acerola (*Malpighia emarginata* DC.) and acai (*Euterpe oleracea* Mart.) by HPLC–PDA–MS/MS. *J. Food Comp. Anal.* 21: 291–9.

Vieira, F.G.K.; Borges, G.S.C.; Copetti, C.; Di Pietro, P.F.; Nunes, E.C.; Fett, R. 2011. Phenolic compounds and antioxidant activity of the apple flesh and peel of eleven cultivars grown in Brazil. *Sci. Hort.* 128: 261–6.

Viscardi, D.; Migliori, C.; Di Cesare, L.F. 2009. Alimentary and nutraceutical quality of "Violetto di Catania" cauliflower cultivars. *Progr. Nutr.* 11: 110–7.

Vitaglione, P. et al. 2007. Protocatechuic acid is the major human metabolite of cyanidin-glucosides. *J. Nutr.* 137: 2043–8.

Vowinkel, E. 1975. Cell wall pigments of peat mosses. 2. Structure of sphagnorubin. *Chem. Ber.* 108: 1166–81.

Wang, S.Y.; Lin, H.S. 2000. Antioxidant activity in fruits and leaves of blackberry, raspberry, and strawberry varies with cultivar and developmental stage. *J. Agric. Food Chem.* 48: 140–6.

Wang, S.Y.; Stretch, A.W. 2001. Antioxidant capacity in cranberry is influenced by cultivar and storage temperature. *J. Agric. Food Chem.* 49: 969–74.

Weidel, E.; Schantz, M.; Richling, E. 2011. Anthocyanin contents in black currant (*Ribes nigrum* L.) juices and fruit drinks. *Fruit Process.* 21: 102–7.

Wiesenborn, D.; Zbikowski, Z.; Nguyen, H. 1995. Process conditions affect pigment quality and yield in extracts of purple sunflower hulls. *J. Am. Oil Chem. Soc.* 72: 183–8.

Wolfe, K.; Wu, X.; Liu, R.H. 2003. Antioxidant activity of apple peels. *J. Agric. Food Chem.* 51, 609–14.

Woods, E.; Clifford, M.N.; Gibbs, M.; Hampton, S.M.; Arendt, J.; Morgan L.M. 2003. Estimation of mean intakes of fourteen classes of dietary phenols in a population of male shift workers. *Proc. Nutr. Soc.* 62: 60A.

Woodward, G.M.; Needs, P.W.; Kay, C.D. 2011. Anthocyanin-derived phenolic acids form glucuronides following simulated gastrointestinal digestion and microsomal glucuronidation. *Mol. Nutr. Food Res.* 55: 378–86.

Wrolstad, R.E.; Heatherbell, D.A. 1968. Anthocyanin pigments of rhubarb, Canada Red. *J. Food Sci.* 33: 592–94.

Wrolstad, R.E.; Heatherbell, D.A. 1974. Identification of anthocyanins and distribution of flavonoids in tamarillo fruit (*Cyphomandra betaceae* (Cav.) Sendt.). *J. Sci. Food Agric.* 25: 1221–8.

Wu, X.; Beecher, G.R.; Holden, J.M.; Haytowitz, D.B.; Gebhardt, S.E.; Prior, R.L. 2006. Concentrations of anthocyanins in common foods in the United States and estimation of normal consumption. *J. Agric. Food Chem.* 54: 4069–75.

Wu, X.; Cao, G.; Prior, R.L. 2002. Absorption and metabolism of anthocyanins in elderly women after consumption of elderberry or blueberry. *J. Nutr.* 132: 1865–71.

Wu, X.; Gu, L.; Prior, R.L.; McKay, S. 2004a. Characterization of anthocyanins and proanthocyanidins in some cultivars of *Ribes*, *Aronia*, and *Sambucus* and their antioxidant capacity. *J. Agric. Food Chem.* 52: 7846–56.

Wu, X.; Pittman, H.E. III; Prior, R.L. 2004b. Pelargonidin is absorbed and metabolized differently than cyanidin after marionberry consumption in pigs. *J. Nutr.* 134: 2603–10.

Wu, X.; Pittman, H.E.; McKay, S.; Prior, R.L. 2005. Aglycones and sugar moieties alter anthocyanin absorption and metabolism following berry consumption in the weanling pig. *J. Nutr.* 135: 2417–24.

Wu, X.; Prior, R.L. 2005. Identification and characterization of anthocyanins by high-performance liquid chromatography-electrospray ionization-tandem mass spectrometry in common foods in the United States: Vegetables, nuts, and grains. *J. Agric. Food Chem.* 53: 3101–13.

Xu, B.; Chang, S.K.C. 2010. Phenolic substances characterization and chemical, cell-based antioxidant activities of 11 lentils grown in Northern United States. *J. Agric. Food Chem.* 58: 1509–17.

Yamaguchi, M.-A.; Kawanobu, S.; Maki, T.; Ino, I. 1996. Cyanidin 3-malonylglucoside and malonyl-coenzyme A: anthocyanidin malonyltransferase in *Lactuca sativa* leaves. *Phytochemistry* 42: 661–3.

Yoshida, K.; Kondo, T.; Kameda, K.; Kawakishi, S.; Lubag, M.A.J.; Mendoza, E.M.T.; Goto, T. 1991. Structures of alatanin A, B and C isolated from edible purple yam *Dioscorea alata. Tetrahedron Lett.* 32: 5575–8.

Yoshida, K.; Mori, M.; Kondo, T. 2009. Blue flower color development by anthocyanins: From chemical structure to cell physiology. *Nat. Prod. Rep.* 26: 884–915.

Yoshida, K.; Sato, Y.; Okuno, R.; Kameda, K.; Isobe, M.; Kondo, T. 1996. Structural analysis and measurement of anthocyanins from colored seed coats of *Vigna*, *Phaseolus*, and *Glycine* legumes. *Biosci. Biotechnol. Biochem.* 60: 589–93.

Zamora-Ros, R.; Andres-Lacueva, C.; Lamuela-Raventos, R.M. et al. 2010. Estimation of dietary sources and flavonoid intake in a Spanish adult population (EPIC-Spain). *J. Am. Diet. Assoc.* 110: 390–98.

Zhang, H.; Wang, L.; Deroles, S.; Bennett, R.; Davies, K. 2006. New insight into the structures and formation of anthocyanic vacuolar inclusions in flower petals. *BMC Plant Biol.* 6: 29.

Zheng, J.; Ding, C.; Wang, L.; Li, G.; Shi, J.; Li, H.; Wang, H.; Suo, Y. 2011. Anthocyanins composition and antioxidant activity of wild *Lycium ruthenicum* Murr. from Qinghai-Tibet Plateau. *Food Chem.* 126: 859–65.

Zheng, Y.; Tian, L.; Liu, H.; Pan, Q.; Zhan, J.; Huang, W. 2009. Sugars induce anthocyanin accumulation and flavanone 3-hydroxylase expression in grape berries. *Plant Growth Regul.* 58: 251–60.

Zimman, A.; Waterhouse, A.L. 2002. Enzymatic synthesis of [3′-*O*-methyl-(3)H]malvidin-3-glucoside from petunidin-3-glucoside. *J. Agric. Food Chem.* 50: 2429–31.

3 Bioavailability and Bioabsorption of Anthocyanins

Tony K. McGhie and David E. Stevenson

CONTENTS

3.1 INTRODUCTION

Anthocyanins, the flavonoid compounds that provide the red, purple, and blue colors of many vegetables and fruits, are consumed as part of a normal human diet. The initial estimates of anthocyanin consumption in the United States were as high as 180–255 mg/day (Kuhnau, 1976), a value that far exceeds the consumption of most other flavonoids. However, more recent studies suggest that the average anthocyanin consumption is around 82 mg/day in Finland and 12.5 mg/day in the United States (Wu et al., 2006a). Berryfruit are particularly rich sources of dietary anthocyanins (Wada and Ou, 2002; Siriwoharn et al., 2004; Wu et al., 2004a; Wu and Prior, 2005), and a single serving of berryfruit can provide many milligrams of anthocyanin. Therefore, dietary choice can have a substantial impact on the amounts and types of anthocyanins consumed by individuals, and potentially an individual's health.

Numerous health benefits have been associated with anthocyanins; for example, anthocyanins have been reported to be strong antioxidants (Rice-Evans et al., 1996; Wang et al., 1997) to inhibit the growth of cancerous cells (Hou, 2003), inhibit inflammation (Seeram et al., 2001), to be vasoprotectors (Xu et al., 2005), and to have antiobesity effects (Tsuda et al., 2003). The therapeutic properties of anthocyanins have been reviewed (Kong et al., 2003; Lila, 2004; Ghosh, 2005; Tsuda, 2012), but have still not been convincingly demonstrated by *in vivo* evidence from animal or human-intervention studies, and much of the supporting evidence is *in vitro* or mechanistic in nature. The therapeutic properties of anthocyanins depend on sufficient cellular or organism bioavailability, either through exposure to cells or through exposure to the whole organism through the diet. Anthocyanins exhibit complex biochemistry and much still remains to be discovered about the biochemical activity of these compounds *in vivo*. In this chapter, we discuss what is currently known about the bioavailability of anthocyanins, and the features of anthocyanin biochemistry and metabolism that affect their potential therapeutic properties.

3.2 BIOCHEMISTRY OF ANTHOCYANINS

Anthocyanins are a subgroup of the large group of compounds known as flavonoids, and more than 700 individual anthocyanin compounds are known (Andersen, 2012). Chemically, anthocyanins are glycosylated, polyhydroxy, or polymethoxy derivatives of the 2-phenyl benzopyrylium moiety and anthocyanins are usually represented as flavylium cations (Figure 3.1) (Mazza and Miniati, 1993). The different chemical structures of anthocyanins arise from the position and number of hydroxyl groups on the molecule, the degree of methylation of these hydroxyl groups, the nature and number of the sugar moieties attached to the phenolic (aglycone) molecule and the position of the attachment, as well as the nature and number of aliphatic or aromatic acids attached to the sugars (Mazza and Miniati, 1993). Anthocyanins occur most frequently in plants as 3-monosides, 3-biosides and 3-triosides as well as 3,5-diglycosides, and more rarely, 3,7-diglycosides associated with the sugars glucose, galactose, rhamnose, arabinose, and xylose (Mazza and Miniati, 1993).

Anthocyanins are reactive compounds; stability is affected by oxygen, temperature, light, and enzymes and they are readily degraded to colorless or brown compounds. Importantly, pH has a marked effect on anthocyanin stability and on the color of anthocyanin-containing solutions (Jackman et al., 1987). Anthocyanin biochemistry is more complex than that of other flavonoid compounds. In aqueous solutions, a number of different molecular forms exist in dynamic equilibrium and these are summarized in Figure 3.1. At pH < 2, the red flavylium cation is the most abundant molecular form. With increasing pH, a proton is rapidly removed to produce the blue quinonoidal structure. Concurrently, there is a slower hydration

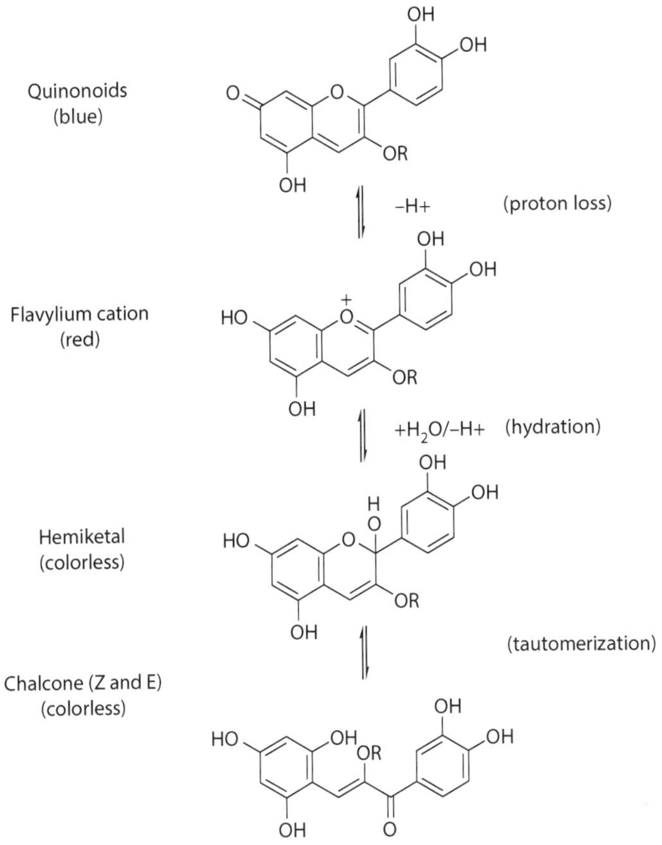

FIGURE 3.1 The molecular structures of anthocyanin that occur at different pH conditions.

of the flavylium cation to produce the colorless hemiketal structure that tautomerizes through an opening of the C-ring to generate the chalcone (*cis* and *trans*) forms. Thus, the relative amounts of the various molecular structures coexisting in an aqueous solution at any given time are determined by pH, temperature, and time in a particular environment. These factors are of significance when considering the preparation of manufactured foods, product shelf life, and passage of anthocyanin-containing foods through the gastrointestinal tract (GI tract). During these processes, anthocyanins are exposed to different pH and temperature environments and the chemical forms and, therefore, the bioactive properties of anthocyanins will vary. Although anthocyanins in red foods are predominantly present as the red flavylium cation, at neutral pHs, the flavylium cation concentration is much lower and it is likely that the other molecular forms will dominate (e.g., quinonoidal bases, hemiketals, and chalcones). This pH-driven structural rearrangement of anthocyanins may alter both their bioactivity and bioavailability. The structural variation caused by anthocyanin biochemistry complicates efforts to understand the health benefits of these flavonoids. For example, the commonly used antioxidant assays are performed at different pHs. The outcomes of bioassays for anthocyanins are likely to depend on the pH of the assay, and on the temperature and time between the dilution of the substance to be tested and the actual assay.

3.3 BIOAVAILABILITY OF ANTHOCYANINS

Bioavailability is defined as the quantity or fraction of the ingested dose that is orally absorbed (Heaney, 2001). Consequently, to assess the efficacy of a bioactive present in the food, it is necessary to know both the amount of the bioactive present in the food *and* the bioavailability of the bioactive. Therefore, the efficacy of a target bioactive can be improved by increasing the bioavailability without increasing the concentration of the bioactive. The concept of bioavailability and developing a mechanistic understanding to manipulate bioavailability are the key aspects of functional food development. Ideally, the measurement of bioavailability requires the measurement of both the unabsorbed and the absorbed portions of the applied dose, and the absorbed portion should include the original bioactive and any metabolite derived from the bioactive, either before or after absorption. Anthocyanin bioavailability, as part of the larger flavonoid group of compounds, has been reviewed (Manach et al., 2005).

3.3.1 ANIMAL STUDIES

Even though the digestive systems of animals differ from that of humans, animals have often been used to investigate the bioavailability of anthocyanins and the reported studies are summarized in Table 3.1. Animal studies (rat/rabbit/pig) have demonstrated the major features of anthocyanin bioavailability, which are: (1) rapid absorption from the GI tract into the plasma, (2) absorption of the intact native anthocyanin glycosides, and (3) methylated and glucuronidated metabolites of the anthocyanin glycosides appear in the urine.

Anthocyanins are detected in the circulatory system within 0.25–2 h of dosing. In rats, the anthocyanin concentrations in the plasma reached a peak (2–3 µg/mL) after only 15 min and then rapidly declined within 2 h following single oral administration of *Vaccinium myrtillus* (bilberry) (Morazzoni et al., 1991). Similarly, following the gastric intubation of a red fruit anthocyanin (320 mg cyanidin-3-glucoside/ kg BW) into rats, the intact anthocyanin glycoside was detected in the plasma (C_{max}: 3.8 µmol/L [1.8 µg/mL] at 15 min (Miyazawa et al., 1999). A study with rats by Tsuda and colleagues detected intact anthocyanin glycosides in the plasma (C_{max}: 0.31 µmol/L [0.14 µg/mL] at 30 min after the oral administration of cyanidin-3-glucoside (400 mg/kg BW) (Tsuda et al., 1999)). This rapid absorption indicates that the stomach is an important site for the bioabsorption of anthocyanins. Anthocyanins are absorbed into gastric tissues and intestinal tissues and bind to the protein in these tissues (He et al., 2009). Interestingly, anthocyanin concentrations are greater in gastric and intestinal tissues than in the plasma (Walton et al., 2008; He et al., 2009) suggesting that the bioabsorption of anthocyanins from the stomach to the GI tract might be relatively high, but that other, currently unknown factors limit the concentrations achieved in the plasma.

TABLE 3.1
Animal Bioavailability Studies

Species	Anthocyanin Source	Dose (per kg Body Weight)[a]	C_{max}[b]	t_{max} (h)[c]	Urinary Excretion (%)[d]	Reference
Rat	Bilberry	400 mg	2–3 µg/mL	0.25		Morazzoni et al. (1991)
Rat	Elderberry Black currant	360 mg	3.80 µmol/L	0.25		Miyazawa et al. (1999)
Rat	Purple corn	400 mg	0.31 µmol/L	0.5		Tsuda et al. (1999)
Rat	Black currant	359 mg Cy-3-glu[e]	0.84 µmol/L	0.5		Matsumoto et al. (2001)
		476 mg Cy-3-rut[f]	0.85 µmol/L	0.5		
		489 mg Dp-3-rut[g]	0.58 µmol/L	2.0		
Rabbit	Black currant	117 mg	780 ng/mL	0.5	0.035 (4 h)	Nielsen et al. (2003)
		164 mg	100 ng/mL	0.25	0.009 (4 h)	
		53 mg	450 ng/mL	0.5	0.023 (4 h)	
Rat	Black currant	100 mg Dp-3-glu[h]	0.4 µmol/L	0.25		Ichiyanagi et al. (2004)
Pig	Marionberry	74 mg	0.103 µmol/L	1	0.088 (24 h)	Wu et al. (2004b)
Rat	Purple black rice	100 mg C3G[e]			0.005 (4 h)	Ichiyanagi et al. (2005a)
Rat	Black currant	100 mg Cy-3-glu[e]	0.18 µmol/L	0.25		Ichiyanagi et al. (2005b)
Rat	Black currant		0.36 µmol/L	3	0.190 (24 h)	Talavéra et al. (2005)
Pig	Chokeberry	229 µmol			0.096 (24 h)	Wu et al. (2005)
	Black currant	140 µmol			0.067 (24 h)	
	Elderberry	228 µmol			0.131 (24 h)	
Pig	Black currant	100 mg	0.09 µg/mL	2–4		Walton et al. (2006)
Pig	Raspberry	50 mg			0.073 (4 h)	Wu et al. (2006b)

[a] Total anthocyanins, if not stated otherwise.

[b] Maximal plasma concentration.

[c] Time to reach C_{max}.

[d] % of intake.

[e] Cy-3-glu, cyanidin-3-glucoside.

[f] Cy-3-rut, cyanidin-3-rutinoside.

[g] Dp-3-rut, delphinidin-3-rutinoside.

[h] Dp-3-glu, delphinidin-3-glucoside.

Urinary excretion is often used to assess bioavailability and as urine contains the end products of metabolism, urine components are useful to investigate the overall metabolism of a compound by an organism. All the studies mentioned above (Tsuda et al., 1999, Morazzoni et al., 1991, Miyazawa et al., 1999) reported intact anthocyanin glycosides in either the plasma or the urine. Furthermore, in rats, after the oral administration of three purified anthocyanins (delphinidin-3-O-β-rutinoside, cyanidin-3-O-β-rutinoside, and cyanidin-3-O-β-glucoside) from black currant, anthocyanin glycosides were detected in the blood (t_{max} at 0.5–2 h) and were excreted in the urine. No other peaks or potential metabolites were detected (Matsumoto et al., 2001). These studies provide clear evidence that anthocyanin glycosides can be absorbed intact from the GI tract and can be ultimately excreted without being metabolized. It appears that anthocyanins are metabolized to a lesser extent than other flavonoids, and this aspect is discussed in more detail in Section 3.4.

As described above, the initial studies on anthocyanin bioavailability detected the intact anthocyanin glycosides in the plasma and in the urine; more recent studies in rats have reported the presence of methylated and glucuronidated anthocyanins. For example, when rats were dosed with 100 mg delphinidin-3-glucoside/kg body weight (BW), the maximum concentration (0.4 μmol/L) in the plasma was reached within 15 min and the methylated metabolite reached a maximum plasma concentration after 1 h (Ichiyanagi et al., 2004). Additional investigations detected anthocyanin glucuronides in rat plasma, and it was suggested that these metabolites were produced in the liver, rather than in the intestinal flora (Ichiyanagi et al., 2005a,b). The glucuronidated and methylated versions of anthocyanin glycosides were reported to be the major types of metabolites present in the urine in pigs (Wu et al., 2004b). The original intact anthocyanin glycosides showed a maximum plasma concentration of 0.103 μmol/L after 1 h. Wu and colleagues (Wu et al., 2005) administered three types of berryfruit with different anthocyanin profiles to pigs, and suggested that the aglycone and the sugar moieties altered the absorption and metabolism of anthocyanins. Talavéra and colleagues were the first to report the presence of the original intact anthocyanins and the methylated and glucuronidated metabolites of anthocyanins in the jejunum (Talavéra et al., 2005).

3.3.2 HUMAN STUDIES

Although animals are often used to study the bioavailability of compounds, studies with humans are important and necessary to confirm that the findings found in animal studies can be transferred to humans. Generally, the features of anthocyanin bioavailability found in animal studies have also been observed in humans. The reported human studies are summarized in Table 3.2.

There exist numerous studies that show conclusively that in humans, intact anthocyanins are absorbed from the digestive tract, transit the circulatory system, and are excreted in the urine. For example, black currant anthocyanins were directly absorbed by humans and were detected in the blood and in the urine as the glycosylated forms (Matsumoto et al., 2001). Similarly, berryfruit anthocyanins from black currant, boysenberry, and blueberry were detected in human urine (McGhie et al., 2003). More recently, the intact glycosides and their metabolites have been detected

TABLE 3.2
Human Anthocyanin Bioavailability Studies

Material	Anthocyanin Dose (Total Intake)[a]	C_{max}[b]	t_{max} (h)[c]	Urinary Excretion (%)[d]	References
Red wine (300 mL)	218 mg			5.10 (12 h)	Lapidot et al. (1998)
Elderberry extract (25 g)	1.5 g	100 ng/mL	0.5		Cao and Prior (1999)
Black currant	236 mg	0.120 µmol/L	1.25–1.75	0.11 (8 h)	Matsumoto et al. (2001)
Elderberry juice (spray-dried capsules)	180 mg	35 ng/mL	1		Murkovic et al. (2000)
Black currant juice (200 mL)	153 mg			0.02–0.05 (5 h)	Netzel et al. (2001)
Red wine (500 mL)	68 mg Mv-3-glu[e]	0.0014 µmol/L	0.8	0.02 (6 h)	Bub et al. (2001)
Dealcoholized red wine	56 mg Mv-3-glu[e]	0.0017 µmol/L	1.5	0.02 (6 h)	
Red grape juice (500 mL)	117 mg Mv-3-glu[e]	0.0028 µmol/L	2.0	0.02 (6 h)	
Elderberry (11 g)	1.9 g			0.003–0.012 (6 h)	Mulleder et al. (2002)
Blueberry powder (100 g)	1.2 g	0.029 µmol/L	4		Mazza et al. (2002)
Elderberry extract (12 g)	720 mg	0.097 µmol/L	1.2	0.06 (24 h)	Milbury et al. (2002)
Elderberry extract (12 g)	720 mg			0.08 (4 h)	Wu et al. (2002)
Blueberry (189 g)	690 mg			0.004 (6 h)	Wu et al. (2002)
Red wine (400 mL)	180 mg	43 ng/mL	1.5	0.23 (7 h)	Frank et al. (2003)
Red grape juice (400 mL)	284 mg	100 ng/mL	0.5	0.18 (7 h)	Frank et al. (2003)
Black currant juice	1.24 g	53 ng/mL	0.75	0.07 (4 h)	Nielsen et al. (2003)
	0.72 g	16 ng/mL	0.75	0.05 (4 h)	
	0.75 g	32 ng/mL	1.5	0.05 (4 h)	
Black currant concentrate (300 mL)	189 mg			0.06 (7 h)	McGhie et al. (2003)
Boysenberry concentrate (300 mL)	345 mg			0.03 (7 h)	McGhie et al. (2003)
Blueberry extract (300 mL)	439 mg			0.02 (7 h)	McGhie et al. (2003)

continued

TABLE 3.2   (continued)
Human Anthocyanin Bioavailability Studies

Material	Anthocyanin Dose (Total Intake)[a]	C_{max}[b]	t_{max} (h)[c]	Urinary Excretion (%)[d]	References
Strawberries (200 g)	76 mg			1.80 (24 h)	Felgines et al. (2003)
Chokeberry extract (7.1 g)	721 mg	0.096 μmol/L	2.8	0.15 (24 h)	Kay et al. (2004)
Blackberries (200 g)	431 mg			0.16 (24 h)	Felgines et al. (2005)
Strawberries (200 g)	222 μmol Pg-glu 13 μmol Pg-rut 6.2 μmol Cy-glu			0.75 (without cream) 1.00 (with cream)	Mullen et al. (2008)
Chokeberry	0.8 mg/kg BW				Wiczkowski et al. (2010)
Strawberries (300 g)	Fresh 9.57 mg Stored 7.19 mg			0.9 0.8	Azzini et al. (2010)
Plum	1117 mg				Netzel et al. (2012)

[a] Total anthocyanins, if not stated otherwise.
[b] Maximal plasma concentration.
[c] Time to reach C_{max}.
[d] % of intake.
[e] Mv-3-glu, malvidin-3-glucoside.

in the urine after the consumption of plum anthocyanins (Netzel et al., 2011). The initial human studies of anthocyanin bioavailability detected only the intact anthocyanin glycosides. For example, after the ingestion of red wine containing 68 mg malvidin-3-glucoside/500 mL or red grape juice containing 117 mg malvidin-3-glucoside/500 mL, malvidin-3-glucoside was found in the plasma (C_{max}: 1.4 nM at 20 min, red wine; C_{max}: 2.8 nM at 180 min, red grape juice) and in the urine of human volunteers. Neither aglycones nor glucuronide or sulfate conjugates were found in the plasma or in the urine samples, indicating that malvidin-3-glucoside is absorbed in its glucosylated form with no indication of metabolism occurring (Bub et al., 2001). However, when anthocyanins from boysenberry were consumed, the intact boysenberry cyanidin-based anthocyanins were detected in the urine, in addition to several additional compounds that were identified by liquid chromatography–mass spectrometry (LC–MS) as methylated and glucuronidated metabolites of boysenberry anthocyanins (Cooney et al., 2004). Felgines and colleagues showed that when strawberry anthocyanin (pelargonidin-3-glucoside) is consumed, monoglucuronide and sulfoconjugated metabolites are excreted in the urine. In this study, the monoglucuronide of pelargonidin accounted for more than 80% of the excretion (Felgines et al., 2003). Similar results were found for cyanidin-3-glucoside from blackberry (Felgines et al., 2005). Very recently, Netzel and colleagues reported that when cyanidin-3-glucoside from plum was consumed, the monoglucuronide and methylated metabolites accounted for 20% and 42%, respectively, of the detected anthocyanin in the urine (Netzel et al., 2012). These studies indicate that the absorbed anthocyanins may be metabolized into conjugates, but the type and the extent of the metabolite formation depend on the type of anthocyanin consumed.

Compared with other polyphenols, the portion of the total anthocyanin dose that can be accounted for by urinary excretion is low and is often <0.1% (Table 3.2). Although Lapidot and colleagues (Lapidot et al., 1998) measured a urinary excretion rate of 5% over a 12-h period following a dose of red wine anthocyanins (218 mg anthocyanin/300 mL), subsequent studies reported much lower rates of excretion. In a study with black currant juice (153 mg of anthocyanin/200 mL), urinary excretion accounted for only 0.02–0.05% of the oral dose (Netzel et al., 2001). In a study with boysenberry concentrate (345 mg anthocyanin), black currant concentrate (189 mg anthocyanin), and blueberry extract (439 mg anthocyanin), urinary excretion accounted for 0.01–0.06% of the consumed dose over a 7-h period after ingestion (McGhie et al., 2003). In another study with malvidin glycosides from red grape juice (284 mg total anthocyanins/400 mL) and red wine (180 mg total anthocyanins/400 mL), urinary excretion of anthocyanins was between 0.18% and 0.23% of the ingested dose (Frank et al., 2003). On the other hand, Felgines and colleagues reported a urinary excretion of 1.80% for pelargonidin-3-glucoside of the anthocyanins dose after the consumption of 200 g of strawberries, which is to date the highest reported recovery (Felgines et al., 2003), with the exception of the Lapidot study (Lapidot et al., 1998). In contrast, the same authors showed a urinary excretion of only 0.16% of the dose with blackberry anthocyanins (mainly cyanidin-3-glucoside) (Felgines et al., 2005).

It is clear from the numerous urinary analysis bioavailability studies discussed above that the fraction of anthocyanins absorbed and excreted intact or as conjugates after oral dosing rarely exceeds 0.1% of the administered dose, with the exception of pelargonidin, which can be excreted to the extent of ~2%. In comparison, many other polyphenols may be excreted to the extent of 2–5% and a few achieve over 10% (Stevenson et al., 2009). In addition, anthocyanin concentrations achieved in human plasma, from an acute dose, are typically in the 10–100 nM range, compared with 0.5–2 µM for many other polyphenols (Manach et al., 2005). These results may be explained in two ways: either the absorption of anthocyanins from the GI tract is much slower than other polyphenols, or their absorption rate is comparable, but they clear from the bloodstream particularly rapidly. The latter explanation could be consistent with the unusually rapid clearance of anthocyanins or their metabolites through the renal system or by rapid disappearance arising from chemical instability, that is, they readily break down or react with other molecules. The consensus arising from recent research is that anthocyanins appear to be absorbed from the GI tract, at least under ideal conditions, at least as fast as other polyphenols. They may even be absorbed faster, as they appear to be among the few polyphenol glycosides that may be absorbed from the stomach (Talavéra et al., 2003). This phenomenon may be linked to the much greater stability of anthocyanins under acidic conditions, as found in the stomach. Potentially, anthocyanins in the plasma may transiently reach concentrations comparable with other polyphenols, but as the clearance of anthocyanins from the bloodstream appears to be rapid, the overall amount of anthocyanins observed passing through the bloodstream could be underestimated. Although there is little evidence from human studies for these unusual properties of anthocyanins, there is a strong support from animal studies. Bilberry anthocyanins fed to mice reached a maximum plasma concentration of 1.2 µM after only 15 min, and then dropped equally quickly (Sakakibara et al., 2009). Cyanidin glucoside administered to rats peaked in the plasma in <1 min, then disappeared with a half-life of also less than a minute, but then reappeared in the liver and kidneys and was rapidly excreted (Vanzo et al., 2011). These absorption and clearance rates measured in minutes compare with the values often measured in hours for other polyphenols (Manach et al., 2005).

The main reason for the rapid disappearance appears to be particularly low chemical stability, resulting in the rapid breakdown of anthocyanins into lower-molecular-weight phenolic compounds, independent of mammalian metabolic enzymes. These properties are discussed in more detail below.

3.4 ANTHOCYANIN METABOLISM AND BIOTRANSFORMATION

3.4.1 ANTHOCYANINS CAN DEGRADE EXTENSIVELY IN PROCESSED FOODS

A variety of anthocyanins is present in plant foods and because anthocyanins are reactive compounds, they can be easily degraded or transformed during the processing and storage/shelf life of manufactured foods. Therefore, in processed foods, the anthocyanins consumed may not actually be the same compounds that were present in the fruits, vegetables, or grains that were harvested and used for their manufacture.

The observation of high losses of anthocyanins during food processing helps to illustrate their relative instability compared with other polyphenols. An extensive analytical survey of processed black currant consumer products, carried out in the United Kingdom, found losses of anthocyanins between ~90% and 99+%, compared with concentrations measured in fresh black currant fruit. Even the commercial individually quick frozen (IQF) whole fruit had lost 30% of its original anthocyanin content (Hollands et al., 2008). A study of polyphenol stability during blueberry processing also found major anthocyanin losses, for example, 40% during juicing, compared with only minor losses for other polyphenols (Lee et al., 2002). The major mechanism of the degradation of anthocyanin in juice from acai fruit appeared to be oligomer/polymer formation (Palencia, 2009). It is well known that the plant enzyme, polyphenol oxidase (PPO), can readily polymerize polyphenolic compounds and anthocyanins appear to be good substrates (Skrede et al., 2000). Furthermore, unlike most other polyphenols, anthocyanins in the solution appear to readily polymerize and/or react with oligomeric procyanidins, even in the absence of PPO (Howard et al., 2012).

3.4.2 METABOLISM IN THE GASTROINTESTINAL TRACT: ANTHOCYANINS APPEAR TO DEGRADE EXTENSIVELY IN THE GI TRACT AND *IN VIVO*

Upon consumption, food components experience numerous different environments whose physicochemical conditions may lead to molecular modification. Within the

GI tract, there are distinct compartments (e.g., stomach, small intestine, and colon), each with its own physicochemical environment (pH, biochemistry, and microbiology). These different environments provide ample opportunity for the degradation and/or metabolism of anthocyanins.

During an investigation of encapsulation to stabilize anthocyanins during *in vitro* digestion, it was found that 80% of unencapsulated anthocyanins had degraded after 2.5 h, compared with around 60% when they were encapsulated (Oidtmann et al., 2011). Cyanidin glucoside administered to rats exhibited a half-life in the intestinal lumen of 2 h (He et al., 2009). In addition, 7.5% of the dose was absorbed into the GI tract tissues, supporting the assertion that anthocyanins can be well absorbed, at least under ideal circumstances. *In vitro* degradation studies on cyanidin and its glucoside at a physiological pH revealed that after 4 h, 57% of the glucoside and 96% of the aglycone had degraded into protocatechuic acid (PCA, 3,4-dihydroxybenzoic acid) and phloroglucinaldehyde (2,3,6-trihydroxybenzaldehyde) (Kay et al., 2009). *In vivo*, glucosides are the only polyphenol glycosides that are extensively hydrolyzed by intestinal glycosidases (Stevenson et al., 2009). Since anthocyanidins (anthocyanins minus the sugar moiety) are substantially less stable than their glycosylated forms at physiological pH (Kay et al., 2009), it appears likely that anthocyanin glucosides are likely to be extensively and rapidly degraded after intestinal deglycosylation and therefore are much less readily absorbed or metabolized. Glycosides resistant to deglycosylation, such as the commonly occurring rutinosides, should have a higher probability of being absorbed intact. In the absence of digestive glycosidase enzymes, that is, during *in vitro* digestion, there was no preferential degradation of glucosides relative to other glycosides, but pelargonidin glucoside was considerably enriched compared with other pelargonidin or cyanidin glycosides (McDougall et al., 2005). This agrees with *in vivo* evidence that pelargonidin is more stable than other anthocyanidins, but the accumulation of the glucoside could not be easily explained.

It appears that anthocyanins differ markedly in their stability in the GI tract, but nevertheless, most anthocyanins are stable enough for a significant proportion to be absorbed into the intestinal tissue. Their stability after absorption, however, appears to be much lower than that of other polyphenols, and this could account for the almost complete loss of the absorbed anthocyanins between the intestinal tissue and the final excretion into the urine.

3.5 ABSORPTION AND TRANSPORT MECHANISMS

3.5.1 ANTHOCYANIN GLYCOSIDES ARE MOSTLY ABSORBED INTACT WITHOUT MODIFICATION

The results of the various stability studies (see Section 3.4.2) suggest that only anthocyanins that are not deglycosylated to the very unstable anthocyanidins are significantly absorbed. The rapid degradation of absorbed anthocyanins in the blood and in tissues would limit the concentration that could accumulate; so, one would expect that they would be less likely to be metabolized by conjugation. There have been many studies that have searched for anthocyanin metabolites, and not all have been successful. Only the original glycosylated forms of anthocyanins were detected in elderly women fed with anthocyanins (Cao et al., 2001), and similar results were found in four other studies looking at the bioavailable forms of berry anthocyanins in humans and/or rats (Miyazawa et al., 1999; Milbury et al., 2002, 2010; McGhie et al., 2003). A greater number of studies, however, have detected metabolites of anthocyanins *in vivo*. Another study on elderly women detected the glucuronide conjugates and the methylation of cyanidin, presumably by the liver enzyme, catechol-*O*-methyl transferase, which produced peonidin; an *O*-methylated form of cyanidin often found in plants (Wu et al., 2002). Similarly, other human and animal studies detected metabolites resulting from methylation and glucuronidation reactions, but also a significant proportion of unmetabolized anthocyanins (Cooney et al., 2004; Kay et al., 2004; Lehtonen et al., 2009; Stalmach et al., 2011). Methylation and glucuronidation were also observed in pigs dosed with berry extracts, with the additional observation that anthocyanins glycosylated with di- or trisaccharides were metabolized to a much lesser extent than those with monosaccharides, and a correspondingly larger proportion of the former was excreted intact via the urine (Wu et al., 2005). Another study in pigs confirmed the latter finding, in that cyanidin and pelargonidin glucosides were extensively deglycosylated and conjugated, whereas cyanidin rutinoside (a disaccharide) was largely recovered intact (Wu et al., 2004b). Delphinidin glucoside was dosed in rats and was detected in the plasma, with two peaks in concentration at 15 and 60 min; the only metabolite detected was the methylated glucoside (Ichiyanagi et al., 2004). Two human studies compared the excretion of metabolites between the consumption of strawberries (predominantly pelargonidin) and black currants (predominantly delphinidin and cyanidin) (Felgines et al., 2003, 2005). Both the studies found a high proportion of glucuronide conjugates in urine and a small proportion of sulfate conjugates. Pelargonidin was more extensively conjugated than black currant anthocyanidins and the proportion excreted (1.8% of the dose) was more than 10 times higher. All the studies mentioned were consistent, in that when glucuronide conjugates were detected, only glucuronides of the anthocyanidin aglycone (i.e., the original glycosyl groups had been removed) were identified. Only one study with rats has reported that glucuronides retained the original plant glycosyl groups (Ichiyanagi et al., 2005a).

An *in vitro* study mentioned earlier (Kay et al., 2009), suggested that cyanidin may break down *in vivo* into PCA (or 3,4-dihydroxybenzoic acid). This suggestion is supported by an *in vivo* study in which rats were dosed with cyanidin glucoside and accumulated eight times more PCA than cyanidin glucoside in their plasma (Tsuda et al., 1999). Similar but less-marked results were observed in people fed with blood orange (*Citrus sinensis*) juice (Vitaglione et al., 2007). Long-term feeding of rats with berry-supplemented diets confirmed that the absorption of anthocyanins is strongly modulated by differences in glycosylation and suggested that acylated anthocyanins are particularly well absorbed (Jian et al., 2006).

Overall, it is clear that anthocyanins can be extensively metabolized after absorption, but the extent of metabolism appears to be dependent on the glycosidic form of the anthocyanin. Bioavailability may be affected by trial variables, such as species tested, plant source of anthocyanins, and dosing format used (i.e., purified compounds, fruit extracts, juice, or whole fruit). In this respect, anthocyanins have much in common with the other classes of flavonoids (Stevenson et al., 2009). The apparent misconception that anthocyanins are not metabolized may have arisen from their relatively low stability after absorption and the correspondingly low concentrations that accumulate in the plasma (generally in the range of ~10–50 nM). This would make anthocyanin metabolites difficult to detect experimentally and may explain why in a number of studies they were not observed.

3.5.2 Are Anthocyanins Absorbed from the Colon as Biotransformation Products?

There is good evidence that, in common with most other polyphenols and in spite of their relatively rapid breakdown, most of a dose of anthocyanins is not absorbed from the GI tract, but finds its way to the colon. In common with other flavonoids, the colon microflora appears to be able to metabolize anthocyanins into smaller molecules such as phenolic acids. Ileostomy patients are often used to assess duodenal absorption/stability of polyphenols and this method resulted in a 40% recovery of dosed anthocyanins in the ileostomy bag, with only 0.1% in the urine (Gonzalez-Barrio et al., 2010). In a similar comparison between apple juice and blueberries, 85% of blueberry anthocyanins were found in the ileostomy bag, compared with 0–33% of apple polyphenols (Kahle et al., 2006). Raspberry anthocyanins were poorly absorbed by rats, most apparently accumulating in the colon (Borges et al., 2007) and large amounts of phenolic acid biotransformation products were detected in urine samples, although these could not be definitively linked to particular polyphenols. When rats were fed a blueberry-enriched diet for 8 weeks, anthocyanins and metabolites were detected in the urine, but not in the plasma or in the organs (Del Bò et al., 2009). There was, however, a large increase in the excretion of hippuric acid (benzoylaminoethanoic acid), the metabolite of benzoic acid, which is the known end product of flavonoid biotransformation by the human colon flora. Two similar studies confirmed these findings (Del Bò et al., 2009; Prior et al., 2010). These studies show that anthocyanins are almost certainly able to reach the colon, although the proportion that degrades to smaller molecules during the passage through the small intestine is unknown. The analysis of anthocyanins in feces from rats fed with berry-enriched diets revealed both high concentrations, of the order of 1–2 mg/g wet feces, and rapid degradation, even in frozen samples, unless the samples were sterilized immediately after collection (He et al., 2005). This finding suggests that both a significant proportion of anthocyanins reaches the colon and that the colon microflora can rapidly degrade them. There are many reports of anthocyanin consumption being associated with increased excretion of phenolic acid biotransformation products. Other studies have demonstrated that isolated colon flora have the ability to biotransform anthocyanins, in a similar manner to other polyphenols. The incubation of purified anthocyanins or metabolites generated using rat liver microsomes, with human fecal bacteria resulted in the rapid degradation and accumulation of hydroxybenzoic acids thought to be derived from the flavonoid β-ring (Fleschhut et al., 2006), that is, similar products to those generated by "chemical" degradation at physiological pH *in vitro*, as mentioned above. Similar results were obtained from raspberry anthocyanins (Gonzalez-Barrio et al., 2011). Other studies linked the biotransformation to the human colon flora of individual anthocyanins with their corresponding β-ring-derived phenolic acids, that is, PCA from cyanidin, gallic acid from delphinidin, and methylated forms of gallic acid (3,4,5-trihydroxybenzoic acid) from malvidin (Aura et al., 2005; Keppler and Humpf, 2005; Avila et al., 2009). No biotransformation study, however, has yet determined the final fate of the remainder of the anthocyanin flavonoid skeleton, although it is likely to be one or more of the commonly observed phenolic acid biotransformation products. It seems likely that the A-ring fragment, in whatever form it may be liberated by biotransformation, would not accumulate in detectable quantities and that it would be quickly biotransformed further to relatively simple benzoic acid derivatives. Human subjects fed with a berry puree containing anthocyanins and a small proportion of plant-derived phenolic acids showed elevated phenolic acid biotransformation product excretion 2–4 h later (Nurmi et al., 2009). Urine analysis of humans on a generally high polyphenol diet found only a tiny proportion of metabolized plant polyphenols; the remainder comprised large amounts of the various known colonic biotransformation products (Rechner et al., 2002). This suggests that microflora-derived phenolic acids have high bioavailability and that many of the biological effects associated with polyphenols in general may actually be attributed to their much more abundant biotransformation products. The link between anthocyanins and biotransformation products is less well defined, however. Although there is much animal and *in vitro* evidence that a high proportion of dietary anthocyanins reaches the colon and that colon bacteria have the capability to biotransform anthocyanins, no report so far has unequivocally demonstrated phenolic acid production from anthocyanins in the absence of other possible sources. The weight of evidence, however, strongly favors the conclusion that anthocyanins are biotransformed in a similar manner to other flavonoids.

3.5.3 Distribution of Anthocyanins after Absorption

Although there is good evidence that anthocyanins can achieve detectable plasma concentrations, albeit much lower than most other polyphenols, there is an emerging consensus that the significant biological effects from any polyphenol are likely to require access to cells and tissues in concentrations sufficient to at least influence the processes such as gene expression and signal transduction (Holst and Williamson, 2008; Milbury and Kalt, 2010). Studies on the tissue distribution of polyphenols are relatively few and are experimentally much more challenging than the plasma or urine measurements. A major problem with tissue measurements of polyphenols is to clearly distinguish between how much is in the tissue and how much is in the entrained blood, the latter being difficult to remove completely (Stevenson et al., 2009). Such measurements do, however, consistently find significant concentrations of polyphenols in tissues, sometimes higher than in the plasma (Stevenson et al., 2009). Polyphenols are particularly abundant in the liver and kidneys, although this is to be expected as these organs actively accumulate xenobiotic compounds for the purposes of metabolism and excretion (Stevenson et al., 2009). Other organs would not necessarily behave similarly. Although it is clearly difficult to quantify polyphenols in organs and tissues reliably, the evidence strongly suggests that polyphenols are transported to them, although to varying extents. The organ of particular interest in the distribution of anthocyanins is the brain because there is accumulating evidence linking anthocyanins with slowing the process of neurodegeneration (Lau et al., 2006; Krikorian et al., 2010; Tsuda, 2012). Five separate studies in rats and pigs all found anthocyanins in brain tissue. Feeding rats with blueberries as 2% of the diet for 10 weeks resulted in the detection of the major blueberry anthocyanins in all parts of the brain, although quantification was not reported (Andres–Lacueva et al., 2005). A similar study on pigs found low nanomolar concentrations of anthocyanins in brain tissues (Milbury and Kalt, 2010). Another pig study found anthocyanins in all the tissues, including the brain, when sampled after all anthocyanins had cleared from the bloodstream, providing strong evidence that their detection in the tissue was genuine (Kalt et al., 2008). The dosing of rats with the pelargonidin aglycone resulted in around 150 nM pelargonidin glucuronide in the brain, lung, and liver tissues, but none was detected in the heart or in the spleen tissues, again strongly suggesting that the relatively high concentrations in some organs were genuine (El Mohsen et al., 2006). Two further rat studies provided additional evidence of the distribution of anthocyanin to organ tissues (Talavéra et al., 2005; Ichiyanagi et al., 2006). The reports of anthocyanins in some organs, but not others, and in organs after all had cleared from bloodstream, are compelling evidence that they are bioavailable to at least some tissues. It remains to be seen whether their apparent neuroprotective effects are mediated directly by the trace amounts detected in the brain, or indirectly, perhaps by a degradation product such as PCA.

3.6 SUMMARY OF CURRENT KNOWLEDGE

There is an accumulation of evidence suggesting that anthocyanins and other phytochemicals possess therapeutic properties. This presents individuals with opportunities for improving their health and well-being through dietary choices and for the development of new food products, both fresh and manufactured, with greater health efficacy. Consequently, it is becoming more important to understand the nature of absorption and metabolism of purported bioactive phytochemicals *in vivo*. Despite many studies, there are still significant gaps in our knowledge about how anthocyanins and compounds derived from them, enter the body, are distributed to tissues, and exert beneficial health effects. A schematic depicting the probable anthocyanin bioabsorption and metabolism is given in Figure 3.2.

Following the ingestion of anthocyanin glycosides in plant-based foods, a portion is rapidly absorbed from the stomach into the systemic circulation after passing through the liver by a process that may involve the transporter bilitranslocase. Anthocyanin glycosides that are not absorbed from the stomach move into the small intestine where they convert into hemiketal, chalcone, and quinonoidal forms as a consequence of the higher physiological pH in the intestine. Further absorption appears to take place in the jejunum. The mechanism for absorption in the jejunum may include the hydrolysis of the glycosides by various hydrolases and the absorption of the phenolic aglycones. The absorbed anthocyanins enter the systemic circulation after the passage through the liver. In the liver, anthocyanins are metabolized by methylation and glucuronidation reactions and some of the metabolites

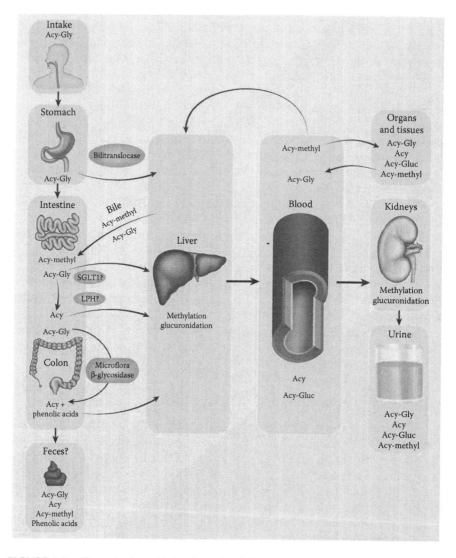

FIGURE 3.2 (See color insert.) A schematic of the current understanding of anthocyanin bioabsoprtion.

are transported back to the intestine as bile. Both unmetabolized and metabolized anthocyanins are available for transport into organs and tissues and are eventually excreted in the urine. Anthocyanins reaching the colon encounter a substantial microbial population and appear to be degraded to sugar and phenolic components, with the phenolic components being further degraded by the disruption of the anthocyanin middle C-ring to yield phenolic acids and aldehydes. These microbial metabolite products may constitute the greatest amount of absorbed anthocyanin-derived compounds and may contribute to the health effect of anthocyanins either directly in the GI tract or after absorption from the colon.

Substantial detail is missing about the biochemical processes by which anthocyanins are absorbed, how the variation of molecular structures consumed in the food, and the forms generated *in vivo* contribute to health benefits. A greater understanding will generate the potential for consumers to gain even more health benefits for high anthocyanin-containing foods such as berryfruit, than is currently the case.

While this manuscript has been in production a recent human study has elegantly confirmed the consensus arising from the research reviewed here that anthocyanins have substantial intrinsic bioavailability, but are difficult to measure reliably, degrade rapidly and are cleared from the bloodstream rapidly (Czank et al., 2013). Using ^{13}C labelled cyanidin-3-glucoside (C3G) 44 ± 26% of the isotope label was recovered from the human subjects, around 5% in urine, 7% in exhaled air and the remainder in faeces, giving a total oral bioavailability of over 12%. Anthocyanin metabolites identified in urine that contained the isotope label included Phase II conjugates of C3G, cyanidin, PCA, phenylpropenoic, phenylacetic and hippuric acids.

3.7   CONCLUDING COMMENTS: ANTHOCYANIN BIOAVAILABILITY—WHERE TO FROM HERE?

The bioavailability properties of anthocyanins have a number of features that are different from other flavonoids such as the flavonols (e.g., quercetin glycosides). These differences are as follows:

- Usually, <0.1% of the ingested dose can be accounted for by urinary excretion and, therefore, bioavailability was previously believed to be very low.
- Bioabsorption occurs rapidly after consumption with the t_{max} in the plasma of 15–60 min and excretion being completed within 6–8 h. These observations have led to the suggestion that anthocyanins are absorbed from the stomach.
- Intact anthocyanin glycosides are absorbed, distributed into the circulatory systems, and excreted in the urine, along with metabolites, but the intact glycosides form an unusually high proportion of species excreted.

The most pressing issue for the bioactivity of anthocyanins is to establish the bioavailability of all the compounds that result from the consumption of anthocyanins. The picture that has emerged so far indicates that many different compounds are produced as a result of anthocyanin consumption. Anthocyanins are present as a number of different molecular species, there is a variety of conjugated metabolites produced once anthocyanins are absorbed *in vivo*, and anthocyanins are extensively metabolized to smaller molecules in the GI tract, particularly the colon. Furthermore, anthocyanin metabolites vary in stability and simple "snapshots" of concentrations may not provide a correct understanding of metabolite dynamics. A true picture of anthocyanin bioavailability will only emerge when all metabolites can be adequately measured, which is a substantial challenge for analytical chemists, and the appropriate experimental designs are implemented.

ACKNOWLEDGMENTS

The authors are grateful to Drs. Janine Cooney and Adam Matich for reviewing the manuscript and for their many useful comments.

REFERENCES

Andersen, Ø.M. 2012. Personal database on anthocyanins.

Andres-Lacueva, C., Shukitt-Hale, B., Galli, R.L., Jauregui, O., Lamuela-Raventos, R.M., and Joseph, J.A. 2005. Anthocyanins in aged blueberry-fed rats are found centrally and may enhance memory. *Nutritional Neuroscience,* 8, 111–120.

Aura, A.M., Martin-Lopez, P., O'leary, K.A., Williamson, G., Oksman-Caldentey, K.M., Poutanen, K., and Santos-Buelga, C. 2005. *In vitro* metabolism of anthocyanins by human gut microflora. *European Journal of Nutrition,* 44, 133–142.

Avila, M., Hidalgo, M., Sanchez-Moreno, C., Pelaez, C., Requena, T., and De Pascual-Teresa, S. 2009. Bioconversion of anthocyanin glycosides by *Bifidobacteria and Lactobacillus. Food Research International,* 42, 1453–1461.

Azzini, E., Vitaglione, P., Intorre, F., Napolitano, A., Durazzo, A., Foddai, M.S., Fumagalli, A. et al. G. 2010. Bioavailability of strawberry antioxidants in human subjects. *British Journal of Nutrition,* 104, 1165–1173.

Borges, G., Roowi, S., Rouanet, J.-M., Duthie, G.G., Lean, M.E.J., and Crozier, A. 2007. The bioavailability of raspberry anthocyanins and ellagitannins in rats. *Molecular Nutrition & Food Research,* 51, 714–725.

Bub, A., Watzl, B., Heeb, D., Rechkemmer, G., and Briviba, K. 2001. Malvidin-3-glucoside bioavailability in humans after ingestion of red wine, dealcoholised red wine and red grape juice. *European Journal of Nutrition,* 40, 113–120.

Cao, G., Muccitelli, H.U., Sanchez-Moreno, C., and Prior, R.L. 2001. Anthocyanins are absorbed in glycated forms in elderly women: A pharmacokinetic study. *American Journal of Clinical Nutrition,* 73, 920–926.

Cao, G. and Prior, R.L. 1999. Anthocyanins are detected in human plasma after oral administration of an elderberry extract. *Clinical Chemistry,* 45, 574–576.

Cooney, J.M., Jensen, D.J., and Mcghie, T.K. 2004. LC–MS identification of anthocyanins in boysenberry extract and anthocyanin metabolites in human urine following dosing. *Journal of the Science of Food and Agriculture,* 84, 237–245.

Czank, C., Cassidy, A., Zhang, Q., Morrison, D.J., Preston, T., Kroon, P.A., Botting, N.P., and Kay, C.D. 2013. Human metabolism and elimination of the anthocyanin, cyanidin-3-glucoside: a 13C-tracer study. *American Journal of Clinical Nutrition,* 97, 995–1003.

Del Bò, C. D., Ciappellano, S., Klimis-Zacas, D., Gardana, C., Riso, P., Porrini, M., and Martini, D. 2009. Anthocyanin absorption, metabolism, and distribution from a wild blueberry-enriched diet (*Vaccinium angustifolium*) is affected by diet duration in the Sprague–Dawley rat. *Journal of Agricultural and Food Chemistry,* 58, 2491–2497.

El Mohsen, M.A., Marks, J., Kuhnle, G., Moore, K., Debnam, E., Srai, S.K., Rice-Evans, C., and Spencer, J.P.E. 2006. Absorption, tissue distribution and excretion of pelargonidin and its metabolites following oral administration to rats. *British Journal of Nutrition,* 95, 51–58.

Felgines, C., Talavera, S., Gonthier, M.P., Texier, O., Scalbert, A., Lamaison, J.L., and Remesy, C. 2003. Strawberry anthocyanins are recovered in urine as glucuro- and sulfoconjugates in humans. *Journal of Nutrition,* 133, 1296–1301.

Felgines, C., Talavera, S., Texier, O., Gil-Izquierdo, A., Lamaison, J.L., and Remesy, C. 2005. Blackberry anthocyanins are mainly recovered from urine as methylated and glucuronidated conjugates in humans. *Journal of Agricultural and Food Chemistry,* 53, 7721–7727.

Fleschhut, J., Kratzer, F., Rechkemmer, G., and Kulling, S. E. 2006. Stability and biotransformation of various dietary anthocyanins *in vitro*. *European Journal of Nutrition*, 45, 7–18.

Frank, T., Netzel, M., Strass, G., Bitsch, R., and Bitsch, I. 2003. Bioavailability of anthocyanins-3-glucosides following consumption of red wine and red grape juice. *Canadian Journal of Physiology and Pharmacology*, 81, 423–435.

Ghosh, D.K. 2005. Anthocyanins and anthocyanin-rich extracts in biology and medicine: Biochemical, cellular and medicinal properties. *Current Topics in Nutraceutical Research*, 3, 113–124.

Gonzalez-Barrio, R., Borges, G., Mullen, W., and Crozier, A. 2010. Bioavailability of anthocyanins and ellagitannins following consumption of raspberries by healthy humans and subjects with an ileostomy. *Journal of Agricultural and Food Chemistry*, 58, 3933–3939.

Gonzalez-Barrio, R., Edwards, C.A., and Crozier, A. 2011. Colonic catabolism of ellagitannins, ellagic acid, and raspberry anthocyanins: *In vivo* and *in vitro* studies. *Drug Metabolism and Disposition*, 39, 1680–1688.

He, J., Magnuson, B.A., and Giusti, M.M. 2005. Analysis of anthocyanins in rat intestinal contents—Impact of anthocyanin chemical structure on fecal excretion. *Journal of Agricultural and Food Chemistry*, 53, 2859–2866.

He, J., Wallace, T.C., Keatley, K.E., Failla, M.L., and Giusti, M.M. 2009. Stability of black raspberry anthocyanins in the digestive tract lumen and transport efficiency into gastric and small intestinal tissues in the rat. *Journal of Agricultural and Food Chemistry*, 57, 3141–3148.

Heaney, R.P. 2001. Factors influencing the measurement of bioavailability, taking calcium as a model. *Journal of Nutrition*, 131, 1344S–1348S.

Hollands, W., Brett, G.M., Radreau, P., Saha, S., Teucher, B., Bennett, R.N., and Kroon, P.A. 2008. Processing blackcurrants dramatically reduces the content and does not enhance the urinary yield of anthocyanins in human subjects. *Food Chemistry*, 108, 869–878.

Holst, B. and Williamson, G. 2008. Nutrients and phytochemicals: From bioavailability to bioefficacy beyond antioxidants. *Current Opinion in Biotechnology*, 19, 73–82.

Hou, D.-X. 2003. Potential mechanisms of cancer chemoprevention by anthocyanins. *Current Molecular Medicine*, 3, 149–159.

Howard, L.R., Prior, R.L., Liyanage, R., and Lay, J.O. 2012. Processing and storage effect on berry polyphenols: Challenges and implications for bioactive properties. *Journal of Agricultural and Food Chemistry*, 60, 6678–6693.

Ichiyanagi, T., Rahman, M.M., Kashiwada, Y., Ikeshiro, Y., Shida, Y., Hatano, Y., Matsumoto, H., Hirayama, M., Tsuda, T., and Konishi, T. 2004. Absorption and metabolism of delphinidin 3-*O*-[beta]-D-glucopyranoside in rats. *Free Radical Biology and Medicine*, 36, 930–937.

Ichiyanagi, T., Shida, Y., Rahman, M.M., Hatano, Y., and Konishi, K. 2006. Bioavailability and tissue distribution of anthocyanins in bilberry (*Vaccinium myrtillus* L.) extract in rats. *Journal of Agricultural and Food Chemistry*, 54, 6578–6587.

Ichiyanagi, T., Shida, Y., Rahman, M.M., Hatano, Y., and Konishi, T. 2005a. Extended glucuronidation is another major path of cyanidin 3-*O*-[beta]-D-glucopyranoside metabolism in rats. *Journal of Agricultural and Food Chemistry*, 53, 7312–7319.

Ichiyanagi, T., Shida, Y., Rahman, M.M., Hatano, Y., Matsumoto, H., Hirayama, M., and Konishi, T. 2005b. Metabolic pathway of cyanidin 3-*O*-beta-D-glucopyranoside in rats. *Journal of Agricultural and Food Chemistry*, 53, 145–150.

Jackman, R.L., Yada, R.Y., Tung, M.A., and Speers, R.A. 1987. Anthocyanins as food colorants—A review. *Journal of Food Biochemistry*, 11, 201–270.

Jian, H., Magnuson, B.A., Lala, G., Qingguo, T., Schwartz, S.J., and Giusti, M.M. 2006. Intact anthocyanins and metabolites in rat urine and plasma after 3 months of anthocyanin supplementation. *Nutrition & Cancer*, 54, 3–12.

Kahle, K., Kraus, M., Scheppach, W., Ackermann, M., Ridder, F., and Richling, E. 2006. Studies on apple and blueberry fruit constituents: Do the polyphenols reach the colon after ingestion? *Molecular Nutrition and Food Research,* 50, 418–423.

Kalt, W., Blumberg, J.B., Mcdonald, J.E., Vinqvist-Tymchuk, M.R., Fillmore, S.A. E., Graf, B.A., O'leary, J.M., and Milbury, P.E. 2008. Identification of anthocyanins in the liver, eye, and brain of blueberry-fed pigs. *Journal of Agricultural and Food Chemistry,* 56, 705–712.

Kay, C.D., Kroon, P.A., and Cassidy, A. 2009. The bioactivity of dietary anthocyanins is likely to be mediated by their degradation products. *Molecular Nutrition and Food Research,* 53, (suppl. 1), S92–S101.

Kay, C.D., Mazza, G., Holub, B. J., and Wang, J. 2004. Anthocyanin metabolites in human urine and serum. *British Journal of Nutrition,* 91, 933–942.

Keppler, K. and Humpf, H.-U. 2005. Metabolism of anthocyanins and their phenolic degradation products by the intestinal microflora. *Bioorganic & Medicinal Chemistry,* 13, 5195–5205.

Kong, J.-M., Chia, L.-S., Goh, N.-K., Chia, T.-F., and Brouillard, R. 2003. Analysis and biological activities of anthocyanins. *Phytochemistry,* 64, 923–933.

Krikorian, R., Shidler, M.D., Nash, T.A., Kalt, W., Vinqvist-Tymchuk, M.R., Shukitt-Hale, B., and Joseph, J.A. 2010. Blueberry supplementation improves memory in older adults. *Journal of Agricultural and Food Chemistry,* 58, 3996–4000.

Kuhnau, J. 1976. The flavonoids. A class of semi-essential food components: Their role in human nutrition. *World Review of Nutrition and Diet,* 24, 117–191.

Lapidot, T., Harel, S., Granit, R., and Kanner, J. 1998. Bioavailability of red wine anthocyanins as detected in human urine. *Journal of Agricultural and Food Chemistry,* 46, 4297–4302.

Lau, F.C., Shukitt-Hale, B., and Joseph, J.A. 2006. Beneficial effects of berry fruit polyphenols on neuronal and behavioral aging. *Journal of the Science of Food and Agriculture,* 86, 2251–2255.

Lee, J., Durst, R.W., and Wrolstad, R.E. 2002. Impact of juice processing on blueberry anthocyanins and polyphenolics: Comparison of two pretreatments. *Journal of Food Science,* 67, 1660–1667.

Lehtonen, H.-M., Rantala, M., Suomela, J.-P., Viitanen, M., and Kallio, H. 2009. Urinary excretion of the main anthocyanin in lingonberry (*Vaccinium vitis-idaea*), cyanidin 3-*O*-galactoside, and its metabolites. *Journal of Agricultural and Food Chemistry,* 57, 4447–4451.

Lila, M. A. 2004. Anthocyanins and human health: An *in vitro* investigative approach. *Journal of Biomedicine and Biotechnology,* 2004, 306–313.

Manach, C., Williamson, G., Morand, C., Scalbert, A., and Remesy, C. 2005. Bioavailability and bioefficacy of polyphenols in humans. 1. Review of 97 bioavailability studies. *American Journal of Clinical Nutrition,* 81, 230S–242S.

Matsumoto, H., Inaba, H., Kishi, M., Tominaga, S., Hirayama, M., and Tsuda, T. 2001. Orally administered delphinidin 3-rutinoside and cyanidin 3-rutinoside are directly absorbed in rats and humans and appear in the blood as the intact forms. *Journal of Agricultural and Food Chemistry,* 49, 1546–1551.

Mazza, G., Kay, C.D., Cottrell, T., and Holub, B.J. 2002. Absorption of anthocyanins from blueberries and serum antioxidant status in human subjects. *Journal of Agricultural and Food Chemistry,* 50, 7731–7737.

Mazza, G. and Miniati, E. 1993. *Anthocyanins in Fruits, Vegetables, and Grains,* Boca Raton, FL, CRC Press.

McDougall, G.J., Dobson, P., Smith, P., Blake, A., and Stewart, D. 2005. Assessing potential bioavailability of raspberry anthocyanins using an *in vitro* digestion system. *Journal of Agricultural and Food Chemistry,* 53, 5896–5904.

McGhie, T.K., Ainge, G.D., Barnett, L.E., Cooney, J.M., and Jensen, D.J. 2003. Anthocyanin glycosides from berry fruit are absorbed and excreted unmetabolized by both humans and rats. *Journal of Agricultural and Food Chemistry,* 51, 4539–4548.

Milbury, P.E., Cao, G., Prior, R.L., and Blumberg, J.B. 2002. Bioavailability of elderberry anthocyanins. *Mechanisms of Ageing and Development,* 123, 997–1006.

Milbury, P.E. and Kalt, W. 2010. Xenobiotic metabolism and berry flavonoid transport across the blood–brain barrier. *Journal of Agricultural and Food Chemistry,* 58, 3950–3956.

Milbury, P.E., Vita, J.A., and Blumberg, J.B. 2010. Anthocyanins are bioavailable in humans following an acute dose of cranberry juice. *Journal of Nutrition,* 140, 1099–1104.

Miyazawa, T., Nakagawa, K., Kudo, M., Muraishi, K., and Someya, K. 1999. Direct intestinal absorption of red fruit anthocyanins, cyanidin-3-glucoside and cyanidin-3,5-diglucoside, into rats and humans. *Journal of Agricultural and Food Chemistry,* 47, 1083–1091.

Morazzoni, P., Livio, S., Scilingo, A., and Malandrino, S. 1991. *Vaccinium myrtillus* anthocyanosides pharmacokinetics in rats. *Drug Research,* 41, 128–131.

Mulleder, U., Murkovic, M., and Pfannhauser, W. 2002. Urinary excretion of cyanidin glycosides. *Journal of Biochemical and Biophysical Methods,* 53, 61–66.

Mullen, W., Edwards, C.A., Serafini, M., and Crozier, A. 2008. Bioavailability of pelargonidin-3-*O*-glucoside and its metabolites in humans following the ingestion of strawberries with and without cream. *Journal of Agricultural and Food Chemistry,* 56, 713–719.

Murkovic, M., Adam, U., and Pfannhauser, W. 2000. Analysis of anthocyane glycosides in human serum. *Fresenius Journal of Analytical Chemistry,* 366, 379–381.

Netzel, M., Fanning, K., Netzel, G., Zabaras, D., Karagianis, G., Treloar, T., Russell, D., and Stanley, R. A. 2012. Urinary excretion of antioxidants in healthy humans following Queen Garnet plum juice ingestion: A new plum variety rich in antioxidant compounds. *Journal of Food Biochemistry,* 36, 159–170.

Netzel, M., Strass, G., Janssen, M., Bitsch, I., and Bitsch, R. 2001. Bioactive anthocyanins detected in human urine after ingestion of blackcurrant juice. *Journal of Environmental Pathology, Toxicology and Oncology,* 20, 89–95.

Nielsen, I.L.F., Dragsted, L.O., Ravin-Haren, G., Freese, R., and Rasmussen, S.E. 2003. Absorption and excretion of black currant anthocyanins in humans and Watanabe heritable hyperlipidemic rabbits. *Journal of Agricultural and Food Chemistry,* 51, 2813–2820.

Nurmi, T., Mursu, J., Heinonen, M., Nurmi, A., Hiltunen, R., and Voutilainen, S. 2009. Metabolism of berry anthocyanins to phenolic acids in humans. *Journal of Agricultural and Food Chemistry,* 57, 2274–2281.

Oidtmann, J., Schantz, M., Mäder, K., Baum, M., Berg, S., Betz, M., Kulozik, U. et al. 2011. Preparation and comparative release characteristics of three anthocyanin encapsulation systems. *Journal of Agricultural and Food Chemistry,* 60, 844–851.

Palencia, L.A.P. 2009. *Chemical Characterization, Bioactive Properties, and Pigment Stability of Polyphenols in Acai (Euterpe oleracea* Mart.*).* PhD thesis, Texas A&M University.

Prior, R.L., Rogers, T.R., Khanal, R.C., Wilkes, S.E., WU, X., and Howard, L.R. 2010. Urinary excretion of phenolic acids in rats fed cranberry. *Journal of Agricultural and Food Chemistry,* 58, 3940–3949.

Rechner, A.R., Kuhnle, G., Bremner, P., Hubbard, G.P., Moore, K.P., and Rice-Evans, C.A. 2002. The metabolic fate of dietary polyphenols in humans. *Free Radical Biology and Medicine,* 33, 220–235.

Rice-Evans, C.A., Miller, N.J., and Paganga, G. 1996. Structure–antioxidant activity relationships of flavonoids and phenolic acids. *Free Radical Biology and Medicine,* 20, 933–956.

Sakakibara, H., Ogawa, T., Koyanagi, A., Kobayashi, S., Goda, T., Kumazawa, S., Kobayashi, H., and Shimoi, K. 2009. Distribution and excretion of bilberry anthocyanines in mice. *Journal of Agricultural and Food Chemistry,* 57, 7681–7686.

Seeram, N.P., Bourquin, L.D., and Nair, M.G. 2001. Degradation products of cyanidin glycosides from tart cherries and their bioactivities. *Journal of Agricultural and Food Chemistry,* 49, 4924–4929.

Siriwoharn, T., Wrolstad, R.E., Finn, C.E., and Pereira, C.B. 2004. Influence of cultivar, maturity, and sampling on blackberry (*Rubus* L. hybrids) anthocyanins, polyphenolics, and antioxidant properties. *Journal of Agricultural and Food Chemistry,* 52, 8021–8030.

Skrede, G., Wrolstad, R.E., and Durst, R.W. 2000. Change in anthocyanin and polyphenolics during juice processing of higbush blueberries (*Vaccinium corymbosum* L.). *Journal of Food Science,* 65, 357–364.

Stalmach, A., Edwards, C.A., Wightman, J.D., and Crozier, A. 2011. Identification of (poly) phenolic compounds in Concord grape juice and their metabolites in human plasma and urine after juice consumption. *Journal of Agricultural and Food Chemistry,* 59, 9512–9522.

Stevenson, D.E., Scheepens, A., and Hurst, R.D. 2009. Bioavailability and metabolism of dietary flavonoids—Much known—much more to discover. In: Keller, R.B. (ed.) *Flavonoids: Biosynthesis, Biological Effects and Dietary Sources.* Nova Science Publishers, Inc. Hauppauge, NY.

Talavéra, S., Felgines, C., Texier, O., Besson, C., Lamaison, J.L., and Remesy, C. 2003. Anthocyanins are efficiently absorbed from the stomach in anesthetized rats. *Journal of Nutrition,* 133, 4178–4182.

Talavéra, S., Felgines, C., Texier, O., Lamaison, J.-L., Besson, C., Gil-Izquierdo, A., and Réme, S.Y.C. 2005. Anthocyanin metabolism in rats and their distribution to digestive area, kidney, and brain. *Journal of Agricultural and Food Chemistry,* 53, 3902–3908.

Tsuda, T. 2012. Dietary anthocyanin-rich plants: Biochemical basis and recent progress in health benefits studies. *Molecular Nutrition and Food Research,* 56, 159–170.

Tsuda, T., Horio, F., and Osawa, T. 1999. Absorption and metabolism of cyanidin 3-*O*-[beta]-glucoside in rats. *FEBS Letters,* 449, 179–182.

Tsuda, T., Horio, F., Uchida, K., Aoki, H., and Osawa, T. 2003. Dietary cyanidin 3-*O*-beta-D-glucoside-rich purple corn color prevents obesity and ameliorates hyperglycemia in mice. *Journal of Nutrition,* 133, 2125–2130.

Vanzo, A., Vrhovsek, U., Tramer, F., Mattivi, F., and Passamonti, S. 2011. Exceptionally fast uptake and metabolism of cyanidin 3-glucoside by rat kidneys and liver. *Journal of Natural Products,* 74, 1049–1054.

Vitaglione, P., Donnarumma, G., Napolitano, A., Galvano, F., Gallo, A., Scalfi, L., and Fogliano, V. 2007. Protocatechuic acid is the major human metabolite of cyanidin-glucosides. *Journal of Nutrition,* 137, 2043–2048.

Wada, L. and Ou, B. 2002. Antioxidant activity and phenolic content of Oregon caneberries. *Journal of Agricultural and Food Chemistry,* 50, 3495–3500.

Walton, M.C., Hendriks, W.H., Mcghie, T.K., and Broomfield, A.M. 2008. A viscous food matrix influences absorption and excretion but not metabolism of blackcurrant anthocyanins in rats. *Journal of Food Science,* 74, H22–H29.

Walton, M.C., Hendriks, W.H., Reynolds, G.W., Kruger, M.C., and Mcghie, T.K. 2006. Anthocyanin absorption and plasma antioxidant status in pigs. *Journal of Agricultural and Food Chemistry,* 54, 7940–7946.

Wang, H., Cao, G., and Prior, R.L. 1997. Oxygen radical absorbing capacity of anthocyanins. *Journal of Agricultural and Food Chemistry,* 45, 304–309.

Wiczkowski, W., Romaszko, E., and Piskula, M. 2010. Bioavailability of cyanidin glycosides from natural chokeberry (*Aronia melanocarpa*) juice with dietary-relevant dose of anthocyanins in humans. *Journal of Agricultural and Food Chemistry,* 58, 12130–12136.

Wu, X., Beecher, G.R., Holden, J.M., Haytowitz, D., Gebhardt, S.E., and Prior, R.L. 2006a. Concentrations of anthocyanins in common foods in the United States and estimation of normal consumption. *Journal of Agricultural and Food Chemistry,* 54, 4069–4075.

Wu, X., Cao, G., and Prior, R.L. 2002. Absorption and metabolism of anthocyanins in elderly women after consumption of elderberry or blueberry. *Journal of Nutrition,* 132, 1865–1871.

Wu, X., Gu, L., Prior, R.L., and Mckay, S. 2004a. Characterization of anthocyanins and pro-anthocyanidins in some cultivars of *Ribes*, *Aronia*, and *Sambucus* and their antioxidant capacity. *Journal of Agricultural and Food Chemistry, 52*, 7846–7856.

Wu, X., Pittman, H.E. III, Mckay, S., and Prior, R.L. 2005. Aglycones and sugar moieties alter anthocyanin absorption and metabolism after berry consumption in weanling pigs. *Journal of Nutrition, 135*, 2417–2424.

Wu, X., Pittman, H.E. III, and Prior, R.L. 2004b. Pelargonidin is absorbed and metabolized differently than cyanidin after marionberry consumption in pigs. *Journal of Nutrition, 134*, 2603–2610.

Wu, X., Pittman, H.E. III, and Prior, R.L. 2006b. Fate of anthocyanins and antioxidant capacity in contents of the gastrointestinal tract of weanling pigs following black raspberry consumption. *Journal of Agricultural and Food Chemistry, 54*, 583–589.

Wu, X. and Prior, R.L. 2005. Systematic identification and characterization of anthocyanins by Hplc–Esi–MS/MS in common foods in the United States: Fruits and berries. *Journal of Agricultural and Food Chemistry, 53*, 2589–2599.

Xu, J.-W., Ikeda, K., and Yamori, Y. 2005. Cyandin-3-glucoside regulates phosphorylation of endothelial nitric oxide synthase. *FEBS Letters, 574*, 176–180.

4 Analysis of Anthocyanins in Biological Samples

Pu Jing and M. Monica Giusti

CONTENTS

4.1 INTRODUCTION

Anthocyanins are water-soluble plant pigments, responsible for the red, purple, and blue colors of many fruits, vegetables, cereal grains, and flowers. They are of interest to nutritionists because of their postulated health benefits, as well as to food processors because of their colorful character as a natural alternative to the use of synthetic dyes. Research concerning their health benefits has become prominent since the late 1990s. Up to now, numerous studies have suggested that anthocyanins are protective against many chronic diseases such as dietary antioxidant [1,2], antimutagenic [3,4], and chemopreventive [5–8] nutraceuticals that contribute to reduced incidences of chronic diseases and the prevention of obesity and diabetes [9,10]. These topics are discussed in detail in the following chapters of this book. Research on the metabolic fate and bioavailability of anthocyanins *in vivo* become crucial for elucidating pharmacokinetics as well as the possible mechanisms of bioactivity. Therefore, accurate and suitable identification and quantification of anthocyanins and their metabolites must be established. This chapter will present an overview of the current knowledge of how anthocyanin chemistry affects their stability and the chemistry of their metabolites, as well as their bioavailability. More specifically, we will discuss some of the challenges of evaluating anthocyanins in biological samples as well as the different methodologies available for the analysis of anthocyanins and their metabolites in biological samples, including blood, urine, and different tissues or fluids.

4.2 ANTHOCYANINS AND THEIR BIOLOGICAL METABOLITES

4.2.1 TYPES OF ANTHOCYANINS

The chemistry of anthocyanins is extensively discussed in Chapter 2 of this book. Their chemical structures, reactivity, and transformations are critical to their fate in the gastrointestinal tract (GIT) after ingestion, as well as their ability to interact with, or pass through tissues lining the GIT. Anthocyanins belong to the class of flavonoid compounds and are commonly known as plant polyphenols. The anthocyanin pigments are composed of two or three main chemical units: the first and the main structure is the flavylium ring or aglycone base (anthocyanidin) that is typically found in nature attached to sugar substitutions (second chemical unit). Sometimes, a third element is present; aliphatic or aromatic acids can be attached to those sugars as acylating groups. Anthocyanidins present the C6–C3–C6 carbon skeleton typical of flavonoids. They are polyhydroxy and polymethoxy derivatives of 2-phenylbenzoprylium or flavylium salts, differing in the number and position of their hydroxyl and methoxyl groups on the β-ring (Figure 4.1). There are 27 known anthocyanidins present in nature [11]. Among them, the six most common anthocyanidins in nature are cyanidin, pelargonidin, peonidin, delphinidin, petunidin, and malvidin. In addition to the anthocyanidins described above, other less-common structures have been reported, the 4-substituted aglycone (Figure 4.2), such as the 5-carboxypryanopelargonidin-3-O-β-glucopyranoside in strawberry (*Fragaria* × *ananassa* Duch) [12]. Anthocyanins, as glycosides of anthocyanidin chromophores, can be in turn linked to aromatic acids, aliphatic acids, and/or ethyl ester derivatives [12].

Anthocyanidins rarely occur in their free form in plant nature because of their high reactivity. Anthocyanins are mainly glycosylated with one or more sugar

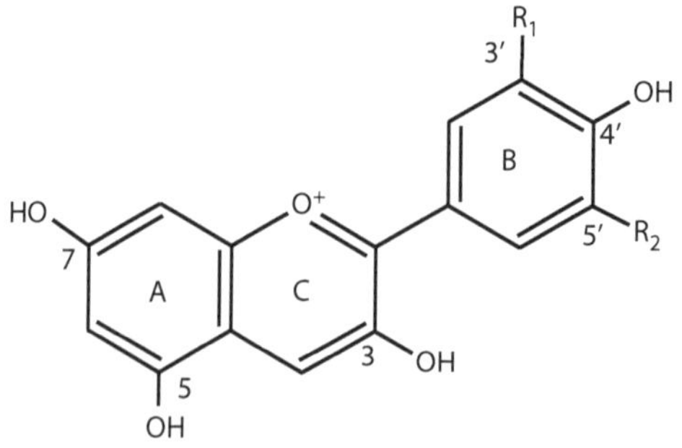

Pelargonidin	$R_1 = H$	$R_2 = H$
Cyanidin	$R_1 = OH$	$R_2 = H$
Delphinidin	$R_1 = OH$	$R_2 = OH$
Peonidin	$R_1 = OCH_3$	$R_2 = H$
Petunidin	$R_1 = OCH_3$	$R_2 = OH$
Malvidin	$R_1 = OCH_3$	$R_2 = OCH_3$

FIGURE 4.1   Chemical structures of anthocyanidins commonly found in nature.

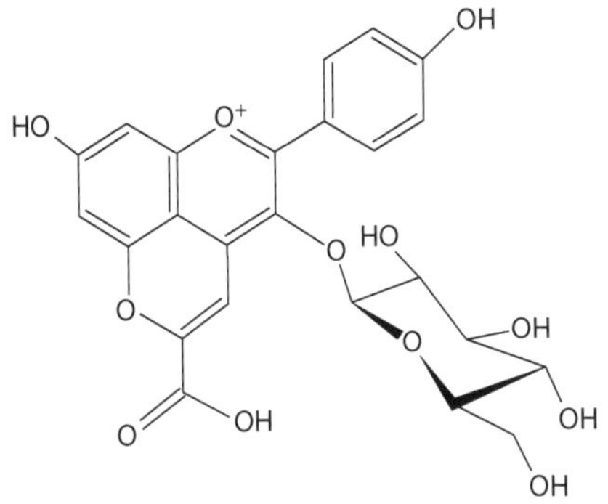

5-Carboxypryanopelargonidin 3-
O-β-glucopyranoside

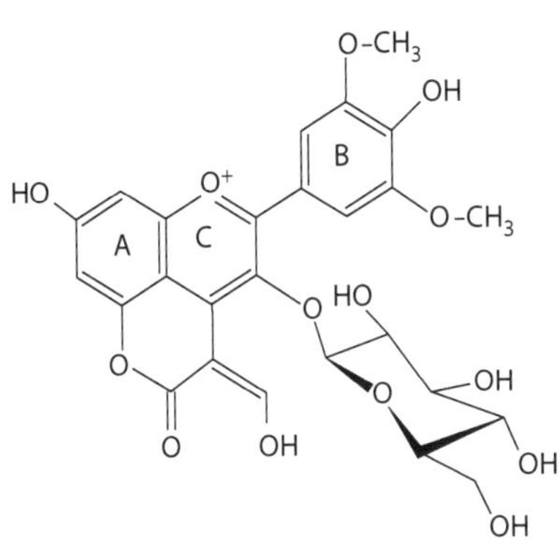

Vitisin A

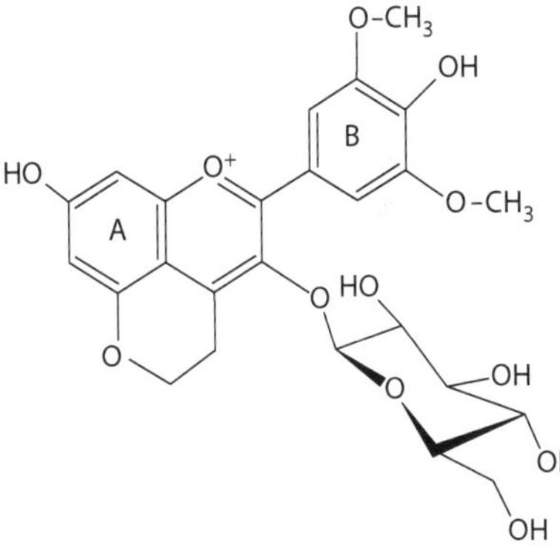

Vitisin B

FIGURE 4.2 Structures of 5-carboxypryanopelargonidin 3-O-β-glucopyranoside and vitisin A/B. (Adapted from Andersen, Ø. et al., *Phytochemistry*, 2004, *65*, 405–410; Bakker, J. et al., *J. Agric. Food Chem.* 1997, *45*, 35–43.)

moieties that enhance their stability and solubility [13–15]. The most common sugar moieties found attached to the aglycone is glucose, followed by galactose, rhamnose, xylose, arabinose as mono-, di-, and triglycosides. Anthocyanins can be glycosylated with one or more sugar units, typically up to three, with some rare exceptions. The C_3-hydroxyl position of the C-ring is the primary place for glycosylation, followed by the C_5 position of the A ring (Figure 4.2). Although not common, sugars can also attach to the anthocyanidin molecule through any one of the hydroxyls at C_7, $C_{3'}$, $C_{5'}$, and even $C_{4'}$ in rare cases [16,17].

These sugars may be acylated [18] with aromatic acids, such as *p*-coumaric, caffeic, ferulic, sinapic, gallic or *p*-hydroxybenzoic acids, or aliphatic acids, such as malonic, acetic, malic, succinic, or oxalic acids (Figure 4.3). Acyl substituents are commonly bound to the C_3 sugar, esterified to the 6-OH, or less frequently to the 4-OH group of the sugars [19]. However, anthocyanins containing rather complicated acylation patterns attached to different sugar moieties have been reported [20–23]. As a result, close to 700 structurally distinct anthocyanins have been identified in nature (see Chapter 1).

Aliphatic acids

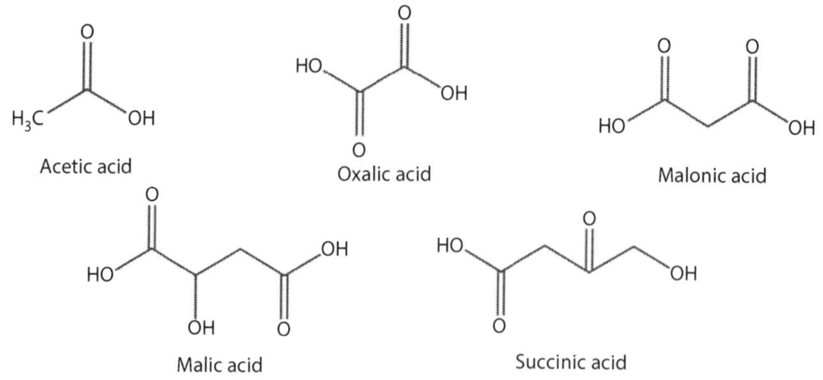

Aromatic acids

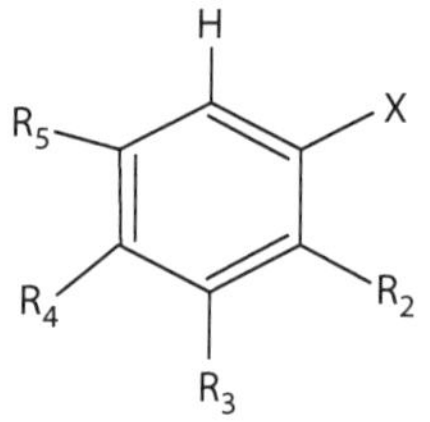

Catalogue	Common Name	R_2	R_3	R_4	R_5
	p-coumaric acid	H	H	–OH	H
Hydroxycinnamic acids (Xa)	Ferulic acid	H	–OCH$_3$	–OH	H
	Sinapic acid	H	–OCH$_3$	–OH	–OCH
	Caffeic acid	H	–OH	–OH	H
Hydroxybenzoic acids (Xb)	p-hydroxybenzoic acid	H	H	–OH	H
	gallic acid	–OH	–OH	–OH	–OH

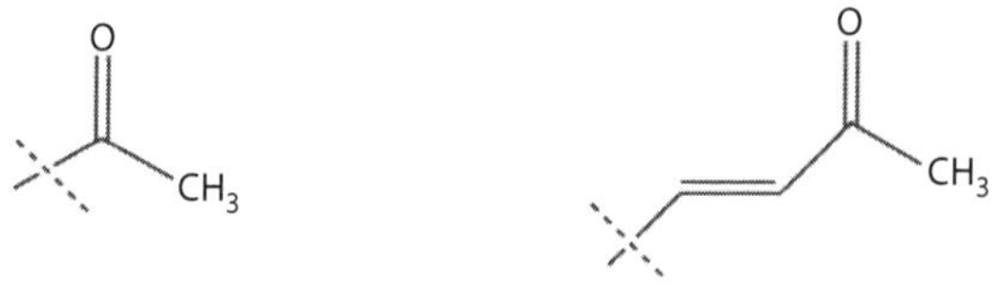

FIGURE 4.3 Common organic acids acylated with sugar moieties on anthocyanins. (Adapted from Robbins, R.J., *J Agric Food Chem* 2003, *51*, 2866–2887.)

4.2.2 ANTHOCYANIN METABOLITES

Anthocyanin compounds are typically very reactive and interact with the surrounding environment. As soon as the anthocyanins enter the body, they are susceptible to modification due to the different and distinct GIT environments (e.g., oral, stomach, small intestine, and colon) and physicochemical conditions [24–32]. Anthocyanins degradation rates and patterns of methylation, glucuronidation, oxidization, sulfo-conjugation, and deglycosylation are somewhat unique and different in the different areas of the GIT as well as in other environments.

Limited information is available on the effect of saliva and the upper GI tract on anthocyanin metabolism. A recent study shows that anthocyanin transformations, metabolism, and degradation start from the first contact with the mouth [33]. All anthocyanins were partially degraded when incubated with saliva. Both the chemical environment of saliva and the oral microbiota were the important factors affecting the stability of anthocyanins in the oral cavity.

The loss of glycosylation has been reported to occur to some degree in the stomach, likely due to the acidic environment [32]. Using an *in vitro* model of pig cecum, significant microbial deglycosylation and degradation of anthocyanins were obtained [31]. Similar results were obtained in studies on the biotransformation of pure anthocyanins, anthocyanin extracts from radish, and the assumed degradation products in models to mimic *in vivo* conditions with human fecal microflora [25]. The results showed that glycosylated and acylated anthocyanins were rapidly degraded by the intestinal microflora after anaerobic incubation with a human fecal suspension with an increase of the degradation product protocatechuic acid.

In weanling pigs that had a single meal with a freeze-dried powder of chokeberry, black currant, or elderberry, anthocyanins were metabolized via methylation, glucuronidation, and sulfoconjugation after absorption [26], and anthocyanins with different aglycons were found to be metabolized differently *in vivo*. The metabolites of delphinidin anthocyanins were not detected, whereas cyanidin derivatives were metabolized by methylation, glucuronidation, or both [26]. In a human study, the metabolites were identified as glucuronide conjugates, as well as methylated and oxidized derivatives of cyanidin-3-galactoside and cyanidin glucuronide in urine after the consumption of chokeberry extracts [34]. The anthocyanin metabolites in humans consuming a meal containing strawberries (providing pelargonidin-3-glucoside) showed that besides pelargonidin-3-glucoside, five metabolites were found in urine samples, including three monoglucuronides of pelargonidin, one sulfoconjugate of pelargonidin with the monoglucuronides accounting for more than 80% of the metabolites [28]. Studies conducted with cyanidin-3-glucoside have found that the intact cyanidin-3-glucoside as well as several anthocyanin metabolites are detected in the human urine: methylated glycosides, glucuronides of anthocyanidins and anthocyanins, a sulfoconjugate of cyanidin, and anthocyanidins after the consumption of blackberries, among which monoglucuronides of anthocyanidins were the major metabolites in urine (>60% of excretion) [29]. It is important to highlight that some studies do not detect the presence of the metabolites described. For example, malvidin-3-glucoside was detected in the plasma and urine after the ingestion of red wine, dealcoholized red wine, and red grape juice whereas other metabolites including the aglycon, sulfate, or glucuronate conjugates were not detected [35]. The reasons why these metabolites are not detected could be related to the chemical structures of the exact anthocyanins evaluated, effects of the matrix or conditions of the experiment, or even the methodologies used for the chemical analyses of the analytes.

Several biotransformation enzymes located in the small intestine, liver, or kidney appear to be the key for anthocyanin metabolism after absorption since the major anthocyanin metabolites recovered in the urine and serum were methylated or/ and glucuronate conjugates. Catechol-*O*-methyltransferase (COMT), which occurs in various tissues, might transfer a methyl group to the aglycone [36–38]. Uridine diphosphoglucose glucuronosyl transferase (UDPGT) and uridine diphosphoglucose glucose dehydrogenase (UDPGD), both abundant in the liver and intestine, were proposed to catalyze the glucuronidation of aglycones [29,34,38]. Cytosolic enzymes phenol sulfotransferases (SULT), widely distributed throughout the body [39], are likely to sulfate aglycones [28]. Oxidized derivatives have been found in human urine and serum [34]. The possible pathways for the formation of anthocyanin metabolites were proposed by Kay et al. [34] and Prior and Wu [40].

4.3 BIOAVAILABILITY

The bioavailability of anthocyanins is generally low, and lower than that of other fla-vonoids based on numerous animal and human studies using anthocyanin-rich foods or extracts. The bioavailability of anthocyanins has been assessed in numerous ani-mal and human studies by plasma concentration and urinary excretion. Anthocyanins are found in the plasma and urine in quite small amounts after the intervention of large amounts of anthocyanin-rich foods or extracts, generally reported at levels lower than ~1% of the dietary intake. Despite their low absorption into the plasma, anthocyanins have been found in different tissues around the body. Berry antho-cyanins and their metabolites were found to transport across the blood–brain bar-rier [41]. Milbury and Kalt (2010) found anthocyanins and their glucuronides in the range of femtomoles per gram of fresh weight of brain tissue at 18 h postprandial. Anthocyanin-rich berries have been proposed as better inhibitors of tumorigenesis in the GIT than in other sites based on different animal studies, suggesting that low bioavailability may cause the difference in biological activities [7,8,42–44]. He et al. (2005) reported that a high anthocyanin concentration in the fecal and cecal con-tent of rats may favor anthocyanin biological activities at the colon site, resulting in potential health benefits [24].

The bioavailability of anthocyanins seems to be related to the chemical struc-ture of the pigments present. Anthocyanins with different patterns of aglycones and glycosylations from different dietary sources (blueberry, boysenberry, black raspberry, and black currant) were detected intact and unmetabolized in the urine of male Sprague–Dawley rats and males [45]. In this study, the total amount of anthocyanin excreted as a percentage of the amount consumed is <0.1% for all anthocyanins. The urinary excretion of anthocyanins from strawberries corre-sponded to 1.8% of pelargonidin-3-glucoside consumed by humans [28], whereas the total urinary excretion of blackberry anthocyanin was 0.16% of the amount of anthocyanins ingested [29]. Four major native anthocyanidin glycosides of black currant juice were detected in the human plasma and urine and the total anthocyanins excreted was <0.133% of the total anthocyanins ingested [46]. The total urinary excretion of anthocyanins (3-monoglucosides of delphinidin, cyani-din, petunidin, peonidin, and malvidin) and metabolites was $0.05 \pm 0.01\%$ of the administered dose within 24 h after consuming a red wine extract [47]. In a study with rats, anthocyanin excretion levels were 1.2% of the ingested amount over a 24-h period after raspberry anthocyanins were fed. Additionally, the trace amount of anthocyanins was detected in the cecum, colon, and feces and they were not found in the liver, kidneys, and brain [48].

Glycosylated anthocyanins have been detected in the plasma and urine after the consumption of 720 mg anthocyanins by elderly women, with anthocyanin concentration reaching up to 168 nmol/L in the plasma [49]. The average concentrations of anthocyanins and their metabolites in human urine reached 17.9 μmol/L within 5 h of consumption of 20 g chokeberry extracts (1.3 g cyanidin-3-glycosides) whereas concentrations in the serum were 591.7 nmol/L within 2 h of consumption [34].

Studies conducted with berry extracts showed that the proportions of individual anthocyanins in the urine appear to be different from that ingested, suggesting that absorption and excretion of anthocyanins were influenced by the type and number of sugars attached to the molecule [45]. Anthocyanins with either a di- or trisaccharide moiety were excreted in the urine in higher proportions than the monoglycoside [26], suggesting that larger proportions of the anthocyanin rutinosides than anthocyanin glucosides were absorbed into the blood stream [50]. The presence of acylating acids also affects anthocyanin bioavailability [27,51]. Charron et al. (2007) found that the recovery of nonacylated anthocyanins was more than fourfold that of acylated anthocyanins in human urine after the consumption of red cabbage. Lately, they found that the peak plasma concentrations of nonacylated anthocyanins were fourfold higher than their acylated counterparts after the consumption of carrot juices.

The reasons why anthocyanins have such a low bioavailability are not clear yet. A possible reason proposed was that anthocyanins could not be efficiently hydrolyzed by β-glucosidase in the GI tract, resulting in a low absorption into the blood stream [52]. The selective decrease of cyanidin-3-glucoside in the small intestinal content likely resulted from β-glucosidase activity [53]. Another possible reason is the degradation of anthocyanins due to the neutral and mild alkaline condition in the intestines. However, a study found that anthocyanins were abundant in rat feces (0.7/1.8/2.0 g/kg wet feces for chokeberry/bilberry/grape, respectively) after metabolism by the intestinal microflora after 14-day intake of anthocyanin-rich extracts from chokeberries, bilberries, or grapes [24].

Nevertheless, studies looking at the mass balance of ingested anthocyanins versus the absorbed and excreted anthocyanins cannot account for the total amount of ingested compounds. Assuming <1% of absorbed anthocyanins, 50–70% being excreted, and 5–10% being uptaken by tissues lining the GIT, there is a significant proportion of anthocyanins still unaccounted for [52]. Some of these anthocyanins are degraded into other phenolics, such as protocatechuic acid (PCA) and phloroglucinol aldehyde (PGA), but some degradation products may be difficult to detect by traditional methods. Anthocyanins are susceptible to degradation in biological systems, but they are also susceptible to degradation during extraction or analyses. Therefore, it is essential to take great care when selecting a methodology for anthocyanin analyses from biological samples. An overview of the methodologies currently in use is presented in the following section. However, it is recommended that whenever possible, the methodologies must be optimized for the particular matrix to be analyzed.

4.4   ISOLATION OF ANTHOCYANINS OR ANTHOCYANIN METABOLITES FROM BIOLOGICAL SAMPLES

4.4.1   PRETREATMENT

The extraction of anthocyanins in biological samples could be a complex task. Anthocyanins are moderately polar structures and should be extracted in relatively polar solvents (i.e., water, methanol, ethanol, or acetone). The extraction of anthocyanins from plant material typically involves the use of organic solvents to help break down plant cell walls and membranes. Similarly, when extracting anthocyanins from biological samples, the composition of the samples as well as the potential localization of the anthocyanins in the matrix should be taken into account. Anthocyanin extraction from cells or tissues will most likely require a mixture of water and organic solvent, most commonly methanol or acetone, whereas the extraction of anthocyanins from feces has been successfully achieved with aqueous solvents. The extraction temperature should be considered since anthocyanins are relatively labile, whereas the anthocyanin aglycones, the anthocyanidins, are highly reactive and can be easily lost during the extraction if proper care is not taken. Cold temperatures will protect the anthocyanins from degradation, and minimize any chemical or enzymatic activity. When testing anthocyanins in biological fluids, such as urine or plasma, the samples are generally acidified before storage or extraction, to convert and maintain the anthocyanin in its flavylium form. Solid anthocyanins containing samples (cells, tissues, or organs) will require washing with distilled water and crushing using a homogenizer before a representative sample can be extracted. Freezing the samples and pulverizing them in the presence of liquid nitrogen have proven to be extremely effective with a variety of samples (e.g., for anthocyanin extraction from tissues). Liquid nitrogen offers a number of advantages including the displacement of oxygen, the low temperatures, and the maximization of the surface area due to the brittle texture of tissues when frozen at such low temperatures.

4.4.2   PURIFICATION

Biological samples can have a complex composition. Sample purification of extracts from biological materials is often necessary for further qualitative and quantitative analyses. An effective preparation of anthocyanins and their metabolites can be usually achieved using solid-phase extraction (SPE) with different commercially available sorbents. The SPE purification has gained popularity because of their ease of use and high efficiency for fractionating anthocyanins and their metabolites. Different resins have been used to clean up or prefractionate anthocyanins prior to isolation or characterization: ion-exchange resins, polyamide powders, and gel materials. Sorbents (C18, HLB, LH-20, and MCX) appeared not to have equal selection on all individual anthocyanins [54,55]. Among them, two sorbents (C18 and HLB) based on reversed-phase interaction are probably the most frequently employed for anthocyanin separation in biological samples [30,35,49,53,56–58]. The selection of the proper SPE resin, solvent combination, and the right internal standards can allow for the proper quantification of anthocyanins and their metabolites.

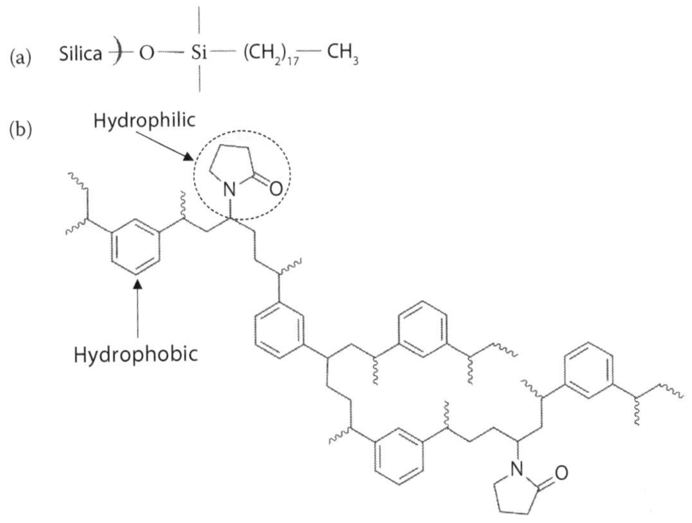

FIGURE 4.4 Functionalized sorbents of (a) C_{18} and (b) HLB .

The C_{18} cartridge is a strong hydrophobic silica-based bonded reversed-phase sorbent (Figure 4.4), which retains even weak hydrophobic organic compounds (e.g., anthocyanins and their metabolites) from aqueous solutions such as serum, plasma, or urine while allowing matrix interference such as sugar and acids to pass through to waste [30,35,49,53,56]. The cartridges should be first activated through the application of at least one sorbent-bed volume of acidified methanol and acidified water [59]. The biological fluid is then loaded onto the SPE cartridge, followed by washing the cartridge with two column volumes acidified water to remove compounds not adsorbed. The anthocyanins and their metabolites can be recovered from the cartridge with acidified methanol. The methanolic eluent can then be dried under a stream of nitrogen gas to remove the residual solvent. Depending on the volume of the material, the solvents can be alternatively removed by a rotary evaporator. The acids commonly used to acidify the samples and solvents used are trifluoroacetic acid [49,53], formic acid [48], oxalic acid [56], acetic acid [60], and phosphoric acid [35] in water. Ideally, the acid used should be volatile, so that it will not become more concentrated during concentration of the sample and should be compatible with the equipment and methodology to be used for the separation and monitoring of the anthocyanins. An environment of low pH is required to maintain the majority of anthocyanins in the flavylium cation, the most stable form of anthocyanins. The use of the spiked standards (internal standard) or additional recovery experiments allows correcting for the loss of anthocyanins during sample preparation and purification. The losses of individual anthocyanins have been found to be different in a series of experiments even following the same SPE purification [54,61]. Therefore, the standards should be chosen carefully to correct accurately for the loss of tested anthocyanins and metabolites.

HLB cartridge is a hydrophilic–lipophilic-balanced reversed-phase sorbent with surface functionality of *m*-divinylbenzene and *N*-vinylpyrrolidone copolymer for acids, bases, and neutrals (Figure 4.4). Such polymers provide stronger reversed-phase interaction than the C_{18} gel, and they maintain better water wettability due to the hydrophilic *N*-vinylpyrrolidone [62]. The polymer structure is more durable and sustains wider pH range than the silica-based gel. Therefore, HLB appeared to promote as a replacement of the C_{18} gel for polyphenolics purification. However, the recovery of anthocyanins was unsatisfactory by using HLB methods compared with the C_{18} method [54]. Ling et al. (2009) used HLB cartridges to extract freeze-dried black raspberry anthocyanins in human plasma or oral mucosa tissue homogenates [58]. Cooke et al. (2006) extracted anthocyanins from the urine or plasma [57]. The purification solvents and procedures for HLB sorbent are similar to those for C_{18} [57,58].

We evaluated C18, HLB, LH-20, and MCX sorbents for anthocyanin isolation from plant mixture [54]. The MCX sorbent features a mixed mode mechanisms of strong cation-exchange and reversed-phase adsorption and has been found to increase purity and efficiency while maintaining excellent recovery rate for isolation of plant anthocyanin mixtures. Therefore, the MCX sorbent has great potential for anthocyanin purification in biological samples in the future.

It is important to highlight that biological samples can be scarce. Any efforts for sample purification and concentration are intended to remove the interfering materials and improve the quality of the results. However, these steps may also cause some losses of the material—therefore, the need to estimate recoveries from these steps. When the amount of the sample is limited, and the concentration of anthocyanins or metabolites is known to be very small, it may be advisable to skip the purification steps and proceed to sample analyses directly.

Prior to liquid chromatography analyses, all samples should be filtered through a 0.45 μm (or smaller pore size) filter; in some cases, centrifugation may also be advisable prior to filtration.

4.5 DETECTION

4.5.1 HPLC–PDA DETECTION

High-performance liquid chromatography (HPLC) has proved to be fast and sensitive for the analyses of phenolic plant constituents, and is especially useful for the analysis of anthocyanins [63]. The first application of HPLC to anthocyanin analyses was in 1975 by Manley and Shubiak [64] and it has now become the method of choice for the separation of mixtures of anthocyanins and anthocyanidins [65]. HPLC is now used for anthocyanin qualitative, quantitative, and preparative work, offering improved resolution compared to chromatographic procedures previously employed (Figure 4.5). It also allows for simultaneous rapid monitoring of the eluting anthocyanins [66,67].

The most popular system is a reversed-phase column (C_{18}), on a silica-based column. However, the use of C_{18} on a polymer-based column has been reported to provide better resolution, especially for the separation of complex anthocyanin mixtures containing acylated pigments [68,69]. Polymer-based columns also show better stability at lower pH-operating conditions. The overall polarity and stereochemistry of anthocyanins are the key factors for their separation. The order of elution will

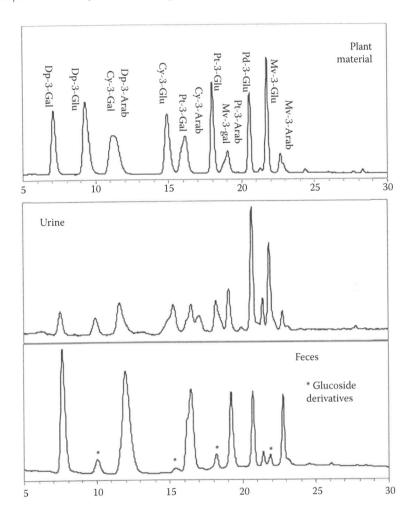

FIGURE 4.5 Chromatographic separation of bilberry anthocyanins found in the berry and in biological samples after dietary supplementation. Detection with photodiode array detector at 520 nm. (Adapted from He, et al., Analysis of anthocyanins in rat intestinal contents—Impact of anthocyanin chemical structure on fecal excretion. *J Agric Food Chem* 2005, *53*, 2859–2866; He, et al., Intact anthocyanins and metabolites in rat urine and plasma after 3 months of anthocyanin supplementation. *Nutr Cancer* 2006, *54*, 3–12.)

be dependent on the hydroxyl or methoxyl substitutions of the pyrylium ring, the number and nature of sugar substituents, and the presence, number, and nature of acylating groups. In general, more glycosylations or hydroxylations increase anthocyanin mobility, whereas *O*-methylations and acylations increase the elution time. Nevertheless, this is not a strict rule; deviations of this behavior may be dependent on the nature and position of the substitutions [65,69]. For instance, 3-rutinosides have longer retention times than the corresponding 3-glucosides because of the nonpolarity imparted by the methyl group of rhamnose at position C6.

The typical mobile phases are composed of gradients of acetic, phosphoric, or formic acid in water–methanol or water–acetonitrile solvents [63] that separate anthocyanins as their flavylium cations and can be easily detected by their absorbance at their visible maximum.

Different detectors are available for monitoring the eluting anthocyanins, including monochromatic or multiple-wavelength spectrophotometers, photodiode array (PDA) detectors, and more recently, mass spectrometers. A PDA detector scans multiple ultraviolet–visible (UV–vis) spectral data of the eluting sample per seconds and allows the spectroscopic interpretation of these data and detection of impurities coeluting with the compounds of interest [65,66,70]. Anthocyanin elution is usually monitored at wavelengths between 480 and 550 nm by comparing the retention times and UV–vis spectra with the respective standards or reference compounds. A careful evaluation of the anthocyanin UV–vis spectra (from ~250 to 650 nm) can provide valuable information about the chemical structure of the pigment. Although not conclusive, the spectral characteristics obtained with a PDA constitute valuable information that when combined with other pieces (such as retention time and molecular mass) can aid in the chemical characterization of the pigment. Each anthocyanidin has a characteristic wavelength of maximum absorbance, the position of the sugar substitutions on the molecule (either position 3 or 5 of the anthocyanin molecule) is reflected by a characteristic shoulder in the 440 nm range, and the presence of cinnamic acid acylations can be revealed by a bathochromic shift on the wavelength of maximum absorbance and an additional spectral peak of absorbance in the 320–340 nm range. The reference materials used can be well-characterized anthocyanin sources, such as strawberries or grapes (before or after alkaline or acid hydrolyses). The reported anthocyanin metabolites, such as glucuronic acid derivatives, methylated anthocyanins, or sulfated anthocyanins should also be detected under similar chromatographic conditions. The elution of the degradation products, such as PGA and PCA, will not be noticed in the 500 nm range as these compounds are not colored. For this reason, simultaneous detection of the 280 nm chromatogram is of great importance. The use of a PDA detector allows for the extraction of chromatograms at any given wavelength of interest once the spectral data have been collected.

HPLC has been a valuable analytical tool for the characterization and quantification of anthocyanins and the metabolites in cells [71,72], urine [26,29,37,38,73–76], blood [26,51,56,74–78], and different organs [30,56,78].

4.5.2 Mass Spectrometer Detection

The use of mass spectroscopic analyses for the characterization of anthocyanins dates from the late 1980s and early 1990s [61,82], and it has increased dramatically over the past decade. Most reports cite the use of HPLC coupled to mass spectrometry (MS) detectors or after the isolation of individual pigments or metabolites prior to the mass spectroscopic analysis [28,47,79–85].

The accurate molecular-weight determination is very important for structure elucidation of complex anthocyanins or metabolites since small components with little UV–vis absorption that show weak or no nuclear magnetic resonance (NMR) signals may be otherwise overlooked [86]. MS can be used for the qualitative and quantitative analyses of anthocyanins by differentiating charged fragments in electronic and magnetic fields according to the mass-to-charge ratios. Generally, MS is composed of three components: ion sources, mass analyzer, and detector. The ionization techniques suitable for anthocyanin chemistry are fast atom bombardment (FAB), electrospray ionization (ESI), matrix-assisted laser desorption ionization (MALDI), and atmospheric pressure chemical ionization (APCI). The most reported ionization techniques for anthocyanin analyses are FAB and ESI. The introduction of FAB–MS constituted an advance in anthocyanin structural identification [63]. FAB is an ionization technique that generates ions by bombarding the sample and liquid matrix by an atom beam. FAB–MS provides molecular ions and various fragmentation ions that give direct information on the structure of a molecule. Only a small amount of the sample is required (1–2 µL may be enough). After solvent evaporation, a liquid matrix (glycerol, thioglycerol, or 3-nitrobenzyl alcohol) is introduced. A neutral atom beam usually of an inert gas such as xenon or argon is then used to bombard the droplet surface from which the material is sputtered. The mass-to-charge ratio of the intact molecular ion is obtained, and the fragmentation of the molecule allows the side chains to be determined although their positions are ambiguous [63,65]. FAB–MS has been used to identify or characterize anthocyanin structures such as cyanidin-3-glucoside acylated with *p*-coumaric acid from *Camellias* [87], nasunin from eggplants [88], cyanidin-3-oxalylglucoside from orchids [89], and delphinidin-3-glucoside, cyanidin-3-glucoside, and petunidin-3-glucoside from seed coats of black soybeans [90].

ESI–MS has emerged as a powerful technique for the characterization of biomolecules, and is the most versatile ionization technique in existence today [80, 91, 92]. The highly sensitive and soft ionization technique allows the mass spectrometric analysis of thermolabile, nonvolatile, and polar compounds and produces intact ions from large and complex species in the solution [93,94]. In addition, it has the ability to introduce liquid samples to a mass detector with minimum manipulation. The volatile acids (such as formic acid and acetic acid) are often added to the mobile phase as well to protonate anthocyanins. A chromatogram with only the base peak for every mass spectrum provides more readily interpretable data because of fewer interference peaks [95]. Cleaner mass spectra are achieved if anthocyanins or metabolites were isolated from mixtures by the use of C_{18} solid-phase purification [53,91,96].

Tandem mass spectrometry (MS–MS) uses more than one mass analyzer for structural and sequencing studies that have been found very useful for anthocyanin or metabolite characterization. The mass analyzers may be of the same type (triple or quadrupole) [80,91] or hybrid such as ion trap [83,97,98] and quadrupole–time of flight (TOF) [99] for anthocyanin structural analysis.

The triple-quadrupole instrument is mostly used in MS–MS, allowing for the formation of low-energy collisionally induced dissociation fragments [91,94]. The mass of the parent ion of interest is scanned in the first quadrupole, *m/z* selected and collisionally activated in the second quadrupole, and the daughter ions are analyzed in the third quadrupole [91,94,100–102]. Consequently, the signal-to-noise ratio was improved greatly by the selection of a particular fragmented ion. This process provides structural information on the components of a mixture and that leads to the unambiguous identification of fragmentation pathways [101]. ESI and MS–MS were used early to structurally characterize anthocyanins from radishes (*Raphanus sativus*), red-fleshed potatoes (*Solanum tuberosum*), red cabbage (*Brassica oleracea*), chokeberries (*Aronia melanocarpa*), concord grapes (*Vitis labrusca*), and roselle (*Hibiscus sabdariffa* L.) anthocyanin extracts [91]. Currently, the ion trap technology appears as a strong alternative to the quadrupole systems for its mass accuracy, mass resolution and high sensitivity (MSn) performance. The ion trap system enables rapid anthocyanin/metabolite screening for the maximum structural information [98,103–105]. With the aid of a linear ion trap system, we tentatively identified 17 radish anthocyanins that were characterized as 3-mono- or dihydroxycinnamoyl (*p*-coumaric, caffeic, and/or ferulic acid)—diglucoside-5-glucoside, 3-mono-, or dihydroxycinnamoyl (*p*-coumaric, caffeic, and/or ferulic acid)-diglucoside-5-malonylglucoside of pelargonidin [98]. Since MS–MS provided clear and characteristic fragmentation patterns for complicated anthocyanins, many have followed to use ESI in combination with tandem MS for the characterization of a wide range of plant materials or anthocyanins found in biological samples.

As every methodology, the powerful MS techniques have limitations. MS analyses will not allow for the differentiation of compounds of the same molecular weights. Therefore, the same anthocyanidin glycosylated with glucose or galactose will produce exactly the same molecular ion. Even more, some acylating groups share the same molecular mass as some typical glycosylations. For example, cyanidin glucoside would produce an *m/z* signal of 449.2. Cyanidin with two glucoses would show a molecular mass of *m/z* 611.4, exactly the same as cyanidin-3-glucoside acylated with caffeic acid. These compounds would not be differentiated by ESI and not even by the more powerful MS–MS. Fortunately, the retention time and spectral characteristics of both pigments are clearly different, with the acylated pigment being characterized by longer retention times on reversed-phase chromatography and a peak of absorption in the 320 nm range not present in the spectra of a nonacylated anthocyanin. For this reason, it is always recommended to combine MS data with other chromatographic and spectral information before a peak is identified.

MS–MS has proven to be very valuable for the characterization of anthocyanin metabolites. The resulting change in the molecular weight of an anthocyanin that has been glucuronidated exactly matches the molecular mass increase after methylation: an increase of 14 units. However, methylation typically occurs on a hydroxyl group of the main anthocyanidin structure, whereas the glucuronidation is a change in the glycosylation. The difference in both molecules can be easily determined by selective fragmentation of the parent ion, by determination of the molecular mass of the daughter ions, through tandem MS (Figure 4.6). Recently, more powerful MS equipment has become available that couples liquid chromatography (high pressure or ultra-high-pressure liquid chromatography) with TOF instruments. The TOF instruments provide very high mass accuracy, allowing for the differentiation of chemical structures with very similar molecular masses. These instruments can be more powerful tools for the elucidation of the chemical structures of anthocyanin metabolites or their degradation products.

It is still recommended, however, to use the MS data in combination with additional chemical data obtained by other analytical techniques (such as retention

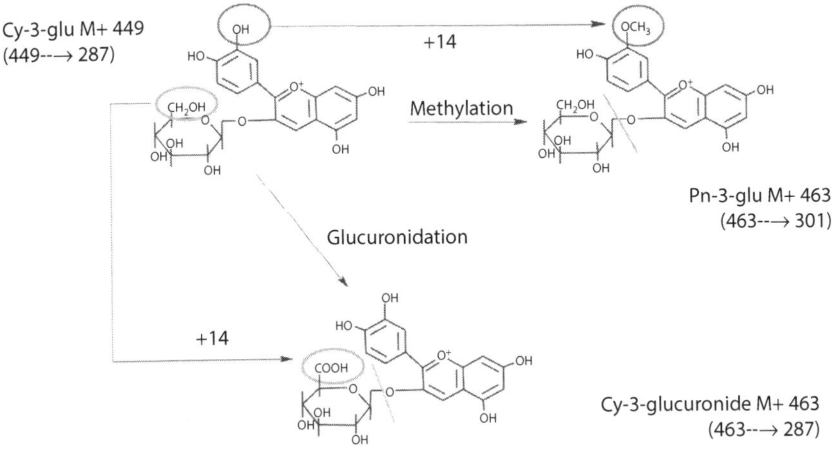

FIGURE 4.6 Reported metabolic transformations of cyanidin-3-glucoside *in vivo* that result on similar molecular mass increase. Simple MS data may not differentiate the compounds while Tandem MS results on clear differentiation of the metabolites.

times and spectral data) to assure the correct interpretation of the results and proper identification of the analytes.

4.5.3 NMR Spectroscopy

NMR spectroscopy is the most powerful method for the structural elucidation in the solution, and advances in NMR techniques have made significant impacts on anthocyanin studies [63,65,106]. The complete structural characterization of anthocyanins is possible with one- and two-dimensional (2D) NMR techniques. However, relatively large quantities of purified materials are required for the resolution of proton signals associated with sugars and positions C6 and C8 of the flavylium nucleus [19].

In the late 1970s, the first successful ^{1}H-NMR recorded on natural anthocyanins by Goto et al. [107] made it possible to determine the complete structures and stereochemistry of complex anthocyanins [86]. ^{13}C-NMR has also been reported. One- and 2-D NMR information allowed the determination of the *trans* configuration of cinnamic acid acylation, the β-glucopyranoside conformation of the sugars present, and the exact position of attachment of acylating groups to the anthocyanin's sugar moieties [86].

The nature of the aglycon and the number of glycosylations and acylating groups present in a molecule can be normally assessed from one-dimensional ^{1}H-NMR. When the one-dimensional spectrum is too complex, it is useful to perform a 2-D *J*-resolved experiment to simplify the spectra and clarify the magnitude of the coupling constants. Coupling constants provides information regarding *trans* versus *cis* configurations and equatorial or axial protonations in the glycosidic moiety, as well as information regarding the configurations of the anomeric carbons in the sugars. Two-dimensional shift correlation and total correlation analyses (2-D NMR COSY and TOCSY) assisted in the assignment of all individual proton signals from the individual sugar moieties. Acylation caused a low field shift of 0.5–1.0 ppm for the geminal proton relative to the nonacylated pigment [63].

HMQC and HMBC experiments enabled to trace connectivities between ^1H and ^{13}C atoms through indirect detection of the low natural abundance nuclei ^{13}C, via ^1H nuclei [108]. The HMQC spectra provide the correlation between directly bonded ^1H and the corresponding ^{13}C. HMBC is a long-range heteronuclear chemical shift correlation technique that provides intraresidue multiple bond correlation; this information is valuable for confirming ^{13}C and/or ^1H assignments. It also provides interresidue multiple bond correlation between the anomeric carbon and the aglycon proton and thus serves to identify the interglycosidic linkages [108]. 2-D NOESY experiments provide valuable information in mapping specific through-space inter-nuclear distances [109] that could be sufficient to determine the molecular three-dimensional structure. Cross-peaks are observed in NOESY spectra between proton pairs that are close in space, typically <5 Å, close enough to allow through-space interactions [108]. The greater the signal, the closer together those hydrogens are in space [110]. It has been proposed that the stacking between the aromatic nuclei of the anthocyanin and the planar ring of the aromatic acid occurs via the formation of π–π hydrophobic interactions [86,111]. Giusti et al. (1998) structurally elucidated two novel diacylated anthocyanins and two monoacylated anthocyanins from radish by one- and 2-D NMR [112]. Jordheim et al. (2006) applied 2-D NMR to charac-terize carboxypyranoanthocyanins [113]. Two 3-deoxyanthocyanins, luteolinidin-5-glucoside, and apigeninidin-5-glucoside were identified by Swinny et al. [114] using ^1H and ^{13}C NMR.

Liquid chromatography coupled with NMR detection (LC–NMR) has arisen as an important technique for analyses of anthocyanins and their metabolites in biological samples, which surpass any previous methodologies in the structure elucidation of anthocyanins or unknown compounds. The LC–NMR has become an important technique for the biomedical, pharmaceutical, environmental, food and natural products analysis, as well as for the identification of drug metabolites [115]. The lack of sensitivity of LC–NMR has been solved with progress in sol-vent suppression, pulse field gradients, probe technology, and high-field magnets [116]. At the current time, LC–NMR methodology is mostly limited to ^1H NMR spectra.

However, whenever the concentration of analyte as eluted from LC column is not sufficient, the sensitivity of LC–NMR is a limiting factor for more sophisti-cated 2-D NMR experiments (COSY, NOESY, HMQC). Recently, a solution to this problem has been developed by inserting an SPE unit between HPLC unit and NMR spectrometer (LC–SPE–NMR), to trap and accumulate the compounds onto SPE cartridges [117,118]. Each one of the trapped compounds in SPE cartridges was eluted into the NMR probe with deuterated solvent, avoiding the need for using deuterated solvents during the LC separation. The hybrid technique, LC/SPE/NMR, has been applied to analyze anthocyanins with very promising results [119]. This technique has not yet been adopted by many labs due to its high cost and the need for highly trained personnel, but may become more popular as the costs of the instruments go down.

4.5.4 OTHER ANALYTICAL TOOLS

There are other analytical tools often used for anthocyanin analyses in plant materials that are not being used for anthocyanin analyses in biological systems. This may be in part due to the limited sample volumes typical of biological samples. Some common analytical tools may not be practical with small samples, as could be the case with acid or alkaline hydrolysis. Spectrophotometric techniques to quantify anthocyanins, such as the pH-differential method [120], become secondary as the HPLC can provide qualitative and quantitative results in just one run and with a very small sample.

Other analytical techniques do not have the limitation of the sample size, but have not become widely used, however can offer great potential for anthocyanin analyses in biological samples. Fourier transform infrared (IR) spectroscopy combined with multivariate statistical analyses has been used by different groups for the rapid discrimination and quantitation of anthocyanins from different sources, mainly in fruits, juices, and wines [121–124]. The infrared spectral signal (Figure 4.7) is the result of molecular vibrations of functional groups (i.e., C = C, C = O, C–H, C–OH) present in the sample, providing unique fingerprinting profiles for characterization of biological materials. Some advantages of this technique is that it is fast and simple, requires minimal sample preparation, such as SPE clean-up of the sample, and very small sample size. These characteristics make IR a promising technique for the analyses of biological samples. The application of IR imaging offers the potential to evaluate localization of anthocyanins in cells, membranes, or other biological materials.

4.6 CONCLUSIONS

Our understanding of anthocyanin stability in the body, bioavailability, metabolic transformations, and biological activity will greatly depend on our ability to

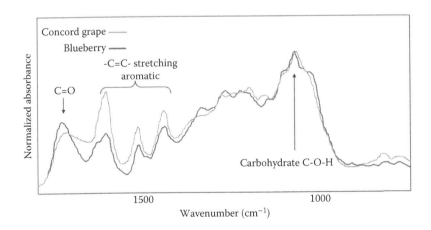

FIGURE 4.7 Fourier transform infrared spectra of concord grape and blueberry phenolic fractions. Bands highlighted are associated with anthocyanin IR signals and their glycosidic substitutions.

efficiently evaluate their presence in different biological tissues. Anthocyanin compounds have a complex chemistry, and are very reactive. Extreme care should be taken when working with anthocyanin containing tissues as the analytes of interest could be degraded during the analyses, or transformed due to the analytical procedure. Many different instruments are available for the analyses of anthocyanins, metabolites, and their degradation products, and it is recommended that different data be combined for better results.

It is possible that our current methodologies for anthocyanin analyses are not capable of extracting, detecting, or quantifying accurately anthocyanins that have become tightly bound to proteins, membranes, or some tissues under biological conditions. This may explain the relatively strong bioactivity attributed to anthocyanin feeding trials as compared to the low bioavailability reported. Anthocyanins may be circulating the body in forms not yet fully understood. The field of anthocyanin analyses in biological tissues will need to continue to evolve to produce new tools that would allow for testing for the presence of anthocyanins without disrupting the tissues. A combination of advanced microscopic techniques with different vibrational spectroscopic techniques may emerge in the future as additional tools for analyses of anthocyanins in biological samples.

REFERENCES

1. Prior, R.L., Fruits and vegetables in the prevention of cellular oxidative damage. *Am J Clin Nutr* 2003, *78*, 570S–578S.
2. Wang, H., Cao, G., Prior, R.L., Oxygen radical absorbing capacity of anthocyanins. *J Agric Food Chem* 1997, *45*, 304–309.
3. Aoki, H., Wada, K., Kuze, N., Ogawa, Y., Koda, T., Inhibitory effect of anthocyanin colors on mutagenicity induced by 2-amino-1-methyl-6-phenylimidazo[4,5-b]pyridine (PhIP). *Foods Food Ingred J Jpn* 2004, *209*, 240–246.
4. Gasiorowski, K., Szyba, K., Brokos, B., Kolaczynska, B., Jankowiak-Wlodarczyk, M., Oszmianski, J., Antimutagenic activity of anthocyanins isolated from *Aronia melanocarpa* fruits. *Cancer Lett* 1997, *119*, 37–46.
5. Hagiwara, A., Miyashita, K., Nakanishi, T., Sano, M., Tamano, S., Kadota, T., Koda, T. et al., Pronounced inhibition by a natural anthocyanin, purple corn color, of 2-amino-1-methyl-6-phenylimidazo[4,5-b]pyridine (PhIP)-associated colorectal carcinogenesis in male F344 rats pretreated with 1,2-dimethylhydrazine. *Cancer Lett* 2001, *171*, 17–25.
6. Lala, G., Malik, M., Zhao, C., He, J., Kwon, Y., Giusti, M.M., Magnuson, B.A., Anthocyanin-rich extracts inhibit multiple biomarkers of colon cancer in rats. *Nutr Cancer* 2006, *54*, 84–93.
7. Casto, B.C., Kresty, L.A., Kraly, C.L., Pearl, D.K., Knobloch, T.J., Schut, H.A., Stoner, G.D., Mallery, S.R., Weghorst, C.M., Chemoprevention of oral cancer by black raspberries. *Anticancer Res* 2002, *22*, 4005–4015.
8. Kresty, L.A., Morse, M.A., Morgan, C., Carlton, P.S., Lu, J., Gupta, A., Blackwood, M., Stoner, G.D., Chemoprevention of esophageal tumorigenesis by dietary administration of lyophilized black raspberries. *Cancer Res* 2001, *61*, 6112–6119.
9. Tsuda, T., Horio, F., Uchida, K., Aoki, H., Osawa, T., Dietary cyanidin 3-*O*-beta-D-glucoside-rich purple corn color prevents obesity and ameliorates hyperglycemia in mice. *J Nutr* 2003, *133*, 2125–2130.
10. McDougall, G.J., Stewart, D., The inhibitory effects of berry polyphenols on digestive enzymes. *Biofactors* 2005, *23*, 189–195.

11. Andersen, Ø.M., Jordheim, M., The anthocyanins. In *Flavonoids: Chemistry, Biochemistry and Applications*, 1st ed., Andersen, Ø.M., Markham, K.R., eds. Boca Raton, FL: CRC Press, 2005.

12. Brouillard, R., Chemical structure of anthocyanins. In *Anthocyanins as Food Colors*, Markakis, P., ed. New York, NY: Academic Press, 1982, pp. 1–40.

13. Harborne, J.B., Correlations between flavonoid chemistry, anatomy and geography in the *Restionaceae* from Australasia and South Africa. *Phytochemistry* 1979, *18*, 1323–1327.

14. Clifford, M.N., Anthocyanins—Nature, occurrence and dietary burden. *J Sci Food Agric* 2000, *80*, 1063–1072.

15. Giusti, M.M., Wrolstad, R.E., Acylated anthocyanins from edible sources and their applications in food systems. *Biochem Eng J* 2003, *14*, 217–225.

16. Brouillard, R., Flavonoids and flower color. In *The Flavonoids: Advances in Research Since 1980*, Harborne, J.B., ed. London: Chapman & Hall, 1988, pp. 525–538.

17. Mazza, G., Miniati, E., *Anthocyanins in Fruits, Vegetables and Grains*. Boca Raton, FL: CRC Press, 1993.

18. Robbins, R.J., Phenolic acids in foods: An overview of analytical methodology. *J Agric Food Chem* 2003, *51*, 2866–2887.

19. Jackman, R.L., Smith, J.L., Anthocyanins and betalains. In *Natural Food Colorants*, 2nd ed., Hendry, G.A.F., Houghton, J.D., eds. Scotland: Blackie & Son Glasgow, 1996.

20. Andersen, Ø.M., Viksund, R.I., Pedersen, A.T., Malvidin 3-(6-acetylglucoside)-5-glucoside and other anthocyanins from flowers of *Geranium sylvaticum*. *Phytochemistry* 1995, *38*, 1513–1517.

21. Andersen, Ø.M., Fossen, T., Anthocyanins with an unusual acylation pattern from stem of *Allium victorialis*. *Phytochemistry* 1995, *40*, 1809–1812.

22. Giusti, M.M., Ghanadan, H., Wrolstad, R.E., Elucidation of the structure and conformation of red radish (*Raphanus sativus*) anthocyanins using one- and two-dimensional nuclear magnetic resonance techniques. *J Agric Food Chem.* 1998, *46*, 4858–4863.

23. Torskangerpoll, K., Noerbaek, R., Nodland, E., Oevstedal, D.O., Andersen, Ø.M., Anthocyanin content of *Tulipa* species and cultivars and its impact on tepal colours. *Biochem Syst Ecol* 2005, *33*, 499–510.

24. He, J., Magnuson, B.A., Giusti, M.M., Analysis of anthocyanins in rat intestinal contents—Impact of anthocyanin chemical structure on fecal excretion. *J Agric Food Chem* 2005, *53*, 2859–2866.

25. Fleschhut, J., Kratzer, F., Rechkemmer, G., Kulling, S.E., Stability and biotransformation of various dietary anthocyanins *in vitro*. *Eur J Nutr* 2006, *45*, 7–18.

26. Wu, X., Pittman, H.E., McKay, S., Prior, R.L., Aglycones and sugar moieties alter anthocyanin absorption and metabolism after berry consumption in weanling pigs. *J Nutr* 2005, *135*, 2417–2424.

27. Charron, C.S., Clevidence, B.A., Britz, S.J., Novotny, J.A., Effect of dose size on bioavailability of acylated and nonacylated anthocyanins from red cabbage (*Brassica oleracea* L. var. *capitata*). *J Agric Food Chem* 2007, *55*, 5354–5362.

28. Felgines, C., Talavera, S., Gonthier, M.P., Texier, O., Scalbert, A., Lamaison, J.L., Remesy, C., Strawberry anthocyanins are recovered in urine as glucuro- and sulfoconjugates in humans. *J Nutr* 2003, *133*, 1296–1301.

29. Felgines, C., Talavera, S., Texier, O., Gil-Izquierdo, A., Lamaison, J.L., Remesy, C., Blackberry anthocyanins are mainly recovered from urine as methylated and glucuronidated conjugates in humans. *J Agric Food Chem* 2005, *53*, 7721–7727.

30. Felgines, C., Texier, O., Garcin, P., Besson, C., Lamaison, J.L., Scalbert, A., Tissue distribution of anthocyanins in rats fed a blackberry anthocyanin-enriched diet. *Mol Nutr Food Res* 2009, *53*, 1098.

31. Keppler, K., Humpf, H.U., Metabolism of anthocyanins and their phenolic degradation products by the intestinal microflora. *Bioorg Med Chem* 2005, *13*, 5195–5205.

32. He, J., Giusti, M.M., Anthocyanins: Natural colorants with health-promoting properties. *Annu Rev Food Sci Technol* 2010, *1*, 163–187.

33. Kamonpatana, K., Giusti, M., Chitchumroonchokchai, C., MorenoCruz, M., Riedl, K., Kumar, P., Failla, M., Susceptibility of anthocyanins to *ex vivo* degradation in human saliva. *Food Chem* 2012, *135*, 738–747.

34. Kay, C.D., Mazza, G., Holub, B.J., Wang, J., Anthocyanin metabolites in human urine and serum. *Br J Nutr* 2004, *91*, 933–942.

35. Bub, A., Watzl, B., Heeb, D., Rechkemmer, G., Briviba, K., Malvidin-3-glucoside bioavailability in humans after ingestion of red wine, dealcoholized red wine and red grape juice. *Eur J Nutr* 2001, *40*, 113–120.

36. Ichiyanagi, T., Shida, Y., Rahman, M.M., Hatano, Y., Matsumoto, H., Hirayama, M., Konishi, T., Metabolic pathway of cyanidin-3-O-β-D-glucopyranoside in rats. *J Agric Food Chem* 2005, *53*, 145–150.

37. Tian, Q., Giusti, M.M., Stoner, G.D., Schwartz, S.J., Urinary excretion of black raspberry (*Rubus occidentalis*) anthocyanins and their metabolites. *J Agric Food Chem* 2006, *54*, 1467–1472.

38. Wu, X., Pittman, H.E., Prior, R.L., Pelargonidin is absorbed and metabolized differently than cyanidin after marionberry consumption in pigs. *J Nutr* 2004, *134*, 2603–2610.

39. Scalbert, A., Williamson, G., Dietary intake and bioavailability of polyphenols. *J Nutr* 2000, *130*, 2073S–2085S.

40. Prior, R.L., Wu, X., Anthocyanins: Structural characteristics that result in unique metabolic patterns and biological activities. *Free Radic Res* 2006, *40*, 1014–1028.

41. Milbury, P.E., Kalt, W., Xenobiotic metabolism and berry flavonoid transport across the blood–brain barrier. *J Agric Food Chem* 2010, *58*, 3950–3956.

42. Carlton, P.S., Kresty, L.A., Siglin, J.C., Morse, M.A., Lu, J., Morgan, C., Stoner, G.D., Inhibition of *N*-nitrosomethylbenzylamine-induced tumorigenesis in the rat esophagus by dietary freeze-dried strawberries. *Carcinogenesis* 2001, *22*, 441–446.

43. Aziz, R.M., Nines, R., Rodrigo, K., Harris, K., Hudson, T., Gupta, A., Morse, M., Carlton, P., Stoner, G.D., The effect of freeze-dried blueberries on *N*-nitrosomethylbenzylamine tumorigenesis in the rat esophagus. *Pharmaceut Biol* 2002, *40(suppl.)*, 43–49.

44. Carlton, P.S., Kresty, L.A., Stoner, G.D., Failure of dietary lyophilized strawberries to inhibit 4-(methylnitrosamino)-1-(3-pyridyl)-1-butanone and benzo[a]pyrene-induced lung tumorigenesis in strain A/J mice. *Cancer Lett* 2000, *159*, 113–117.

45. McGhie, T.K., Ainge, G.D., Barnett, L.E., Cooney, J.M., Jensen, D.J., Anthocyanin glycosides from berry fruit are absorbed and excreted unmetabolized by both humans and rats. *J Agric Food Chem* 2003, *51*, 4539–4548.

46. Rechner, A.R., Kuhnle, G., Hu, H., Roedig-Penman, A., van den Braak, M.H., Moore, K.P., Rice-Evans, C.A., The metabolism of dietary polyphenols and the relevance to circulating levels of conjugated metabolites. *Free Radic Res* 2002, *36*, 1229–1241.

47. Garcia-Alonso, M., Minihane, A.M., Rimbach, G., Rivas-Gonzalo, J.C., de Pascual-Teresa, S., Red wine anthocyanins are rapidly absorbed in humans and affect monocyte chemoattractant protein 1 levels and antioxidant capacity of plasma. *J Nutr Biochem* 2009, *20*, 521–529.

48. Borges, G., Roowi, S., Rouanet, J.-M., Duthie, G.G., Lean, M.E.J., Crozier, A., The bioavailability of raspberry anthocyanins and ellagitannins in rats. *Mol Nutr Food Res* 2007, *51*, 714–725.

49. Cao, G., Muccitelli, H.U., Sánchez-Moreno, C., Prior, R.L., Anthocyanins are absorbed in glycated forms in elderly women: A pharmacokinetic study. *Am J Clin Nutr* 2001, *73*, 920–926.

50. Nielsen, I.L., Dragsted, L.O., Ravn-Haren, G., Freese, R., Rasmussen, S.E., Absorption and excretion of black currant anthocyanins in humans and Watanabe heritable hyperlipidemic rabbits. *J Agric Food Chem* 2003, *51*, 2813–2820.

51. Charron, C.S., Kurilich, A.C., Clevidence, B.A., Simon, P.W., Harrison, D.J., Britz, S.J., Baer, D.J., Novotny, J.A., Bioavailability of anthocyanins from purple carrot juice: Effects of acylation and plant matrix. *J Agric Food Chem* 2009, *57*, 1226–1230.

52. Nemeth, K., Plumb, G.W., Berrin, J.G., Juge, N., Jacob, R., Naim, H.Y., Williamson, G., Swallow, D.M., Kroon, P.A., Deglycosylation by small intestinal epithelial cell beta-glucosidases is a critical step in the absorption and metabolism of dietary flavonoid glycosides in humans. *Eur J Nutr* 2003, *42*, 29–42.

53. He, J., Wallace, T.C., Keatley, K.E., Failla, M.L., Giusti, M.M., Stability of black raspberry anthocyanins in the digestive tract lumen and transport efficiency into gastric and small intestinal tissues in the rat. *J Agric Food Chem* 2009, *57*, 3141–3148.

54. He, J., Giusti, M.M., High-purity isolation of anthocyanins mixtures from fruits and vegetables—A novel solid-phase extraction method using mixed mode cation-exchange chromatography. *J Chromatogr A* 2011, *1218*, 7914–7922.

55. Kraemer-Schafhalter, A., Fuchs, H., Pfannhauser, W., Solid-phase extraction (SPE)—A comparison of 16 materials for the purification of anthocyanins from *Aronia melanocarpa* var Nero. *J Sci Food Agric* 1998, *78*, 435–440.

56. Tsuda, T., Horio, F., Osawa, T., Absorption and metabolism of cyanidin 3-*O*-β-D-glucoside in rats. *FEBS Lett* 1999, *449*, 179–182.

57. Cooke, D.N., Thomasset, S., Boocock, D.J., Schwarz, M., Winterhalter, P., Steward, W.P., Gescher, A.J., Marczylo, T.H., Development of analyses by high-performance liquid chromatography and liquid chromatography/tandem mass spectrometry of bilberry (*Vaccinium myrtilus*) anthocyanins in human plasma and urine. *J Agric Food Chem* 2006, *54*, 7009–7013.

58. Ling, Y., Ren, C., Mallery, S.R., Ugalde, C.M., Pei, P., Saradhi, U.V.R.V., Stoner, G.D., Chan, K.K., Liu, Z., A rapid and sensitive LC–MS/MS method for quantification of four anthocyanins and its application in a clinical pharmacology study of a bioadhesive black raspberry gel. *J Chromatogr B* 2009, *877*, 4027–4034.

59. Rodriguez-Saona, L.E., Wrolstad, R.E., Extraction, isolation, and purification of anthocyanins. In *Current Protocols in Food Analytical Chemistry*, 1st ed., Wrolstad, R.E., Acree, T.E., An, H., Decker, E.A., Penner, M.H., Reid, D.S., Schwartz, S.J., Shoemaker, C.F., Sporns, P., eds. New York, NY: John Wiley & Sons, Inc., 2001, vol. 1, pp. F1.1.1–F1.1.11.

60. Rechner, A.R., Kuhnle, G., Bremner, P., Hubbard, G.P., Moore, K.P., Rice-Evans, C.A., The metabolic fate of dietary polyphenols in humans. *Free Radic Biol Med* 2002, *33*, 220–235.

61. Woodward, G., Kroon, P., Cassidy, A., Kay, C., Anthocyanin stability and recovery: Implications for the analysis of clinical and experimental samples. *J Agric Food Chem* 2009, *57*, 5271–5278.

62. Masqué, N., Marcé, R.M., Borrull, F., Molecularly imprinted polymers: New tailor-made materials for selective solid-phase extraction. *TrAC Trends Anal Chem* 2001, *20*, 477–486.

63. Strack, D., Wray, V., Anthocyanins. In *Methods in Plant Biochemistry (vol. I. Plant Phenolics)*, Dey, P.M., Harborne, J.B., eds. San Diego, CA: Academic Press, 1989.

64. Manley, C.H., Shubiak, P., High pressure liquid chromatography of anthocyanins. *Can Inst Food Sci Technol* 1975, *8*, 33–39.

65. Strack, D., Wray, V., The anthocyanins. In *The Flavonoids: Advances in Research Since 1986*, Harborne, J.B., ed. London: Chapman & Hall, 1994.

66. Hong, V., Wrolstad, R.E., Use of HPLC separation/photodiode array detection for characterization of anthocyanins. *J Agric Food Chem* 1990, *38*, 708–715.

67. Hong, V., Wrolstad, R.E., Characterization of anthocyanin-containing colorants and fruit juices by HPLC/photodiode array detection. *J Agric Food Chem* 1990, *38*, 698–708.

68. Hale, M.L., Francis, F.J., Fagerson, I.S., Detection of endocyanin in cranberry juice cocktail by HPLC anthocyanin profile. *J Food Sci* 1986, *51*, 1511–1513.

69. Wrolstad, R.E., Hong, V., Boyles, M.J., Durst, R.W., Use of anthocyanin pigment analysis for detecting adulteration in fruit juices. In *Methods to Detect Adulteration in Fruit Juice and Beverages, vol. I*, Nagy, S., Wade, R.L., eds. Auburndale, FL: AgScience Inc., 1995.

70. Andersen, Ø.M., Chromatographic separation of anthocyanins in cowberry (lingonberry) *Vaccinium vites-idaea* L. *J Food Sci* 1985, *50*, 1230–1232.

71. Yi, W., Akoh, C.C., Fischer, J., Krewer, G., Absorption of anthocyanins from blueberry extracts by caco-2 human intestinal cell monolayers. *J Agric Food Chem* 2006, *54*, 5651–5658.

72. Youdim, K.A., Martin, A., Joseph, J.A., Incorporation of the elderberry anthocyanins by endothelial cells increases protection against oxidative stress. *Free Radic Biol Med* 2000, *29*, 51–60.

73. Mullen, W., Edwards, C.A., Serafini, M., Crozier, A., Bioavailability of pelargonidin-3-*O*-glucoside and its metabolites in humans following the ingestion of strawberries with and without cream. *J Agric Food Chem* 2008, *56*, 713–719.

74. Wiczkowski, W., Romaszko, E., Piskula, M.K., Bioavailability of cyanidin glycosides from natural chokeberry (*Aronia melanocarpa*) juice with dietary-relevant dose of anthocyanins in humans. *J Agric Food Chem* 2010, *58*, 12130–12136.

75. Milbury, P.E., Cao, G., Prior, R.L., Blumberg, J., Bioavailability of elderberry anthocyanins. *Mech Ageing Dev* 2002, *123*, 997–1006.

76. He, J., Magnuson, B.A., Lala, G., Tian, Q., Schwartz, S.J., Giusti, M.M., Intact anthocyanins and metabolites in rat urine and plasma after 3 months of anthocyanin supplementation. *Nutr Cancer* 2006, *54*, 3–12.

77. Felgines, C., Texier, O., Besson, C., Vitaglione, P., Lamaison, J.L., Fogliano, V., Scalbert, A., Vanella, L., Galvano, F., Influence of glucose on cyanidin 3-glucoside absorption in rats. *Mol Nutr Food Res* 2008, *52*, 959–964.

78. Vanzo, A., Terdoslavich, M., Brandoni, A., Torres, A.M., Vrhovsek, U., Passamonti, S., Uptake of grape anthocyanins into the rat kidney and the involvement of bilitranslocase. *Mol Nutr Food Res* 2008, *52*, 1106–1116.

79. Chirinos, R., Campos, D., Betalleluz, I., Giusti, M.M., Schwartz, S.J., Tian, Q., Pedreschi, R., Larondelle, Y., High-performance liquid chromatography with photodiode array detection (HPLC–DAD)/HPLC–mass spectrometry (MS) profiling of anthocyanins from Andean mashua tubers (*Tropaeolum tuberosum* Ruíz and Pavón) and their contribution to the overall antioxidant activity. *J Agric Food Chem* 2006, *54*, 7089–7097.

80. Tian, Q., Giusti, M.M., Stoner, G.D., Schwartz, S.J., Screening for anthocyanins using high-performance liquid chromatography coupled to electrospray ionization tandem mass spectrometry with precursor–ion analysis, product–ion analysis, common-neutral-loss analysis, and selected reaction monitoring. *J Chromatogr A* 2005, *1091*, 72–82.

81. Wu, X., Gu, L., Prior, R.L., McKay, S., Characterization of anthocyanins and proanthocyanidins in some cultivars of *Ribes, Aronia,* and *Sambucus* and their antioxidant capacity. *J Agric Food Chem* 2004, *52*, 7846–7856.

82. Hvattum, E., Determination of phenolic compounds in rose hip (*Rosa canina*) using liquid chromatography coupled to electrospray ionisation tandem mass spectrometry and diode-array detection. *Rapid Commun Mass Spectrom* 2002, *16*, 655–662.

83. Montoro, P., Tuberoso, C.I.G., Perrone, A., Piacente, S., Cabras, P., Pizza, C., Characterisation by liquid chromatography–electrospray tandem mass spectrometry of anthocyanins in extracts of *Myrtus communis* L. berries used for the preparation of *Myrtle liqueur*. *J Chromatogr A* 2006, *1112*, 232–240.

84. Ohnishi, R., Ito, H., Kasajima, N., Kaneda, M., Kariyama, R., Kumon, H., Hatano, T., Yoshida, T., Urinary excretion of anthocyanins in humans after cranberry juice ingestion. *Biosci Biotechnol Biochem* 2006, *70*, 1681–1687.

85. Oki, T., Suda, I., Terahara, N., Sato, M., Hatakeyama, M., Determination of acylated anthocyanin in human urine after ingesting a purple-fleshed sweet potato beverage with various contents of anthocyanin by LC–ESI–MS/MS. *Biosci Biotechnol Biochem* 2006, *70*, 2540–2543.

86. Goto, T., Structure, stability and color variation of natural anthocyanins. *Fortsch Chem Org Natur* 1987, *52*, 113–158.

87. Saito, N., Yokoi, M., Yamaji, M., Honda, T., Cyanidin 3-*p*-coumaroylglucoside in *Camellia* species and cultivars. *Phytochemistry* 1987, *26*, 2761–2762.

88. Kazuya, H., Hiroko, S., Atsuko, S., Akio, T., Photo-isomerization of the *Nasunin*, the major eggplant anthocyanins. *Food Sci Technol Int,* Tokyo, 1998, *4*, 25–28.

89. Strack, D., Busch, E., Wray, V., Grotjahn, L., Klein, E., Cyanidin 3-oxalylglucoside in orchids. *J Biosci* 1986, *41*, 707–711.

90. Choung, M.-G., Baek, I.-Y., Kang, S.-T., Han, W.-Y., Shin, D.-C., Moon, H.-P., Kang, K.-H., Isolation and determination of anthocyanins in seed coats of black soybean (*Glycine max* (L.) Merr.). *J Agric Food Chem* 2001, *49*, 5848–5851.

91. Giusti, M.M., Rodríguez-Saona, L.E., Griffin, D., Wrolstad, R.E., Electrospray and tandem mass spectroscopy as tools for anthocyanin characterization. *J Agric Food Chem* 1999, *47*, 4657–4664.

92. Covey, T., Analytical Characteristics of the electrospray ionization process. In *Biochemical and Biotechnological Applications of Electrospray Ionization Mass Spectrometry, American Chemical Society* 1996, *619*, 21–59.

93. Black, G.E., Fox, A., Liquid chromatography with electrospray ionization tandem mass spectrometry: Profiling carbohydrates in whole bacterial cell hydrolysates. In *Biochemical and Biotechnological Applications of Electrospray Ionization Mass Spectrometry*, Snyder, A.P., ed. ACS Symposium Series: 1995.

94. Hutton, T., Major, H.J., Characterizing biomolecules by electrospray ionization as spectrometry coupled to liquid chromatography and capillary electrophoresis. *Biochem Soc Trans* 1995, *23*, 924–927.

95. Sagesser, M., Deinzer, M., HPLC–ion spray–tandem mass spectrometry of flavonol glycosides in hops. *J Am Soc Brew Chem* 1996, *54*, 129–134.

96. Skrede, G., Wrolstad, R.E., Durst, R.W., Changes in anthocyanins and polyphenolics during juice processing of Highbush Blueberries (*Vaccinium corymbosum* L.). *J Food Sci* 2000, *65*, 357–364.

97. Pati, S., Losito, I., Gambacorta, G., Notte, E.L., Palmisano, F., Zambonin, P.G., Simultaneous separation and identification of oligomeric procyanidins and anthocyanin-derived pigments in raw red wine by HPLC-UV-ESI-MSn. *J Mass Spectrom* 2006, *41*, 861–871.

98. Jing, P., Zhao, S.-J., Ruan, S.-Y., Xie, Z.-H., Dong, Y., Yu, L., Anthocyanin and glucosinolate occurrences in the roots of Chinese red radish (*Raphanus sativus* L.), and their stability to heat and pH. *Food Chem* 2012, *133*, 1569–1576.

99. Mazzuca, P., Ferranti, P., Picariello, G., Chianese, L., Addeo, F., Mass spectrometry in the study of anthocyanins and their derivatives: Differentiation of *Vitis vinifera* and hybrid grapes by liquid chromatography, electrospray ionization mass spectrometry and tandem mass spectrometry. *J Mass Spectrom* 2005, *40*, 83–90.

100. Jennings, K.R., MS/MS instrumentation. In *Applications of Modern Mass Spectrometry in Plant Science Research*, Newton, R.P., Walton, T.J., eds. New York: Clarendon Press, 1996.

101. Lawson, G., Ostah, N., Woolley, J.G., MS/MS studies of tropane alkaloids: Detection and determination of structure. In *Applications of Modern Mass Spectrometry in*

Plant Science Research, Newton, R.P., Walton, T.J., eds. New York: Clarendon Press, 1996.

102. McLafferty, F.W., Turecek, F., *Interpretation of Mass Spectra*. 4 ed. Sausalito, CA: University Science Books, 1993.

103. Carkeet, C., Clevidence, B.A., Novotny, J.A., Anthocyanin excretion by humans increases linearly with increasing strawberry dose. *J Nutr* 2008, *138*, 897–902.

104. Truong, V.-D., Deighton, N., Thompson, R.T., McFeeters, R.F., Dean, L.O., Pecota, K.V., Yencho, G.C., Characterization of anthocyanins and anthocyanidins in purple-fleshed sweetpotatoes by HPLC-DAD/ESI-MS/MS. *J Agric Food Chem* 2009, *58*, 404–410.

105. Wu, X., Prior, R.L., Systematic identification and characterization of anthocyanins by HPLC-ESI-MS/MS in common foods in the United States: Fruits and berries. *J Agric Food Chem* 2005, *53*, 2589–2599.

106. Andersen, O.M., Fossen, T., Characterization of anthocyanins by NMR. In *Handbook of Food Analytical Chemistry: Pigments, Colorants, Flavors, Texture, and Bioactive Food Components*, Wrolstad, R.E., ed. New York: John Wiley & Sons, 2005.

107. Goto, T., Takase, S., Kondo, T., PMR spectra of natural acylated anthocyanins determination of stereostructure of awobanin, shisonin and violanin (*Commelina communis, Viola tricolor, Perilla orimoides*). *Tetrahedron Lett* 1978, *27*, 2413–2416.

108. Agrawal, P.K., NMR spectroscopy in the structural elucidation of oligosaccharides and glycosides. *Phytochemistry* 1992, *31*, 3307–3330.

109. Keepers, J.W., James, T.L., A theoretical study of distance determinations from NMR. Two-dimensional nuclear overhauser effect spectra. *J Magn Reson (1969)* 1984, *57*, 404–426.

110. Kemp, W., *Organic Spectroscopy*. 2 ed.; Macmillan: the University of Michigan, 1987.

111. Figueiredo, P., Elhabiri, M., Saito, N., Brouillard, R., Anthocyanin intramolecular interactions: A new mathematical approach to account for the remarkable colorant properties of the pigments extracted from *Matthiola incana*. *J Am Chem Soc* 1996, *118*, 4788–4793.

112. Giusti, M.M., Ghanadan, H., Wrolstad, R.E., Elucidation of the structure and conformation of red radish (*Raphanus sativus*) anthocyanins using one- and two-dimensional nuclear magnetic resonance techniques. *J Agric Food Chem* 1998, *46*, 4858–4863.

113. Jordheim, M., Fossen, T., Andersen, Ø.M., Preparative isolation and NMR characterization of carboxypyranoanthocyanins. *J Agric Food Chem* 2006, *54*, 3572–3577.

114. Swinny, E.E., Bloor, S.J., Wong, H., ^{1}H and ^{13}C NMR assignments for the 3-deoxyanthocyanins, luteolinidin-5-glucoside and apigeninidin-5-glucoside. *Magn Reson Chem* 2000, *38*, 1031–1033.

115. Albert, K., Liquid chromatography–nuclear magnetic resonance spectroscopy. *J Chromatogr A* 1999, *856*, 199–211.

116. Welch, C.R., Wu, Q., Simon, J.E., Recent advances in anthocyanin analysis and characterization. *Curr Anal Chem* 2008, *4*, 75–101.

117. Godejohann, M., Preiss, A., Mügge, C., Wünsch, G., Application of on-Line HPLC–^{1}H NMR to environmental samples: Analysis of groundwater near former ammunition plants. *Anal Chem* 1997, *69*, 3832–3837.

118. de Koning, J.A., Hogenboom, A.C., Lacker, T., Strohschein, S., Albert, K., Brinkman, U.A.T., On-line trace enrichment in hyphenated liquid chromatography–nuclear magnetic resonance spectroscopy. *J Chromatogr A* 1998, *813*, 55–61.

119. Acevedo De la Cruz, A., Hilbert, G., Rivière, C., Mengin, V., Ollat, N., Bordenave, L., Decroocq, S. et al., Anthocyanin identification and composition of wild *Vitis* spp. accessions by using LC–MS and LC–NMR. *Anal Chim Acta* 2011, 10.1016/j.aca.2011.11.060.

120. Giusti, M.M., Jing, P., Analysis of pigments and colorants: Analysis of anthocyanins. In *Food Colorants: Chemical and Functional Properties. Chemical & Functional*

Properties of Food Components, Socaciu, C., ed. CRC Press: Boca Raton, FL, 2007, pp. 479–506.

121. Giust, I.M., Atnip, A., Sweeney, C., Rodriguez-Saona, L. In *Rapid Authentication of Fruit Juices by Infrared Spectroscopic Techniques*, ACS Symposium Series, Washington, DC, 2011, Ebeler, S.E., Takeoka, G.R., Winterhalter, P., eds. Washington, DC: Oxford University Press, 2011, pp 275–299.

122. He, J., Rodriguez-Saona, L., Giusti, M., Midinfrared spectroscopy for juice authentication-rapid differentiation of commercial juices. *J Agric Food Chem* 2007, *55*, 4443–4452.

123. Soriano, A., Pérez-Juan, P.M., Vicario, A., González, J.M., Pérez-Coello, M.S., Determination of anthocyanins in red wine using a newly developed method based on Fourier transform infrared spectroscopy. *Food Chem* 2007, *104*, 1295–1303.

124. Edelmann, A., Diewok, J., Schuster, K.C., Lendl, B., Rapid method for the discrimination of red wine cultivars based on mid-infrared spectroscopy of phenolic wine extracts. *J Agr Food Chem* 2001, *49*, 1139–1145.

125. Andersen, Ø.M., Fossen, T., Torskangerpoll, K., Fossen, A., Hauge, U., Anthocyanin from strawberry (*Fragaria x ananassa*) with the novel aglycone, 5-carboxypyranopelargonidin. *Phytochemistry* 2004, *65*, 405–410.

126. Bakker, J., Timberlake, C.F., Isolation, identification, and characterization of new color-stable anthocyanins occurring in some red wines. *J. Agric. Food Chem.* 1997, *45*, 35–43.

127. He, et al., Analysis of anthocyanins in rat intestinal contents—Impact of anthocyanin chemical structure on fecal excretion. *J Agric Food Chem* 2005, *53*, 2859–2866.

128. He, et al., Intact anthocyanins and metabolites in rat urine and plasma after 3 months of anthocyanin supplementation. *Nutr Cancer* 2006, *54*, 3–12.

5 Antioxidant Activities of Anthocyanins

Xianli Wu

CONTENTS

5.1 INTRODUCTION

Anthocyanins (ACNs) are water-soluble plant secondary metabolites responsible for the blue, purple, and red color of many plant tissues. They occur primarily as glycosides of their respective aglycone anthocyanidin chromophores (Figure 5.1), with the sugar moiety mainly attached at the 3-position on the C-ring or the 5, 7-position on the A-ring. Glycosylation at 3', 4', 5'-position of the B-ring, although very rare, has also been observed [1]. Glucose, galactose, arabinose, rhamnose, and xylose are the most common sugars that are attached to anthocyanidins as mono-, di-, or trisaccharide forms. Aglycones are rarely found in fresh plant materials. There are about 25 different anthocyanidins found in nature [2], whereas only six of them, cyanidin, delphinidin, petunidin, peonidin, pelargonidin, and malvidin, are ubiquitously distributed (Figure 5.1). The differences in chemical structure of these six common anthocyanidins occur at the 3' and 5' positions of the B-ring

141

Six common anthocyanidins

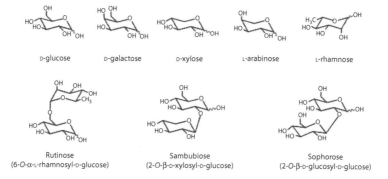

Anthocyanidin	R_1	R_2	R_3	MW
Pelargonidin	H	OH	H	271
Cyanidin	OH	OH	H	287
Delphinidin	OH	OH	OH	303
Peonidin	OMe	OH	H	301
Petunidin	OMe	OH	OH	317
Malvidin	OMe	OH	OMe	331

Common mono- and disaccharides

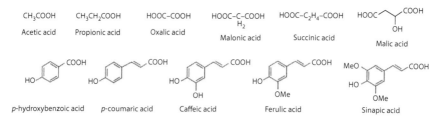

| D-glucose | D-galactose | D-xylose | L-arabinose | L-rhamnose |

| Rutinose | Sambubiose | Sophorose |
| (6-O-α-L-rhamnosyl-D-glucose) | (2-O-β-D-xylosyl-D-glucose) | (2-O-β-D-glucosyl-D-glucose) |

Common aliphatic and aromatic acids

CH_3COOH CH_3CH_2COOH $HOOC-COOH$ $HOOC-C-COOH$ $HOOC-C_2H_4-COOH$
                                                                                              H_2

Acetic acid   Propionic acid   Oxalic acid   Malonic acid   Succinic acid   Malic acid

p-hydroxybenzoic acid   p-coumaric acid   Caffeic acid   Ferulic acid   Sinapic acid

FIGURE 5.1 Common ACN structures. Sugar moieties are generally on position 3 of the C-ring. Selected sugars that commonly occur in ACN structures.

(Figure 5.1). The sugar moieties may also be acylated by a range of aromatic or aliphatic acids. Common acylating agents are cinnamic acids (Figure 5.1). For most naturally occurring ACNs, their molecular weights (MW) are within the range of 400–1200, depending on the type of anthocyanidin and the type/number of sugar and acylated groups. Two ACNs, which represent simple or complex ACNs isolated from strawberry or red cabbages, are shown in Figure 5.2. 3-Deoxyanthocyanins are another group of anthocyanidins that do not have a hydroxyl group on 3-position

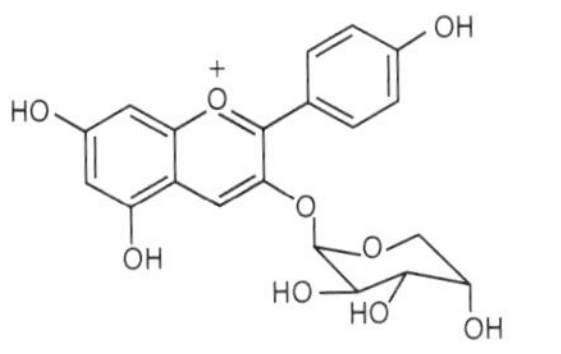

Pelargonidin 3-arabinoside (MW = 403)

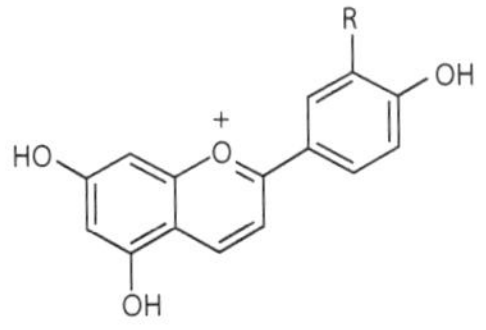

3-Deoxyanthocyanin

Apigeninidin   R = H (MW = 255)
Luteolinidin   R = OH (MW = 271)

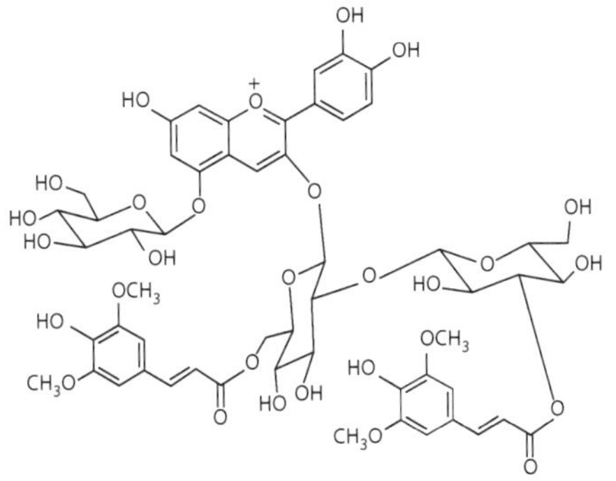

Cyanidin 3-(sinapoyl)diglucoside-5-(sinapoyl)glucoside (MW = 1185)

FIGURE 5.2   Representatives of simple and complicated ACNs and 3-deoxyanthocyanidins.

(Figure 5.2). Unlike other anthocyanidins with a 3-hydroxyl group, they can occur naturally in the aglycone forms in limited dietary sources such as certain varieties of sorghum [3,4]. They were also reported to be very stable in acidic solutions compared to other anthocyanidins. Over 700 naturally occurring ACNs have been reported [2] and they are known to vary in: (1) the number and position of hydroxyl and methoxyl groups on the basic anthocyanidin skeleton; (2) the identity, number, and positions at which the sugars are attached; and (3) the extent of sugar acylation and the identity of the acylated agent [5].

ACNs play an important role in plant physiology in both pollination and seed dispersal [6]. Because of their intense color, ACNs are also regarded as potential candidates for natural colorants in the food industry [7,8]. However, currently attention has been paid on their antioxidant activities and possible health effects and disease prevention properties. ACNs have long been considered as strong antioxidants [9,10]. They effectively quench free radicals and terminate the chain reaction, which have been demonstrated in a number of *in vitro* assays [11,12]. The protective effects of ACNs on oxidative stress-induced damage have also been shown in various *in vivo* models [13–16]. Most recently, emerging evidence indicated that ACNs may be able to regulate redox-sensitive signaling molecules [17,18]. ACNs exert a wide range of health benefits including prevention of cardiovascular diseases (CVDs), anticancer, and improving cognition functions, which are believed to occur, at least partly through the antioxidant mechanisms [19,20].

The aim of chapter is to summarize the current knowledge on antioxidant activities of ACNs. To facilitate the discussion, a brief overview of oxidative stress, its role in disease etiology and the action of antioxidant was provided in the first part of the chapter. The antioxidant activities of ACNs were then discussed based on the testing models and targets, with emphasis on the emerging new research areas. Finally, the disease preventive effects of ACNs, which were more or less associated with their antioxidant activities, were also discussed.

5.2 REACTIVE OXIDANTS, ANTIOXIDANTS, AND DISEASES

All animals need O_2 for production of energy in mitochondria [21]. When cells utilize oxygen to generate energy, free radicals are created as a consequence of ATP (adenosine triphosphate) production by mitochondria. A free radical is any species containing unpaired electron. With unpaired electron in its atomic orbital, the free radical is highly reactive. There are many types of free radicals and non-radical reactive species in living system [21], nonetheless, reactive oxygen species (ROS) and reactive nitrogen species (RNS) are the primary reactive species in the human body (Figure 5.3). These species play a dual role as both deleterious and beneficial agents. Thus maintaining the "redox balance" or "homeostasis" is considered as a fundamental theme of aerobic life. If this balance shifts in favor of the oxidants, a status call oxidative stress is generated. Oxidative stress is often defined as "a serious imbalance between the generation of reactive oxygen species and antioxidant protection in favor of the former, causing excessive oxidative damage" [22, p. 125]. Excess ROS and RNS can damage cellular lipids, proteins, or DNA to impair their normal function. Thus, oxidative stress has been implicated in the etiology of many chronic and degenerative diseases, and it was also believed to play a major role in the aging process [23–27] (Figure 5.4). According to Valko et al. [25], these diseases fall into two groups: (a) diseases characterized by pro-oxidants shifting the thiol/disulfide redox state and impairing glucose tolerance—the so-called "mitochondrial oxidative stress" conditions (cancer and diabetes mellitus); (b) diseases characterized by "inflammatory oxidative conditions" and enhanced activity of either NAD(P)H oxidase (leading to atherosclerosis and chronic inflammation) or xanthine oxidase-induced formation of ROS (implicated in ischemia and reperfusion injury). The process of aging is to a large extent due to the damaging consequence of free radical action (lipid peroxidation, DNA damage, protein oxidation).

The implication of oxidative stress in the etiology of chronic and degenerative diseases suggests that antioxidants, both endogenously and exogenously, are necessary to protect cellular components from oxidative damage. The term "antioxidant" has

A. Radicals (superoxide anion and free radical):

Superoxide: $O_2^{\cdot}$	Alkoxyl radical: $LO^{\cdot}$
Hydroxyl radical: $OH^{\cdot}$	Nitric oxide: $^{\cdot}NO$
Peroxyl radical: $RO_2^{\cdot}$	Nritrogen dioxide: $NO_2^{\cdot}$

B. Nonradicals (molecule and ion):

Singlet Oxygen: $^1O_2$	Lipid hydroperoxide: $LOOH$
Hydrogen peroxide: H_2O_2	Hypochlorite: $OCI-$
Hypochlorite: $HOCI$	Hydroxide anion: $OH-$
Peroxynitrite: $ONOO^{-}$	

FIGURE 5.3 Common ROS and RNS in human body.

Pathological roles of ROS

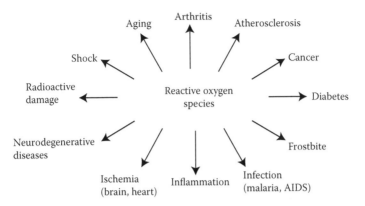

ROS attack lipids, sugars, proteins and DNA induce their oxidation, which may result in oxidative damage such as membrane dysfunction, protein modification, enzyme inactivation, and break of DNA strands and modification of its bases.

FIGURE 5.4 Oxidative stress and human chronic diseases.

different connotations to different audiences and there is thus far no authentic definition [28]. The early well-known definition was given by Halliwell and Guteridge [29] as "any substance that when present at low concentrations compared with those of an oxidized substrate significantly delays or prevents oxidation of that substrate" (p. 125). Later the same authors redefined antioxidant as "any substance that delays, prevents, or removes oxidative damage to a target molecule" [30, p. 1147]. And most recently, it was defined again by Halliwell as "a molecule that protects a biological target against oxidative damage" [22, p. 125]. The definition evolved with broader and deeper the knowledge we have gained toward better understanding the true meaning of antioxidant. The endogenous antioxidant defense system is believed to play a primary role in modulating the redox balance and reducing oxidative stress in the human body. It can be largely divided into three groups: chain-breaking antioxidant, antioxidant enzymes, and metal binding proteins [31,32]. Some of the key components of the endogenous antioxidant defense system are summarized in Figure 5.5. The exogenous antioxidants, mostly through dietary sources, may act as important supplements to endogenous antioxidants. In the last two decades or so, intense research efforts have been made to identify antioxidants from the human diet that are capable of reducing oxidative stress. Of all the exogenous dietary antioxidants, flavonoids have received primary attention. As a homogenous group, there are over 8000 flavonoids being described. They are ubiquitously distributed in the plant kingdom, particularly abundant in fruits and vegetables [33]. Flavonoids are part of our diet with a daily intake estimated to be 1 g/day, higher than any other dietary phytochemicals [34]. Epidemiological studies indicated that a high intake of fruit and vegetables rich in flavonoids was linked to lower incidence of various human chronic diseases, which were thought to be at least partly related to their antioxidant activities [35–39].

Antioxidant Defense System

Enzyme antioxidants	Metal binding proteins
• Superoxide dismutases	• Transferrin
• Catalase	• Ferritin
• Glutathione peroxidase	• Lactoferrin
• Caeruloplasmin	

Chain breaking antioxidants

• Tocopherols	• Ascorbate
• Ubiquinol	• Urate
• Carotenoids	

FIGURE 5.5   Endogenous antioxidant defense system.

As an important subgroup of flavonoid, ACNs have long been considered as strong dietary antioxidants [9,10].

On the basis of recent findings, dietary flavonoids including ACNs may exert their antioxidant activities and attenuate oxidative stress through either direct or indirect mechanisms. Early studies focused on their direct reaction with free radicals—direct scavenging free radicals [40] or chelating metal ions [41]. However, recent studies suggested that in addition to the ability in directly quenching free radicals, flavonoids were able to stimulate the activities of important antioxidant enzymes and regulate key redox cell signaling such as nuclear factors erythroid-2-related factor (Nrf2) and nuclear factor kappa B (NF-κB) pathway [42], thus exerting antioxidant activity through indirect mechanisms.

5.3   ANTIOXIDANT CAPACITY OF ACNS: DIRECT RADICAL SCAVENGING/REDUCING CAPACITY

The early studies of antioxidant activities of ACNs focused on their direct free radical scavenging effects. The idea was probably adopted from oxidation control and spoilage retardation by synthetic antioxidants such as tert-butylhydroquinone (TBHQ) and butylated hydroxytoluene (BHT) in the food industry [43]. So it was not surprising to see the comparisons of antioxidant capacities between ACNs and TBHQ and/or BHT in early publications [44–46]. Various chemical-based *in vitro* assays were developed and used to evaluate the direct antioxidant activities of dietary flavonoids [47]. Furthermore, assessment of direct scavenging activities of ACNs and ACN-rich foods could also be conducted by cell-based assays or *in vivo* models, which were considered to be more biologically relevant.

Another direct antioxidant effect of ACNs is through chelating metal ions such as iron (Fe) and copper (Cu). These redox active metals undergo redox cycling reactions and possess the ability to produce reactive radicals such as superoxide anion radical and nitric oxide in biological systems. A major aspect of *in vivo* antioxidant defense is to minimize the availability of transition metal ions that are catalytically available for Fenton reaction and other free-radical reactions [48]. Many low-molecular weight antioxidants including ACNs are capable of chelating metal ions thus reducing their catalytic activity to form ROS [49].

5.3.1 CHEMICAL-BASED ASSAYS

A broad variety of chemical-based assays have been developed for assessing antioxidant capacities of dietary antioxidants [47], and many of which, including the most commonly used ORAC [50], DPPH [51], ABTS [52], TEAC [53], and TRAP [54], have been widely applied in evaluating antioxidant activities of ACNs and/ or ACN-rich foods. In these methods, the capacity of antioxidant has been measured by competition method using a probe as a reference compound. The capacity of scavenging free radicals has been assessed by the extent of suppression of probe consumption, measured by UV/visible absorption, fluorescence, chemiluminescence, and EPR spectroscopy. Owing to their chemical structures (Figure 5.1) [55,56], ACNs are excellent hydrogen donors and thus show high antioxidant capacity in the above-mentioned chemical-based bioassays involving hydrogen transfer reactions (HAT) [47,57]:

$$ROO^{\bullet} + AH \rightarrow ROOH + A^{\bullet}$$

However, according to a recent press release by USDA Agriculture Research Service (http://www.ars.usda.gov/services/docs.htm?docid=15866), these assays are based on discrete underlying mechanisms that use different radical or oxidant sources and therefore generate distinct values and cannot be compared directly. Another major criticism for chemical-based assay is that they lack biological relevance, since they are basically test tube assays and many of the free radicals used in these assays do not exist in the human body.

The elevated levels of low-density lipoprotein (LDL) cholesterol constitute a major risk factor for atherosclerotic disease. Oxidatively modified LDL (Ox-LDL) has been positively correlated with the progression of atherosclerosis [58,59]. The concept of inhibiting LDL oxidation by antioxidants was used to develop the LDL oxidation assay. This assay was usually performed on isolated LDL oxidized by copper, and lipid oxidation products such as malondialdehyde (MDA), conjugated diene, lipid peroxides, and thiobarbituric acid-reactive substances (TBARS) were measured as an index of oxidation. The antioxidant activity of the six common anthocyanidins and their glycosidic forms was evaluated in three lipid-containing models including copper-induced human LDL [10]. Most ACNs and their aglycones acted as strong antioxidants in reducing LDL oxidation and showed an activity comparable to the well-known antioxidants such as α-tocopherol, Trolox, catechin, and quercetin. The aglycons showed in general higher activities than the glycosides. Studies on other ACNs, such as cyanidine 3-O-β-glucoside [60]; or ACN-rich fractions from *Hibiscus* ACN-rich extract [61], mulberry ACN extract [62], showed similar effects, indicating that they had strong protection for LDL oxidation.

Delphinidin-3-(p-coumaroyl)-rutinoside-5-glucoside (nasunin) is an ACN isolated as purple-colored crystals from eggplant peels. A spectrophotometric study showed that nasunin formed an iron complex with a molar ratio of nasunin:Fe^{3+} of 2:1.

Therefore, it suggested that hydroxyl radical scavenging by this ACN was not due to direct radical scavenging but inhibition of OH˙ generation by chelating iron [63].

5.3.2 Cell-Based Assays

Since chemical-based assays have been consistently criticized for their lack of biological relevance, cell-based assays were developed to evaluate protective effects of dietary antioxidants including ACNs. The advantage of using cultured cells is that various different stressors and cell types including model systems for some specific disease can be used for evaluation of the antioxidant effects [64]. In the last 5 years or so, several cell-based assays were developed and used to assess antioxidant activities of ACNs or ACN-rich extracts/foods. The principle and methodology of these assays were well summarized in a recent review [65].

The cellular antioxidant activity (CAA) assay in HepG2 cell line was used to determine the CAA of 25 fruits commonly consumed in the United States [66]. The data were compared with chemical-based ORAC values. Similar to ORAC data [67], ACN-rich pomegranate and berries (wild blueberry, blackberry, raspberry, and blueberry) had the highest CAA values [68]. Another cell-based assay was developed using erythrocytes (RBC) and polymorphonuclear (PMN) cells [40]. The data were also compared with ORAC values. However, in this case the magnitude of the effect was not directly correlated to the ORAC value, indicating the different mechanisms involved. This assay was further used to evaluate antioxidant properties of a juice blend (JB) containing a mixture of fruits and berries. The phytochemical antioxidants in the JB are primarily in the form of ACNs, predominantly cyanidin 3-rutoside, cyanidin 3-diglycoside, and cyanidin 3-glucoside. The cell-based antioxidant protection of erythrocytes (CAP-e) assay showed that antioxidants in the JB penetrated and protected cells from oxidative damage, whereas polymorphonuclear cells showed reduced formation of ROS [69]. Antioxidant assessment of crude extract and ACN-enriched blackberry extract was determined by both ORAC and cell-based (INT-407 intracellular) assays [70]. The results suggested that the antioxidant activities, assessed either by ORAC and in INT-407 cells, are much higher in ACN-enriched extract, indicating the ability of ACN in blackberry to suppress both peroxyl radical-induced chemical and intracellular oxidation. In another study conducted by the same group [71], the efficacy of crude blackberry ACN protection against peroxyl radical (AAPH)-induced oxidative damage in Caco-2 colon cancer cells was evaluated. The ACN extract suppressed AAPH-initiated Caco-2 intracellular oxidation in a concentration-dependent manner, with an IC_{50} value of 6.5 ± 0.3 mg/mL. The results showed that the antioxidant activity of ACNs principally attributed to cyanidin-3-O-glucoside (C3G) and common to blackberry are effective at inhibiting peroxyl radical-induced apoptosis in cultured Caco-2 cells. Cyanidin-3-rhamnoglucoside (C3R), a major ACN in fresh figs, was evaluated by various *in vitro* assays and correlated with the protection afforded to cultured NIH-3T3 fibroblast cells. C3R inhibited lipid peroxidation from producing peroxyl radicals (ROO˙). C3R also showed a strong chelating activity toward the Fe^{2+} ion. Pretreatment with C3R inhibited proapoptotic processes that were initiated by the oxidation of lysosome membranes in fibroblast cells [72].

In one of the most recent studies [73], the antioxidant activity of ACN fractions from bilberry and blueberry was examined using CAA assay in different cell lines: human colon cancer (Caco-2), human hepatocarcinoma (HepG2), human endothelial (EA.hy926), and rat vascular smooth muscle (A7r5). These cell lines were chosen because they represent standard experimental models for the study of the bioactivity of the extract. The results showed that ACNs had intracellular antioxidant activity if applied at very low concentrations (<1 μg/L), thereby providing a long-sought rationale for their health-protecting effects in spite of their unfavorable pharmacokinetic properties.

When conducting cell-based antioxidant assays, one must be aware that ACNs exist as the flavylium cation at pH < 3, but at pH 3–6 they may exist as a quinoidal-base and at pH 7–8 they may convert into the chalcone [11]. Seeram et al. [45] found that cyanidin glycosides from tart cherries spontaneously degraded to protocatechuic acid, 2,4-dihydroxybenzoic acid, and 2,4,6-trihydroxybenzoic acid in solution at pH 7. Thus, in any cell or tissue culture study at neutral pH, ACNs may degrade.

The physiological relevancy of cell-based assays is upgraded from simple chemical-based assays. However, the results of cell culture experiments with antioxidants strongly depended on the cell type, the cell culture medium, and the absence or the presence of other antioxidants [74]. To quote the author of a recent paper "the fact that the effects of 'antioxidants' such as ascorbate, lycopene, epigallocatechin gallate and several other polyphenols on cells in culture are often artifacts—reflecting the reactions of these compounds with components of cell culture media and/or their rapid decomposition into other bioactive agents in cell culture" [22, p. 129]. A simple one-to-one extrapolation of *in vitro* data to *in vivo* behavior must therefore be misleading: "What happen in a Petri dish or in preclinical assays may not happen in people" [74, p. 166, 75].

5.3.3 *In Vivo* Assessment

Studies on the antioxidant effects in animals and human subjects following consumption of ACNs are not definitive. Much of the early work on ACNs has resulted from studies of concentrated forms of ACNs [76], and the primary outcomes were antioxidant capacity and biomarker changes in plasma, serum, or other biological fluid.

A small but significant increase was observed in plasma hydrophilic and lipophilic antioxidant capacity following the consumption of a single meal of 189 g of blueberries (10 mg ACNs/kg) [77,78]. Others [79] reported an increase in plasma antioxidant capacity (acetone fraction) after the consumption of approximately 1.2 g of ACNs (15 mg ACNs/kg) from blueberry. Matsumoto et al. [80] observed a rapid increase in plasma antioxidant activity, as indicated by monitoring chemiluminescence intensity, after oral administration of black currant ACNs (0.573 mg/kg). What is not known is whether ACNs are accumulated in tissues if consumed over an extended period of time. In another comprehensive study, changes in plasma antioxidant capacity (AOC) following consumption of a single meal of berries/fruits were studied in five clinical trials [15]. Consumption of blueberry in two studies and of mixed grape powder increased hydrophilic AOC. Lipophilic AOC increased following a meal of blueberry containing 12.5 mmol TE AOC. Consumption of 280 g of cherries (4.5 mmol TE AOC) increased plasma L-AOC but not H-AOC. These results suggested that consumption of ACN-rich berries and fruits, such as blueberries and mixed grape, was associated with increased plasma AOC in the postprandial state. Consumption of an energy source of macronutrients containing no antioxidants was associated with a decline in plasma AOC.

Pawlowicz [81] found that ACNs from chokeberry may decrease the generation of autoantibodies to oxidized LDLs in pregnancies complicated by intrauterine growth retardation (IUGR). These results suggested that ACNs could be useful in controlling oxidative stress during pregnancies complicated by IUGR [81].

In rats, oral pretreatment with *Hibiscus* anthocyanins (HAs) (100 and 200 mg/kg) for 5 days before a single dose of t-butyl hydroperoxide (t-BHP) (0.2 mmol/kg, ip) significantly lowered the serum levels of alanine and aspartate aminotransferase, enzyme markers of liver damage, and also reduced oxidative liver damage in rats. Histopathological evaluation of the liver revealed that HAs reduced the incidence of liver lesions including inflammatory, leukocyte infiltration, and necrosis induced by t-BHP in rats. Based on these results, the authors suggested that HAs may play a role in the prevention of oxidative damage in living systems [82]. In another study [83], ACNs obtained from the petals of *H. rosa-sinensis* were shown to prevent carbon tetrachloride(CCl_4)-induced acute liver damage in the rat. Treatment of separate groups of rats with 2.5 mL of 1%, 5%, and 10% ACN extract in 5% aqueous ethanol/kg body weight, 5 days/week for 4 weeks before giving 0.5 mL/kg CCl_4, resulted in significantly less hepatotoxicity than with CCl_4 alone, as measured by serum aspartate- and alanine-aminotransferase activities 18 h after CCl_4 [83].

The decreased food intake and body weight gain, and increased lung weight and atherogenic index observed in rats in which paraquat was used to induce oxidative stress were clearly suppressed by supplementing the paraquot diet with acylated ACNs from red cabbage [84]. Paraquat feeding increased the concentration of TBARS in liver lipids, and decreased the liver triacylglycerol level. These effects tended to be suppressed by supplementing the paraquot diet with acylated ACNs. In addition, catalase activity in the liver mitochondrial fraction was markedly decreased by feeding the paraquat diet; this decrease was partially suppressed by supplementing the paraquat diet with acylated ACNs. An increase in the NADPH-cytochrome-P450-reductase activity in the liver microsome fraction by paraquat was suppressed by supplementing the paraquat diet with acylated ACNs. These results suggest that acylated ACNs from red cabbage acted to prevent oxidative stress *in vivo* that may have been due to active oxygen species formed through the action of paraquat [84].

Feeding C3G significantly suppressed changes caused by hepatic ischemia–reperfusion (I/R) in rats fed 2 g/kg diet of C3G for 14 days. I/R treatment elevated liver TBARS and serum activities of glutamic oxaloacetic transaminase, glutamic pyruvic transaminase, and lactate dehydrogenase, marker enzymes for liver injury, and lowered liver reduced glutathione concentration. Although liver ascorbic acid concentrations were also lowered by hepatic I/R, concentrations were restored more quickly in C3G fed rats compared to control rats. Feeding C3G also resulted in a significant decrease in generation of TBARS during serum formation, and serum also showed a significantly lower susceptibility to further lipid peroxidation provoked by AAPH or Cu^{2+} than that of the control group [85]. Under these feeding and oxidative stress conditions, C3G functioned as a potent *in vivo* antioxidant [86,87].

In rats fed a vitamin E-deficient diet for 12 weeks and then replaced with a diet containing a highly purified ACN-rich extract (1 g/kg), a significant improvement in plasma antioxidant capacity and a decrease in the vitamin E deficiency-enhanced hydroperoxides and 8-oxo-deoxyguanosine concentrations in the liver were observed [16]. The ACN extract consisted of a mixture of the 3-glucoside forms of delphinidin, cyanidin, petunidin, peonidin, and malvidin. Thus, it appears that ACNs can be effective *in vivo* antioxidants when included in the diet at 1 or 2 g/kg. These levels in the diet provide 20–40 mg/day, which are much higher amounts on a body weight basis than found in the typical diet of humans.

In a most recent study, rats were given purified ACN extract from red grape skin orally for 10 days in either normal physiological conditions or exposed to a pro-oxidant chemical (CCl_4). The oral administration of the ACN extract significantly enhanced the ORAC value of the de-proteinized serum of about 50% ACN administration was also able to completely reverse the decrease in the serum ORAC activity induced by the CCl_4 treatment [88].

While total antioxidant capacity assays have been commonly used to assess total antioxidant capacity in human fluids (plasma and serum) in intervention studies, it has been understood among the scientific community that the results of the assays obtained from intervention studies might not simply be resulted from the absorbed flavonoids since their concentrations are too low compared to the other existing antioxidants at much larger concentrations. Further, reductions in biomarkers of oxidative stress may not simply be ascribed to the direct antioxidant effect of ACNs found in animal studies.

5.3.4 *In Vitro* vs. *In Vivo* Antioxidant Activity of ACNs: Impact of Bioavailability

Some of the key factors that impact *in vivo* antioxidant effects of ACNs and other flavonoids include (1) quantities consumed; (2) quantities absorbed or metabolized; and (3) plasma and/or tissue concentrations. Bioavailability plays a key role to link the *in vitro* antioxidant assays and the possible *in vivo* antioxidant activities. Although ACNs display antioxidant effects in cell culture and other *in vitro* systems at relatively high concentrations, it is not clear whether their concentrations can be reached *in vivo* at the tissue level to produce antioxidant effects as intact forms. Because of the instabilities of ACNs in the neutral pH range, whether ACNs remain intact in tissues long enough to act as antioxidants is an open question.

As being discussed earlier, many studies on antioxidant activities of ACNs were carried out in various *in vitro* models. To validate the prominent *in vivo* antioxidant effects revealed in these *in vitro* models, the bioavailability of ACNs must be taken into consideration. Evidence accumulated in the last decade indicated that the levels of ACNs detected in plasma and urine, as intact forms, are in general very low. Though the results of absorption of ACNs from different studies have not always agreed, we now know from most studies that the total recovery of ACNs from urine is lower than 0.1% [5]. Therefore, the doses reported in some *in vitro* studies might have insignificant relevance to *in vivo* conditions given that the level of intact ACNs exposed to tissues (except GIT luminal side tissues) could be very limited owing to the observed low concentration in blood [11]. Another important issue is the metabolites of ACNs that are present in the tissues. Some metabolites, especially the catabolites generated by gut microflora from native ACNs such as protocatechuic acid and the other simple phenolic compounds [89], may be absorbed at a fairly high concentration and have comparable or even more potent antioxidant activities than the precursors.

5.4 ANTIOXIDANT ACTIVITIES OF ACNS: INDIRECT EFFECTS

Over recent years, a new concept of the antioxidant effects of dietary flavonoid has emerged regarding their indirect antioxidant activity. In addition to direct quenching free radicals, the antioxidant function of flavonoids may also be fulfilled by regulating expression of multiple genes and certain signaling pathways in responding to oxidative stress, which could further lead to the activation or fortification of the antioxidant defense system.

5.4.1 ACNs and Antioxidant/Oxidant Enzymes

In the human body, multiple enzyme systems use different substrates as electron source of electron to produce a variety of ROS. Under physiological conditions, ROS formation and elimination are delicately balanced. However, enhanced activity of oxidant enzymes and/or reduced activity of antioxidant enzymes lead to oxidative stress [90]. Modulation of antioxidant/oxidant enzymes has emerged in recent years as another important mechanism of antioxidant activity of ACNs. Owing to the central role of superoxide in generating other free radicals *in vivo* [21], enzymes related to superoxide formation and degradation received primary attention (Figure 5.6). Administration of ACN-rich extract from black rice (AEBR) (500 mg/kg) along with alcohol in rats showed a better profile of the antioxidant system with normal glutathione peroxidase (GSH-Px), superoxide dismutase (SOD) and glutathione *S*-transferase (GST) activities [91]. The effects of the daily intake of ACNs and ellagitannins (ET) extracted from blackberries on the markers of oxidative status in healthy rats were

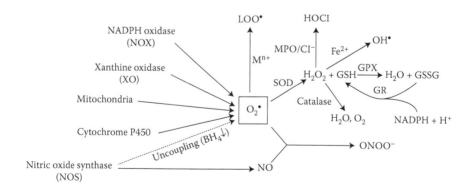

FIGURE 5.6 Antioxidant and oxidant enzymes related to superoxide.

also reported. In this study, the phenolic compounds were administered from three different extracts: an aqueous extract of blackberry (BJ) and its two derived fractions: ACN-enriched (AF) and ET-enriched (EF) fractions. After 35 days of administration, a significant increase in the catalase activity was observed only in the plasma of the groups administered ACN-containing extracts, which were the BJ and AF groups [92]. Another recent study was designed to investigate the antioxidant and protective effects of strawberry extracts against ethanol-induced gastric mucosa damage in a rat model, and to test whether strawberry extracts affect antioxidant enzyme activities in the gastric mucosa. Male Wistar rats received 40 mg/day/kg body weight of strawberry crude extracts for 10 days. Gastric damage was induced by ethanol, which caused significant activity reduction of both catalase and SOD. Strawberry consumption was able to restore the activity of these two enzymes to normal [93]. Protective effects of ACN-rich blueberries (BB) against atherosclerosis and potential underlying mechanisms in reducing oxidative stress were examined in apoE-deficient (apoE(−/−)) mice. ApoE(−/−) mice were fed an AIN-93G diet (CD) or CD formulated to contain 1% freeze-dried whole BB for 20 week. The mean lesion area for apoE(−/−) mice fed BB was significantly reduced in the aorta sinus and in the descending aorta compared with CD-fed mice. Genes analyzed by RT-PCR array showed that four major antioxidant enzymes in the aorta (superoxide dismutase (SOD) 1, SOD2, glutathione reductase (GSR), and thioredoxin reductase 1) were upregulated in BB-fed mice. Enzyme activities of SOD and GSR were greater in the liver and/or serum of BB-fed mice than those of CD-fed mice. In addition, serum paraoxonase 1 activity in serum of BB-fed mice was also increased than that of the CD-fed mice. These results suggested the potential mechanisms of protective effectiveness of BB against atherosclerosis may involve reduction in oxidative stress by enhancement of antioxidant defense [94].

The ability of ACN in modulating antioxidant/oxidant enzymes was also studied in cell culture models. ACN-rich fraction prepared from Hull blackberries were found to upregulate the expression of catalase, MnSOD, Gpx1/2, and Gsta1 antioxidant enzymes [95].

5.4.2 ACNs and Redox Cell Signaling Pathways

In recent years, amounting evidence revealed that antioxidant effects of dietary polyphenols may be achieved by regulating redox-sensitive signaling molecules [96]. One good example is the activation of nuclear factor-erythroid-2-related factor 2 (Nrf2). Nrf2 is referred to as the "master regulator" of the antioxidant response, modulating the expression of hundreds of genes, including antioxidant enzymes such as glutathione peroxidase (GPx), glutathione S-transferase, catalase, NAD(P)H: quinone oxidoreductase-1 (NQO1), and/or phase II enzymes [97–99]. The ARE-Nrf2 has also proven to be a primary means of regulating the production of proteins that regulate intracellular iron [28]. Owing to its critical role, the Nrf2 system was proposed as a potential target for the development of indirect antioxidants including ACNs [100] (Figure 5.7).

The mechanisms of the multifunctional effects of cyanidin on regulating antioxidant enzymes and oxidative stress-induced hepatotoxicity were investigated in

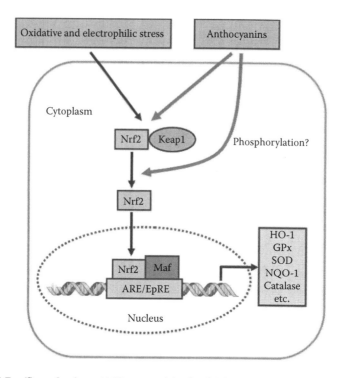

FIGURE 5.7   **(See color insert.)** The potential role of ACNs in regulating factor-erythroid-2-related factor 2 (Nrf2) pathway.

a recent paper [17]. The data indicated that cyanidin-mediated antioxidant enzyme expression involved the extracellular signal-regulated kinase (ERK) and c-Jun N-terminal kinase (JNK) pathways and Nrf2 activation. The synergistic effect of cyanidin and the PPAR agonist, troglitazone, on Nrf2-PPAR activation, was also observed. Treatment of cyanidin and troglitazone abolished H_2O_2-induced down-regulation of genes involved in lipid metabolism and ameliorated H_2O_2-mediated cytotoxicity. Cyanidin and PPAR agonists also showed synergistic benefits against metabolic dysfunction-related oxidative damage, which was found to be associated with the regulatory roles of mitogen-activated protein kinases (MAPKs) and the transcription factor Nrf2.

Another study supported the role of Nrf2 in inducing antioxidants after treatment of ACN fraction (AF) from purple sweet potato in rats. AF attenuated the DMN-induced increased serum alanine aminotransferase and aspartate aminotransferase activities. AF was also found to increase the expression of Nrf2, NADPH:quinine oxidoreductase-1, heme oxygenase-1, and GSTalpha, which were reduced by DMN, and decrease the expression of cyclooxygenase-2 and inducible nitric oxide synthase. An increase of NF-κB was observed in the DMN-induced liver injury group, while AF inhibited this translocation. These data implied that ACNs were able to induce antioxidant defense via the Nrf2 pathway and reduce inflammation via NF-κB inhibition [18].

Rats fed a diet containing an ACN-rich potato cultivar (SQ) showed an increase in the hepatic superoxide dismutase (SOD)-2 mRNA level. The possible signaling pathway was investigated in HepG2 cell line. SQ would directly increase the hepatic SOD-2 mRNA level. The induction of SOD-2 mRNA expression was blocked by an inhibitor of the extracellular signal-related kinase (ERK) 1/2 pathway. Furthermore, an extract of SQ increased the phosphorylation of ERK1/2 after 15 or 30 min of stimulation. These indicated that ACNs could directly induce hepatic SOD-2 mRNA expression via activation of ERK1/2 pathway in HepG2 cells [101].

Treatment of human HepG2 cells with the ACN C3G significantly reduced ROS levels induced by high glucose. C3G incubation increased glutamate-cysteine ligase expression, which in turn mediated the reduction in ROS levels. The upregulation of glutamate-cysteine ligase catalytic subunit (Gclc) expression by C3G occurred independent of the Nrf1/2 transcription factors. The cAMP-response element binding protein (CREB) was identified as the target transcription factor involved in the C3G-mediated upregulation of Gclc expression. C3G also increased phosphorylation of CREB through protein kinase A (PKA) activation, which induced a CREB-mediated upregulation of Gclc transcription. Furthermore, treatment with C3G increased the GSH synthesis in the liver of diabetic db/db mice through PKA-CREB-dependent induction of Gclc expression. This study suggested that ACN C3G has an effect of activating GSH synthesis through a novel antioxidant defense mechanism against excessive ROS production, contributing to the prevention of hyperglycemia-induced hepatic oxidative damage [102].

5.5 ANTIOXIDANT ACTIVITIES IN DISEASE PREVENTION OF ACNS

Oxidative stress and ROS/RNS have been suggested to cause almost every chronic disease. But according to recent data, their role in many diseases is likely to be peripheral, only in some diseases it is fundamental. The diseases in which ROS/RNS play a fundamental role may include cancer, neurodegenerative diseases, and probably atherosclerosis [22,48]. ACNs or ACN-rich food have demonstrated a broad spectrum of biomedical functions. These include prevention of CVD, cancer, Alzheimer's disease (AD), inflammatory responses, and metabolic diseases, all of which are more or less associated with their potent antioxidant property [11,19]. Considering the biological relevance, only the data from human or animal studies that linked to possible antioxidant activities were discussed in this section.

5.5.1 ATHEROSCLEROSIS

Atherosclerosis is a multifactorial disease, with several risk factors that affect the likelihood and severity of the clinical outcome. The oxidation theory is now an established hypothesis of atherogenesis and suggests that accumulation of oxidized LDL in the vessel wall is an early event in disease progression [103]. Epidemiological studies have examined the relationship between foods rich in ACNs (such as red wine and several species of berries) and CVD as well as the relationship between total ACN intake and risk of developing CVD [89]. Human intervention studies using ACN-rich berries (chokeberries, cranberries, blueberries, and strawberries), or purified ACN extracts have demonstrated significant improvements in LDL oxidation, lipid peroxidation, total plasma antioxidant capacity, dyslipidemia, and glucose metabolism [104]. Hemodialysis patients face an elevated risk of cancer, arteriosclerosis, and other diseases, ascribed in part to increased oxidative stress. Red fruit juice with high ACN/polyphenol content had been shown to reduce oxidative damage in healthy probands [105].

Benefits of ACNs in cardiovascular functions have also been demonstrated in many animal studies. Watanabe heritable hyperlipidemic rabbits were fed purified ACN fraction from black currants, black currant juice, probucol, or control diet for 16 weeks. Purified ACNs significantly increased plasma cholesterol and LDL cholesterol. Intake of black currant juice had no effect on total plasma cholesterol, but lowered very-low-density lipoprotein (VLDL) cholesterol significantly. The erythrocyte antioxidant enzyme glutathione peroxidase was significantly increased by purified ACNs and superoxide dismutase was increased by both ACN-containing treatments [106]. Male Wistar rats were fed the ACN-rich or the ACN-free diet for a period of 8 weeks. The hearts of rats fed the ACN-rich diet were more resistant to regional ischemia and reperfusion insult. Cardioprotection was associated with increased myocardial glutathione levels, suggesting that dietary ACNs might modulate cardiac antioxidant defenses [107].

5.5.2 CANCER

The development of cancer in humans is a complex process including cellular and molecular changes mediated by diverse endogenous and exogenous stimuli. It is well established that oxidative DNA damage is responsible for cancer development [25]. Cancer initiation and promotion are associated with chromosomal defects and oncogene activation induced by free radicals. A common form of damage is the formation of hydroxylated bases of DNA, which are considered an important event in chemical carcinogenesis [24,25].

The studies on the cancer-preventive effects of the ACNs, including results from *in vitro* cell culture and *in vivo* animal models, have been well documented [108,109]. The antioxidant effects of ACNs *in vitro*, such as directly scavenging ROS, increasing the oxygen-radical absorbing capacity of cells, reducing the formation of oxidative adducts in DNA, have been demonstrated using several cancer cell culture systems [109]. However, in most animal or human studies, the mechanisms of cancer prevention of ACNs were not fully understood. In the azoxymethane (AOM)-induced model of colon cancer in F344 rats, diets containing 2.5%, 5%, and 10% lyophilized black raspberries significantly decreased total tumors (adenomas and adenocarcinomas) by 42%, 45%, and 71%, and urinary 8-hydroxy-2′-deoxyguanosine (8-OHdG) levels by 73%, 81%, and 83%, respectively. These indicated that berries reduce ROS-induced DNA damage in animals [110]. Afaq et al. [111] investigated the photo-chemopreventive effect of delphinidin, a major anthocyanidin present in many pigmented fruits and vegetables, on UVB-induced biomarkers of skin cancer development in the SKH-1 hairless mouse. Topical application of delphinidin (1 mg/application) to mouse skin inhibited apoptosis and markers of DNA damage such as cyclobutane pyrimidine dimers and 8-OHdG. These results suggested that delphinidin inhibited UVB-mediated oxidative stress and reduced DNA damage, thereby protecting the cells from UVB-induced apoptosis. In a most recent study, the underlying antioxidant mechanisms of black currant ACNs in the chemopreventive effects were elucidated in diethylnitrosamine (DENA)-initiated hepatocarcinogenesis in rats. Dietary ACN-rich black currant skin extract (BCSE) (100 and 500 mg/kg) administered 4 weeks before and 18 weeks after DENA challenge decreased abnormal lipid peroxidation, protein oxidation, and expression of inducible nitric oxide synthase (iNOS) and 3-nitrotyrosine (3-NT) in a dose-responsive fashion. Mechanistic studies revealed that BCSE upregulated the gene expression of a number of hepatic antioxidant and carcinogen detoxifying enzymes through activation of the Nrf2-regulated antioxidant pathway, which led to the upregulation of a variety of housekeeping genes [112].

Unlike the *in vivo* animal model studies, epidemiological studies in humans have not provided convincing evidence of the anticancer effects of ACNs [109]. Although epidemiological studies have not shown that ACN intake reduces cancer risk in humans, they suggest that ACN intake may reduce certain parameters of oxidative damage [109]. A study in Germany showed that individuals who consumed an ACN/polyphenol-rich fruit juice had reduced oxidative DNA damage and a significant increase in reduced glutathione when compared to controls [113]. In addition, in an investigation of patients with Barrett's esophagus, the oral administration of 45 or 32 g (males and females, respectively) of lyophilized black raspberry powder (which contains about 5–7% ACNs) in a slurry of water daily for 6 months reduced levels of 8-epi-prostaglandin F2α (8-Iso-PGF2) and 8-OHdG in urine [114].

5.5.3 ALZHEIMER'S DISEASE

Oxidative stress has been strongly implicated in the pathophysiology of neurodegenerative disorders such as AD. In recent years, antioxidants—especially those of dietary origin—have been suggested as possible agents useful for the prevention and treatment of AD [115]. The protective effect of ACNs on the cerebral oxidative stress was studied using the whiskers cut model. In mice, such treatment causes psychological or emotional distress leading to oxidative stress in tissues. To investigate the *in vivo* antioxidant activity of ACNs, an extract of *Vaccinium myrtillis* L., an ACN mixture, was orally administered (100 mg/kg of body weight) to mice for 7 days, and then psychological stress was assessed by cutting off their whiskers. Whisker removal increased both protein carbonyl formation and lipid peroxidation in the brain, heart, kidney, and liver. The levels of oxidative markers showed regional differences in the brain. Orally administered ACNs were also active in the brain, suppressing stress-induced cerebral oxidative stress and dopamine abnormalities in distressed mice. These effects of ACN treatment suggest their possible usefulness for the treatment of cerebral disorders related to oxidative stress [116]. Amyloid-beta peptide (Abeta) is known to induce the redox imbalance, mitochondrial dysfunction and caspase activation, resulting in neuronal cell death. Treatment with antioxidants provided a new therapeutic strategy for AD patients. Purple sweet potato anthocyanins (PSPA) were used to treat Abeta toxicity in PC12 cells. The results showed that pretreatment of PC12 cells with PSPA reduced Abeta-induced toxicity, intracellular

ROS generation, and lipid peroxidation dose dependently. The study indicates that PSPA may be a promising approach for the treatment of AD and other oxidative-stress-related neurodegenerative diseases [117].

5.6 CONCLUDING REMARKS

As a major group of dietary polyphenols, ACNs have received considerable attention on their potential antioxidant activities. Though showing high antioxidant potential with various *in vitro* assays, ACNs are unlikely to act as antioxidants *in vivo* by directly reacting with ROS/RNS, due to their extremely low absorption and fast excretion rates. Exceptions to this may be the direct antioxidant effects in the gastrointestinal tract.

In human intervention studies, total antioxidant capacity assays have been commonly used to assess total antioxidant capacity in biological fluids. However, the results of the assays might not be simply from the absorbed ACNs, and the reductions in biomarkers of oxidative stress may not be simply ascribed to the direct antioxidant effect of ACNs found in animal studies. It is possible that the metabolites of ACNs, especially the catabolites generated by the gut microflora, are absorbed and exert antioxidant activity *in vivo*.

Emerging evidence has shown that ACNs are able to regulate expression of multiple genes and certain key signaling pathways in responding to the oxidative stress *in vivo*, thus fulfilling the antioxidant function through indirect ways. These indirect antioxidant activities are likely to be the major mechanisms underlying the disease prevention effects of ACNs.

While there has been a major focus on the antioxidant properties of ACNs in preventing chronic diseases, the evidence has been rapidly accumulated that ACNs and their *in vivo* metabolites exert modulation action far beyond antioxidant activities, which may include regulating gene expression, certain signaling pathways, enzyme activties, and so on.

Last but not the least, one must be aware of the fact that the majority of the published studies related to antioxidant or other bioactivities, along with the disease prevention effects of ACNs, have relied on crude ACN extracts or fruits/berries rich in ACNs. Therefore, the data should be interpreted with caution as the observed bioactivities and/or health effects may not be solely from ACNs. And synergistic effects with other bioactive compounds may also exist.

REFERENCES

1. Mazza, G., Miniati, E., *Anthocyanins in Fruits, Vegetables, and Grains*. Boca Raton, FL: CRC Press, 1993.
2. Andersen, O.M. 2012. Personal database on anthocyanins.
3. Wu, X., Prior, R.L., Systematic identification and characterization of anthocyanins by HPLC-ESI-MS/MS in common foods in the United States: Fruits and berries. *J Agric Food Chem* 2005, *53*, 2589–99.
4. Awika, J.M., Rooney, L.W., Waniska, R.D., Properties of 3-deoxyanthocyanins from sorghum. *J Agric Food Chem* 2004, *52*, 4388–94.
5. Prior, R.L., Wu, X., Anthocyanins: Structural characteristics that result in unique metabolic patterns and biological activities. *Free Radic Res* 2006, *40*, 1014–28.

6. Stintzing, F.C., Carle, R., Functional properties of anthocyanins and betalains in plants, food, and in human nutrition. *Trends Food Sci Tech* 2004, *15*, 19–38.
7. Giusti, M.M., Jing, P., *Natural Pigments of Berries: Functionality and Application.* Boca Raton, FL: CRC Press, 2007; pp. 105–146.
8. Wilska-Jeszka, J., *Food Colorants.* Boca Raton, FL: CRC Press, 2007; pp. 245–274.
9. Duthie, G., Gardner, P., Kyle, J., McPhail, D., Antioxidant activity of anthocyanins *in vitro* and *in vivo*. *ACS Symposium Series* 2004, *871*, 90–102.
10. Kahkonen, M.P., Heinonen, M., Antioxidant activity of anthocyanins and their aglycons. *J Agric Food Chem* 2003, *51*, 628–33.
11. He, J., Giusti, M.M., Anthocyanins: Natural colorants with health-promoting properties. *Annu Rev Food Sci Technol* 2010, *1*, 163–87.
12. Mazza, G.J., Anthocyanins and heart health. *Ann Ist Super Sanita* 2007, *43*, 369–74.
13. Han, K.H., Matsumoto, A., Shimada, K., Sekikawa, M., Fukushima, M., Effects of anthocyanin-rich purple potato flakes on antioxidant status in F344 rats fed a cholesterol-rich diet. *Br J Nutr* 2007, *98*, 914–21.
14. Ke, Z., Liu, Y., Wang, X., Fan, Z., Chen, G., Xu, M., Bower, K.A., Frank, J.A., Ou, X., Shi, X., Luo, J., Cyanidin-3-glucoside ameliorates ethanol neurotoxicity in the developing brain. *J Neurosci Res* 2011, *89*, 1676–84.
15. Prior, R.L., Gu, L., Wu, X., Jacob, R.A., Sotoudeh, G., Kader, A.A., Cook, R.A., Plasma antioxidant capacity changes following a meal as a measure of the ability of a food to alter *in vivo* antioxidant status. *J Am Coll Nutr* 2007, *26*, 170–81.
16. Ramirez-Tortosa, C., Andersen, O.M., Gardner, P.T., Morrice, P.C., Wood, S.G., Duthie, S.J., Collins, A.R., Duthie, G.G., Anthocyanin-rich extract decreases indices of lipid peroxidation and DNA damage in vitamin E-depleted rats. *Free Radic Biol Med* 2001, *31*, 1033–7.
17. Shih, P.H., Hwang, S.L., Yeh, C.T., Yen, G.C., Synergistic effect of cyanidin and PPAR agonist against nonalcoholic steatohepatitis-mediated oxidative stress-induced cytotoxicity through MAPK and Nrf2 transduction pathways. *J Agric Food Chem* 2012, *60*, 2924–33.
18. Hwang, Y.P., Choi, J.H., Yun, H.J., et al., Anthocyanins from purple sweet potato attenuate dimethylnitrosamine-induced liver injury in rats by inducing Nrf2-mediated antioxidant enzymes and reducing COX-2 and iNOS expression. *Food Chem Toxicol* 2011, *49*, 93–9.
19. Zafra-Stone, S., Yasmin, T., Bagchi, M., Chatterjee, A., Vinson, J.A., Bagchi, D., Berry anthocyanins as novel antioxidants in human health and disease prevention. *Mol Nutr Food Res* 2007, *51*, 675–83.
20. Domitrovic, R., The molecular basis for the pharmacological activity of anthocyans. *Curr Med Chem* 2011, *18*, 4454–69.
21. Halliwell, B., Reactive species and antioxidants. Redox biology is a fundamental theme of aerobic life. *Plant Physiol* 2006, *141*, 312–22.
22. Halliwell, B., Free radicals and antioxidants—Quo vadis? *Trends Pharmacol Sci* 2011, *32*, 125–30.
23. Haulica, I., Boisteanu, D., Bild, W., Free radicals between health and disease. *Rom J Physiol* 2000, *37*, 15–22.
24. Pham-Huy, L.A., He, H., Pham-Huy, C., Free radicals, antioxidants in disease and health. *In J Biomed Sci* 2008, *4*, 89–96.
25. Valko, M., Leibfritz, D., Moncol, J., Cronin, M.T., Mazur, M., Telser, J., Free radicals and antioxidants in normal physiological functions and human disease. *Int J Biochem Cell Biol* 2007, *39*, 44–84.
26. Finkel, T., Holbrook, N.J., Oxidants, oxidative stress and the biology of ageing. *Nature* 2000, *408*, 239–47.
27. Spector, A., Review: Oxidative stress and disease. *J Ocul Pharmacol Ther* 2000, *16*, 193–201.

28. Finley, J.W., Kong, A.N., Hintze, K.J., Jeffery, E.H., Ji, L.L., Lei, X.G., Antioxidants in foods: State of the science important to the food industry. *J Agric Food Chem* 2011, *59*, 6837–46.
29. Halliwell, B., Gutteridge, J.M., The definition and measurement of antioxidants in biological systems. *Free Radic Biol Med* 1995, *18*, 125–6.
30. Halliwell, B., Biochemistry of oxidative stress. *Biochem Soc Trans* 2007, *35*, 1147–50.
31. Halliwell, B., Free radicals and antioxidants: A personal view. *Nutr Rev* 1994, *52*, 253–65.
32. Young, I.S., Woodside, J.V., Antioxidants in health and disease. *J Clin Pathol* 2001, *54*, 176–86.
33. Harborne, J.B., Anthocyanins and other flavonoids. *Natural Product Reports* 2001, *18*, 310–33.
34. Grassi, D., Desideri, G., Ferri, C., Flavonoids: Antioxidants against atherosclerosis. *Nutrients* 2010, *2*, 889–902.
35. Arts, I.C., Hollman, P.C., Polyphenols and disease risk in epidemiologic studies. *Am J Clin Nutr* 2005, *81*, 317S–325S.
36. Esfahani, A., Wong, J.M., Truan, J., Villa, C.R., Mirrahimi, A., Srichaikul, K., Kendall, C.W., Health effects of mixed fruit and vegetable concentrates: A systematic review of the clinical interventions. *J Am Coll Nutr* 2011, *30*, 285–94.
37. Fraga, C.G., Plant polyphenols: How to translate their *in vitro* antioxidant actions to *in vivo* conditions. *IUBMB Life* 2007, *59*, 308–15.
38. Scalbert, A., Manach, C., Morand, C., Remesy, C., Jimenez, L., Dietary polyphenols and the prevention of diseases. *Crit Rev Food Sci Nutr* 2005, *45*, 287–306.
39. Yao, L.H., Jiang, Y.M., Shi, J., Tomas-Barberan, F.A., Datta, N., Singanusong, R., Chen, S.S., Flavonoids in food and their health benefits. *Plant Foods Hum Nutr* 2004, *59*, 113–22.
40. Honzel, D., Carter, S.G., Redman, K.A., Schauss, A.G., Endres, J.R., Jensen, G.S., Comparison of chemical and cell-based antioxidant methods for evaluation of foods and natural products: Generating multifaceted data by parallel testing using erythrocytes and polymorphonuclear cells. *J Agric Food Chem* 2008, *56*, 8319–25.
41. Kuhnau, J., The flavonoids. A class of semi-essential food components: Their role in human nutrition. *World Rev Nutr Dietet* 1976, *24*, 117–91.
42. Speciale, A., Chirafisi, J., Saija, A., Cimino, F., Nutritional antioxidants and adaptive cell responses: An update. *Curr Mol Med* 2011, *11*, 770–89.
43. Finley, J.W., Given Jr, P., Technological necessity of antioxidants in the food industry. *Food Chem Toxicol* 1986, *24*, 999–1006.
44. Wang, H., Nair, M.G., Strasburg, G.M., Chang, Y.C., Booren, A.M., Gray, J.I., DeWitt, D.L., Antioxidant and antiinflammatory activities of anthocyanins and their aglycon, cyanidin, from tart cherries. *J Nat Prod* 1999, *62*, 294–6.
45. Seeram, N.P., Bourquin, L.D., Nair, M.G., Degradation products of cyanidin glycosides from tart cherries and their bioactivities. *J Agric Food Chem* 2001, *49*, 4924–9.
46. Seeram, N.P., Cichewicz, R.H., Chandra, A., Nair, M.G., Cyclooxygenase inhibitory and antioxidant compounds from crabapple fruits. *J Agric Food Chem* 2003, *51*, 1948–51.
47. Prior, R.L., Wu, X., Schaich, K., Standardized methods for the determination of antioxidant capacity and phenolics in foods and dietary supplements. *J Agric Food Chem* 2005, *53*, 4290–302.
48. Halliwell, B., Free radicals and antioxidants: Updating a personal view. *Nutr Rev* 2012, *70*, 257–65.
49. Jomova, K., Valko, M., Advances in metal-induced oxidative stress and human disease. *Toxicology* 2011, *283*, 65–87.
50. Wang, H., Cao, G., Prior, R.L., Oxygen radical absorbing capacity of anthocyanins. *J Agric Food Chem* 1997, *45*, 304–309.

51. Espin, J.C., Soler-Rivas, C., Wichers, H.J., Garcia-Viguera, C., Anthocyanin-based natural colorants: A new source of antiradical activity for foodstuff. *J Agric Food Chem* 2000, *48*, 1588–92.

52. Bao, J., Cai, Y., Sun, M., Wang, G., Corke, H., Anthocyanins, flavonols, and free radical scavenging activity of Chinese bayberry (*Myrica rubra*) extracts and their color properties and stability. *J Agric Food Chem* 2005, *53*, 2327–32.

53. Blando, F., Gerardi, C., Nicoletti, I., Sour cherry (*Prunus cerasus* L.) Anthocyanins as ingredients for functional foods. *J Biomed Biotechnol* 2004, *2004*, 253–58.

54. Philpott, M., Lim, C.C., Ferguson, L.R., Dietary protection against free radicals: A case for multiple testing to establish structure–activity relationships for antioxidant potential of anthocyanic plant species. *Int J Mol Sci* 2009, *10*, 1081–103.

55. Rice-Evans, C.A., The relative antioxidant activities of plant-derived polyphenolic flavonoids. *Free Radic Res* 1995, *22*, 375–83.

56. Cao, G., Sofic, E., Prior, R.L., Antioxidant and prooxidant behavior of flavonoids: Structure–activity relationships. *Free Radic Biol Med* 1997, *22*, 749–60.

57. Huang, D., Ou, B., Prior, R.L., The chemistry behind antioxidant capacity assays. *J Agric Food Chem* 2005, *53*, 1841–56.

58. Parthasarathy, S., Steinberg, D., Witztum, J.L., The role of oxidized low-density lipoproteins in the pathogenesis of atherosclerosis. *Annu Rev Med* 1992, *43*, 219–25.

59. Jialal, I., Devaraj, S., The role of oxidized low density lipoprotein in atherogenesis. *J Nutr* 1996, *126*, 1053S–7S.

60. Amorini, A.M., Fazzina, G., Lazzarino, G., Tavazzi, B., Di Pierro, D., Santucci, R., Sinibaldi, F., Galvano, F., Galvano, G., Activity and mechanism of the antioxidant properties of cyanidin-3-*O*-beta-glucopyranoside. *Free Radic Res* 2001, *35*, 953–66.

61. Chang, Y.C., Huang, K.X., Huang, A.C., Ho, Y.C., Wang, C.J., Hibiscus anthocyanins-rich extract inhibited LDL oxidation and oxLDL-mediated macrophages apoptosis. *Food Chem Toxicol* 2006, *44*, 1015–23.

62. Liu, L.K., Lee, H.J., Shih, Y.W., Chyau, C.C., Wang, C.J., Mulberry anthocyanin extracts inhibit LDL oxidation and macrophage-derived foam cell formation induced by oxidative LDL. *J Food Sci* 2008, *73*, H113–21.

63. Noda, Y., Kaneyuki, T., Igarashi, K., Mori, A., Packer, L., Antioxidant activity of nasunin, an anthocyanin in eggplant. *Res Commun Mol Pathol Pharmacol* 1998, *102*, 175–87.

64. Niki, E., Assessment of antioxidant capacity *in vitro* and *in vivo*. *Free Radic Biol Med* 2010, *49*, 503–15.

65. Cheli, F., Baldi, A., Nutrition-based health: Cell-based bioassays for food antioxidant activity evaluation. *J Food Sci* 2011, *76*, R197–205.

66. Wolfe, K.L., Liu, R.H., Cellular antioxidant activity (CAA) assay for assessing antioxidants, foods, and dietary supplements. *J Agric Food Chem* 2007, *55*, 8896–907.

67. Wu, X., Beecher, G.R., Holden, J.M., Haytowitz, D.B., Gebhardt, S.E., Prior, R.L., Lipophilic and hydrophilic antioxidant capacities of common foods in the United States. *J Agric Food Chem* 2004, *52*, 4026–37.

68. Wolfe, K.L., Kang, X., He, X., Dong, M., Zhang, Q., Liu, R.H., Cellular antioxidant activity of common fruits. *J Agric Food Chem* 2008, *56*, 8418–26.

69. Jensen, G.S., Wu, X., Patterson, K.M., Barnes, J., Carter, S.G., Scherwitz, L., Beaman, R., Endres, J.R., Schauss, A.G., *In vitro* and *in vivo* antioxidant and anti-inflammatory capacities of an antioxidant-rich fruit and berry juice blend. Results of a pilot and randomized, double-blinded, placebo-controlled, crossover study. *J Agric Food Chem* 2008, *56*, 8326–33.

70. Elisia, I., Hu, C., Popovich, D.G., Kitts, D.D., Antioxidant assessment of an anthocyanin-enriched blackberry extract *Food Chem* 2007, *101*, 1052–1058.

71. Elisia, I., Kitts, D.D., Anthocyanins inhibit peroxyl radical-induced apoptosis in Caco-2 cells. *Mol Cell Biochem* 2008, *312*, 139–45.

72. Solomon, A., Golubowicz, S., Yablowicz, Z., Bergman, M., Grossman, S., Altman, A., Kerem, Z., Flaishman, M.A., Protection of fibroblasts (NIH-3T3) against oxidative damage by cyanidin-3-rhamnoglucoside isolated from fig fruits (*Ficus carica* L.). *J Agric Food Chem* 2010, *58*, 6660–5.

73. Bornsek, S.M., Ziberna, L., Polak, T., Vanzo, A., Ulrih, N.P., Abram, V., Tramer, F., Passamonti, S., Bilberry and blueberry anthocyanins act as powerful intracellular antioxidants in mammalian cells. *Food Chem* 2012, *134*, 1878–1884.

74. Berger, R.G., Lunkenbein, S., Strohle, A., Hahn, A., Antioxidants in food: Mere myth or magic medicine? *Crit Rev Food Sci Nutr* 2012, *52*, 162–71.

75. Bjelakovic, G., Gluud, C., Surviving antioxidant supplements. *J Natl Cancer Inst* 2007, *99*, 742–3.

76. Morazzoni, P., Bombardelli, E., Vaccinium myrtillus L. *Fitoterapia* 1996, *67*, 3–29.

77. Wu, X., Beecher, G.R., Holden, J.M., Haytowitz, D.B., Gebhardt, S.E., Prior, R.L., Concentrations of anthocyanins in common foods in the United States and estimation of normal consumption. *J Agric Food Chem* 2006, *54*, 4069–75.

78. Prior, R.L., Fruits and vegetables in the prevention of cellular oxidative damage. *Am J Clin Nutr* 2003, *78*, 570S–578S.

79. Mazza, G., Kay, C.D., Cottrell, T., Holub, B.J., Absorption of anthocyanins from blueberries and serum antioxidant status in human subjects. *J Agric Food Chem* 2002, *50*, 7731–7.

80. Matsumoto, H., Nakamura, Y., Hirayama, M., Yoshiki, Y., Okubo, K., Antioxidant activity of black currant anthocyanin aglycons and their glycosides measured by chemiluminescence in a neutral pH region and in human plasma. *J Agric Food Chem* 2002, *50*, 5034–7.

81. Pawlowicz, P., Zastosowanie naturalnych antocyjanow pochodzacych z soku aronii czarnoowocowej (aronia melanocarpa) w leczeniu idiopatycznego oraz powiklanego preeclampsja wewnatrzmacicznego zahamowania wzrostu plodu. Wplyw na metabolizm utlenowanych lipoprotein osocza—rola przeciwcial anty-*o*-LDL (olab). *Ginekologia Polska* 2000, *71*, 848–53.

82. Wang, C.J., Wang, J.M., Lin, W.L., Chu, C.Y., Chou, F.P., Tseng, T.H., Protective effect of Hibiscus anthocyanins against tert-butyl hydroperoxide-induced hepatic toxicity in rats. *Food Chem Toxicol* 2000, *38*, 411–6.

83. Obi, F.O., Prevention of carbon tetrachloride-induced hepatotoxicity in the rat by *H. rosa-sinensis* anthocyanin extract administered in ethanol. *Toxicology* 1998, *131*, 93–8.

84. Igarashi, K., Kimura, Y., Takenaka, A., Preventive effects of dietary cabbage acylated anthocyanins on paraquat-induced oxidative stress in rats. *Biosci Biotechnol Biochem* 2000, *64*, 1600–7.

85. Tsuda, T., Horio, F., Osawa, T., Dietary cyanidin 3-*O*-beta-D-glucoside increases ex vivo oxidation resistance of serum in rats. *Lipids* 1998, *33*, 583–8.

86. Tsuda, T., The role of anthocyanins as an antioxidant under oxidative stress in rats. *Biofactors* 2000, *13*, 133–9.

87. Tsuda, T., Horio, F., Kitoh, J., Osawa, T., Protective effects of dietary cyanidin 3-*O*-beta-D-glucoside on liver ischemia-reperfusion injury in rats. *Arch Biochem Biophys* 1999, *368*, 361–6.

88. Lionetto, M.G., Giordano, M.E., Calisi, A., Erroi, E., De Nuccio, F., Schettino, T., Effect of the daily ingestion of a purified anthocyanin extract from grape skin on rat serum antioxidant capacity. *Physiol Res* 2011, *60*, 637–45.

89. Wallace, T.C., Anthocyanins in cardiovascular disease. *Adv Nutr* 2011, *2*, 1–7.

90. Wassmann, S., Wassmann, K., Nickenig, G., Modulation of oxidant and antioxidant enzyme expression and function in vascular cells. *Hypertension* 2004, *44*, 381–6.

91. Hou, Z., Qin, P., Ren, G., Effect of anthocyanin-rich extract from black rice (*Oryza sativa* L. *Japonica*) on chronically alcohol-induced liver damage in rats. *J Agric Food Chem* 2010, *58*, 3191–6.

92. Hassimotto, N.M., Lajolo, F.M., Antioxidant status in rats after long-term intake of anthocyanins and ellagitannins from blackberries. *J Sci Food Agric* 2011, *91*, 523–31.

93. Alvarez-Suarez, J.M., Dekanski, D., Ristic, S., Radonjic, N.V., Petronijevic, N.D., Giampieri, F., Astolfi, P. et al. Strawberry polyphenols attenuate ethanol-induced gastric lesions in rats by activation of antioxidant enzymes and attenuation of MDA increase. *PLoS One* 2011, *6*, e25878.

94. Wu, X., Kang, J., Xie, C., Burris, R., Ferguson, M.E., Badger, T.M., Nagarajan, S., Dietary blueberries attenuate atherosclerosis in apolipoprotein E-deficient mice by upregulating antioxidant enzyme expression. *J Nutr* 2010, *140*, 1628–32.

95. Murapa, P., Dai, J., Chung, M., Mumper, R.J., D'Orazio, J., Anthocyanin-rich fractions of blackberry extracts reduce UV-induced free radicals and oxidative damage in keratinocytes. *Phytother Res* 2012, *26*, 106–12.

96. Rahman, I., Biswas, S.K., Kirkham, P.A., Regulation of inflammation and redox signaling by dietary polyphenols. *Biochem Pharmacol* 2006, *72*, 1439–52.

97. Hu, M.L., Dietary polyphenols as antioxidants and anticancer agents: More questions than answers. *Chang Gung Med J* 2011, *34*, 449–60.

98. Hybertson, B.M., Gao, B., Bose, S.K., McCord, J.M., Oxidative stress in health and disease: The therapeutic potential of Nrf2 activation. *Mol Aspects Med* 2011, *32*, 234–46.

99. Kaspar, J.W., Niture, S.K., Jaiswal, A.K., Nrf2:INrf2 (Keap1) signaling in oxidative stress. *Free Radic Biol Med* 2009, *47*, 1304–9.

100. Jung, K.A., Kwak, M.K., The Nrf2 system as a potential target for the development of indirect antioxidants. *Molecules* 2010, *15*, 7266–91.

101. Hashimoto, N., Noda, T., Kim, S.J., Yamauchi, H., Takigawa, S., Matsuura-Endo, C., Suzuki, T., Han, K.H., Fukushima, M., Colored potato extracts induce superoxide dismutase-2 mRNA via ERK1/2 pathway in HepG2 cells. *Plant Foods Hum Nutr* 2010, *65*, 266–70.

102. Zhu, W., Jia, Q., Wang, Y., Zhang, Y., Xia, M., The anthocyanin cyanidin-3-*O*-beta-glucoside, a flavonoid, increases hepatic glutathione synthesis and protects hepatocytes against reactive oxygen species during hyperglycemia: Involvement of a cAMP-PKA-dependent signaling pathway. *Free Radic Biol Med* 2012, *52*, 314–27.

103. Mashima, R., Witting, P.K., Stocker, R., Oxidants and antioxidants in atherosclerosis. *Curr Opin Lipidol* 2001, *12*, 411–8.

104. Basu, A., Rhone, M., Lyons, T.J., Berries: Emerging impact on cardiovascular health. *Nutr Rev* 2010, *68*, 168–77.

105. Spormann, T.M., Albert, F.W., Rath, T., Dietrich, H., Will, F., Stockis, J.P., Eisenbrand, G., Janzowski, C., Anthocyanin/polyphenolic-rich fruit juice reduces oxidative cell damage in an intervention study with patients on hemodialysis. *Cancer Epidemiol Biomarkers Prev* 2008, *17*, 3372–80.

106. Finne Nielsen, I.L., Elbol Rasmussen, S., Mortensen, A., Ravn-Haren, G., Ma, H.P., Knuthsen, P., Hansen, B.F. et al. Anthocyanins increase low-density lipoprotein and plasma cholesterol and do not reduce atherosclerosis in Watanabe heritable hyperlipidemic rabbits. *Mol Nutr Food Res* 2005, *49*, 301–8.

107. Toufektsian, M.C., de Lorgeril, M., Nagy, N., Salen, P., Donati, M.B., Giordano, L., Mock, H.P. et al., Chronic dietary intake of plant-derived anthocyanins protects the rat heart against ischemia–reperfusion injury. *J Nutr* 2008, *138*, 747–52.

108. Hou, D.X., Potential mechanisms of cancer chemoprevention by anthocyanins. *Curr Mol Med* 2003, *3*, 149–59.

109. Wang, L.S., Stoner, G.D., Anthocyanins and their role in cancer prevention. *Cancer Lett* 2008, *269*, 281–90.

110. Harris, G.K., Gupta, A., Nines, R.G., Kresty, L.A., Habib, S.G., Frankel, W.L., LaPerle, K., Gallaher, D.D., Schwartz, S.J., Stoner, G.D., Effects of lyophilized black raspberries

on azoxymethane-induced colon cancer and 8-hydroxy-2'-deoxyguanosine levels in the Fischer 344 rat. *Nutr Cancer* 2001, *40*, 125–33.

111. Afaq, F., Syed, D.N., Malik, A., Hadi, N., Sarfaraz, S., Kweon, M.H., Khan, N., Zaid, M.A., Mukhtar, H., Delphinidin, an anthocyanidin in pigmented fruits and vegetables, protects human HaCaT keratinocytes and mouse skin against UVB-mediated oxidative stress and apoptosis. *J Invest Dermatol* 2007, *127*, 222–32.

112. Thoppil, R.J., Bhatia, D., Barnes, K.F., Haznagy-Radnai, E., Hohmann, J., Darvesh, A.S., Bishayee, A., Black currant anthocyanins abrogate oxidative stress through Nrf2-Mediated antioxidant mechanisms in a rat model of hepatocellular carcinoma. *Curr Cancer Drug Targets* 2012, *12*, 1244–57.

113. Weisel, T., Baum, M., Eisenbrand, G., Dietrich, H., Will, F., Stockis, J.P., Kulling, S., Rufer, C., Johannes, C., Janzowski, C., An anthocyanin/polyphenolic-rich fruit juice reduces oxidative DNA damage and increases glutathione level in healthy probands. *Biotechnol J* 2006, *1*, 388–97.

114. Kresty, L.A., Frankel, W.L., Hammond, C.D., Baird, M.E., Mele, J.M., Stoner, G.D., Fromkes, J.J., Transitioning from preclinical to clinical chemopreventive assessments of lyophilized black raspberries: Interim results show berries modulate markers of oxidative stress in Barrett's esophagus patients. *Nutr Cancer* 2006, *54*, 148–56.

115. Darvesh, A.S., Carroll, R.T., Bishayee, A., Geldenhuys, W.J., Van der Schyf, C.J., Oxidative stress and Alzheimer's disease: Dietary polyphenols as potential therapeutic agents. *Expert Rev Neurother* 2010, *10*, 729–45.

116. Rahman, M.M., Ichiyanagi, T., Komiyama, T., Sato, S., Konishi, T., Effects of anthocyanins on psychological stress-induced oxidative stress and neurotransmitter status. *J Agric Food Chem* 2008, *56*, 7545–50.

117. Ye, J., Meng, X., Yan, C., Wang, C., Effect of purple sweet potato anthocyanins on beta-amyloid-mediated PC-12 cells death by inhibition of oxidative stress. *Neurochem Res* 2010, *35*, 357–65.

6 Anthocyanins in Cardiovascular Disease Prevention

Taylor C. Wallace

CONTENTS

6.1 INTRODUCTION

Cardiovascular disease (CVD) is the number one cause of death worldwide. An estimated 17.3 million people died from CVD in 2008, representing 30% of all global deaths according to the World Health Organization (WHO). Of these deaths, an estimated 7.3 million were thought to be due to coronary heart disease (CHD) and 6.2 million were due to stroke (WHO 2011). Despite aggressive management of CVD risk factors and improved outcomes through pharmaceutical and medical research, the societal burden from CVD remains high. The number of annual CVD-related deaths is on the rise, particularly in lower-income countries, and is expected to reach 23.6 million people per year by 2030 (WHO 2011).

Throughout the past decade, there has been increased interest in lifestyle and dietary approaches to reducing cardiovascular risk. Consumption of colorful fruits and vegetables has been inversely associated with a decreased risk of CVD (Nöthlings et al. 2008), most likely due to the abundance and variety of bioactive compounds present. Interest in the biochemistry and biological effects of anthocyanins, as well as their breakdown and/or metabolic products, has substantially increased because of growing evidence that supports potential preventative and therapeutic effects of these compounds toward the onset and progression of CVD. It is widely accepted that the preventive and therapeutic activities of anthocyanins involve modulation of multiple pathways that are crucial to the pathogenesis of CVD. This chapter seeks to assemble the totality of current evidence related to the CVD preventative and therapeutic activities of anthocyanins from *in vitro* cell culture studies to randomized controlled trials, all contributing to a greater body of scientific knowledge that collectively supports the concept that anthocyanins may exert significant effects toward long-term cardiovascular health and homeostasis.

6.2 RISK FACTORS FOR CVD AND THE ROLE OF OXIDATIVE STRESS

The WHO defines CVD as disorders of the heart and blood vessels, including CHD, cerebrovascular disease, peripheral arterial disease, rheumatic heart disease, congenital heart disease, deep-vein thrombosis, and pulmonary embolism (WHO 2011). Atherosclerosis, a chronic inflammatory disease, is the primary origin of CVD. Atherosclerosis is caused by plaque rupture or erosion that leads to acute formation of platelet-rich thrombi that occlude or partially occlude the arterial lumen, leading to CVD clinical events such as myocardial infarction, unstable angina, or cerebrovascular accident. The process is initiated through accumulation and uptake of low-density lipoproteins (LDL) by macrophages in the intima, often resulting in oxidation and aggregation of the LDL. The entrapped oxidized LDL initiates an inflammatory response by the endothelial cells, which attract monocytes to the area. The monocytes then adhere to the endothelium, cross into the intima, and differentiate into macrophages, which engulf the oxidized LDL and cause unregulated accumulation of the LDL. These lipid-rich macrophages are known as "foam cells" and are characteristic of fatty streaks (early atherosclerotic lesions). Foam cells are incapable of escaping the intima and undergo apoptosis, further contributing to the inflammatory process. The progression of the fatty streaks and atherosclerotic lesions involves smooth muscle cells, collagen, and platelets in addition to further lipid accumulation. The advanced lesion may develop a fibrous cap, reduce endothelial functions, limit the effective diameter of the vessels, and thereby restrict blood flow and the supply of oxygen to tissues. As the plaque grows larger, it has an increased tendency to rupture and cause thrombosis and sudden death through myocardial infarction.

Behavioral risk factors such as an unhealthy diet, lack of physical activity, smoking, and excessive use of alcohol account for approximately 80% of CVD incidences.

Behavioral risk factors may promptly lead to intermediate risk factors of developing CVD such as obesity, and may result in elevated blood pressure, glucose, and lipid levels. Comprehensive and integrated action is the means to prevent and control CVD (WHO 2011).

6.3 DIETARY ANTHOCYANINS AND CVD RISK: EPIDEMIOLOGICAL EVIDENCE

Epidemiological data have clearly shown an inverse relationship between CVD risk and consumption of plant-based foods such as fruits, vegetables, and legumes (Joshipura et al. 2001; Ness and Powles 1997; Nöthlings et al. 2008). It has further been hypothesized that bioactive components present in these foods greatly contribute to a lower incidence of CVD. A few large prospective cohort studies have examined both the relationship between CVD and foods rich in anthocyanins (such as red wine and several species of berries) as well as the relationship between total anthocyanin intake and risk of developing CVD. The existing epidemiological evidence advocates that anthocyanins have a positive influence on the cardiovascular system. However, these studies are limited by one or more common nutritional epidemiological challenges such as reliance on self-reporting, inadequate adjudication of CVD-related events, potential confounding factors, and weaknesses in study design (i.e., most were not set up to specifically examine cardiovascular outcomes). The most promising evidence stems from the analyses conducted by Jennings et al. (2012), McCullough et al. (2012), and Mink et al. (2007). Most recently a cross-sectional study of 1898 women showed that increased anthocyanin consumption, calculated from food-frequency questionnaires and an updated/extended USDA database, showed that higher anthocyanin intake is associated with lower arterial stiffness and central blood pressure. The intakes of anthocyanins associated with these findings could be incorporated into the diet by the consumption of 1–2 portions of berries daily and are, therefore, relevant for public health strategies to reduce CVD risk. Direct effects on arterial stiffness and atherosclerosis included decreases in central systolic blood pressure (-3.0 ± 1.4 mmHg), mean arterial pressure (-2.3 ± 1.2 mmHg), and pulse wave velocity (-0.4 ± 0.2) for quintile 5 versus quintile 1 (Jennings et al. 2012). Another recent analysis of 98,469 men and women in the Cancer Prevention Study II Nutrition Cohort provides another accurate estimation of actual anthocyanin intakes and analysis of their impact on CVD mortality given the many confounding factors that exist when studying intake of these compounds (McCullough et al. 2012). The study examined men and women (mean age 70 y and 69 y, respectively) during 7 y of follow-up and reported that the intakes of anthocyanidins in all quintiles were associated with a decreased risk of CVD mortality in men, women, and men and women combined. These data illustrate a dose relationship for the first time. As intake of anthocyanins increases, the relative risk (RR) seems to inversely decrease in the age-adjusted model (Table 6.1). Men and women combined in the highest intake quintile showed the most significant decrease in CVD mortality (RR, 0.70; 95% confidence interval [95% CI], 0.62–0.79) (McCullough et al. 2012).

Another recent 2007 analysis of 34,489 postmenopausal women participating in the Iowa Women's Health Study also showed significant reductions in CVD mortality

TABLE 6.1

RR and 95% CI for CVD Mortality by Quintile of Energy-Adjusted Anthocyanin Intake in Men and Women in the Cancer Prevention Study II Nutrition Cohort

Quintile	Median Intake, mg (Range)	Age-Adjusted RR (95% CI)		
		Men ($n = 38,180$)	Women ($n = 60,289$)	Men and Women ($n = 98,469$)
1	3.8 (<5.5)	1.00 (—)	1.00 (—)	1.00 (—)
2	6.8 (5.5–8.1)	0.83 (0.70–0.97)	0.74 (0.62–0.87)	0.78 (0.70–0.88)
3	9.8 (8.2–11.4)	0.82 (0.71–0.96)	0.71 (0.60–0.84)	0.77 (0.68–0.86)
4	13.7 (11.5–16.6)	0.78 (0.67–0.91)	0.63 (0.53–0.76)	0.72 (0.64–0.80)
5	22.2 (>16.7)	0.72 (0.62–0.84)	0.70 (0.58–0.84)	0.70 (0.62–0.79)
p-Value		0.0002	0.0002	0.0001

Source: From McCullough ML et al. 2012. Am J Clin Nutr 95: 454–64.
Note: CI, confidence interval; CVD, cardiovascular disease; RR, relative risk.

associated with the consumption of total anthocyanins and anthocyanin-rich foods during a 16-year follow-up period using an age- and energy-adjusted model (Mink et al. 2007). This cohort study reported that a mean intake of 0.2 mg/d of anthocyanins was associated with reduced risk of CVD mortality (RR, 0.85; 95% CI, 0.78–0.92), CHD mortality (RR, 0.81; 95% CI, 0.73–0.91), and total mortality (RR, 0.86; 95% CI, 0.81–0.90) in postmenopausal women (Mink et al. 2007). Anthocyanin consumption was estimated using a food frequency questionnaire and food composition data obtained from three U.S. Department of Agriculture databases. Blueberries and strawberries also showed a significant decrease in CVD mortality ($RR_{Blueberries}$, 0.85; 95% CI, 0.75–0.96) ($RR_{Strawberries}$, 0.82; 95% CI, 0.74–0.89) and CHD mortality ($RR_{Blueberries}$, 0.81; 95% CI, 0.69–0.95) ($RR_{Strawberries}$, 0.84; 95% CI, 0.74–0.95). A significant reduction in RR was associated with consumption of blueberries and strawberries at least once a week (Mink et al. 2007). A recent analysis of 87,242 female participants in the Nurse's Health Study II, 46,672 female participants Nurse's Health Study I, as well as 23,043 male participants in the Health Professionals Follow-Up Study showed that the intake of anthocyanins helps to significantly reduce the development of hypertension in a multivariate adjusted model of the pooled cohorts (RR, 0.87; 95% CI, 0.81–0.92) (Cassidy et al. 2011). Intake of all six predominant anthocyanidins (cyanidin, delphinidin, malvidin, pelargonidin, peonidin, and petunidin) showed significant reductions in hypertension in participants aged >60 y, except for delphinidin, which had borderline significance. Anthocyanin intake across all quintiles of each of the three cohorts showed slight but significant reductions in hypertension among participants. Food intake data were also collected and anthocyanin content was estimated using the U.S. Department of Agriculture databases (Cassidy et al. 2011).

Other intake data of anthocyanin-rich foods are noteworthy in assessing the role of anthocyanins in cardiovascular maintenance and CVD prevention. Female health professionals enrolled in the Women's Health Study ($n = 38,176$) showed a significant risk reduction of C-reactive protein levels among women consuming >2 servings of strawberries per week (RR, 0.86; 95% CI, 0.75–0.99). Cardiovascular death also seemed to be influenced in a positive manner; however, these findings were not significant or dose dependent (Sesso et al. 2007). Several epidemiological studies have shown that CVD mortality can be decreased by moderate consumption of red wine. A meta-analysis of wine consumption among 209,418 participants pooled from 26 studies showed a decreased RR of CVD associated with light-to-moderate wine intake (RR, 0.68; 95% CI, 0.59–0.77) (Di Castelnuovo et al. 2002). The "French Paradox" first drew attention to the CVD protective effects of red wine after epidemiological data collected by the WHO revealed a discord in CVD mortality in a cohort of subjects from Toulouse, France, compared with other cohorts from 17 Western countries including the United States and the United Kingdom (Colling et al. 1989; Keil and Kuulasmaa 1989; Renaud and de Lorgeril 1992). The French cohort had a lower risk of CVD mortality despite higher consumption of saturated fat (Colling et al. 1989). The Kuopio Ischaemic Heart Disease Risk Factor Study showed a positive influence of anthocyanidin consumption as related to CVD; however, these results were not significant or dose dependent (large fluctuations between intake quartiles), nor was the ascertainment conducted according to clinical protocol (sole reliance on computer linkage to the national death registry) (Mursu et al. 2008).

The hypotheses generated by epidemiological data have initiated numerous laboratory and clinical studies devoted to understanding the potential role that anthocyanins may play in the prevention of CVD.

6.4 EFFECTS OF ANTHOCYANIN CHEMISTRY ON CVD BIOMARKERS

The chemical structure (position, number, and types of substitutions) of an anthocyanin plays an important role in the biological activity exerted. The potential of some specific fruits, vegetables, or legumes to prevent CVD induced by oxidative stress may be attributed to the unique ability of certain dietary components, such as anthocyanins, to influence a variety of cellular activities related to cardiovascular health, as many researchers have observed. Anthocyanins have long been known to show potent antioxidant activity *in vitro*; however, the extent to which individual anthocyanins successfully scavenge reactive oxygen species (ROS) varies. Erlejman et al. (2004) suggested three important structures present within flavonoids that are responsible for their antioxidant ability: (1) catechol group in the B-ring, (2) double bonds with 4-oxo group in the C-ring, and (3) the presence of hydroxyl groups in the third and fifth positions. The number of hydroxide groups in total or on the B-ring, a 3′,4′-ortho-dihydroxyl group on the B-ring, and a 3-hydroxyl group on the C-ring seemed to be the main structural requirements of anthocyanins for the inhibitory effects on oxidative injury in endothelial cells (Yi et al. 2010). Methylation of the 4′-hydroxyl group is associated with a significant decrease in radical scavenging activity *in vitro*; however, methylation of other positions on the B-ring has a large effect (Rahman et al. 2006). An additional hydroxyl group on the B-ring (e.g., delphinidin vs. cyanidin) does not influence the *in vitro* radical scavenging activity of anthocyanins (Rahman et al.

2006). Interestingly, a direct comparison of the corresponding anthocyanins with fla-vonoids (cyanidin-3-galactoside vs. quercetin-3-galactoside) showed better efficacy for the former, although the aglycons (cyanidin and quercetin) had similar activity (Kähkönen and Heinonen 2003; Yan et al. 2002).

6.5 RANDOMIZED CONTROLLED TRIALS OF PURIFIED ANTHOCYANINS

Randomized controlled trials were introduced into clinical medicine when strepto-mycin was evaluated in the treatment of tuberculosis and have since become inte-grated in nutrition science as the "gold standard" for assessing the effectiveness of nutrients and nonnutrient dietary components such as anthocyanins. Table 6.2 illus-trates the outcomes of randomized controlled trials for pharmacologic interventions with purified anthocyanins. Although pharmacological clinical models of purified anthocyanins are useful in generating dietary guidance and public policy, other randomized controlled trials of anthocyanin-rich foods and extracts exist and con-tribute to the knowledge of potential synergistic interactions that anthocyanins may have with other phytonutrients and food components. The most compelling evidence from the studies shown in Table 6.2 demonstrates the positive effect of anthocyanins on lipid profiles and inhibition of cholesteryl ester transfer protein in dyslipidemic patients (Qin et al. 2009), as well as improvement in endothelial function (i.e., flow-mediated dilation), vascular cell adhesion molecule-1 (VCAM-1), and lipid profiles among hypercholesterolemic patients (Zhu et al. 2011). Similarly, administration of purified anthocyanins demonstrated significant decreased levels of high sensitiv-ity C-reactive protein, VCAM-1 and plasma interleukin-1β (IL-1β) among adults with hypercholesterolemia. The study further found significant decreases in LDL-cholesterol and increases in HDL-cholesterol (Zhu et al. 2012). The other two stud-ies included in Table 6.2, while showing minor to statistically insignificant effects, were of healthy patients whose markers of CVD were normal. Studying subjects with an absence of elevated CVD markers causes difficulty in the ability to show actual significant effects, especially when the effects are more than likely incre-mental, span a lifetime of consumption, and are dependent on other healthy lifestyle factors. Anthocyanins may have an active role in helping to maintain homeostasis in the cardiovascular system. For this reason, large long-term prospective cohort trials with healthy individuals and randomized controlled trials of individuals with elevated risk biomarkers offer equally valuable information on the actual effect of anthocyanins in relation to CVD prevention and treatment.

6.6 RANDOMIZED CONTROLLED TRIALS OF ANTHOCYANIN-RICH EXTRACTS AND FOODS

6.6.1 STUDIES INVOLVING SUBJECTS WITH ELEVATED CVD RISK BIOMARKERS

Other randomized controlled trials of anthocyanin-rich fruits and vegetables, as well as their extracts, support the findings of trials with purified anthocyanins and offer additional valuable information on potential mechanisms by which anthocyanins may

TABLE 6.2
Summary of Clinical Intervention Trials That Examine Purified Anthocyanins and CVD Markers

Study, Year (Reference)	Participants	Source Intervention	Duration	Main Findings
Zhu et al. (2012)	150 hypercholesterolemic individuals (aged 40–65 y)	320 mg anthocyanins (administered as 160 mg twice daily) in the form of cyanidin-3-glucoside and delphinidin-3-glucoside	24 wk	Anthocyanin supplementation significantly decreased the levels of serum high sensitivity C-reactive protein (−21.6% vs. −2.5%), soluble VCAM-1 (−12.3% vs. 0.4%) and plasma interleukin-1β (−12.8% vs. −1.3%) in the treatment vs. the placebo group. Significant differences in LDL cholesterol (−10.4% vs. 0.3%) and HDL cholesterol (14.0% vs. 0.9%) were also observed in the treatment vs. placebo group
Zhu et al. (2011)	150 hypercholesterolemic individuals (aged 40–65 y)	*Short-term study:* 320 mg anthocyanins purified from bilberry and black currant were administered once daily *Long-term study:* 320 mg anthocyanins (administered as 160 mg twice daily) purified from bilberry and black currant	12 wk	In this short-term crossover study, 12 individuals showed significant increases in flow-mediated dilation from 8.3% at baseline to 11.0% after 1 h and 10.1% after 2 h ($p < 0.05$). In the long-term study, participants in the anthocyanin intervention group showed significant increases in flow-mediated dilation (24.4% vs. 2.2%), cGMP (12.6% vs. −1.2%), and HDL cholesterol concentrations ($p < 0.05$). Decreases in VCAM-1 and LDL cholesterol were also significant ($p < 0.05$). The changes in cGMP and HDL cholesterol were positively correlated with flow-mediated dilation in the anthocyanin group ($p < 0.05$)

continued

TABLE 6.2   (continued)
Summary of Clinical Intervention Trials That Examine Purified Anthocyanins and CVD Markers

Study, Year (Reference)	Participants	Source Intervention	Duration	Main Findings
Hassellund et al. (2011)	31 healthy men with a high normal blood pressure (aged 35–51 y)	640 mg anthocyanins (administered as 320 mg twice daily) purified from bilberry and black currant	4 wk	No significant effect on blood pressure
Qin et al. (2009)	120 dyslipidemic subjects (aged 40–65 y)	160 mg anthocyanins (administered as 80 mg twice daily) purified from bilberry and black currant	12 wk	Anthocyanin supplementation increased HDL cholesterol concentrations by 13.7% ($p < 0.001$) and decreased LDL cholesterol concentrations by 13.6% ($p < 0.001$) in the treatment group compared with 2.8% and –0.6% in the placebo group Cellular cholesterol efflux to serum increased more in the anthocyanin group than in the placebo group (20.0% and 0.2%; $p < 0.001$) Anthocyanin supplementation decreased the mass and activity of plasma cholesteryl ester transfer protein by 10.4% and 6.3% in the treatment group and –3.5% and 1.1% in the placebo group ($p < 0.001$)
Karlsen et al. (2007)	120 healthy subjects (aged 40–74 y)	300 mg anthocyanins (administered as 150 mg twice daily) purified from bilberry and black currant	3 wk	Anthocyanin supplementation improved NF-κB CXCL8, RANTES, and IFNα (an inducer of NF-κB activation) significantly in the treatment group compared with the placebo group

Note: cGMP, cyclic guanosine monophosphate; CVD, cardiovascular disease; CXCL8, controlled inflammatory chemokines of interleukin-8; HDL, high-density lipoprotein; ICAM, intercellular adhesion molecule; IFNα, interferon alpha; LDL, low-density lipoprotein; NF-κB, nuclear factor kappa B; RANTES, regulated upon activation, normal T-cell expressed and secreted; VCAM, vascular cell adhesion molecule.

positively influence CVD. Erlund et al. (2008) found that consumption of moderate amounts of berries by 72 middle-aged individuals with cardiovascular risk factors resulted in favorable changes in platelet function, HDL cholesterol, and blood pressure. The subjects were given whole bilberries (100 g) and one nectar containing 50 g of crushed lingonberries every other day as well as black currant or strawberry puree (100 g, containing 80% black currants) and cold-pressed chokeberry and raspberry juice (0.7 dL juice, containing 80% chokeberry) on the opposite days for 8 weeks in a single-blind, randomized, placebo-controlled intervention trial. Serum HDL cholesterol increased significantly more in the berry treatment group versus the control group (5.2% and 0.6%, respectively; $p < 0.006$). Platelet function was inhibited significantly more in the berry versus the placebo group (11% and –1.4%; $p = 0.018$). Systolic blood pressure decreased significantly, and the decrease mostly occurred in subjects with high baseline blood pressure (7.3 mm in the highest tertile; $p = 0.024$) (Erlund et al. 2008). A study of 18 patients suffering from hypertriglyceridemia examined the effects of blond and red grapefruits on serum triglycerides and lipid levels. The study showed that red grapefruits versus blonde grapefruits showed a significantly greater decrease compared with the control group in total cholesterol (15.5% vs. 7.6%), LDL cholesterol (20.3% vs. 10.7%), and triglycerides (17.2% vs. 5.6%) (Gorinstein et al. 2006). Pomegranate juice consumption effectively reduced the progression of carotid lesions, blood pressure, and LDL oxidation in 19 patients with carotid artery stenosis. In the control group that did not consume pomegranate juice, common carotid intima-media thickness increased by 9% during 1 y, whereas pomegranate juice consumption of 50 mL/d resulted in a significant reduction by up to 30% after 1 y. Serum levels of antibodies against oxidized LDL cholesterol were decreased by 19%; in parallel, serum total antioxidant status increased by 130% after 1 y of pomegranate juice consumption. Systolic blood pressure was also reduced by 21%; however, no further reduction was noted past 1 y of consumption (Aviram et al. 2004). Pomegranate juice consumption of 240 mL/d also reduced the extent of stress-induced ischemia in 45 patients who had CHD and myocardial ischemia ($p < 0.05$). Study participants underwent electrocardiographic-gated myocardial perfusion single-photon emission computed tomographic technetium-99 m tetrofosmin scintigraphy at rest and during stress at baseline and 3 months. A blinded independent nuclear cardiologist performed visual scoring of images using standardized segmentation and nomenclature (17 segments, 0–4 scale). The benefit was observed without changes in cardiac medications (Sumner et al. 2005). Chokeberry extracts given three times daily (255 mg flavonoids) for 6 weeks significantly enhanced the reduction of cardiovascular risk markers in 44 patients who survived myocardial infarction. The study showed reduced serum 8-isoprostans and oxidized LDL cholesterol levels by 38% and 29% ($p < 0.001$), as well as reductions in C-reactive protein ($p < 0.007$) and monocyte chemoattractant protein-1 (MCP-1) ($p < 0.0001$) levels by 23% and 29%, respectively. In addition, systolic and diastolic blood pressures were decreased by a mean average of 11 and 7.2 mmHg, respectively (Naruszewicz et al. 2007). Cranberry juice (double strength; 835 mg total polyphenols and 94 mg anthocyanins) consumption over 4 weeks decreased central aortic stiffness in a chronic crossover study of 44 subjects with CHD. Carotid-femoral pulse wave velocity decreased after cranberry juice consumption (8.3 ± 2.3 to 7.8 ± 2.2 m/s) in contrast with an increase after administration of the placebo (8.0 ± 2.0 to 8.4 ± 2.8 m/s) ($p < 0.003$) (Dohadwala et al. 2011). Short-term ingestion of purple grape juice improved flow-mediated dilation and reduced LDL cholesterol susceptibility to oxidation in patients ($n = 15$) with coronary artery disease. At baseline, flow-mediated dilation was impaired ($2.2\% \pm 2.9\%$) but increased ($6.4\% \pm 4.7\%$) ($p = 0.003$) after ingestion of purple grape juice, demonstrating that endothelium-dependent vasodilation and prevention of LDL cholesterol oxidation are important mechanisms in the prevention of cardiovascular events (Stein et al. 1999). Sixteen female participants with metabolic syndrome showed significant improvements in lipid profiles and lipid peroxidation after consumption of 25 g/d of freeze-dried strawberry pomace for 4 weeks. Total cholesterol and LDL cholesterol levels were significantly lower at 4 weeks versus baseline (-5% and -6%, respectively; $p < 0.05$), as was lipid peroxidation in the form of malondialdehyde and hydroxynonenal (-14%; $p < 0.01$) (Basu et al. 2009).

These studies, as well as other pertinent clinical studies involving anthocyanin-rich extracts and individuals with elevated CVD risk biomarkers, are summarized in Table 6.3.

6.6.2 STUDIES WITH HEALTHY SUBJECTS

Many well-designed clinical trials of healthy individuals have been conducted; however, the evidence of a positive effect is not as concrete due to the absence of elevated CVD risk biomarkers and other potential confounding healthy habits exhibited by these individuals. A study of 34 healthy volunteers showed a small but not statistically significant positive change in total cholesterol concentrations among the treatment group taking 400 mg of spray-dried elderberry powder capsules (10% anthocyanins) versus the control (Murkovic et al. 2004), suggesting a protective and/or preventative effect. The treatment also had minor but insignificant effects on serum antioxidant capacity (Murkovic et al. 2004). On the contrary, a study of 52 healthy volunteers showed no significant effects of a 12-week intervention consisting of 500 mg/d of anthocyanins administered in the form of elderberry extract capsules on inflammatory markers, platelet reactivity, lipids, and glucose (Curtis et al. 2009). Consumption of Bing sweet cherries did seem to alter concentrations of inflammatory markers but not cholesterol and triglyceride levels in 18 healthy subjects over a 28-day period. After cherries were consumed for the 28-day interval, circulating concentrations of C-reactive protein, RANTES, and nitric oxide decreased by 25% ($p < 0.05$), 21% ($p < 0.05$), and 18% ($p = 0.07$), respectively (Kelley et al. 2006). Four-week ingestion of blood orange juice on the contrary did not significantly affect cellular markers related to CVD risk (Giordano et al. 2012). Table 6.4 summarizes the totality of evidence with regard to clinical intervention trials involving anthocyanin-rich extracts and healthy individuals.

6.7 MECHANISMS BY WHICH ANTHOCYANINS INFLUENCE THE CARDIOVASCULAR SYSTEM

In the absence of strong conclusive clinical data, animal models and *in vitro* cell line work have become critical to understanding the relationships between anthocyanin-rich diets and CVD prevention at physiologic, metabolic, biochemical, and molecular

TABLE 6.3

Summary of Clinical Intervention Trials That Examine Anthocyanin-Rich Foods/Extracts in Individuals with Elevated CVD Risk Biomarkers

Study, Year (Reference)	Participants	Source Intervention	Duration	Main Findings
Dohadwala et al. (2011)	44 individuals (aged 62 ± 8 y) with CHD	Cranberry juice (835 mg total polyphenols and 94 mg anthocyanins) daily	4 wk	Carotid-femoral pulse wave velocity significantly decreased in the treatment group and increased in the placebo group
Basu et al. (2010)	48 individuals (aged 50 ± 3 y) with metabolic syndrome	Beverage containing 50 g of freeze-dried blueberries	8 wk	Decreases in systolic and diastolic blood pressure were significantly greater in the blueberry supplement group versus the control Decreases in plasma oxidized LDL cholesterol and malondialdehyde and hydroxynonenal concentrations were greater in the blueberry group versus the control
Burton-Freeman et al. (2010)	24 hyperlipidemic individuals (aged 50.9 ± 15 y)	Strawberry beverage containing 10 g freeze-dried fruit daily accompanied by a high-fat meal	6 wk	Triglycerides and oxidized LDL cholesterol were lower in the strawberry treatment group after the high-fat meal was consumed Decreased lipid levels were noted after 6 wk in the treatment group
Basu et al. (2009)	16 women with metabolic syndrome (aged 51 ± 9.1 y)	50 g of freeze-dried strawberry pomace as a beverage (25 g twice daily)	4 wk	Decrease in total and LDL cholesterol and lipid peroxidation at 4 wk versus baseline ($p < 0.05$)
Erlund et al. (2008)	72 subjects with CVD risk factors (aged 57.5 ± 6.3 y)	2 portions of berries daily (one portion after lunch and the other after dinner) were administered; every other day 100 g whole bilberries and a nectar containing 50 g crushed lingonberries were consumed; 100 g of black currant or strawberry puree and 0.7 dL cold-pressed chokeberry and raspberry juice were consumed on alternate days	8 wk	Berry consumption inhibited platelet function by 11% in the treatment compared with −1.4% in the placebo group ($p < 0.02$). Serum HDL cholesterol increased by 5.2% in the treatment group compared with a 0.6% increase in the placebo group ($p < 0.01$). Total serum cholesterol and triacylglycerol were not significantly affected Systolic blood pressure decreased significantly ($p < 0.05$)

continued

TABLE 6.3 (continued)

Summary of Clinical Intervention Trials That Examine Anthocyanin-Rich Foods/Extracts in Individuals with Elevated CVD Risk Biomarkers

Study, Year (Reference)	Participants	Source Intervention	Duration	Main Findings
Lee et al. (2008)	30 subjects with type 2 diabetes (aged 65 ± 1 y)	1500 mg of cranberry extract powder administered as 3×500-mg capsules daily	12 wk	Decrease in total cholesterol, LDL cholesterol, and total cholesterol/HDL cholesterol ratio ($p < 0.05$). No effects on glucose or glycated hemoglobin
Naruszewicz et al. (2007)	44 individuals (aged 65.87 ± 8.3 y) who survived myocardial infarction and received statin therapy	3×85 mg of chokeberry extract daily	6 wk	Chokeberry extract significantly reduced serum 9-isoprostans and oxidized LDL cholesterol by 38% and 29%, respectively C-reactive protein and MCP-1 decreased by 23% and 29%, respectively A reduction in systolic and diastolic blood pressure by an average of 11 and 7.2 mmHg was also noted
Gorinstein et al. (2006)	57 hyperlipidemic subjects (aged 39–72 y) with previous coronary bypass surgery	1 blond or red Israeli grapefruit daily	30 d	Red grapefruit consumption decreased total cholesterol 15.5% ($p < 0.01$), LDL cholesterol by 20.3% ($p < 0.005$), and triglycerides by 17.2% ($p < 0.005$) Red grapefruit consumption had a more positive impact on lipid profiles than did blond grapefruit
Summer et al. (2005)	45 individuals who had CHD and myocardial ischemia	240 mL pomegranate juice daily	3 mo	Pomegranate juice consumption significantly reduced the extent of stress-induced ischemia

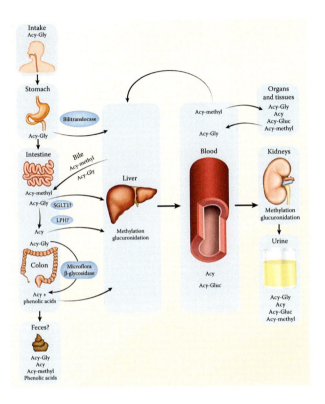

FIGURE 3.2 See text for caption.

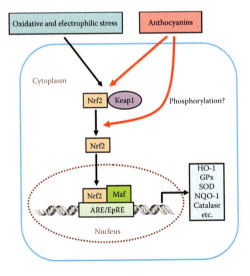

FIGURE 5.7 The potential role of ACNs in regulating factor-erythroid-2-related factor 2 (Nrf2) pathway.

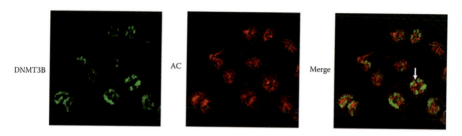

FIGURE 8.2 Cellular uptake of black raspberry-derived anthocyanins (AC) and colocalization of AC with DNMT3B in HCT116 cells. AC uptake was observed in HCT116 cells treated with AC at 25 μg/mL for 1 day. AC in red presented in both cytoplasm and nuclei. Same cells were stained with DNMT3B shown in green. Colocalization of AC with DNMT3B appeared in yellowish as indicated by arrowhead.

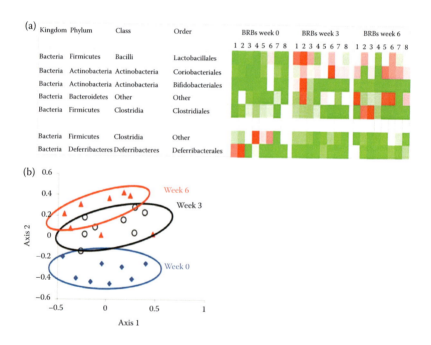

FIGURE 8.4 (a) Heatmap analyses of the fecal microbiota. Heatmaps were generated for the BRB-treated rats by normalizing the percent abundance of each bacterial order within each sample to the average number of sequences for all 24 samples in each treatment group. Green-to-red shading indicates relatively lower-to-higher number of sequences. (b) Bray–Curtis analysis of the intestinal microbiota in response to diet. Nonmetric multidimensional scaling (NMDS) ordination of the Bray–Curtis distances showed that the samples at 3 and 6 weeks were progressively distant (i.e., different in biodiversity) from the pretreatment samples for BRB diet supplementation. The R package Ecodist 1.2.3 was used to calculate Bray–Curtis distances and perform NMDS.

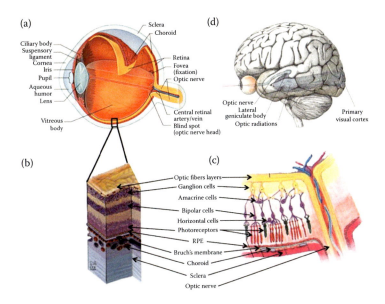

FIGURE 9.1 Anatomy of the primate visual system. (a) Macroanatomy of the eyeball, which includes the transparent media (cornea, lens, aqueous humor filling the space between the lens and the cornea, and the vitreous filling the posterior part of the eye) the ciliary body and suspensory ligaments responsible for accommodation, and the retina lying at the back of the eye. (http://visionandeyecare.wordpress.com/tag/retina/) (b) The retina is composed of several layers of cells, with the photoreceptors facing the RPE that is separated from the choroidal vascular plexus by the Bruch's membrane. (Modified from http://www.mcponline.org/content/9/6/1031.full.pdf + html) (c) The retina is composed of various types of cells. Photoreceptors, bipolar, and ganglion cells are glutamatergic excitatory cells, whereas horizontal (GABAergic) and amacrine (mostly GABAergic and glycinergic) are providing inhibitory inputs. Axons from the ganglion cells gather in the nerve fiber layer (NFL) to reunite at the optic nerve. (Modified from http://www.cidpusa.org/senses.htm) (d) Ganglion cell axons in the optic nerve synapse mainly in the lateral geniculate body, a small relay thalamic structure that conveys the information to the primary visual cortex via the optic radiations.

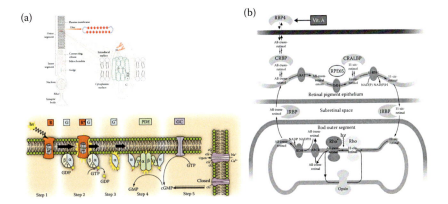

FIGURE 9.2 See text for caption.

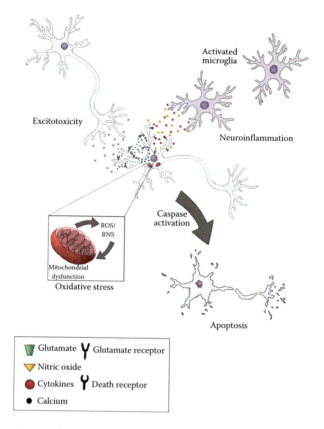

FIGURE 10.1 See text for caption.

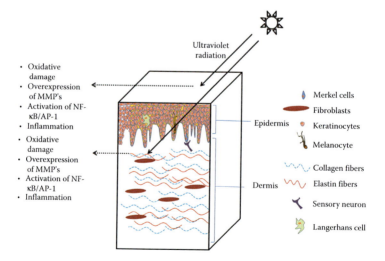

FIGURE 11.1 Schematic representation of the skin architecture and mechanisms of skin damage.

McAnulty et al. (2005)	20 smokers (aged 26 ± 4.2 y)	Acute or daily consumption of 250 g blueberries	3 wk or postprandial	Decrease in lipid hydroperoxides in the blueberry group versus the control group at 3 wk ($p < 0.001$)
Murkovic et al. (2004)	34 healthy subjects (aged 24–36 y)	400 mg spray-dried elderberry juice (containing approximately 100 mg anthocyanins) administered in the form of gelatinous capsules daily	2 wk (14 subjects continued for 1 additional wk)	A small but insignificant decrease in total cholesterol, triglycerides, and LDL cholesterol as well as an insignificant increase in HDL cholesterol in the treatment group was noted Postprandial triglyceride concentrations were not significantly different when the six subjects were investigated with and without the elderberry juice gelatin capsules
Aviram et al. (2004)	19 individuals (aged 65–75 y) with carotid artery stenosis	50 mL pomegranate juice daily	3 y	Consumption of pomegranate juice resulted in a significant reduction of carotid intima-media thickness by 30% compared with an increase of 9% in the control group after 1 y Serum levels of antibodies against oxidized LDL cholesterol decreased by 19% and serum total antioxidant status increased by 130% after 1 y Systolic blood pressure was reduced by 21% after 1 y
Chambers and Camire (2003)	27 individuals with type 2 diabetes (aged 56 ± 13 y)	Cranberry juice concentrate powder (6 capsules daily)	12 wk	No effect on fasting glucose, lipids, or HbA_{1C} levels ($p > 0.05$)
Simeonov et al. (2002)	62 patients with type 1 or type 2 diabetes (aged 46.2 ± 4 y)	200 mL chokeberry juice daily	3 mo	Decrease in lipid profiles, fasting glucose and HbA_{1C} levels ($p < 0.001$)
Stein et al. (1999)	15 individuals (aged 62.5 ± 12.7 y) with coronary artery disease	Patients were given 7.7 ± 1.2 mL/kg purple grape juice daily	14 d	At baseline flow-mediated dilation was impaired ($2.9\% \pm 2.2\%$) but increased ($6.4\% \pm 4.7\%$) after ingestion of purple grape juice Purple grape juice reduced LDL cholesterol susceptibility to oxidation

Note: CHD, coronary heart disease; CVD, cardiovascular disease; HDL, high-density lipoprotein; LDL, low-density lipoprotein; MCP-1, monocyte chemoattractant protein-1.

TABLE 6.4
Summary of Clinical Intervention Trials That Examine Anthocyanin-Rich Foods/Extracts in Healthy Individuals

Study, Year (Reference)	Participants	Source Intervention	Duration	Main Findings
Giordano et al. (2012)	18 healthy men and women (aged 23–44 y)	1 L of blood orange juice per day	4 wk	Cellular markers such as platelets and leukocytes were not significantly affected
Jin et al. (2011)	20 healthy subjects (aged 44.5 ± 13.3 y)	250 mL of a 20% black currant juice drink	120 min (vascular reactivity; 8 h plasma and urine sample; 24 h urine sample)	No significant effects on acute measures of vascular reactivity, biomarkers of endothelial function, or lipid risk factors
Curtis et al. (2009)	52 healthy postmenopausal women (aged 58.2 ± 5.6 y)	500 mg anthocyanins (administered as 250 mg twice daily) as elderberry extract capsules daily	12 wk	No significant changes in CVD biomarkers were found in clinically acceptable ranges Postprandial metabolism increased ($p < 0.02$)
Jensen et al. (2008)	12 healthy subjects (aged 19–52 y)	120 mL juice blend containing acaí berry, cranberry, blueberry, wolfberry, and bilberry in addition to other fruit juices daily	Postprandial	Increase in serum antioxidant status and inhibition of lipid peroxidation versus placebo ($p < 0.03$)
Ruel et al. (2008)	30 healthy men (aged 51 ± 10 y)	125, 250, and 500 mL/d cranberry juice cocktail	16 wk	Decrease in plasma oxidized LDL cholesterol, adhesion molecules (ICAM and VCAM) and systolic blood pressure at 12 and 16 wk ($p < 0.05$)

Reference	Subjects	Dose	Duration	Results
Kelley et al. (2006)	18 healthy subjects (aged 45–61 y)	280 g Bing sweet cherries daily	4 wk	Circulating concentrations of C-reactive protein, RANTES, and nitric oxide decreased by 25 ($p < 0.05$), 21 ($p < 0.05$), and 18% ($p < 0.07$), respectively. Plasma lipid profiles were not significantly affected
Duthie et al. (2006)	20 healthy women (aged 28 ± 7 y)	750 mL (250 mL, 3 × daily) cranberry juice	2 wk	No effect on blood or cellular antioxidant status, lipid status, or oxidative DNA damage ($p > 0.05$)
Ruel et al. (2006)	30 healthy men (aged 51 ± 10 y)	Increasing doses of cranberry juice cocktail (125, 250, 500 mL/d) during 3 successive 4-week periods	Four successive 4 wk phases	Increase in HDL cholesterol in the 250 mL/d treatment group
Marniemi et al. (2000)	60 healthy individuals (mean age 60 y)	100 g deep-frozen berries (bilberries, lingonberries, or black currants); 240 g berries in postprandial study	8 wk and postprandial	Decrease in LDL cholesterol oxidation and slight increase in serum antioxidant capacity

Note: HDL, high-density lipoprotein; ICAM, intercellular adhesion molecule; LDL, low-density lipoprotein; RANTES, regulated upon activation, normal T-cell expressed and secreted; VCAM, vascular cell adhesion molecule.

levels. As our understanding of gene homology has improved, confidence has grown that what is learned in common animal models has a good likelihood of being relevant to human biology. *In vitro* gastrointestinal tract simulator and animal models of defined microbiota have been and will likely continue to be extremely relevant in characterizing the breakdown and metabolic derivatives of anthocyanins, which may have a greater potential to cross the basolateral membrane into the circulation and positively influence CVD compared with the parent molecule. There are also caveats to using these models. Humans and animals have many important dissimilarities. In addition, the relevance of many animal and *in vitro* studies to the clinical *in vivo* situation must be confirmed because many of these studies apply high concentrations of anthocyanins (not taking into account breakdown and/or metabolic derivatives) that far exceed the level observed clinically *in vivo*. However, some studies do achieve results at comparable levels.

Mechanistic studies support the beneficial effects of flavonoids, including anthocyanins, on the established biomarkers of CVD risk including nitric oxide, inflammation, and endothelial dysfunction (Loke et al. 2008; Pergola et al. 2006; Steffen et al. 2008). Inflammation defined by calor (heat), rubor (redness), and tumor (swelling) plays a major role in the development of CVD. The role of anthocyanins in CVD prevention is linked to protection against oxidative stress. Several mechanisms of action have been proposed to explain the *in vivo* anti-inflammatory actions of anthocyanins and other flavonoids. Anthocyanin isolates and anthocyanin-rich mixtures of flavonoids may provide protection from DNA cleavage, estrogenic activity (altering the development of hormone-dependent disease symptoms), enzyme inhibition, boosting production of cytokines (thus regulating immune responses), anti-inflammatory activity, lipid peroxidation, decreasing capillary permeability and fragility, and membrane strengthening (Acquaviva et al. 2003; Lazze et al. 2003; Lefevre et al. 2004; Ramirez-Tortosa et al. 2001; Rossi et al. 2003). Dietary anthocyanins have been shown to accumulate in the tissues of pigs during long-term feeding and have a longer residence time in tissues than in the bloodstream (Kalt et al. 2008). Whether anthocyanins accumulate in the cardiac or vascular tissues during long-term feeding is still unknown; however, anthocyanins do affect vascular reactivity in animal studies (Kalea et al. 2009). A diet rich in foods high in anthocyanins such as blueberries may protect the myocardium from induced ischemic damage and may have the potential to attenuate the development of postmyocardial infarction heart failure (Ahmet et al. 2009). Taking these comments into account, the overwhelming body of evidence obtained from these studies suggests that anthocyanins influence many mechanisms involved with the onset and progression of CVD.

6.7.1 DIRECT SCAVENGING EFFECTS

LDL cholesterol accumulated in the extracellular subendothelial space of the arteries is mildly oxidized to a minimally modified LDL cholesterol. This form of LDL cholesterol induces local vascular cells to produce molecules such as MCP-1 as well as granulocyte and macrophage colony-stimulating factors, which stimulate monocyte recruitment and differentiation to macrophages in arterial walls. These monocytes and macrophages stimulate further peroxidation of LDL cholesterol and the formation of foam cells. In addition to the creation of foam cells, the now fully oxidized LDL cholesterol has the ability to bind monocytes to the endothelium, thus initiating the formation of atherosclerotic lesions. Several important mechanisms may underlie the role of antioxidants (such as anthocyanins) in preventing CVD. Incorporation of antioxidants into LDL cholesterol protects the LDL cholesterol against oxidation and leads to the reduced formation of oxidized LDL cholesterol. In addition, incorporation of antioxidants may reduce vascular cell oxidation of LDL cholesterol and the cellular responses to oxidized LDL cholesterol, resulting in less monocyte adhesion, foam cell formation, cytotoxicity to vascular cells, and vascular dysfunction (Diaz et al. 1997) (Figure 6.1).

There is a very large amount of evidence that suggests that anthocyanins, like other flavonoids, possess direct scavenging activity against free radicals or ROS in general, thus allowing them to potentially prevent the oxidation of LDL cholesterol, vascular cell oxidation of LDL cholesterol, and/or cellular responses to oxidized LDL cholesterol at least *in vitro*. The direct ROS-scavenging properties and activity of anthocyanins have been assessed by many scientists using various ROS/oxidative stress inducers such as DPPH (1,1-diphenyl-2-picrylhydrazyl) among others. Even though results of these studies are not always in agreement, in general the larger number of free hydroxyl groups corresponds to a greater scavenging effect. However, as stated previously, the position of the hydroxyl groups present is crucial to their exerted effects. Yi et al. (2010) found that the number of hydroxyl groups on the B-ring, a 3′,4′-orthodihydroxyl group on the B-ring, and a 3-hydroxyl group on the C-ring seemed to be the main structural requirements of anthocyanins for the inhibitory effects on the oxidative injury to endothelial cells induced by oxidized LDL cholesterol. Delphinidin has been shown to effectively protect human umbilical

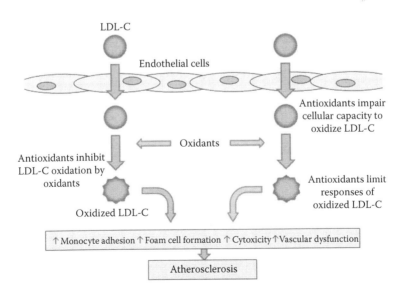

FIGURE 6.1 LDL-specific (left) and tissue-specific (right) mechanisms of antioxidant action. Amended from Diaz et al. (1997). *Note:* LDL-C, low-density lipoprotein cholesterol.

vascular endothelial cells against oxidative stress induced by oxidized LDL cholesterol through decreasing the concentration of intracellular ROS, suppressing the production of lipid peroxidation, and restoring the activities of endogenous antioxidants (Chen et al. 2010). Nicotinamide adenine dinucleotide phosphate (NADPH) oxidases are a major source of ROS generation in endothelial cells. Delphinidin attenuated oxidized LDL cholesterol-induced generation of ROS, of p38 mitogen-activated protein kinase (p38 MAPK) protein expression, nuclear factor kappa B (NF-κB) transcription activity and protein expression, inhibitor of kappa B alpha (IκB-α) degradation, and NADPH oxidase subunit (Nox2 and p22phox) protein and mRNA expression in endothelial cells in a dose-dependant manner (Chen et al. 2011). Accordingly, four anthocyanins isolated from elderberries were incorporated into the plasma and cytosol of endothelial cells to directly examine the protective roles. The results demonstrated that not only could anthocyanins be directly incorporated into the endothelial cells, but also that significant oxidative stress protection was a result (Youdim et al. 2000). *Hibiscus* anthocyanin-rich extracts decreased oxidized LDL cholesterol-mediated foam cell formation. A dramatic increase in lipid accumulation was observed in mouse macrophage J774A.1 cells treated with only oxidized LDL cholesterol (control group) compared with those treated with oxidized LDL cholesterol and *Hibiscus* anthocyanin-rich extracts (treatment group), which largely prevented the lipid accumulation (Kao et al. 2009). The theory of oxidative stress protection by anthocyanins is supported by many rat and clinical studies that show a positive correlation between total serum anthocyanins and antioxidant capacity of the serum.

6.7.2 INHIBITION OF ROS-FORMING ENZYMES

Anthocyanins seem to inhibit ROS-forming enzymes such as xanthine oxidase, lipoxygenases, and NADPH oxidase. NADPH oxidase is a multi-subunit enzymatic complex responsible for the production of superoxide due to the monoelectronic reduction of oxygen using NADPH or NADH. This enzyme complex is considered to be the most relevant source of superoxide in the vessel wall. Mazza et al. (2002) found that the concentration of anthocyanins in the serum was directly correlated with the serum antioxidant capacity using the oxygen radical absorbance capacity (ORAC) assay when men were supplemented with 1.2 g of anthocyanins from freeze-dried blueberries. This change in antioxidant capacity suggests that anthocyanins and their secretory products may play an important role in decreasing superoxide production of NADPH oxidase in addition to other possible mechanisms. A decrease in NADPH oxidase activity typically leads to an increase in serum antioxidant capacity. It has been proposed that eNOS metabolism, rather than general antioxidant activity, is a major target of flavanols, a similar class of flavonoids, and that NADPH oxidase activity is a crucial site of action (Choi et al. 2008). The same theory could hold true for anthocyanins because of their similarity in chemistry and demonstrated positive correlation to serum antioxidant capacity *in vivo*. Particularly, the benzene ring with adjacent methoxy-hydroxyl group structure of anthocyanidins such as malvidin, peonidin, and petunidin seems to be significant for the inhibition of NADPH oxidase. This structure is also significant in other drugs and natural compounds that have shown inhibitory activity (Figure 6.2).

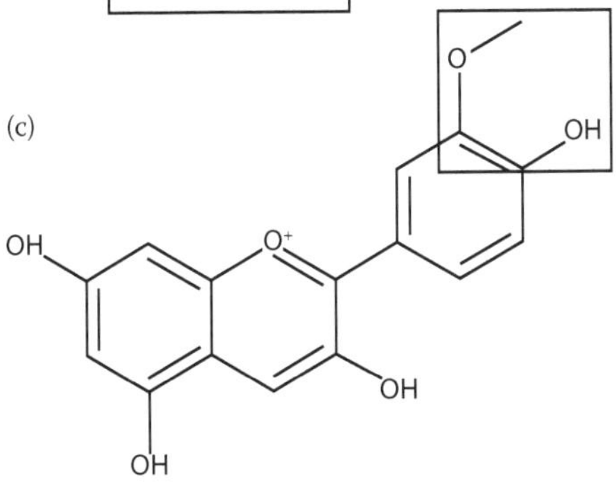

FIGURE 6.2 Compounds inhibiting NADPH oxidase. The most important feature seems to be a benzene ring with adjacent methoxy-hydroxyl groups. (a) Isorhamnetin, a known metabolite of quercetin and an active NADPH oxidase inhibitor; (b) the known drug apocynin; and (c) peonidin, an anthocyanidin.

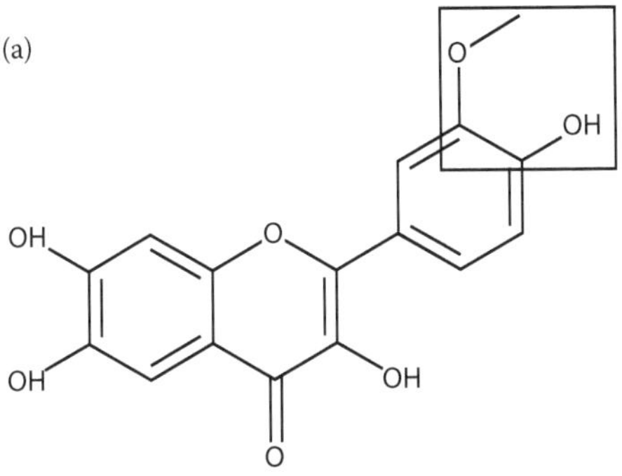

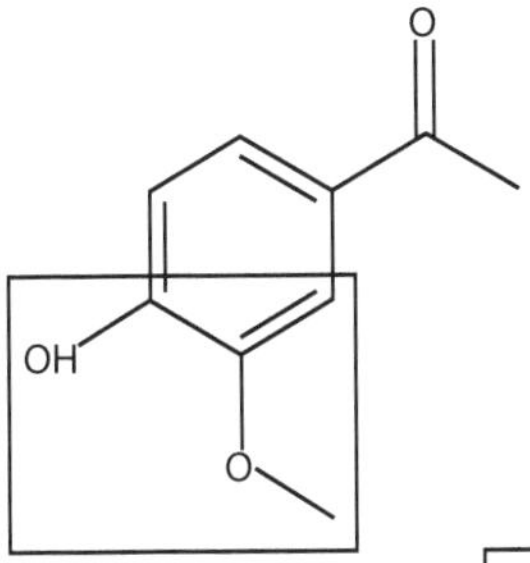

In contrast to other flavonoids, anthocyanidins with free hydroxyl groups are more active when it comes to the competitive inhibition of xanthine oxidase, compared with corresponding substances without hydroxyl groups (Costantino et al. 1995). Knaup et al. (2009) studied the inhibitory potential of delphinidin, cyanidin, peonidin, and malvidin glycosides, as well as their aglycons, on soybean lipoxygenases-1 and human neutrophil granulocyte 5-lipoxygenase. Delphinidin-3-glucoside and delphinidin-3-galactoside were found to be the most effective inhibitors of lipoxygenases, although all anthocyanins showed inhibitory activity. The study concluded that glycosylation of the anthocyanidins with at least two hydroxyl groups in an *ortho*

position improves their efficacy with regard to inhibition of lipoxygenases (Knuap et al. 2009).

6.7.3 DECREASE IN EXPRESSION OF INFLAMMATORY SIGNALING MOLECULES

6.7.3.1 Cyclooxygenases

Cyclooxygenase-2 (COX-2) is expressed very little in resting macrophages; however, inflammation upregulates the expression. Likewise with iNOS inhibition, the effect does not seem to be primarily caused by direct inhibition of enzymes, although this may participate in the overall effect, and it is likely mediated by the inhibition of the NF-κB pathway. Abnormal upregulation of COX-2 inflammatory proteins is common among patients and laboratory animals with CVD. Interestingly, through their ability to inhibit mRNA and/or protein expression levels of COX-2, NF-κB, and various interleukins, anthocyanins have exhibited anti-inflammatory effects *in vitro* (Zhang et al. 2008). All flavonoids, including anthocyanins, are relatively poor direct inhibitors of cyclooxygenase-1 (COX-1) although they may slightly decrease its expression (Ha et al. 2008).

6.7.3.2 Nitric Oxide

Vascular endothelial cells play an important role in maintaining cardiovascular health by producing nitric oxide, a compound that promotes arterial relaxation (vasodilation). Nitric oxide is a signaling molecule that influences the development of atherosclerosis and many aspects of inflammation, ranging from its own production to immunocompetent cells to the recruitment of leukocytes (Bogdan 2001). Nitric oxide is produced from L-arginine by three nitric oxide synthase enzymes: endothelial nitric oxide synthase (eNOS), neuronal nitric oxide synthase (nNOS), and inducible nitric oxide synthase (iNOS). Nitric oxide is a potent vasodilator with antihypertensive, antithrombotic, antiatherogenic, and antismooth muscle proliferative properties (Vallence 2000). Although it is generally considered a protective substance in low concentrations, it may be toxic in higher levels, especially with regard to its excessive production by iNOS. Arterial vasodilation resulting from endothelial production of nitric oxide is termed *endothelium-dependent vasodilation*. Several reports have shown that anthocyanins and other flavonoids are able to inhibit excessive release of iNOS.

The eNOS protein has been shown to be impaired in conditions associated with atherosclerosis, hypertension, diabetes, and ischemia–reperfusion injury (Dillon and Vito 2000). These conditions are also associated with the production of ROS that can chemically quench nitric oxide and/or damage the endothelium, impairing nitric oxide production by eNOS. Anthocyanin concentrations in the bloodstream have been thought to be too low to directly contribute to *in vivo* quenching of ROS, even though they exhibit superior antioxidant potential to classic antioxidants such as butylated hydroxyanisole. They may, however, be adequate to improve endothelial function by influencing nitric oxide levels. Chokeberry and bilberry anthocyanin-rich extracts have the ability to prevent loss of endothelium-dependent nitric oxide-mediated relaxation in porcine arteries *in vitro* at a level that roughly reflects that seen in several studies to exist in the human plasma after consumption of these compounds (Bell and Gochenaur 2006). Four anthocyanins isolated from elderberries were incorporated into the plasma, lemma, and cytosol of endothelial cells *in vitro* to directly examine their role. The results from this study indicate that anthocyanins can be directly incorporated into endothelial cells and produce significant oxidative stress protection (Youdim et al. 2002). Delphinidin provided endothelium-dependent vasorelaxation in the rat aorta, comparable with that of red wine polyphenols (Andriambeloson et al. 1998). A similar finding with black currant concentrate was reported in rat aorta rings *in vitro* (Nakamura et al. 2002). Treatment of bovine artery endothelial cells with cyanidin-3-glucoside for 8 h enhanced eNOS protein expression in a dose- and time-dependent manner, determined by Western blot analysis. Longer incubation (12, 16, and 24 h) with 0.1 μmol/L of cyanidin-3-glucoside caused a further increase in eNOS expression, and subsequently increased nitric oxide output twofold (24 h). In just 12 min of exposure of bovine endothelial cells to cyanidin-3-glucoside, NOS activity was enhanced because of the phosphorylating effects on eNOS (Xu et al. 2004a,b). Other fruit pigment preparations have been shown to produce endothelium-dependent relaxation of arteries; however, these effects have largely been confined to the pigments of red wine and grapes (Andriambeloson et al. 1998; Dell'Agli et al. 2004; Mendes et al. 2003; Ndiaye et al. 2003; Stein et al. 1999). Protection from heart attacks through administration of grape juice and red wine was strongly tied to the ability of anthocyanins to reduce inflammation, inhibit platelet formation, and enhance nitric oxide release (Schewe et al. 2008).

Consequences from excessive production of nitric oxide may be positively influenced by flavonoids due to scavenging of peroxynitrite, a product of nitric oxide and superoxide (Kim et al. 2004). More detailed analysis has shown that active flavonoids inhibit excessive iNOS synthesis through the prevention of IκB degradation and the consequent inhibition of NF-κB activation (Liang et al. 1999). Pelargonidin has been shown to inhibit iNOS and mRNA expression, as well as the production of nitric oxide in a dose-dependent manner in macrophages exposed to the inflammatory stimulus lipopolysaccharide (LPS) (Hämäläinen et al. 2007). Some anthocyanins may inhibit iNOS expression through the NF-κB pathway, but additional studies are required for establishment of the fundamental structure (Garcia-Alonso et al. 2004; Wang et al. 2008). Black rice extract containing known proportions of cyanidin-3-glucoside and peonidin-3-glucoside significantly reduced the formation of nitric oxide by suppressing iNOS expression in murine macrophage RAW 264.7 cells without inducing cell toxicity (Hu et al. 2003).

6.7.4 Vasodilatory Action

A variety of anthocyanins have shown the ability to dilate blood vessels. The studies present in the current literature utilize a wide range of methodological approaches and various types of arteries. Fruit pigment preparations have been shown to produce endothelium-dependent relaxation of arteries, but these determinations have been largely confined to pigments contained in red wine or grapes (Andriambeloson et al. 1998; Dell'Agli et al. 2004; Mendes et al. 2003; Ndiaye et al. 2003; Stein et al. 1999). Several studies have shown that anthocyanin-rich foods can activate endothelial

NOS and improve endothelial function *in vitro* and in animal models (Agewall et al. 2000; Iwasaki-Kurashige et al. 2006). The totality of evidence indicates that anthocyanins seem to induce endothelium-dependent vasorelaxation by activating endothelial NOS and consequently increasing production of the vasorelaxant agent nitric oxide. However, it is not yet clear if anthocyanin-induced vasorelaxation starts with their interaction with plasma membrane receptors in the extracellular compartment or with their membrane transport toward intracellular molecular targets. The mechanism of vasodilatory action has not been fully elucidated. For this reason, vascular reactivity was assessed in thoracic aortic rings obtained from male Wistar rats. Pretreatment of aortic rings with antisequence bilitranslocase antibodies targeting the carrier decreased vasodilation induced by cyanidin 3-glucoside and bilberry anthocyanins, showing for the first time that bilitranslocase mediates a critical step in vasodilation induced by anthocyanins (Ziberna et al. 2011). This research offers new insights into molecular mechanisms involved in endothelium-dependent vasorelaxation by flavonoids as well as the importance of their specific membrane carriers. In addition, anthocyanin-rich extracts from chokeberry, bilberry, and elderberry have shown endothelium-dependent relaxation capacity in porcine coronary arteries. All extracts produced a dose-dependent effect, with chokeberry exhibiting the highest potency. It is clear that berry extracts in this study significantly attenuated the loss of A23187-mediated relaxation of coronary arteries exposed to pyrogallol. It is most likely that the beneficial effects of these extracts are due to their ORAC, which would tend to reduce the effective concentration of O_2 or ROS from pyrogallol reaching the arterial endothelium (Bell and Gochenaur 2006). Along these lines, eNOS in bovine vascular endothelial cells is upregulated after 6-h exposure to 0.1 M cyanidin-3-glucoside. Twelve-minute exposure of these cells to cyanidin-3-glucoside phosphorylates nitric oxide synthase and enhances its activity (Xu et al. 2004a,b). Chronic ingestion of anthocyanins increases cardiac glutathione concentrations in rats (Toufektsian et al. 2008). Alcohol-free red wine showed a protective effect in rat hearts by repressing hypertrophy-associated increased phosphorylation of protein kinase C α/βII and by activating Akt/protein kinase B (Palfi et al. 2009). The synthesized anthocyanin-derivative HK-008 increases endothelium-induced relaxation by suppressing oxidative stress and modulating prostanoid signaling (Kamata et al. 2006).

A recent study evaluating the effects of purified anthocyanins from bilberries and black currants on endothelial function in humans showed promising effects of the compounds with both short- and long-term use. During the initial short-term crossover study, 12 hypercholesterolemic patients showed significant increases in flow-mediated dilation from 8.3% ± 0.6% at baseline to 11.0% ± 0.8% at 1 h and 10.1% ± 0.9% at 2 h after oral administration of 320 mg of purified anthocyanins from bilberry and black currant during a 4-h intervention trial. For the long-term, 12-week, double-blind, placebo-controlled parallel intervention trial, the 150 hypercholesterolemic patients were randomized to receive 320 mg of purified anthocyanins from bilberry and black currant or a placebo. The anthocyanin treatment group showed significant increases in flow-mediated dilation compared with the placebo group (28.4% vs. 2.2%) at the end of the 12-week study. Changes in the cGMP and HDL cholesterol concentrations positively correlated with flow-mediated dilation

(FMD) in the anthocyanin group ($p < 0.05$). In the presence of nitric oxide-cGMP inhibitors, the effects of anthocyanins on endothelial function were abolished in human participants and in a rat aortic ring model (Zhu et al. 2011). This study was the first to examine the clinical benefits of the consumption of purified anthocyanins on endothelial functions.

Anthocyanin-rich extracts have also been found to affect vasodilation in patients with coronary artery disease, as previously mentioned. Mean carotid-femoral pulse wave velocity, a measure of central aortic stiffness, decreased after cranberry juice consumption; however, brachial artery flow-mediated dilation did not change in a chronic crossover study of 15 individuals with coronary artery disease. Brachial artery flow-mediated dilation did show improvement during the uncontrolled pilot study after consumption of a 480-mL portion of cranberry juice (Dohadwala et al. 2011). Purple grape juice improved flow-mediated dilation in patients with coronary artery disease. The effects of purple grape juice remained significant ($p < 0.01$) in a linear regression model that included age, artery diameter, lipid values, and use of lipid-lowering and antioxidant therapies (Stein et al. 1999).

6.7.5 PLATELET AGGREGATION

Activated blood platelets play a central role in the development of CVD, comparable with hypertension or diabetes, because they contribute to plaque formation within blood vessels in the early stages of atherogenesis (Assmann et al. 1999). During this process, platelets become activated upon binding to collagen and von Willebrand factor (vWF) multimers, which are secreted in response to inflammatory stimuli from damaged endothelial cells (Ruggeri 2002). Collagen and vWF bind to receptors and the receptor complex glycoprotein on the platelet surface (Davi and Patrono 2007). The activated platelets then secrete a range of adhesion molecules and bind fibrinogen from the plasma. In addition, they synthesize and secrete agonists such as adenosine diphosphate and thromboxane A_2 (TXA_2), which also induce platelet aggregation and thus amplify and maintain the initial platelet response (Davi and Patrono 2007). Activated platelets further intensify atherogenesis by contributing to the recruitment and binding of leukocytes (Ruggeri 2002).

For many years, it has been known that some flavonoids may decrease platelet aggregation caused by an agonist of the thromboxane receptor, arachidonic acid, or collagen, but not by thrombin (Guerrero et al. 2005). The effects of individual polyphenols and anthocyanin-rich plant extracts on platelet function have been studied *in vitro*. However, the results of these studies are lacking in their physiological relevance because they are not always applicable to the *in vivo* state (Kroon et al. 2004; Scalbert and Williamson 2000). Several studies have shown that foods rich in anthocyanins, as well as their *in vivo* metabolites, are able to inhibit platelet function (Rechner and Kroner 2005). Anthocyanins and phenolic acids seem to exhibit a synergistic effect *in vitro* at physiologically relevant levels (Rechner and Kroner 2005). From an *in vitro* perspective, co-occurring flavonoids working synergistically to antagonize hydrogen peroxide formation are most effective in depressing platelet function (Pignatelli et al. 2000). Using an *in vivo* rat model of dyslipidemia induced by a high-fat diet, dietary supplementation with anthocyanin-rich extracts from black rice reduced platelet hyperactivity and facilitated the maintenance of optimal platelet function, in addition to other reductions in hypertriglyceridemia and weight gain by means of restoring calmodulin and P-selectin levels, TXA_2 production, and the TXA_2/prostacyclin (PGI_2) ratio back to that of the controls (Yang et al. 2011). In comparing their chemical structures, it is not surprising that flavonoids such as anthocyanins interact with the receptor for thromboxane. Collagen-induced *ex vivo* whole blood platelet aggregation decreased by $49\% \pm 9\%$ in seven canines with coronary stenosis after gastric administration of 10 mL/kg of purple grape juice. The same amounts of orange juice and grapefruit juice were not effective in decreasing platelet aggregation (Osman et al. 1998). Because *in vitro* studies have suggested that flavonoids bind to cell membranes and thus may have an accumulative or tissue-loading effect over time, the same authors moved to further test this hypothesis by administering 5 mL/kg of purple grape juice per day orally to five cynomolgus monkeys for 7 days. The results mimicked those of the previous study with canines, suggesting that purple grape juice containing flavonoids such as anthocyanins may exert positive effects toward platelet aggregation (orange and grapefruit juices showed to be poor inhibitors of platelet aggregation) (Osman et al. 1998).

As mentioned in previous sections, the positive effects of anthocyanins and anthocyanin-rich extracts on platelet function have been shown in human models (Erlund et al. 2008; Freedman et al. 2001) although one clinical trial showed no effect (Curtis et al. 2009). Incubation of platelets with diluted purple grape juice *in vitro* led to inhibition of aggregation, enhanced release of platelet-derived nitric oxide, and decreased superoxide production. The *in vivo* relevance of these findings was confirmed when 20 healthy volunteers consumed 7 mL/kg of purple grape juice daily for 2 weeks (Freedman et al. 2001).

6.7.6 ADHESION MOLECULES, CYTOKINES, AND CHEMOKINES

Cytokines are mediators of local and intercellular communications required for an integrated response to a variety of stimuli in immune and inflammatory processes. Different cytokines are associated with inflammatory disease, with a clinical outcome partially determined by the balance between proinflammatory cytokines (i.e., interleukin [IL]-1β, IL-2, and IL-6; interferon [IFN]-γ; or tumor necrosis factor [TNF]-α) and anti-inflammatory cytokines (i.e., IL-10, TGF-β) (Santangelo et al. 2007). A number of anthocyanins and other flavonoids have been reported to inhibit the expression of proinflammatory cytokines. Chemokines are small cytokines that play a significant role in controlling leukocyte migration. Endothelial cells recruit leukocytes by selectively expressing adhesion molecules on their surface as a response to proinflammatory stimuli such as TNF-α and LPS. Flavonoids including anthocyanins generally seem to modulate this type of monocyte adhesion during the inflammatory process by decreasing their expression by endothelial cells. Analysis of structure–activity relationships among flavonoids suggests that four hydroxylations at positions 5, 7, 3′, and 4′ together with a bond at the C2–C3 and the B-ring attachment at the C2 position seem necessary for the highest proinflammatory cytokine expression (Comalada et al. 2006). Mimicking inflammation through the *in vitro* stimulation of vascular endothelial cells with cytokines and with or without LPS leads to an increased adherence of leukocytes to the endothelium due to the increased expression of adhesion molecules such as VCAM-1, intracellular adhesion molecule-1 (ICAM-1), and endothelial cell selectin (E-selectin) (Iiyama et al. 1999). Pretreatment with anthocyanins has shown to reversibly decrease adhesion due to the inhibition of expression of induced adhesion molecules. The chemokine MCP-1, known to mediate in the recruitment of macrophages to sites of inflammation, has been shown to have a protective effect against TNF-α-induced MCP-1 secretion in primary human endothelial cells. Rats fed a high-fat diet and administered 4% freeze-dried whole blueberries showed significant decreases in TNF-α and MCP-1 (Prior et al. 2009). Treatment of endothelial cells with cyanidin-3-glucoside and peonidin-3-glucoside has been reported to inhibit the production of cytokines and matrix metalloproteinases (Chen et al. 2006). Vascular endothelial growth factor is a proatherosclerotic factor in which anthocyanins (delphinidin and cyanidin) have been shown to prevent its expression simulated by platelet-derived growth factor (AB) in vascular smooth muscle cells by preventing the activation of p38 MAPK and c-Jun N-terminal kinase. The hydroxyl residue at the 3 position on the B-ring was shown to be crucial as inhibitory effects were not shared by other anthocyanin compounds such as malvidin and peonidin (Oak et al. 2006). Treatment of human umbilical vein endothelial cells with anthocyanins interferes with the recruitment of TNF receptor-associated factors (TRAF)-2 in lipid rafts, thereby inhibiting CD40-induced proinflammatory signaling (Xia et al. 2007). The anthocyanin delphinidin has been shown to decrease the extent of apoptotic and necrotic cell death in cultured cardiomyocytes and reduces infarct size after ischemia in rats. This is mediated by the inhibition of signal transducers and activators of transcription 1 (STAT1) (Oak et al. 2006).

There are several critical steps in which anthocyanins can modulate molecular events leading to the overexpression of inflammatory mediators. The most widely studied inflammatory mediator is NF-κB, an oxidative stress-sensitive transcription factor that plays a critical role in the regulation of a variety of genes important to cellular responses, including inflammation. In several studies, the suppression of proinflammatory chemokines, growth factors, and adhesion molecules was associated with an inhibition of NF-κB activation (Atalay et al. 2003; Bagchi et al. 2004; Cimino et al. 2006). Targeting of NF-κB can contribute to the inhibition of COX-2 expression by anthocyanidins in RAW 264.7 cells (Banerjee et al. 2002). NF-κB-related proinflammatory chemokines, cytokines, and mediators of inflammatory responses decreased in the plasma of healthy adult subjects postsupplementation with anthocyanins in parallel-designed, placebo-controlled, clinical trials suggesting mediated inhibition of NF-$\kappa\beta$ activation by anthocyanins *in vivo* (Blanco-Colio et al. 2000; Karlsen et al. 2007; Kelley et al. 2006). Direct inhibition of LPS-induced NF-$\kappa\beta$ transactivation by anthocyanins was observed in human monocytes (Karlsen et al. 2007). Mechanisms whereby anthocyanins inhibit NF-κB activation are not fully understood. One possible mechanism is that anthocyanins, their breakdown products, or metabolites serve as redox buffers capable of suppressing oxidative stress and thereby dampen the inflammatory response by direct ROS scavenging (Karlsen et al. 2007).

6.8 CONCLUSION AND FUTURE RESEARCH FOCUS

In conclusion, the current evidence supports that anthocyanins have some potentially positive effects on the cardiovascular system, although pharmacological targets somewhat differ among various studies and individual anthocyanins. The available literature describing the role in modulation and reduction of cardiovascular risk factors and prevention of cardiovascular-related health complications is consistent; however, much work remains to achieve definitive conclusions about the potential of anthocyanins as either preventative or therapeutic agents. The need for future research in this area is clearly evident. All aspects of the evidence-based nutrition paradigm (i.e., randomized controlled trials, animal and cell mechanistic studies, and epidemiological data) give unique insight into how these unique compounds may help the vascular system maintain homeostasis or positively influence elevated risk biomarkers in the case of individuals already suffering from CVD. Larger long-term controlled trials are needed to study the effects of anthocyanins on maintaining vascular homeostasis in which the traditional risk biomarkers are not elevated. Conversely, additional intervention studies involving purified anthocyanins are needed to confirm their effect on validated risk biomarkers (e.g., lipid profiles) in individuals with CVD. A large prospective study of the cardioprotective effects of anthocyanins should be designed specifically for cardiovascular outcomes with cardiovascular events being ascertained by expert medical review on an individual basis. The relevance of many *in vitro* studies to the *in vivo* situation needs to be confirmed because many *in vitro* studies apply high concentrations of anthocyanins that far exceed the level observed clinically *in vivo*; however, some *in vitro* studies do achieve results at comparable levels.

The use of isotopic labeling of anthocyanins would help generate greater knowledge about the way in which these phytonutrients are metabolized and absorbed in the gut, and/or in which tissues they may accumulate in the body. This type of labeling gives researchers insight into which compounds, metabolites, and decretory products cross the basolateral membrane and are most active in influencing the cardiovascular system. Along these lines, a better understanding of metabolism by the gut microbiota is needed to better understand the bioavailability and bioactivity of anthocyanins and their derived products in relation to CVD. Finally, there is a need for better understanding of the synergistic effects of anthocyanins with other polyphenols and food components so effective public policy may be developed and implemented.

REFERENCES

Acquaviva R, Russo A, Galvano F, et al. 2003. Cyanidin and cyanidin 3-O-beta-D-glucoside as DNA cleavage protectors and antioxidants. *Cell Biol Toxicol* 19: 243–52.

Agewall S, Wright S, Doughty RN, Whalley GA, Duxbury M, Sharpe N. 2000. Does a glass of red wine improve endothelial function? *Eur Heart J* 21: 74–8.

Ahmet I, Spangler E, Shukitt-Hale B, Juhaszova M, Sollott SJ, Joseph JA, Ingram DK, Talan M. 2009. Blueberry enriched diet protects rat heart from ischemic damage. *PLos ONE* 4(6): e5954.

Andriambeloson E, Magnier C, Hann-Archipoff G, et al. 1998. Natural dietary polyphenolic compounds cause endothelium-dependent vasorelaxation in rat thoracic aorta. *J Nutr* 128: 2324–33.

Assmann G, Cullen P, Jossa F, Lewis B, Mancini M. 1999. Coronary heart disease: Reducing the risk: The scientific background to primary and secondary prevention of coronary heart disease. A worldwide view. International Task Force for the Prevention of Coronary Heart Disease. *Arterioscler Thromb Vasc Biol* 19: 1819–24.

Atalay M, Gordillo G, Roy S, et al. 2003. Anti-angiogenic property of edible berry in a model of hemangioma. *FEBS Lett* 544: 252–7.

Aviram M, Rosenblat M, Gaitini D, et al. 2004. Pomegranate juice consumption for 3 years by patients with carotid artery stenosis reduces common carotid intima-media thickness, blood pressure and LDL oxidation. *Clin Nutr* 23: 423–33.

Bagchi D, Sen CK, Bagchi M, Atalay M. 2004. Anti-angiogenic, antioxidant, and anti-carcinogenic properties of a novel anthocyanin-rich berry extract formula. *Biochemistry (Mosc)* 69: 75–80.

Banerjee T, Valacchi G, Ziboh VA, van der Vliet A. 2002. Inhibition of TNFalpha-induced cyclooxygenase-2 expression by amentoflavone through suppression of NF-kappaB activation in A549 cells. *Mol Cell Biochem* 238: 105–10.

Basu A, Du M, Leyva MJ, et al. 2010. Blueberries decrease cardiovascular risk factors in obese men and women with metabolic syndrome. *J Nutr* 140: 1582–7.

Basu A, Wilkinson M, Penugonda K, Simmons B, Betts NM, Lyons TJ. 2009. Freeze-dried strawberry powder improves lipid profile and lipid peroxidation in women with metabolic syndrome: Baseline and post intervention effects. *Nutr J* 8: 43.

Bell DR, Gochenauer K. 2006. Direct vasoactive and vasoprotective properties of anthocyanin-rich extracts. *J Appl Physiol* 100: 1164–70.

Bogdan C. 2001. Nitric oxide and the immune response. *Nat Immunol* 2: 907–16.

Blanco-Colio LM, Valderrama M, Alvarez-Sala LA, Bustos C, Ortego M, Hernandez-Presa MA, Cancelas P, Gomez-Gerique J, Millan J. 2000. Red wine intake prevents nuclear factor-kappa beta activation in peripheral blood mononuclear cells of healthy volunteers during postprandial lipemia. *Circulation* 102: 1020–6.

Burton-Freeman B, Linares A, Hyson D, Kappagoda T. 2010. Strawberry modulates LDL oxidation and postprandial lipemia in response to high-fat meal in overweight hyperlipidemic men and women. *J Am Coll Nutr* 29: 46–54.

Cassidy A, O'Reilly EJ, Kay C, et al. 2011. Habitual intake of flavonoid subclasses and incident hypertension in adults. *Am J Clin Nutr* 93: 338–47.

Chambers BK, Camire ME. 2003. Can cranberry supplementation benefit adults with type 2 diabetes? *Diabetes Care* 26: 2695–6.

Chen C, Yi L, Jin X, et al. 2011. Inhibitory effect of delphinidin on monocyte-endothelial cell adhesion induced by oxidized low-density lipoprotein via ROS/p28MAPK/NF-κB pathway. *Cell Biochem Biophys* 61: 337–48.

Chen CY, Yi L, Jin X, et al. 2010. Delphinidin attenuates stress injury induced by oxidized low-density lipoprotein in human umbilical vein endothelial cells. *Chem Biol Interact* 183: 105–12.

Chen PN, Kuo WH, Chiang CL, Chiou HL, Hsieh YS, Chu SC. 2006. Black rice anthocyanins inhibit cancer cells invasion via repressions of MMPs and u-PA expression. *Chem Biol Interact* 163: 218–29.

Choi JS, Choi YJ, Shin SY, et al. 2008. Dietary flavonoids differentially reduce oxidized LDL-induced apoptosis in human endothelial cells: Role of MAPK- and JAK/STAT-signaling. *J Nutr* 138: 983–90.

Cimino F, Ambra R, Canali R, Saija A, Virgili F. 2006. Effect of cyanidin-3-O-glucoside on UVB-induced response in human keratinocytes. *J Agric Food Chem* 54: 4041–7.

Colling M, Weggemann S, Doring A, Keil U, Wolfram G. 1989. Nutrition survey of adults using a 7-day protocol—A pilot study in the Augsburg MONICA project. [article in German]. *Offentl Gesundheitswes* 51: 94–7.

Comalada M, Ballester I, Bailon E, et al. 2006. Inhibition of pro-inflammatory markers in primary bone marrow-derived mouse macrophages by naturally occurring flavonoids: Analysis of the structure–activity relationship. *Biochem Pharmacol* 72: 1010–21.

Costantino L, Rastelli G, Albasini A. 1995. Anthocyanidines as inhibitors of xanthine oxidase. *Pharmazie* 50: 573–4.

Curtis PJ, Kroon PA, Hollands WJ, et al. 2009. Cardiovascular disease risk biomarkers and liver and kidney function are not altered in postmenopausal women after ingesting an elderberry extract rich in anthocyanins for 12 weeks. *J Nutr* 139: 2266–71.

Davi G, Patrono C. 2007. Mechanisms of disease: Platelet activation and atherothrombosis. *N Engl J Med* 357: 1282–94.

Dell'Agli M, Busciala A, Bosisio E. 2004. Vascular effects of wine polyphenols. *Cardiovasc Res* 63: 593–602.

Di Castelnuovo A, Rotondo S, Iacoviello L, Donati MB, DeGaetano G. 2002. Meta-analysis of wine and beer consumption in relation to vascular risk. *Circulation* 105: 2836–44.

Diaz MN, Frei B, Vita JA, Keaney JF Jr. 1997. Antioxidants and atherosclerotic heart disease. *N Engl J Med* 337: 408–16.

Dillon GA, Vito JA. 2000. Nitric oxide and endothelial dysfunction. In: *Nitric Oxide and the Cardiovascular System*, edited by Loscalzo J and Vito JA. Totawa, NJ: Humana, pp. 207–25.

Dohadwala MM, Holbrook M, Hamburg NM, et al. 2011. Effects of cranberry juice consumption on vascular function in patients with coronary artery disease. *Am J Clin Nutr* 93: 934–40.

Duthie SJ, Jenkinson AM, Crozier A. 2006. The effects of cranberry juice consumption on antioxidant status and biomarkers relating to heart disease and cancer in healthy human volunteers. *Eur J Nutr* 45: 113–22.

Erlejman AG, Verstraeten SV, Fraga CG, Oteiza PI. 2004. The interaction of flavonoids with membranes: Potential determinant of flavonoid antioxidant effects. *Free Radic Res* 38: 1311–20.

Erlund I, Koli R, Alfthan G, et al. 2008. Favorable effects of berry consumption on platelet function, blood pressure, and HDL cholesterol. *Am J Clin Nutr* 87: 323–31.

Freedman JE, Parker C, 3rd, Li L, et al. 2001. Select flavonoids and whole juice from purple grapes inhibit platelet function and enhance nitric oxide release. *Circulation* 103: 2792–8.

Garcia-Alonso M, Rimbach G, Rivas-Gozalo JC, De Pascual-Teresa S. 2004. Antioxidant and cellular activities of anthocyanins and their corresponding vitisins A—Studies in platelets, monocytes, and human endothelial cells. *J Agric Food Chem* 52: 3378–84.

Giordano L, Coletta W, Tamburrilli C, et al. 2012. Four-week ingestion of blood orange juice results in measurable anthocyanin urinary levels but does not affect cellular markers related to cardiovascular risk: A randomized cross-over study in healthy volunteers. *Eur J Nutr* 51:541–48.

Gorinstein S, Caspi A, Libman I, et al. 2006. Red grapefruit positively influences serum triglyceride level in patients suffering from coronary atherosclerosis: Studies *in vitro* and in humans. *J Agric Food Chem* 54: 1887–92.

Guerrero JA, Lozano ML, Castillo J, Benavente-Garcia O, Vicente V, Rivera J. 2005. Flavonoids inhibit platelet function through binding to the thromboxane A2 receptor. *J Thromb Haemost* 3: 369–76.

Ha SK, Lee P, Park JA, Ph HR, Lee SY, Park JH, Lee EH, Ryu JH, Lee KR, Kim SY. 2008. Apigenin inhibits the production of NO and PGE2 in microglia and inhibits neuronal cell death in middle cerebral artery occlusion-induced focal ischemia mice model. *Neurochem Int* 52: 878–86.

Hämäläinen M, Nieminen R, Vuorela P, Heinonen M, Moilanen E. 2007. Anti-inflammatory effects of flavonoids: Genistein, kaempferol, quercetin, and daidzein inhibit STAT-1 and NF-kappaB activations, whereas flavones, isorhamnetin naringenin, and pelargonidin inhibit only NF-kappaB activation along with their inhibitory effect on iNOS expression and NO production in activated macrophages. *Mediators Inflamm* 45673: 1–10.

Hassellund SS, Flaa A, Sandvik L, Kjeldsen SE, Rostrup M. 2011. Effects of anthocyanins on blood pressure and stress reactivity: A double-blind randomized placebo-controlled crossover study. *J Hum Hypertens* Published online ahead of print May 5.

Hu C, Zawistowski J, Ling W, Kitts DD. 2003. Black rice (*Oryza sativa* L. *indica*) pigmented fraction suppresses both reactive oxygen species and nitric oxide in chemical and biological model systems. *J Agric Food Chem* 51: 5271–7.

Iiyama K, Hajra L, Iiyama M, Li H, DiChiara M, Medoff BD, Cybulsky MI. 1999. Patterns of vascular cell adhesion molecule-1 and intercellular adhesion molecule-1 expression in rabbit and mouse atherosclerotic lesions and at sites predisposed to lesion formation. *Circ Res* 85(2): 199–207.

Iwasaki-Kurashige K, Loyaga-Rendon RY, Matsumoto H, Tokunaga T, Azuma H. 2006. Possible mediators involved in decreasing peripheral vascular resistance with blackcurrant concentrate (BC) in hind-limb perfusion model of the rat. *Vascul Pharmacol* 44: 215–23.

Jennings A, Welch AA, Fairweather-Tait SJ, Kay C, Minihane A-M, Chowienczyk P, Jiang B, Cecelja M, Spector T, Macgregor A, Cassidy A. 2012. *Am J Clin Nutr.* (published ahead of print).

Jensen GS, Wu X, Patterson KM, et al. 2008. *In vitro* and *in vivo* antioxidant and anti-inflammatory capacities of an antioxidant-rich fruit and berry juice blend. Results of a pilot and randomized, double-blind, placebo-controlled cross-over study. *J Agric Food Chem* 56: 8326–33.

Jin Y, Alimbetov D, George T, Gordon MH, Lovegrove JA. 2011. A randomized trial to investigate the effects of acute consumption of a blackcurrant juice drink on markers of vascular reactivity and bioavailability of anthocyanins in human subjects. *Eur J Clin Nutr* 65: 849–56.

Joshipura KJ, Hu FB, Manson JE, et al. 2001. The effect of fruit and vegetable intake on risk for coronary heart disease. *Ann Intern Med* 134: 1106–14.

Kähkönen MP, Heinonen M. 2003. Antioxidant activity of anthocyanins and their aglycons. *J Agric Food Chem* 51: 628–33.

Kalea AZ, Clark K, Schuschke DA, Klimis-Zacas DJ. 2009. Vascular reactivity is affected by dietary consumption of wild blueberries in the Sprague–Dawley rat. *J Med Food* 12: 21–8.

Kalt W, Blumberg JB, McDonald JE, et al. 2008. Identification of anthocyanins in the liver, eye, and brain of blueberry-fed pigs. *J Agric Food Chem* 56: 705–12.

Kamata K, Makino A, Kanie N, et al. 2006. Effects of anthocyanidin derivative (HK-008) on relaxation in rat perfused mesenterial bed. *J Smooth Muscle Res* 42: 75–88.

Kao ES, Tseng TH, Lee HJ, Chan KC, Wang CJ. 2009. Anthocyanin extracted from *Hibiscus* attenuate oxidized LDL-mediated foam cell formation involving regulation of CD36 gene. *Chem Biol Interact* 179: 212–8.

Karlsen A, Retterstol L, Laake P, et al. 2007. Anthocyanins inhibit nuclear factor-κB activation in monocytes and reduce plasma concentrations of pro-inflammatory mediators in healthy adults. *J Nutr* 137: 1951–4.

Keil U, Kuulasmaa K. 1989. WHO MONICA Project: Risk factors. *Int J Epidemiol* 18: S46–55.

Kelley DS, Rasooly R, Jacob RA, Kader AA, Mackey BE. 2006. Consumption of Bing sweet cherries lowers circulating concentrations of inflammation markers in healthy men and women. *J Nutr* 136: 981–6.

Kim JY, Jung KJ, Choi JS, Chung HY. 2004. Hesperetin: A potent antioxidant against peroxynitrite. *Free Radic Res* 38: 761–9.

Knaup B, Oehme A, Valotis A, Schreier P. 2009. Anthocyanins as lipoxygenase inhibitors. *Mol Nutr Food Res* 53: 617–24.

Kroon PA, Clifford MN, Crozier A, et al. 2004. How should we assess the effects of exposure to dietary polyphenols in vitro? *Am J Clin Nutr* 80: 15–21.

Lazze M, Pizzala R, Savio M, Stivala L, Prosperi E, Bianchi L. 2003. Anthocyanins protect against DNA damage induced by tert-butyl-hydroperoxide in rat smooth muscle and hepatoma cells. *Mutat Res* 535: 103–15.

Lee IT, Chan YC, Lin CW, Lee WJ, Sheu WH. 2008. Effect of cranberry extracts on lipid profiles in subjects with type 2 diabetes. *Diabet Med* 25: 1473–7.

Lefevre M, Howard L, Most M, Ju Z, Delany J. 2004. Microarray analysis of the effects of grape anthocyanins on hepatic gene expression in mice. *FASEB J* 18: A851.

Liang YC, Huang YT, Tsai SH, Lin-Shiau SY, Chen CF, Lin JK. 1999. Suppression of inducible cyclooxygenase and inducible nitric oxide synthase by apigenin and related flavonoids in mouse macrophages. *Carcinogenesis* 20: 1945–52.

Loke WM, Hodgson JM, Proudfoot JM, McKinley AJ, Puddey IB, Croft KD. 2008. Pure dietary flavonoids quercetin and (-)-epicatechin augment nitric oxide products and reduce endothelin-1 acutely in healthy men. *Am J Clin Nutr* 88: 1018–25.

Marniemi J, Hakala P, Maki J, Ahotupa M. 2000. Partial resistance of low density lipoprotein to oxidation *in vivo* after increased intake of berries. *Nutr Metab Cardiovasc Dis* 10: 331–7.

Mazza G, Kay CD, Cottrell T, Holub BJ. 2002. Absorption of anthocyanins from blueberries and serum antioxidant status in human subjects. *J Agric Food Chem* 50: 7731–7.

McAnulty SR, McAnulty LS, Morrow JD, et al. 2005. Effect of daily fruit ingestion on angiotensin converting enzyme activity, blood pressure and oxidative stress in chronic smokers. *Free Radic Res* 39: 1241–8.

McCullough ML, Peterson JJ, Patel R, Jacques PF, Shah R, Dwyer JT. 2012. Flavonoid intake and cardiovascular disease mortality in a prospective cohort of US adults. *Am J Clin Nutr* 95: 454–64.

Mendes A, Desgranges C, Cheze C, Vercauteren J, Freslon JL. 2003. Vasorelaxant effects of grape polyphenols in rat isolated aorta. Possible involvement of a purinergic pathway. *Fundam Clin Pharmacol* 17: 673–81.

Mink PJ, Scrafford CG, Barraj LM, et al. 2007. Flavonoid intake and cardiovascular disease mortality: A prospective study in postmenopausal women. *Am J Clin Nutr* 85: 895–909.

Murkovic M, Abuja PM, Bergmann AR, et al. 2004. Effects of elderberry juice on fasting and postprandial serum lipids and low-density lipoprotein oxidation in healthy volunteers: A randomized, double-blind, placebo-controlled study. *Eur J Clin Nutr* 58: 244–9.

Mursu J, Voutilainen S, Nurmi T, Tuomainen TP, Kurl S, Salonen JT. 2008. Flavonoid intake and the risk of ischaemic stroke and CVD mortality in middle-aged Finnish men: The Kuopio Ischaemic Heart Disease Risk Factor Study. *Br J Nutr* 100: 890–5.

Nakamura Y, Matsumoto H, Todoki K. 2002. Endothelium-dependent vasorelaxation induced by black currant concentrate in rat thoracic aorta. *Jpn J Pharmacol* 89: 29–35.

Naruszewicz M, Laniewska I, Millo B, Dluzniewski M. 2007. Combination therapy of statin with flavonoids rich extract from chokeberry fruits enhanced reduction of cardiovascular risk markers in patients after myocardial infarction. *Atherosclerosis* 194: e179–84.

Ndiaye M, Chataigneau T, Andriantsitohaina R, Stoclet JC, Schini-Kerth VB. 2003. Red wine polyphenols cause endothelium-dependent EDHF-mediated relaxations in porcine coronary arteries via a redox-sensitive mechanism. *Biochem Biophys Res Commun* 310: 371–7.

Ness AR, Powles JW. 1997. Fruit and vegetables, and cardiovascular disease: A review. *Int J Epidemiol* 26: 1–13.

Nöthlings U, Schulze M, Weikert C, et al. 2008. Intake of vegetables, legumes, and fruit, and risk of all-cause cardiovascular, and cancer mortality in European diabetic population. *J Nutr* 138: 775–81.

Oak MH, Bedoui JE, Maderia SVF, Chalupsky K, Schini-Kerth VB. 2006. Delphinidin and cyanidin inhibit PDGFAB-induced VEGF release in vascular smooth muscle cells by preventing activation of p38 MAPK and JNK. *Br J Pharmacol* 149: 283–90.

Osman HE, Maalej N, Shanmuganayagam D, Folts JD. 1998. Grape juice but not orange or grapefruit juice inhibits platelet activity in dogs and monkeys. *J Nutr* 128: 2307–12.

Palfi A, Bartha E, Copf L, et al. 2009. Alcohol-free red wine inhibits isoproterenol-induced cardiac remodeling in rats by the regulation of Akt1 and protein kinase C alpha/beta II. *J Nutr Biochem* 20: 418–25.

Pergola C, Rossi A, Dugo P, Cuzzocrea S, Sautebin L. 2006. Inhibition of nitric oxide biosynthesis by anthocyanin fraction of blackberry extract. *Nitric Oxide* 15: 30–9.

Pignatelli P, Pulcinelli FM, Celestini A, et al. 2000. The flavonoids quercetin and catechin synergistically inhibit platelet function by antagonizing the intracellular production of hydrogen peroxide. *Am J Clin Nutr* 72: 1150–5.

Prior RL, Wu X, Gu L, et al. 2009. Purified berry anthocyanins but not whole berries normalize lipid parameters in mice fed an obesogenic high fat diet. *Mol Nutr Food Res* 53: 1406–18.

Qin Y, Xia M, Ma J, et al. 2009. Anthocyanin supplementation improves serum LDL- and HDL-cholesterol concentrations associated with the inhibition of cholesteryl ester transfer protein in dyslipidemic subjects. *Am J Clin Nutr* 90: 485–92.

Rahman MM, Ichiyanagi T, Komiyama T, Hatano Y, Konishi T. 2006. Superoxide radical- and peroxynitrite-scavenging activity of anthocyanins; structure–activity relationship and their synergism. *Free Radic Res* 40: 993–1002.

Ramirez-Tortosa C, Andersen ØM, Gardner PT, et al. 2001. Anthocyanin-rich extract decreases indices of lipid peroxidation and DNA damage in vitamin E depleted rats. *Free Radic Biol Med* 31: 1033–7.

Rechner AR, Kroner C. 2005. Anthocyanins and colonic metabolites of dietary polyphenols inhibit platelet function. *Thromb Res* 116: 327–34.

Renaud S, de Lorgeril M. 1992. Wine, alcohol, platelets, and the French paradox for coronary heart disease. *Lancet* 339: 1523–6.

Rossi A, Serraino I, Dugo P, et al. 2003. Protective effects of anthocyanins from blackberry in a rat model of acute lung inflammation. *Free Radic Res* 37: 891–900.

Ruel G, Pomerleau S, Couture P, Lemieux S, Lamarche B, Couillard C. 2006. Favourable impact of low-calorie cranberry juice consumption on plasma HDL-cholesterol concentrations in men. *Br J Nutr* 96: 357–64.

Ruel G, Pomerleau S, Couture P, Lemieux S, Lamarche B, Couillard C. 2008. Low-calorie cranberry juice supplementation reduces plasma oxidized LDL and cell adhesion molecule concentrations in men. *Br J Nutr* 99: 352–9.

Ruggeri ZM. 2002. Platelets in atherothrombosis. *Nat Med* 8: 1227–34.

Santangelo C, Vari R, Scazzocchio B, Di Benedetto R, Filesi C, Masella R. 2007. Polyphenols, intracellular signaling and inflammation. *Ann Ist Super Sanita* 43: 394–405.

Scalbert A, Williamson G. 2000. Dietary intake and bioavailability of polyphenols. *J Nutr* 130: 2073S–85S.

Schewe T, Steffen Y, Sies H. 2008. How do dietary flavanols improve vascular function? A position paper. *Arch Biochem Biophys* 476: 102–6.

Sesso HD, Gaziano JM, Jenkins DJ, Buring JE. 2007. Strawberry intake, lipids, C-reactive protein, and the risk of cardiovascular disease in women. *J Am Coll Nutr* 26: 303–10.

Simeonov SB, Botushanov NP, Karahanian EB, Pavlova MB, Husianitis HK, Troev DM. 2002. Effects of *Aronia melanocarpa* juice as part of the dietary regimen in patients with diabetes mellitus. *Folia Med (Plovdiv)* 44: 20–3.

Steffen Y, Gruber C, Schewe T, Sies H. 2008. Mono-o-methylated flavanols and other flavonoids as inhibitors of endothelial NADPH oxidase. *Arch Biochem Biophys* 469: 209–19.

Stein JH, Keevil JG, Weibe DA, Aeschlimann S, Folts JD. 1999. Purple grape juice improves endothelial function and reduces susceptibility of LDL cholesterol to oxidation in patients with coronary artery disease. *Circulation* 100: 1050–5.

Sumner MD, Elliott-Eller M, Weidner G, et al. 2005. Effects of pomegranate juice consumption on myocardial perfusion in patients with coronary heart disease. *Am J Cardiol* 96: 810–4.

Toufektsian MC, de Lorgeril M, Nagy N, et al. 2008. Chronic intake of plant-derived anthocyanins protects the rat heart against ischemia-reperfusion injury. *J Nutr* 138: 747–52.

Vallence P. 2000. Vascular nitric oxide in health and disease. In: *Nitric Oxide Biology and Pathobiology*, edited by Ignarro L. San Diego, CA: Academic, pp. 921–30.

Wang Q, Xia M, Liu C, et al. 2008. Cyanidin-3-O-beta-glucoside inhibits iNOS and COX-2 expression by inducing liver X receptor alpha activation in THP-1 macrophages. *Life Sci* 83: 176–84.

World Health Organization. 2011. Cardiovascular diseases (CVDs). Fact sheet no. 317. Available at: http://www.who.int/mediacentre/factsheets/fs317/en/index.html. Accessed March 25, 2012.

Xia M, Ling W, Zhu H, Wang Q, Ma J, Hou M, Tang Z, Li L, Ye Q. 2007. Anthocyanin prevents CD40-activated proinflammatory signaling in endothelial cells by regulating cholesterol distribution. *Arterioscler Thromb Vasc Biol* 27: 519–24.

Xu J, Ikeda K, Yamori Y. 2004a. Cyanidin-3-glucoside regulates phosphorylation of endothelial nitric oxide synthase. *FEBS Lett* 574: 176–80.

Xu JW, Ikeda K, Yamori Y. 2004b. Upregulation of endothelial nitric oxide synthase by cyanidin-3-glucoside, a typical anthocyanin pigment. *Hypertension* 44: 217–22.

Yan X, Murphy BT, Hammond GB, Vinson JA, Neto CC. 2002. Antioxidant activities and antitumor screening of extracts from cranberry fruit (*Vaccinium macrocarpon*). *J Agric Food Chem* 50: 5844–9.

Yang Y, Andrews MC, Hu Y, et al. 2011. Anthocyanin extract from black rice significantly ameliorates platelet hyperactivity and hypertriglyceridemia in dyslipidemic rats induced by high fat diets. *J Agric Food Chem* 59: 6759–64.

Yi L, Chen CY, Jin X, et al. 2010. Structural requirements of anthocyanins in relation to inhibition of endothelial injury induced by oxidized low-density lipoprotein and correlation with radical scavenging activity. *FEBS Lett* 584: 583–90.

Youdim KA, Martin A, Joseph JA. 2000. Incorporation of the elderberry anthocyanins by endothelial cells increases protection against oxidative stress. *Free Radic Biol Med* 29: 51–60.

Youdim KA, McDonald J, Kalt W, Joseph JA. 2002. Potential role of dietary flavonoids in reducing microvascular endothelium vulnerability to oxidative and inflammatory insults (small star, filled). *J Nutr Biochem* 13: 282–8.

Zhang Y, Seeram NP, Lee R, Feng L, Heber D. 2008. Isolation and identification of strawberry phenolics with antioxidant and human cancer cell antiproliferative properties. *J Agric Food Chem.* 56:670–675.

Zhu Y, Ling W, Guo H, et al. 2012. Anti-inflammatory effect of purified dietary anthocyanin in adults with hypercholesterolemia: A randomized controlled trial. *Nutr Metab Cardiovasc Dis* (published ahead of print).

Zhu Y, Xia M, Yang Y, et al. 2011. Purified anthocyanin supplementation improves endothelial function via NO-cGMP activation in hypercholesterolemic individuals. *Clin Chem* 57: 1524–33.

Ziberna L, Lunder M, Tramer F, Drevensek G, Passamonti S. 2011. The endothelial plasma membrane transporter bilitranslocase mediates rat aortic vasodilation induced by anthocyanins. *Nutr Metab Cardiovacs Dis*. Published online ahead of print May 4.

7 Anthocyanins and Metabolic Syndrome

Diana E. Roopchand, Leonel E. Rojo,
David Ribnicky, and Ilya Raskin

CONTENTS

7.1 OVERVIEW

The 2003–2006 National Health and Nutrition Examination Survey (NHANES) data found that 34% of adults in the United States at that time met the criteria for metabolic syndrome (MetS; discussed below) and were at heightened risk for developing type II diabetes and/or cardiovascular disease (CVD) (Ervin 2009). Current therapies include lifestyle changes and pharmaceuticals to treat the various clinical conditions present in individuals with MetS. Diet and lifestyle interventions to promote weight loss were shown to be more effective than drug-based therapy alone for treatment of MetS or type II diabetes (Knowler et al. 2002; Orchard et al. 2005). As edible plants contain an abundance of polyphenols that are associated with health maintenance and reduced incidence of chronic diseases (Cherniack 2011; Lila 2007), public health agencies in the United States and worldwide have recommended having at least five servings of fruits and vegetables per day. In particular, anthocyanins have been investigated for their beneficial effects in animal models of MetS and in human subjects with one or more features of MetS. In this chapter, we summarize the major clinical and molecular characteristics that are present in MetS and review human clinical trials and *in vivo* studies that investigate the effects of pure anthocyanins, anthocyanin-rich extracts/formulations, or anthocyanin-rich foods on endpoints relevant to MetS. Red grapes, cranberry, and pomegranate are not discussed because a relatively low content of anthocyanins versus other polyphenols in these fruits suggests that their beneficial effects are primarily derived from other polyphenols, with some contribution of anthocyanins.

7.2 CHARACTERISTICS OF METS

MetS is characterized by central obesity, hyperglycemia, hyperinsulinemia, dyslipidemia, and hypertension. Individuals with any three of these five clinical factors are considered positive for the syndrome and at increased risk for developing type II diabetes and CVD (Alberti et al. 2009). In addition, inflammation, elevated biomarkers of lipid oxidation, and a pro-thrombotic state are associated with MetS (Alberti et al. 2006, 2009). Regardless of the ongoing academic debate on the most adequate clinical definition of MetS, it is widely accepted that the presence of these clinical conditions is highly prognostic for developing type II diabetes and CVD and thus serves as a tool for early diagnostic and therapeutic interventions.

Over the past 50 years scientific publications have increased our understanding of the grave consequences of MetS; however, the molecular basis of MetS remains a conundrum. This is not unexpected, since MetS does not result from a single physiologic alteration, but from a cluster of underlying conditions. There is no central altered pathway that can fully account for the development of MetS, which involves alteration of a constellation of complex and interconnected pathways. Central obesity, which is highly correlated with insulin resistance, develops due to a combination of sedentary lifestyle, excessive consumption of calories and a genetic predisposition to weight gain. Nutrient overload results in increased storage of triglycerides and decreased lipolysis, which leads to increased adipocyte size (Arner et al. 2011; Cinti et al. 2005; Spalding et al. 2008). Obesity in humans and rodents is also accompanied by increased adipocyte cell death that seems to occur by both apoptotic (Alkhouri et al. 2010) and necrotic (Cinti et al. 2005) mechanisms, which signal the recruitment of scavenging macrophages that surround dead cells in "crown-like" structures (Cinti et al. 2005). Resident adipose tissue macrophages (ATM) in lean individuals secrete cytokines that predominantly confer an anti-inflammatory M2 phenotype; however, in obesity, ATM profiles are a mix of pro-inflammatory M1 and anti-inflammatory M2 phenotypes (Bourlier et al. 2008; Lumeng et al. 2007a; Morris et al. 2011; Zeyda et al. 2007) that promote a pro-inflammatory microenvironment. Adipose tissue also secretes a variety of cytokines (i.e., adipokines) into the circulation. The adipokine secretion profile in obese subjects with metabolic dysfunction is distinct from lean individuals and shows increased expression of pro-inflammatory adipokines (e.g., tumor necrosis factor alpha (TNFα), interleukin 6 (IL6), C reactive protein (CRP), leptin, retinoid binding protein-4 (RBP4), IL-18) and chemokines (e.g., monocyte chemoattractant protein-1 (MCP-1), plasminogen-activator inhibitor-1 (PAI-1), CC chemokine ligand-5 (CCL5)/RANTES and chemerin), while anti-inflammatory adipokines, such as adiponectin and secreted frizzled-related protein 5 (SFRP5), are reduced (Lumeng et al. 2007a,b; Ouchi et al. 2011; Sun et al. 2012).

The systemic low-grade chronic inflammation in adipose tissue of obese subjects promotes insulin resistance, decreases glucose tolerance and enhances atherogenic processes by mechanisms that are not yet completely understood (Sun et al. 2012). Functional insulin signaling results in decreased release of free fatty acids from adipose tissue, inhibition of hepatic gluconeogenesis and induction of glucose uptake by skeletal muscle; however, in obese subjects, these tissues become insulin resistant due to impaired insulin signaling, leading to ectopic fat accumulation, increased hepatic glucose output, and decreased glucose uptake by muscle (Zeyda and Stulnig 2009). Early insulin resistance is compensated by increased insulin release from pancreatic beta cells to maintain euglycemia. In individuals that are genetically pre-disposed to type II diabetes or have advanced type II diabetes, pancreatic beta cells eventually deteriorate by various mechanisms, including apoptosis (Kasuga 2006; Muoio and Newgard 2008) and complete pancreatic failure occurs. As beta cell mass decreases, compensatory hyperinsulinemia becomes unsustainable, leading to hyperglycemia and overt type II diabetes (Kasuga 2006; Muoio and Newgard 2008).

Decreased glucose uptake results in diminished production of glycerol, the backbone of triglycerides, and decreased storage of fatty acids in adipocytes. Hormone-sensitive lipase is usually inhibited by insulin and prevents intracellular lipolysis of stored triglycerides, but insulin-resistant adipocytes have increased levels of hormone-sensitive lipase leading to increased release of fatty acids (Ginsberg et al. 2006). In addition to increased levels of triglycerides, MetS is also characterized by increased low-density lipoprotein (LDL) particles, and decreased levels of high-density lipoprotein (HDL) in plasma (Ginsberg et al. 2006). Plasma cholesterol levels are heavily influenced by genetic and nutritional factors that regulate cholesterol absorption and synthesis, but not by dietary cholesterol intake (Lecerf and de Lorgeril 2011). HDL particles are antiatherogenic as they participate in reverse cholesterol transport (RCT) by transferring cholesterol from peripheral tissues, including cholesterol-laden macrophages and foam cells in the arterial wall, to the liver for excretion as biliary cholesterol or bile acids (Wang and Peng 2011).

The mechanisms by which obesity promotes hypertension are partially known and involve leptin-mediated activation of the sympathetic nervous system (SNS) and the renin–angiotensin–aldosterone system (RAAS) (Rahmouni 2010; Rahmouni et al. 2005). Obesity results in selective leptin resistance (Correia et al. 2002; Mark et al. 2002; Rahmouni et al. 2002). Leptin signaling is impaired in the arcuate nucleus of the hypothalamus, which is responsible for appetite suppression (Munzberg et al. 2004), but remains effective in the ventromedial and dorsomedial hypothalamus where it mediates activation of the SNS and increases blood pressure by causing peripheral vasoconstriction and increased renal tubular sodium reabsorption (Marsh et al. 2003). Angiotensin II is produced in adipose tissue and acts locally to regulate adipocyte growth and differentiation and systemically to regulate blood pressure

(Ailhaud et al. 2000; Massiera et al. 2001). In obesity, visceral fat produces high levels of angiotensin II, a powerful vasoconstrictor, that is released into the blood stream where it can stimulate renal sodium reabsorption causing hypertension and glomerular damage (Boustany et al. 2004). Obesity and insulin resistance are also correlated with higher levels of oxidative stress. The postprandial state results in increased oxidative stress, especially in obese individuals as they have lower levels of antioxidant enzymes (Furukawa et al. 2004; Tinahones et al. 2009). Obese subjects with high levels of insulin resistance have higher levels of oxidative stress compared with obese persons with lower levels of insulin resistance, both before and after a high fat meal; however, whether the insulin resistance results in the increased oxidative stress or vice versa is unknown (Tinahones et al. 2009). Early stage atherosclerosis involves dysfunction of vascular endothelial cells, which normally release nitric oxide (NO), a powerful vasodilator and inhibitor of platelet aggregation that prevents thrombus formation. Fat accumulation promotes increased reactive oxygen species (ROS) in blood vessels that can inactivate NO by reacting with it to form peroxynitrite causing endothelial dysfunction, vasoconstriction, and elevated blood pressure (Furukawa et al. 2004; Luft 2010).

7.3 EVIDENCE FOR ANTHOCYANINS AS AN INTERVENTION FOR METS

Studies indicate that dietary intake of anthocyanins may delay or reduce clinical manifestations of MetS and thereby delay or prevent the onset of type II diabetes and CVD. Several recent clinical trials (Table 7.1) and animal studies have investigated the effects of anthocyanin-rich extracts or anthocyanin-rich fruit formulations on subjects with MetS or type II diabetes. A caveat of many of these studies is that, although the anthocyanin concentrations may be quantified, there are scores of other compounds present in the test material that may also have contributed to any observed effects. However, studies using highly purified anthocyanins indicate that anthocyanins themselves have antidiabetic activity (Grace et al. 2009). Another consideration is that different types of naturally occurring anthocyanins may have varying degrees of effectiveness. Furthermore, anthocyanins are not well absorbed and have low bioavailability, based on concentrations detected in blood plasma (McGhie and Walton 2007). Consequently, the exact mechanism of how anthocyanins mediate their anti-MetS effects, whether directly or via microbial-derived metabolites produced in the gut, remains to be elucidated. As such, it is critical that a thorough phytochemical characterization of extracts be performed prior to biological interventions. Because of relatively poor understanding of absorption, distribution, metabolism, and excretion (ADMET) of anthocyanins in humans, *in vitro* data produced using intact anthocyanins should be interpreted with caution as anthocyanin concentrations used *in vitro* may not be realistically achieved *in vivo* and are more reliable when presented in conjunction with corroborating *in vivo* evidence. Bearing these limitations and caveats in mind, data from studies investigating the effects of anthocyanins from various food sources on the clinical manifestations of MetS in humans and animal models are reviewed below.

TABLE 7.1

Summary of Clinical Studies Investigating Effects of Anthocyanins on Endpoints Related to MetS, Type II Diabetes, and CVD

Subjects	Anthocyanin Dose	Anthocyanin Source	Results	Reference
Blueberry				
32 insulin-resistant men and women (DBPC)	668 mg/day for 6 weeks in a beverage	Whole blueberry powder formulated into a smoothie	↑ Insulin sensitivity	Stull et al. (2010)
48 obese men and women with MetS (SBPC)	742 mg/day for 8 weeks in a beverage	50 g freeze-dried blueberries (equivalent to 350 g or 2.3 cups fresh blueberries)	↓ BP, ↓ plasma oxidized LDL. No changes in serum glucose, lipid peroxidation, body weight, waist circumference	Basu et al. (2010a)
8 healthy men (SB-X)	Single dose of 1.2 g	100 g of freeze-dried blueberry powder (equivalent to 500–650 g whole blueberries) mixed in water	↑ In *ex vivo* serum antioxidant status	Kay and Holub (2002)
Chokeberry				
35 men with mild hypercholesterolemia	Not stated	250 mL of chokeberry juice/day for two 6 wk periods	↓ TC, ↓ LDL, ↓ TG; ↑ serum NO; ↑ flow mediated dilatation	Poreba et al. (2009)
58 men with mild hypercholesterolemia	Not stated	250 mL of chokeberry juice/day for two 6 wk periods	↓ TC, ↓ LDL, ↓ TG, ↑ HDL$_2$; ↓ serum glucose, ↓ homocysteine, ↓ fibrinogen, ↓BP	Skoczynska et al. (2007)
38 men and women with MetS and 14 healthy subjects	60 mg/day	300 mg of chokeberry extract/day for 2 months	↓ TC, ↓ LDL, ↓ TG ↓ platelet aggregation, ↓ clot formation and lysis	Sikora et al. (2012)
16 subjects with insulin-dependent diabetes and 25 patients with noninsulin-dependent diabetes	Not stated	200 mL of chokeberry juice per day for 3 months	↓ fasting blood glucose, ↓ HbA1c, ↓ TC, ↓ lipids	Simeonov et al. (2002)
25 subjects with MetSyn	300 mg/day for 2 months	Chokeberry extract	↓ BP, ↓ TC, ↓ LDL, ↓ TG, ↓ ET-1, ↓erythrocyte cholesterol	Broncel et al. (2007); Broncel et al. (2010)

continued

TABLE 7.1 (continued)
Summary of Clinical Studies Investigating Effects of Anthocyanins on Endpoints Related to MetS, Type II Diabetes, and CVD

Subjects	Anthocyanin Dose	Anthocyanin Source	Results	Reference
		Strawberry		
24 overweight subjects	81.6 mg per beverage	10 g freeze-dried strawberry powder formulated into a beverage	↓ Postprandial IL6, CRP and insulin	Edirisinghe et al. (2011)
20 obese subjects	ND	Freeze-dried strawberry powder equivalent to 320 g of frozen strawberries formulated into milkshake, yogurt, cream cheese, or water-based beverage for 3 wks	↓ TC, ↓ small HDL particles, ↑ LDL particle size	Zunino et al. (2011)
16 women with MetS	ND	50 g of freeze-dried strawberry powder per day mixed into water for 4 weeks	↓ TC, ↓ LDL, ↓ malondialdehyde, ↓ 4-hydroxynonenal	Basu et al. (2009)
27 subjects with MetS	ND	50 g of freeze-dried strawberry powder per day mixed into water for 8 weeks	↓ TC, ↓ LDL, ↓ small LDL particles, ↓ VCAM-1	Basu et al. (2010b)
		Bilberry and Black Currant		
27 men with BP >140/90 mm Hg (DBPC-X)	640 mg/day in capsule form for 4 weeks with 4-wk washout and crossover	Extract of bilberry and black currant (Medox®)	No differences in BP or stress reactivity; ↑ HDL; ↑ blood glucose; ↑ von Willebrand factor; ↑ in plasma levels of phenolic acids detected	Hassellund et al. (2013); Hassellund et al. (2012)

Subjects	Dose	Source	Outcome	Reference
120 dyslipidemic men and women (DBPC)	320 mg/day for 12 weeks in capsule form	Extract of bilberry and black currant (Medox®)	↑ HDL, ↓ LDL, enhances cellular cholesterol efflux to serum	Qin et al. (2009)
120 healthy men and women (PDPC)	300 mg/day for 3 weeks	Extract of bilberry and black currant (Medox®)	↓ NF-κB transactivation; ↓ plasma concentrations of pro-inflammatory factors	Karlsen et al. (2007)
Elderberry				
52 post-menopausal women	500 mg/day for 12 weeks	Elderberry extract	No change in CVD markers; normal liver and kidney enzyme profiles	Curtis et al. (2009)
Cherry				
19 women with T2D	720 mg/day for 6 weeks	Sour cherry juice	↓ Body weight, ↓ BP, ↓ HbA1c, ↓ total cholesterol, ↓ LDL-C	Ataie-Jafari et al. (2008)
18 healthy men and women	99.7 mg/day for 28 days	Bing sweet cherries (280 g fresh wt./day)	↓ CRP, ↓ RANTES, ↓ NO	Kelley et al. (2006)
Black Rice				
60 men with CVD	ND	10 g/d of black rice bran fraction for 6 months	Enhanced plasma antioxidant capacity; ↓ sVCAM-1, ↓ sCD40 L, ↓ hs-CRP	Wang et al. (2007)

Note: ND, not determined.

7.3.1 STUDIES WITH BLUEBERRY

Vaccinium angustifolium Aiton (lowbush or wild blueberry) and *Vaccinium corymbosum* L. (highbush blueberry) are the most commercially important North American blueberry crops. Blueberry fruits contain up to 27 different anthocyanins (Wu and Prior 2005) and have been used in traditional medicine for the complications of type II diabetes (Martineau et al. 2006). Recent human clinical trials and animal studies have investigated the effects of blueberries on factors related to MetS.

In a recent clinical study, obese and insulin-resistant subjects consumed a beverage containing 45 g of whole blueberry powder (1.3 g of anthocyanins) or a beverage without the blueberry powder for 6 weeks (Stull et al. 2010). Compared to baseline, subjects consuming blueberry bioactives had a significant $22.2 \pm 5.8\%$ improvement in insulin sensitivity whereas the placebo group had only a $4.9 \pm 4.5\%$ improvement, as measured using the hyperinsulinemic–euglycemic clamp technique. No significant changes in blood pressure, adiposity, body weight, energy intake, or inflammatory biomarkers between groups or within groups were observed (Stull et al. 2010). In a second study MetS subjects that consumed 50 g of blueberries (742 mg of anthocyanins) for 8 weeks had decreased blood pressure and markers of lipid oxidation such as oxidized LDL cholesterol and malondialdehyde compared to the control group (Basu et al. 2010a). Changes in body weight, waist circumference, HbA1c, insulin resistance, blood glucose, and lipid profiles were not observed; however, subjects in the blueberry group already had normal baseline levels of lipids (except for low HDL cholesterol), blood glucose, and insulin (Basu et al. 2010a). In a single-blinded crossover study, eight healthy subjects first consumed a control supplement mixed in water with a high-fat meal and a week later consumed a water solution of 100 g of freeze-dried blueberry powder (1.2 g of anthocyanins) with the same high-fat meal (Kay and Holub 2002). Test subjects had higher serum antioxidant status when the high-fat diet was supplemented with blueberry actives indicating that blueberries can reduce postprandial oxidative stress (Kay and Holub 2002).

Spontaneously hypertensive rats (SHR) and spontaneously hypertensive stroke-prone rats (SHRSP) are used as models of human hypertension and hypertension and stroke, respectively (Doggrell and Brown 1998) and both have been used to examine the effects of blueberry actives on hypertension. Compared to control-fed animals, SHRSP fed a diet containing 3% freeze-dried blueberries had lower systolic blood pressure at 4 weeks (19% lower) and 6 weeks (30% lower), although the effect was attenuated at 8 weeks (Shaughnessy et al. 2009). The decreased blood pressure may be at least partially due to inhibition of angiotensin converting enzyme (ACE) activity, which was decreased in SHRSP consuming the 3% blueberry diet, but not normotensive rats (Wiseman et al. 2011). The kidneys of SHRSP fed the 3% blueberry diet also had 40% less nitrites than control diet-fed rats, indicating a decrease in renal oxidative stress (Shaughnessy et al. 2009). In a separate study, a diet enriched with 8% wild blueberry was found to modify vasomotor control and improved vascular tone in adult SHR compared to a control diet (Kristo et al. 2010).

ApoE-deficient (apoE$^{-/-}$) mice spontaneously develop atherosclerosis when placed on a standard chow diet and closely model the pathogenesis of atherosclerotic lesions in humans (Meir and Leitersdorf 2004; Zhang et al. 1992). When fed a diet containing 1% freeze-dried whole blueberries, apoE$^{-/-}$ mice showed decreased atherosclerotic lesions accompanied by increased gene expression of aortal antioxidant enzymes and decreased F(2)-isoprostane, a marker of lipid peroxidation (Wu et al. 2010).

Compared to Zucker Fatty rats fed a high-fat diet (45% of kcal fat; HF45) with added sugar, animals fed a HF45 diet with 2% freeze-dried blueberry powder showed decreased TG, fasting insulin, homeostasis model index of insulin resistance (HOMA-IR), and improved glucose metabolism (Seymour et al. 2011). In addition, rats on blueberry-supplemented HF45 diet had in reduced adiposity and increased adipose and skeletal muscle peroxisome proliferator-activated receptor (PPAR) activity, which is correlated with increased glucose and fatty acid oxidation (Dumasia et al. 2005; Seymour et al. 2011). In contrast, Zucker Lean rats fed a diet supplemented with blueberry had higher body weight and lower TG, but all other measures were stable (Seymour et al. 2011).

C57BL/6J mice develop many of the clinical characteristics of humans with MetS when fed a high-fat diet (Surwit et al. 1988) and have been used to test for actives that decrease hyperglycemia, insulin resistance, weight gain, and alter molecular aspects of obesity and inflammation. Interestingly C57Bl/6J mice fed a high-fat diet with 45% kcal fat (HF45) supplemented with 10% freeze-dried blueberry powder (containing 27.8 mg/g anthocyanins) had increased body weight, more adiposity, and less lean body mass compared to mice on just the HF45 diet. Glucose area under the curve (AUC) measured after oral glucose tolerance testing was also not changed by blueberry powder supplementation of HF45 (Prior et al. 2008). In contrast, when the anthocyanins extracted from blueberry powder was added to the drinking water of mice on a high-fat diet with 60% kcal fat (HF60), their body weights were lower than controls (Prior et al. 2008), indicating that blueberry components could counteract the effects of the HF45 diet. In another study, C57Bl/6J mice fed a HF60 diet containing 4% blueberry powder did not show differences in body weight, adiposity or metabolic rate, but had decreased inflammatory gene expression in adipose tissue, adipocyte death, macrophage infiltration of adipose tissue, and improved insulin sensitivity compared to mice fed the HF60 alone (DeFuria et al. 2009). Blueberry powder supplementation of the HF60 diet did not reduce hepatic glucose output as fasting blood glucose levels in the mice were only modestly decreased compared to mice fed the HF60 diet alone (DeFuria et al. 2009). In a third study, C57Bl/6J fed a HF45 diet along with 0.2 mg/mL purified blueberry anthocyanins in drinking water (0.49 mg anthocyanins/mouse/day) for 72 days showed improved β cell function (HOMA-BCF) scores, lower serum leptin and lower body fat gain (Prior et al. 2010). For unknown reasons, mice on the high-fat diet did not show the same effects when given a higher dose of anthocyanins (1.0 mg/mL) in drinking water, suggesting that higher doses of anthocyanins may not be an advantage for improving/correcting certain clinical endpoints (Prior et al. 2010). Finally, blueberry juice consumption with the high-fat diet was less effective than addition of purified anthocyanins to drinking water (0.2 mg/mL) in preventing increased body weight (Prior et al. 2010). In contrast to biotransformed blueberry juice, regular blueberry juice did not lower body weight gain, food consumption, or improve oral glucose tolerance in KK-A^y mice (Vuong et al. 2009), which is in agreement with the above results obtained in the C57BL/6J fed whole blueberry powder or juice. An anthocyanin-enriched extract of lowbush blueberry (*Vaccinium angustifolium* Aiton) formulated with Labrasol® was found to be hypoglycemic after a single 500 mg/kg dose administration in diabetic C57BL/6J (Grace et al. 2009). At a dose of 300 mg/kg malvidin 3-*O* glucoside formulated in Labrasol®, but not delphinidin 3-*O*-glucoside, showed a hypoglycemic effect similar to that obtained with the anthocyanin-enriched extract, suggesting the latter is not active acutely (Grace et al. 2009). Overall, these data suggest that whole blueberry powder or juice has beneficial effects in the C57BL/6J model of MetS; however, certain blueberry components, such as sugars and/or lipids, can counteract the antiobesity and possibly other beneficial effects of blueberry anthocyanins.

7.3.2 STUDIES WITH STRAWBERRY

In addition to vitamin C, strawberry (*Fragaria* × *ananassa*) polyphenols, mainly ellagitannins and anthocyanins contributed most to its antioxidant capacity (Aaby et al. 2007). Overweight adults (*n* = 24) that consumed a high-carbohydrate, moderate-fat meal with a strawberry beverage decreased postprandial inflammatory and insulin response compared with the consumption of a placebo beverage (Edirisinghe et al. 2011). In a trial of 20 obese adults, consumption of four servings of strawberries per day for 3 weeks lowered plasma cholesterol and improved the lipid particle profile (Zunino et al. 2011). Women diagnosed with MetS had lower total and LDL cholesterol levels and decreased lipid peroxidation after 4 weeks of strawberry supplementation compared to pre-intervention measurements (Basu et al. 2009). In a placebo-controlled study of 27 MetS subjects, supplementation with a beverage containing freeze-dried strawberry powder for 8 weeks resulted in improvement of dyslipidemia and decreased serum levels of vascular cell adhesion molecule-1 (VCAM-1), a marker associated with CVD (Basu et al. 2010b).

Freeze-dried strawberry powder decreased blood glucose levels in obese and lean C57BL/6J mice and lower CRP in lean mice (Parelman et al. 2012). Polyphenol-rich strawberry pomace was found to reduce the serum insulin, serum FFAs, and total hepatic cholesterol in rats fed a fructose-rich diet (Jaroslawska et al. 2011). The addition of purified strawberry anthocyanins to the drinking water of C57BL/6J mice fed a high-fat diet decreased body weight gain and reduced dyslipidemia compared to control mice (Prior et al. 2009).

7.3.3 STUDIES WITH MAQUI BERRY

Maqui berry (*Aristotelia chilensis*), also known as Chilean blackberry, is an edible berry that grows wild in central and southern Chile. A standardized anthocyanin-rich formulation from Maqui berry (*Aristotelia chilensis*) administered at a dose of 500 mg/kg decreased fasting blood glucose and improved glucose tolerance in hyperglycemic C57/Bl6J mice fed a HF diet (Rojo et al. 2012). The application of the anthocyanin-rich formulation from Maqui berry decreased glucose production in H4IIE cells and enhanced both noninsulin and insulin-mediated glucose uptake in L6 muscle cells (Rojo et al. 2012). Similar antidiabetic effects were observed when the aforementioned mice or cells were treated with pure delphinidin

3-sambubioside-5-glucoside (D3S5G), an anthocyanin characteristic of Maqui berry (Rojo et al. 2012).

7.3.4 STUDIES WITH CHOKEBERRY

Black chokeberry (*Aronia melanocarpa* Elliot), common to Eastern Europe and North America and processed mainly into juice and wine, contains about 1.5 g of anthocyanins per 100 g fresh weight (Wu et al. 2004). Daily consumption of chokeberry juice by diabetic patients for 3 months resulted in lower fasting blood glucose, decreased HbA1c, TC, and lipid levels (Simeonov et al. 2002). Compared to baseline measures, daily treatment of MetS patients with 300 mg of anthocyanins from chokeberry resulted in decreased levels of blood pressure, total, LDL and erythrocyte membrane cholesterol, TG, the vasoconstrictor endothelin-1 (ET-1), and oxidative stress (Broncel et al. 2007, 2010). Subjects with mild hypercholesterolemia that consumed chokeberry juice daily for a two 6-week period with a 6-week break in between showed improved lipid metabolism and endothelial function (Poreba et al. 2009). A second study with the same regimen of chokeberry juice treatment showed that subjects with mild hypercholesterolemia had improved lipid and glucose metabolism, lower blood pressure, and decreased serum markers of thrombosis (Skoczynska et al. 2007). MetS patients treated with 300 mg/day of chokeberry extract for 2 months had a reduction in blood pressure, serum endothelin, lipids, and oxidative stress markers (Broncel et al. 2010) as well as inhibition of platelet aggregation, clotting, and fibrinolysis (Sikora et al. 2012).

Oral administration of black chokeberry fruit juice (5–20 mL/kg) having an anthocyanin concentration of 1068 mg/L for 30 days decreased plasma total cholesterol (TC), LDL cholesterol (LDL-C), and triglycerides (TG) in Wistar rats fed a diet containing 4% cholesterol (4% ChD), but did not alter plasma lipids in rats fed a standard diet (Valcheva-Kuzmanova et al. 2007a). Administration of black chokeberry fruit juice (10–20 mL/kg) did not alter plasma glucose or lipid parameters of normal rats, but caused reductions in plasma glucose TG and total cholesterol in streptozotocin-induced diabetic rats as compared to control-treated rats (Valcheva-Kuzmanova et al. 2007b). In a different study, Wistar rats received chokeberry extract in drinking water for 6 weeks while being fed a fructose-rich diet (FRD) to induce insulin resistance, hyperinsulinemia, dyslipidemia, and hypertension (Qin and Anderson 2012). Compared to the control group that received only water, the group receiving the Chokeberry extract had lower levels of epididymal fat, blood glucose, TAG, cholesterol, plasma IL-6, and TNFα while plasma adiponectin was increased (Qin and Anderson 2012). Likewise, the mRNA levels of positive insulin signaling components (Irs1, Irs2, Pi3 k, Glut1, Glut4, and Gys1) were increased while a negative signaling component (Gsk3β) was decreased. Protein and gene expression of adiponectin and PPARγ mRNA were upregulated while fatty acid binding protein 4 (FABp4), fatty acid synthase (Fas), and lipoprotein lipase (Lpl) mRNA levels were inhibited (Qin and Anderson 2012). mRNA levels of inflammatory cytokines IL1β, IL-6, and TNFα were decreased (Qin and Anderson 2012). In a related study, Wistar rats were fed a high-fructose diet and injected with a low dose of streptozotocin to reduce β-cell mass and developed features of MetS (Jurgonski et al. 2008).

Compared to rats receiving control diet, animals fed the high fructose diet supplemented with 0.2% chokeberry extract (containing 404 mg anthocyanins/g) had lower blood glucose and improved antioxidant status with an insignificant decrease in cholesterol after 4 weeks (Jurgonski et al. 2008).

7.3.5 STUDIES WITH BILBERRY AND BLACK CURRANT

Bilberry (*Vaccinium myrtillus* L.; European blueberry) fruits contain higher levels of anthocyanins (3.7 mg anthocyanins/g fresh weight) than commercially available North American blueberry species (Kalt et al. 1999) and have recently been studied for their antidiabetic and atheroprotective activities.

AMP-activated protein kinase (AMPK) is important for cellular energy balance in liver, muscle, and adipose tissue. Activation of AMPK leads to phosphorylation of several targets (e.g., acetyl-CoA carboxylase (ACC), HMG-coA reductase, and mTOR) that leads to fatty acid oxidation and inhibition of gluconeogenesis and lipogenesis, the result being activation of catabolic pathways and inactivation of anabolic pathways (Viollet et al. 2006; Zhang et al. 2009). Hepatic AMPK activation downregulates gluconeogenic enzymes such as phosphoenolpyruvate carboxykinase (PEPCK) and glucose-6-phosphatase (G6Pase) (Viollet et al. 2006). Compared to control diet, KK-A^y mice fed a diet containing 1% bilberry anthocyanins for 5 weeks had decreased serum glucose, serum and liver lipid concentrations, improved insulin sensitivity and decreased glucose flux as measured by pyruvate tolerance; however, there were no inter-group differences in body weight or food intake (Takikawa et al. 2010). Compared to control, bilberry-treated mice showed increased AMPK phosphorylation (Thr172) and thus activation in white adipose tissue (WAT), skeletal muscle, and liver tissue accompanied by increased Glut4 protein expression in WAT and muscle (Takikawa et al. 2010). Bilberry treatment resulted in decreased WAT gene expression of RBP4, an adipokine associated with insulin resistance; however, RBP4 protein in WAT and serum were similar in control and blueberry-treated groups (Takikawa et al. 2010). Hepatic tissue from blueberry-treated mice also showed decreased PEPCK and G6Pase gene expression consistent with decreased gluconeogenesis while increased ACC phosphorylation/inactivation and increased gene expression of PPARα, acyl-CoA oxidase (ACO), and carnitine palmitoyltranaferase-1A (CPT1A) were indicative of increased fatty acid oxidation (Takikawa et al. 2010).

Development of atherosclerotic plaques was diminished in apoE$^{-/-}$ mice after 16 weeks of consuming a diet supplemented with 0.02% bilberry extract comprised of 52% anthocyanins (Mauray et al. 2009). Microarray analysis of liver and aorta tissues taken from the apoE$^{-/-}$ mice fed control or 0.02% bilberry-supplemented diets for 2 weeks revealed that the antiatherogenic effects of bilberry supplementation correlated with downregulation of pro-inflammatory genes and increased expression of genes involved in cholesterol metabolism in the liver while genes implicated in oxidative stress, inflammation, transendothelial migration were modulated in the aorta (Mauray et al. 2010, 2012).

Four anthocyanins (the 3-*O*-glucosides and the 3-*O*-rutinosides of delphinidin and cyanidin) comprise over 97% of the total anthocyanin content of black currant

(*Ribes nigrum* L.) (Slimestad and Solheim 2002). An anthocyanin-enriched fraction from black currant juice administered to hyperlipidemic rabbits *ad libitum* in drinking water did not show any of the improvements in preventing atherosclerotic plaque development or decreasing plasma lipids, cholesterol, or markers of oxidative stress, as was observed with treatment using the positive control drug, probucol (0.5% w/w), compared to the control (Finne Nielsen et al. 2005). Rather the black currant anthocyanin-enriched fraction increased LDL cholesterol, while black currant juice did not, suggesting that the juice contained some component that offset the undesirable effect of the anthocyanin fraction (Finne Nielsen et al. 2005).

Three clinical studies have been performed with Medox® capsules, a proprietary and commercial bilberry and black currant anthocyanin extract containing 75 or 80 mg of 17 different purified anthocyanins and traces of other polyphenols. The first was a parallel designed, placebo-controlled trial where 120 healthy subjects received control or Medox® capsules delivering 300 mg anthocyanin/day for 3 weeks after which the levels 21 cytokines and chemokines were measured in both groups (Karlsen et al. 2007). Compared to baseline, the group that received the anthocyanin extract had greater reductions in plasma concentrations of pro-inflammatory cytokines IL-8, IFN-α, and RANTES after treatment compared to the decrease observed in the placebo-treated group, while changes were not observed in the remaining 18 cytokines/chemokines (Karlsen et al. 2007). The second study was a double-blind, randomized, placebo-controlled trial where 120 dyslipidemic subjects received placebo treatment or Medox® capsules providing 160 mg anthocyanins twice daily for 12 weeks (Qin et al. 2009). Subjects in the anthocyanin group had a 13.7% increase in HDL cholesterol, a 13.6% decrease in LDL cholesterol and enhanced cholesterol efflux into serum, which correlated with inhibition of the plasma cholesteryl ester transfer protein (CETP), thus demonstrating that anthocyanins can improve the cholesterol profile in dyslipidemic subjects (Qin et al. 2009). More recently, a 4-week treatment of Medox® capsules (640 mg anthocyanins/day) or placebo was assessed in 27 pre-hypertensive, nondyslipidemic men in a double-blind, randomized, placebo-controlled crossover study (Hassellund et al. 2013; Hassellund et al. 2012). Compared to placebo, anthocyanin treatment did not show effects on blood pressure, heart rate, or measures of stress reactivity (i.e., plasma noradrenaline and adrenaline) (Hassellund et al. 2012). Analysis of blood chemistry markers related to CVD showed that subjects receiving the anthocyanin treatment had a modest, but significant, increase in HDL cholesterol, consistent with the Qin et al. (2009); however, LDL cholesterol was unchanged (Hassellund et al. 2013). The anthocyanin treatment group showed a small, but significant increase in blood glucose levels, which the authors speculate may be due to the high dose of anthocyanins. Higher levels of Von Willebrand factor, a predictor of adverse cardiac events, were also detected in the anthocyanin treatment group indicating potential unfavorable effects of anthocyanin treatment. No differences were observed in triglycerides, total cholesterol, HbA1c, lipoprotein a, insulin, homeostasis model assessment of insulin resistance (HOMA-IR), homocysteine, or albumin/creatinine ratio. Study subjects were not characterized as having an increased inflammatory state or showing increased signs of oxidative stress prior to treatment. After the intervention period, markers of oxidative stress, endothelial dysfunction, and inflammation remained similar in both groups (Hassellund et al. 2013).

7.3.6 STUDIES WITH ELDERBERRY

European elderberry (*Sambucus nigra*) and American elderberry (*Sambucus canadensis*) contain a range of anthocyanins and other antioxidant polyphenols (Thole et al. 2006; Wu et al. 2004). CVD risk markers and safety were assessed in healthy post-menopausal women in a randomized, parallel design, placebo-controlled trial of a European elderberry extract (Curtis et al. 2009). After treatment with placebo ($n = 26$) or capsules of elderberry extract containing a daily anthocyanin dose of 500 mg ($n = 26$) for 12 weeks, liver and kidney enzymes in both groups were within normal physiological ranges; however, subjects showed no differences in plasma levels of inflammatory biomarkers, vascular activity (endothelin-1, platelet reactivity, blood pressure), plasma lipids, lipoproteins, or glucose (Curtis et al. 2009).

7.3.7 STUDIES WITH CHERRY

Anthocyanins are responsible for the red color in cherries and contribute to their antioxidant, anti-inflammatory and antidiabetic activities (Mulabagal et al. 2009). The effect of consumption of sour cherry (*Prunus cerasus* L., also called tart cherry) juice on blood glucose and CVD risk factors was assessed in 19 women with type II diabetes (Ataie-Jafari et al. 2008). Subjects consumed 40 g of sour cherry juice per day for 6 weeks containing 720 mg of anthocyanins. Compared to pretreatment baseline measurements, subjects had decreased body weight, body mass index (BMI) blood pressure, and HbA1c after 6 weeks of treatment (Ataie-Jafari et al. 2008). In another study, 18 healthy men and women consumed 280 g of fresh bing sweet cherries containing 100 mg of anthocyanins per day for 28 days (Kelley et al. 2006). Serum concentrations of CRP, NO, and RANTES were decreased after 28 days of cherry consumption and RANTES levels continued to decline 28 days after the cherry consumption was discontinued. Plasma lipids, lipoproteins, blood glucose, and insulin levels remained similar between the groups (Kelley et al. 2006).

Insulin resistant and hyperglycemic Dahl salt-sensitive rats were fed a control diet or a diet supplemented with 1% (wt/wt) freeze-dried whole tart cherry (*P. cerasus* L., also called sour cherry) for 90 days (Seymour et al. 2008). Compared to rats on the control diet, rats fed the diet containing tart cherry had decreased hyperlipidemia, hyperinsulinemia, fasting blood glucose, reduced fatty liver, enhanced hepatic PPAR-α mRNA, enhanced acyl-coenzyme A oxidase mRNA and activity and increased plasma antioxidant capacity (Seymour et al. 2008). Another study by the same group using the Zucker fatty rat model of obesity and MetS showed that, compared to control fed animals, rats fed a HF diet supplemented with 1% freeze-dried whole tart cherry powder had decreased hyperlipidemia, percentage adiposity and abdominal fat (retroperitoneal) weight (Seymour et al. 2009). Mice fed the HF diet with tart cherry also reduced retroperitoneal IL-6 and TNFα protein and mRNA, decreased plasma IL-6 and TNFα, and decreased NF-κB activity (Seymour et al. 2009).

Anthocyanins and ursolic acid, the most abundant bioflavonoids in Cornelian cherry (*Cornus mas*), were extracted and purified from the fresh fruit and then formulated into individual HF diets (Jayaprakasam et al. 2006). Compared to C57/Bl6 mice fed the HF diet alone, mice fed a HF diet supplemented with anthocyanins (1 g/kg diet) for 8 weeks showed improved glucose tolerance, a 24% decrease in weight gain and decreased lipid accumulation and triacylglycerol concentration in liver (Jayaprakasam et al. 2006). Compared to mice fed the HF diet, mice fed the HF diet supplemented with ursolic acid (0.5 g/kg diet) showed similar improvements in all endpoints similar to mice on the anthocyanin-supplemented diet; however, the reduction in weight gain was not observed with ursolic acid indicating that this effect is specific to anthocyanins (Jayaprakasam et al. 2006).

7.3.8 STUDIES WITH PURPLE CORN

The main anthocyanin that gives color to purple corn (*Zea mays* L.) is cyanidin-3-glucoside (Prior et al. 2009). The addition of purple corn color (PCC) to a HF diet (11 g/kg diet) for 12 weeks was shown to prevent the obesity, adipocyte hypertrophy, hyperglycemia, hyperinsulinemia, and hyperleptinemia that occurred in mice fed the HF diet alone (Tsuda et al. 2003). The HF diet induced increase in TNF-α mRNA and mRNA of enzymes involved in fatty acid and triacylglycerol synthesis was also decreased by PCC supplementation of the HF diet (Tsuda et al. 2003). In a second study, compared to when fed a control diet, KK-A^y mice fed a diet containing 0.2% anthocyanins from PCC (purified to 96% purity) for 5 weeks had reduced blood glucose and enhanced insulin sensitivity which was correlated with the upregulation of Glut4 mRNA and protein and a decrease of RBP4, TNF-α and monocyte chemotactic protein-1 (MCP-1) mRNA in WAT (Sasaki et al. 2007). In contrast to these results, another study found that C57BL/6J mice fed a high-fat diet supplemented with PCC did not prevent obesity or lower serum lipids, cholesterol, insulin, leptin, or cytokine concentrations compared to mice fed a high-fat diet alone (Prior et al. 2009). The reason for the discrepancy in results is not known and may be due to differences in preparation of the PCC material.

7.3.9 STUDIES WITH BLACK RICE

The bran portion of black rice (*Oryza sativa* L. *indica*) is rich in anthocyanins, mainly cyanidin-3-glucoside (>90%) and peonidin-3-glucoside (Ichikawa et al. 2001; Jang et al. 2012; Zhang et al. 2010). Studies in NZW rabbit models of atherosclerosis have shown that high cholesterol diets (HCD) supplemented with 30% black or red rice (Ling et al. 2001) or 5% black rice bran, also called black rice pigment fraction (Ling et al. 2002; Xia et al. 2003) decreased atherosclerotic plaque formation and increased blood antioxidant status compared to HCD diet supplemented with 30% white rice or 5% white rice bran (also called white rice pigment fraction). Studies in apolipoprotein (apo) E-deficient mice, which spontaneously develop atherogenic plaques, showed that mice fed a diet supplemented with 5% black rice bran had 46–48% less atherosclerotic plaque formation as well as lower cholesterol, oxidative stress, and inflammatory lymphocytes compared to mice fed control diet or control diet containing 5% white rice bran (Xia et al. 2003). Aging apoE$^{(-/-)}$ mice develop advanced atherosclerotic plaques that are prone to rupture, erosion, and thrombosis formation. These features model advanced atherosclerotic lesions in humans and therapy involves plaque stabilization to prevent plaque rupture and thrombosis. Compared to controls, dietary supplementation with an anthocyanin-rich black rice bran extract (300 mg/kg/d; 43.2% anthocyanins) for 20 weeks inhibited progression of atherosclerotic plaques in aged (apo) E-deficient mice as effectively as a diet supplemented with simvastin (50 mg/kg/d), a hypolipidemic drug (Xia et al. 2006). Black rice bran extract was also used for treating hepatic steatosis, or associated with obesity. Mice fed a HF diet supplemented with 1% (w/w) of black rice extract for 7 weeks had diminished hepatic steatosis or nonalcoholic fatty liver disease (NAFLD), decreased serum triglyceride and total cholesterol levels and increased expression of fatty acid metabolism-related genes relative to controls (Jang et al. 2012).

In a study of 60 coronary heart disease patients, subjects were randomly assigned to have their diet supplemented with 10 g of white rice bran or 10 g of black rice bran for 6 months. At the end of the 6-month intervention, subjects in the black rice bran group showed improved plasma total antioxidant capacity, decreased plasma concentrations of soluble vascular cell adhesion molecule-1 (sVCAM-1), soluble CD40 ligand (sCD40 L) and high-sensitive C-reactive protein (hs-CRP) compared to the white rice bran control group (Wang et al. 2007). Therefore, the presence of the anthocyanin pigments from black rice seemed to confer cardio-protective effects.

7.3.10 Studies with Black Soybean

Black soybean seed coats (*Glycine max* (L.) Merr.) are rich in proanthocyanidins and anthocyanins (Kanamoto et al. 2011; Lee et al. 2005) and are a common food staple in the Asian diet. Compared to rats fed a HF diet, rats fed a HF diet supplemented with 10% black soybean or supplemented with 0.037% black soybean anthocyanins (equivalent to that in the 10% soybean diet) showed less weight gain, decreased triglycerides and total cholesterol, increased HDL cholesterol and lower adipose tissue weights (Kwon et al. 2007). In a related study, C57BL/6 mice fed a high-fat diet supplemented with a hydroethanolic extract of black soybean seed coat for 14 weeks had less adiposity, decreased glucose levels, increased insulin sensitivity, decreased expression of WAT inflammatory genes. The animals also showed an increase in the amount of brown adipose tissue (BAT) and in the WAT expression of uncoupling protein 1 (UCP1) and UCP2, which are involved in energy expenditure (Kanamoto et al. 2011).

7.3.11 Studies with Purple Carrot

Anthocyanins are the major antioxidant pigments in purple carrots, while orange carrots contain high levels of α- and β-carotene (Sun et al. 2009). Purple carrot juice and β-carotene were compared for their ability to reverse the histological and metabolic changes in rats fed a high-fat and high-carbohydrate (HF-HC) diet (Poudyal et al. 2010). Rats fed the HF-HC diet supplemented with purple carrot juice showed improvement or attenuation of the negative effects observed in rats fed the HF-HC diet alone, which included hypertension, impaired glucose tolerance, increased abdominal fat deposition, dyslipidemia, liver and cardiac fibrosis, cardiac stiffness endothelial dysfunction, inflammatory cell infiltration, and oxidative stress (Poudyal et al. 2010). Rats fed the HF-HC diet supplemented with β-carotene showed positive effects in several endpoints, but did show reduced levels of oxidative stress, cardiac stiffness, or hepatic fat accumulation (Poudyal et al. 2010).

7.3.12 STUDIES WITH PURPLE SWEET POTATO

Purple sweet potato cultivars contain a variety of anthocyanins (Truong et al. 2010). Compared to mice fed a diet without supplementation, apoE$^{(-/-)}$ deficient mice fed a cholesterol- and fat-enriched diet supplemented with 1% purple sweet potato (*Ipomoea batatas* L.) for 4 weeks had a 46% reduction in atherosclerotic plaque development, lower levels of oxidative stress, and decreased plasma levels of soluble vascular cell adhesion molecule-1 (sVCAM-1) (Miyazaki et al. 2008). The 1% purple sweet potato intervention diet did not alter body weight or dyslipidemia compared to control (Miyazaki et al. 2008). In contrast, mice fed a HF diet supplemented with an anthocyanin fraction of sweet potato (200 mg/kg/day) for 4 weeks showed reduced weight gain without decreased food consumption, decreased hepatic triglycerides and improved serum lipid parameters compared to mice fed the HF diet alone (Hwang et al. 2011b). Liver tissues of mice fed the sweet potato anthocyanin fraction showed increased phosphorylation and activity of AMPK, which was accompanied by increased phosphorylation and inactivation of its target, ACC (Hwang et al. 2011a), indicating decreased fatty acid synthesis. The same molecular events were also observed in HepG2 hepatocytes treated with the sweet potato anthocyanin fraction, along with decreased levels of sterol regulatory element-binding protein 1 (SREBP-1) and its target fatty acid synthase (FAS).

7.4 SORPTION, CONCENTRATION, AND STABILIZATION OF ANTHOCYANINS

Animal studies suggest that anthocyanins would be more effective for decreasing weight gain and consequential complications if they were separated from other components of the whole fruit or juice, such as sugars and/or lipids (Prior et al. 2008, 2009, 2010). The extraction of anthocyanins typically requires the use of costly organic solvents and affinity resins which removes the product from the context and category of natural foods. By leveraging the natural affinity of polyphenols for proteins, protein-rich flours, such as defatted soybean flour (DSF), can be used as natural affinity resins to efficiently sorb mid-polarity range polyphenols, such as anthocyanins from juiced or extracted plant material. The DSF-bound polyphenols can then be separated from highly polar sugars and nonpolar oils/fats, which remain the liquid phase (Roopchand et al. 2012a). Using this natural method of polyphenol sorption, we have shown that anthocyanins and other polyphenols from blueberry, cranberry, and grape juices can be successfully sorbed, concentrated, and stabilized in a DSF matrix (Roopchand et al. 2012a,b). Compared to control treatments, single dose administration of blueberry polyphenol-enriched DSF or grape polyphenol-enriched DSF formulated in Labrasol was found to be hypoglycemic in diabetic C57Bl/6J mice, thus indicating that the sorbed anthocyanins retain their antidiabetic activity (Roopchand et al. 2012a).

7.5 CONCLUSION

Human clinical and preclinical animal studies performed thus far suggest that anthocyanin-rich foods and anthocyanin-enriched extracts are generally beneficial for improving some of the clinical manifestations of MetS. Studies done with highly purified single anthocyanin compounds confirm these findings. The mode of action by which anthocyanins convey benefit for MetS, however, remains unclear and a focus for future research. One of the complexities to discovering a mode of action is that anthocyanins are poorly bioavailable. In addition, they are largely metabolized in the gut into a variety of metabolites that are not clearly defined. The metabolites of anthocyanins, rather than intact parent molecule, may therefore mediate the physiological effects observed *in vivo* from anthocyanin treatment. Studies with ^{13}C or ^{14}C isotope-labeled anthocyanins can go a long way in elucidating the bioavailability, metabolism, tissue distribution, and mode of action of these compounds. Unfortunately, costs associated with the production of these isotopes in quantities necessary for comprehensive *in vivo* and clinical investigation put these studies outside the reach of most investigators. In addition, several types of anthocyanin compounds exist and may have different efficacy for a given biological endpoint (Grace et al. 2009). It is also possible that different anthocyanins may act additively or synergistically among themselves or together with other polyphenols present in fruits and vegetables.

Future directions of anthocyanin research should focus on determining whether intact anthocyanins or specific anthocyanin metabolites are responsible for the physiological outcomes that have been observed *in vivo*. In addition, anthocyanin research would benefit from studies designed to determine whether the general antioxidant activity of anthocyanins is responsible for the anti-MetS activities or whether anthocyanins or their metabolites interact with specific cellular targets to produce their physiological effects, such as decreased expression of inflammatory cytokines. Recent research data suggest a connection between low-grade inflammation and development of MetS, therefore it is reasonable to hypothesize that the anti-inflammatory activity of anthocyanins exert a protective effect against development of MetS. Furthermore, a reduction in ROS and other free radicals may nonspecifically inhibit inflammation, therefore it is tempting to speculate that anthocyanins, being strong antioxidants, may protect against MetS without directly interacting with a specific molecular target. Furthermore, this may explain why many structurally diverse antioxidant polyphenols, in addition to anthocyanins, are also effective against MetS.

REFERENCES

Aaby, K., Ekeberg, D., and Skrede, G. 2007. Characterization of phenolic compounds in strawberry (*Fragaria × ananassa*) fruits by different HPLC detectors and contribution of individual compounds to total antioxidant capacity. *J Agric Food Chem 55*: 4395–4406.

Ailhaud, G., Fukamizu, A., Massiera, F., Negrel, R., Saint-Marc, P., and Teboul, M. 2000. Angiotensinogen, angiotensin II and adipose tissue development. *Int J Obes Relat Metab Disord 24*(Suppl 4): S33–S35.

Alberti, K.G., Eckel, R.H., Grundy, S.M., Zimmet, P.Z., Cleeman, J.I., Donato, K.A., Fruchart, J.C., James, W.P., Loria, C.M., and Smith, S.C., Jr. 2009. Harmonizing the metabolic syndrome: A joint interim statement of the International Diabetes Federation Task Force on Epidemiology and Prevention; National Heart, Lung, and Blood Institute; American Heart Association; World Heart Federation; International Atherosclerosis Society; and International Association for the Study of Obesity. *Circulation 120*: 1640–1645.

Alberti, K.G., Zimmet, P., and Shaw, J. 2006. Metabolic syndrome—A new world-wide definition. A Consensus Statement from the International Diabetes Federation. *Diabet Med 23*: 469–480.

Alkhouri, N., Gornicka, A., Berk, M.P., Thapaliya, S., Dixon, L.J., Kashyap, S., Schauer, P.R., and Feldstein, A.E. 2010. Adipocyte apoptosis, a link between obesity, insulin resistance, and hepatic steatosis. *J Biol Chem 285*: 3428–3438.

Arner, P., Bernard, S., Salehpour, M. et al. 2011. Dynamics of human adipose lipid turnover in health and metabolic disease. *Nature 478*: 110–113.

Ataie-Jafari, A., Hosseini, S., Karimi, F., and Pajouhi, M. 2008. Effects of sour cherry on blood glucose and some cardiovascular risk factor imporovement in diabetic women. *Nutr Food Sci 38*: 355–360.

Basu, A., Du, M., Leyva, M.J., Sanchez, K., Betts, N.M., Wu, M., Aston, C.E., and Lyons, T.J. 2010a. Blueberries decrease cardiovascular risk factors in obese men and women with metabolic syndrome. *J Nutr 140*: 1582–1587.

Basu, A., Fu, D.X., Wilkinson, M., Simmons, B., Wu, M., Betts, N.M., Du, M., and Lyons, T.J. 2010b. Strawberries decrease atherosclerotic markers in subjects with metabolic syndrome. *Nutr Res 30*: 462–469.

Basu, A., Wilkinson, M., Penugonda, K., Simmons, B., Betts, N.M., and Lyons, T.J. 2009. Freeze-dried strawberry powder improves lipid profile and lipid peroxidation in women with metabolic syndrome: Baseline and post intervention effects. *Nutr J 8*: 43.

Bourlier, V., Zakaroff-Girard, A., Miranville, A. et al. 2008. Remodeling phenotype of human subcutaneous adipose tissue macrophages. *Circulation 117*: 806–815.

Boustany, C.M., Bharadwaj, K., Daugherty, A., Brown, D.R., Randall, D.C., and Cassis, L.A. 2004. Activation of the systemic and adipose renin–angiotensin system in rats with diet-induced obesity and hypertension. *Am J Physiol Regul Integr Comp Physiol 287*: R943–R949.

Broncel, M., Kozirog-Kolacinska, M., Andryskowski, G., Duchnowicz, P., Koter-Michalak, M., Owczarczyk, A., and Chojnowska-Jezierska, J. 2007. Effect of anthocyanins from *Aronia melanocarpa* on blood pressure, concentration of endothelin-1 and lipids in patients with metabolic syndrome. *Pol Merkur Lekarski 23*: 116–119.

Broncel, M., Kozirog, M., Duchnowicz, P., Koter-Michalak, M., Sikora, J., and Chojnowska-Jezierska, J. 2010. *Aronia melanocarpa* extract reduces blood pressure, serum endothelin, lipid, and oxidative stress marker levels in patients with metabolic syndrome. *Med Sci Monit 16*: CR28–CR34.

Cherniack, E.P. 2011. Polyphenols: Planting the seeds of treatment for the metabolic syndrome. *Nutrition 27*: 617–623.

Cinti, S., Mitchell, G., Barbatelli, G., Murano, I., Ceresi, E., Faloia, E., Wang, S., Fortier, M., Greenberg, A.S., and Obin, M.S. 2005. Adipocyte death defines macrophage localization and function in adipose tissue of obese mice and humans. *J Lipid Res 46*: 2347–2355.

Correia, M.L., Haynes, W.G., Rahmouni, K., Morgan, D.A., Sivitz, W.I., and Mark, A.L. 2002. The concept of selective leptin resistance: Evidence from agouti yellow obese mice. *Diabetes 51*: 439–442.

Curtis, P.J., Kroon, P.A., Hollands, W.J., Walls, R., Jenkins, G., Kay, C.D., and Cassidy, A. 2009. Cardiovascular disease risk biomarkers and liver and kidney function are not altered in postmenopausal women after ingesting an elderberry extract rich in anthocyanins for 12 weeks. *J Nutr 139*: 2266–2271.

DeFuria, J., Bennett, G., Strissel, K.J., Perfield, J.W., 2nd, Milbury, P.E., Greenberg, A.S., and Obin, M.S. 2009. Dietary blueberry attenuates whole-body insulin resistance in high fat-fed mice by reducing adipocyte death and its inflammatory sequelae. *J Nutr 139*: 1510–1516.

Doggrell, S.A., and Brown, L. 1998. Rat models of hypertension, cardiac hypertrophy and failure. *Cardiovasc Res 39*: 89–105.

Dumasia, R., Eagle, K.A., Kline-Rogers, E., May, N., Cho, L., and Mukherjee, D. 2005. Role of PPAR-gamma agonist thiazolidinediones in treatment of pre-diabetic and diabetic individuals: A cardiovascular perspective. *Curr Drug Targets Cardiovasc Haematol Disord 5*: 377–386.

Edirisinghe, I., Banaszewski, K., Cappozzo, J., Sandhya, K., Ellis, C.L., Tadapaneni, R., Kappagoda, C.T., and Burton-Freeman, B.M. 2011. Strawberry anthocyanin and its association with postprandial inflammation and insulin. *Br J Nutr 106*: 913–922.

Ervin, R.B. 2009. Prevalence of metabolic syndrome among adults 20 years of age and over, by sex, age, race and ethnicity, and body mass index: United States, 2003–2006. In National Health Statistics Report, N.C.f.H. Statistics, ed. (Hyattsville, MD).

Finne Nielsen, I.L., Elbol Rasmussen, S., Mortensen, A., et al., 2005. Anthocyanins increase low-density lipoprotein and plasma cholesterol and do not reduce atherosclerosis in Watanabe Heritable Hyperlipidemic rabbits. *Mol Nutr Food Res 49*: 301–308.

Furukawa, S., Fujita, T., Shimabukuro, M., Iwaki, M., Yamada, Y., Nakajima, Y., Nakayama, O., Makishima, M., Matsuda, M., and Shimomura, I. 2004. Increased oxidative stress in obesity and its impact on metabolic syndrome. *J Clin Invest 114*: 1752–1761.

Ginsberg, H.N., Zhang, Y.L., and Hernandez-Ono, A. 2006. Metabolic syndrome: Focus on dyslipidemia. *Obesity (Silver Spring) 14*(Suppl 1): 41S–49S.

Grace, M.H., Ribnicky, D.M., Kuhn, P., Poulev, A., Logendra, S., Yousef, G.G., Raskin, I., and Lila, M.A. 2009. Hypoglycemic activity of a novel anthocyanin-rich formulation from lowbush blueberry, *Vaccinium angustifolium* Aiton. *Phytomedicine 16*: 406–415.

Hassellund, S.S., Flaa, A., Sandvik, L., Kjeldsen, S.E., and Rostrup, M. 2012. Effects of anthocyanins on blood pressure and stress reactivity: a double-blind randomized placebo-controlled crossover study. *J Hum Hypertens 26*: 396–404.

Hassellund, S.S., Flaa, A., Kjeldsen, S.E., Seljeflot, I., Karlsen, A., Erlund, I., and Rostrup, M. 2013. Effects of anthocyanins on cardiovascular risk factors and inflammation in pre-hypertensive men: a double-blind randomized placebo-controlled crossover study. *J Hum Hypertens 27*:100–106.

Hwang, Y.P., Choi, J.H., Han, E.H., et al., 2011a. Purple sweet potato anthocyanins attenuate hepatic lipid accumulation through activating adenosine monophosphate-activated protein kinase in human HepG2 cells and obese mice. *Nutr Res 31*: 896–906.

Hwang, Y.P., Choi, J.H., Yun, H.J. et al. 2011b. Anthocyanins from purple sweet potato attenuate dimethylnitrosamine-induced liver injury in rats by inducing Nrf2-mediated antioxidant enzymes and reducing COX-2 and iNOS expression. *Food Chem Toxicol 49*: 93–99.

Ichikawa, H., Ichiyanagi, T., Xu, B., Yoshii, Y., Nakajima, M., and Konishi, T. 2001. Antioxidant activity of anthocyanin extract from purple black rice. *J Med Food 4*: 211–218.

Jang, H.H., Park, M.Y., Kim, H.W., Lee, Y.M., Hwang, K.A., Park, J.H., Park, D.S., and Kwon, O. 2012. Black rice (*Oryza sativa* L.) extract attenuates hepatic steatosis in C57BL/6J mice fed a high-fat diet via fatty acid oxidation. *Nutr Metab (Lond) 9*: 27.

Jaroslawska, J., Juskiewicz, J., Wroblewska, M., Jurgonski, A., Krol, B., and Zdunczyk, Z. 2011. Polyphenol-rich strawberry pomace reduces serum and liver lipids and alters gastrointestinal metabolite formation in fructose-fed rats. *J Nutr 141*: 1777–1783.

Jayaprakasam, B., Olson, L.K., Schutzki, R.E., Tai, M.H., and Nair, M.G. 2006. Amelioration of obesity and glucose intolerance in high-fat-fed C57BL/6 mice by anthocyanins and ursolic acid in Cornelian cherry (*Cornus mas*). *J Agric Food Chem 54*: 243–248.

Jurgonski, A., Juskiewicz, J., and Zdunczyk, Z. 2008. Ingestion of black chokeberry fruit extract leads to intestinal and systemic changes in a rat model of prediabetes and hyperlipidemia. *Plant Foods Hum Nutr 63*: 176–182.

Kalt, W., McDonald, J.E., Ricker, R.D., and Lu, X. 1999. Anthocyanin content and profile within and among blueberry species. *Can J Plant Sci 79*: 617–623.

Kanamoto, Y., Yamashita, Y., Nanba, F., Yoshida, T., Tsuda, T., Fukuda, I., Nakamura-Tsuruta, S., and Ashida, H. 2011. A black soybean seed coat extract prevents obesity and glucose intolerance by up-regulating uncoupling proteins and down-regulating inflammatory cytokines in high-fat diet-fed mice. *J Agric Food Chem 59*: 8985–8993.

Karlsen, A., Retterstol, L., Laake, P., Paur, I., Kjolsrud-Bohn, S., Sandvik, L., and Blomhoff, R. 2007. Anthocyanins inhibit nuclear factor-kappaB activation in monocytes and reduce plasma concentrations of pro-inflammatory mediators in healthy adults. *J Nutr 137*: 1951–1954.

Kasuga, M. 2006. Insulin resistance and pancreatic beta cell failure. *J Clin Invest 116*: 1756–1760.

Kay, C.D., and Holub, B.J. 2002. The effect of wild blueberry (*Vaccinium angustifolium*) consumption on postprandial serum antioxidant status in human subjects. *Br J Nutr 88*: 389–398.

Kelley, D.S., Rasooly, R., Jacob, R.A., Kader, A.A., and Mackey, B.E. 2006. Consumption of Bing sweet cherries lowers circulating concentrations of inflammation markers in healthy men and women. *J Nutr 136*: 981–986.

Knowler, W.C., Barrett-Connor, E., Fowler, S.E., Hamman, R.F., Lachin, J.M., Walker, E.A., and Nathan, D.M. 2002. Reduction in the incidence of type 2 diabetes with lifestyle intervention or metformin. *N Engl J Med 346*: 393–403.

Kristo, A.S., Kalea, A.Z., Schuschke, D.A., and Klimis-Zacas, D.J. 2010. A wild blueberry-enriched diet (*Vaccinium angustifolium*) improves vascular tone in the adult spontaneously hypertensive rat. *J Agric Food Chem 58*: 11600–11605.

Kwon, S.H., Ahn, I.S., Kim, S.O., Kong, C.S., Chung, H.Y., Do, M.S., and Park, K.Y. 2007. Anti-obesity and hypolipidemic effects of black soybean anthocyanins. *J Med Food 10*: 552–556.

Lecerf, J.M., and de Lorgeril, M. 2011. Dietary cholesterol: From physiology to cardiovascular risk. *Br J Nutr 106*: 6–14.

Lee, J., Durst, R.W., and Wrolstad, R.E. 2005. Determination of total monomeric anthocyanin pigment content of fruit juices, beverages, natural colorants, and wines by the pH differential method: Collaborative study. *J AOAC Int 88*: 1269–1278.

Lila, M.A. 2007. From beans to berries and beyond: Teamwork between plant chemicals for protection of optimal human health. *Ann N Y Acad Sci 1114*: 372–380.

Ling, W.H., Cheng, Q.X., Ma, J., and Wang, T. 2001. Red and black rice decrease atherosclerotic plaque formation and increase antioxidant status in rabbits. *J Nutr 131*: 1421–1426.

Ling, W.H., Wang, L.L., and Ma, J. 2002. Supplementation of the black rice outer layer fraction to rabbits decreases atherosclerotic plaque formation and increases antioxidant status. *J Nutr 132*: 20–26.

Luft, F.C. 2010. New insights into angiotensin, reactive oxygen and endothelial function. *Nephrol Dial Transplant 25*: 2099–2101.

Lumeng, C.N., Bodzin, J.L., and Saltiel, A.R. 2007a. Obesity induces a phenotypic switch in adipose tissue macrophage polarization. *J Clin Invest 117*: 175–184.

Lumeng, C.N., Deyoung, S.M., Bodzin, J.L., and Saltiel, A.R. 2007b. Increased inflammatory properties of adipose tissue macrophages recruited during diet-induced obesity. *Diabetes 56*: 16–23.

Mark, A.L., Correia, M.L., Rahmouni, K., and Haynes, W.G. 2002. Selective leptin resistance: A new concept in leptin physiology with cardiovascular implications. *J Hypertens 20*: 1245–1250.

Marsh, A.J., Fontes, M.A., Killinger, S., Pawlak, D.B., Polson, J.W., and Dampney, R.A. 2003. Cardiovascular responses evoked by leptin acting on neurons in the ventromedial and dorsomedial hypothalamus. *Hypertension 42*: 488–493.

Martineau, L.C., Couture, A., Spoor, D. et al. 2006. Anti-diabetic properties of the Canadian lowbush blueberry *Vaccinium angustifolium* Ait. *Phytomedicine 13*: 612–623.

Massiera, F., Bloch-Faure, M., Ceiler, D. et al. 2001. Adipose angiotensinogen is involved in adipose tissue growth and blood pressure regulation. *Faseb J 15*: 2727–2729.

Mauray, A., Felgines, C., Morand, C., Mazur, A., Scalbert, A., and Milenkovic, D. 2010. Nutrigenomic analysis of the protective effects of bilberry anthocyanin-rich extract in apo E-deficient mice. *Genes Nutr 5*: 343–353.

Mauray, A., Felgines, C., Morand, C., Mazur, A., Scalbert, A., and Milenkovic, D. 2012. Bilberry anthocyanin-rich extract alters expression of genes related to atherosclerosis development in aorta of apo E-deficient mice. *Nutr Metab Cardiovasc Dis 22*: 72–80.

Mauray, A., Milenkovic, D., Besson, C., Caccia, N., Morand, C., Michel, F., Mazur, A., Scalbert, A., and Felgines, C. 2009. Atheroprotective effects of bilberry extracts in apo E-deficient mice. *J Agric Food Chem 57*: 11106–11111.

McGhie, T.K., and Walton, M.C. 2007. The bioavailability and absorption of anthocyanins: Towards a better understanding. *Molecular Nutrition & Food Research 51*: 702–713.

Meir, K.S., and Leitersdorf, E. 2004. Atherosclerosis in the apolipoprotein-E-deficient mouse: A decade of progress. *Arterioscler Thromb Vasc Biol 24*: 1006–1014.

Miyazaki, K., Makino, K., Iwadate, E., Deguchi, Y., and Ishikawa, F. 2008. Anthocyanins from purple sweet potato *Ipomoea batatas* cultivar Ayamurasaki suppress the development of atherosclerotic lesions and both enhancements of oxidative stress and soluble vascular cell adhesion molecule-1 in apolipoprotein E-deficient mice. *J Agric Food Chem 56*: 11485–11492.

Morris, D.L., Singer, K., and Lumeng, C.N. 2011. Adipose tissue macrophages: Phenotypic plasticity and diversity in lean and obese states. *Curr Opin Clin Nutr Metab Care 14*: 341–346.

Mulabagal, V., Lang, G.A., DeWitt, D.L., Dalavoy, S.S., and Nair, M.G. 2009. Anthocyanin content, lipid peroxidation and cyclooxygenase enzyme inhibitory activities of sweet and sour cherries. *J Agric Food Chem 57*: 1239–1246.

Munzberg, H., Flier, J.S., and Bjorbaek, C. 2004. Region-specific leptin resistance within the hypothalamus of diet-induced obese mice. *Endocrinology 145*: 4880–4889.

Muoio, D.M., and Newgard, C.B. 2008. Mechanisms of disease: Molecular and metabolic mechanisms of insulin resistance and beta-cell failure in type 2 diabetes. *Nat Rev Mol Cell Biol 9*: 193–205.

Orchard, T.J., Temprosa, M., Goldberg, R., Haffner, S., Ratner, R., Marcovina, S., and Fowler, S. 2005. The effect of metformin and intensive lifestyle intervention on the metabolic syndrome: the diabetes prevention program randomized trial. *Ann Intern Med 142*: 611–619.

Ouchi, N., Parker, J.L., Lugus, J.J., and Walsh, K. 2011. Adipokines in inflammation and metabolic disease. *Nature Reviews Immunology 11*: 85–97.

Parelman, M.A., Storms, D.H., Kirschke, C.P., Huang, L., and Zunino, S.J. 2012. Dietary strawberry powder reduces blood glucose concentrations in obese and lean C57BL/6 mice, and selectively lowers plasma C-reactive protein in lean mice. *Br J Nutr 108*: 1789–1799.

Poreba, R., Skoczynska, A., Gac, P., Poreba, M., Jedrychowska, I., Affelska-Jercha, A., Turczyn, B., Wojakowska, A., Oszmianski, J., and Andrzejak, R. 2009. Drinking of chokeberry juice from the ecological farm Dzieciolowo and distensibility of brachial artery in men with mild hypercholesterolemia. *Ann Agric Environ Med 16*: 305–308.

Poudyal, H., Panchal, S., and Brown, L. 2010. Comparison of purple carrot juice and beta-carotene in a high-carbohydrate, high-fat diet-fed rat model of the metabolic syndrome. *Br J Nutr 104*: 1322–1332.

Prior, R.L., S, E.W., T, R.R., Khanal, R.C., Wu, X., and Howard, L.R. 2010. Purified blueberry anthocyanins and blueberry juice alter development of obesity in mice fed an obesogenic high-fat diet. *J Agric Food Chem 58*: 3970–3976.

Prior, R.L., Wu, X., Gu, L., Hager, T., Hager, A., Wilkes, S., and Howard, L. 2009. Purified berry anthocyanins but not whole berries normalize lipid parameters in mice fed an obesogenic high fat diet. *Mol Nutr Food Res 53*: 1406–1418.

Prior, R.L., Wu, X., Gu, L., Hager, T.J., Hager, A., and Howard, L.R. 2008. Whole berries versus berry anthocyanins: Interactions with dietary fat levels in the C57BL/6J mouse model of obesity. *J Agric Food Chem 56*: 647–653.

Qin, B., and Anderson, R.A. 2012. An extract of chokeberry attenuates weight gain and modulates insulin, adipogenic and inflammatory signalling pathways in epididymal adipose tissue of rats fed a fructose-rich diet. *Br J Nutr 108*: 581–587.

Qin, Y., Xia, M., Ma, J., Hao, Y., Liu, J., Mou, H., Cao, L., and Ling, W. 2009. Anthocyanin supplementation improves serum LDL- and HDL-cholesterol concentrations associated with the inhibition of cholesteryl ester transfer protein in dyslipidemic subjects. *Am J Clin Nutr 90*: 485–492.

Rahmouni, K. 2010. Leptin-induced sympathetic nerve activation: signaling mechanisms and cardiovascular consequences in obesity. *Curr Hypertens Rev 6*: 104–209.

Rahmouni, K., Correia, M.L., Haynes, W.G., and Mark, A.L. 2005. Obesity-associated hypertension: New insights into mechanisms. *Hypertension 45*: 9–14.

Rahmouni, K., Haynes, W.G., Morgan, D.A., and Mark, A.L. 2002. Selective resistance to central neural administration of leptin in agouti obese mice. *Hypertension 39*: 486–490.

Rojo, L.E., Ribnicky, D., Logendra, S., Poulev, A., Rojas-Silva, P., Kuhn, P., Dorn, R., Grace, M.H., Lila, M.A., and Raskin, I. 2012. *In vitro* and *in vivo* anti-diabetic effects of anthocyanins from Maqui Berry (*Aristotelia chilensis*). *Food Chem 131*: 387–396.

Roopchand, D.E., Grace, M.H., Kuhn, P., Cheng, D.M., Plundrich, N., Pouleva, A., Howell, A., Fridlender, B., Lila, M.A., and Raskin, I. 2012a. Efficient sorption of polyphenols to soybean flour enables natural fortification of foods. *Food Chem 131*: 1193–1200.

Roopchand, D.E., Kuhn, P., Poulev, A., Oren, A., Lila, M.A., Fridlender, B., and Raskin, I. 2012b. Biochemical analysis and *in vivo* hypoglycemic activity of a grape polyphenol-oybean flour complex. *J Agric Food Chem 60*: 8860–8865.

Sasaki, R., Nishimura, N., Hoshino, H. et al. 2007. Cyanidin 3-glucoside ameliorates hyperglycemia and insulin sensitivity due to downregulation of retinol binding protein 4 expression in diabetic mice. *Biochem Pharmacol 74*: 1619–1627.

Seymour, E.M., Lewis, S.K., Urcuyo-Llanes, D.E., Tanone, II, Kirakosyan, A., Kaufman, P.B., and Bolling, S.F. 2009. Regular tart cherry intake alters abdominal adiposity, adipose gene transcription, and inflammation in obesity-prone rats fed a high fat diet. *J Med Food 12*: 935–942.

Seymour, E.M., Singer, A.A., Kirakosyan, A., Urcuyo-Llanes, D.E., Kaufman, P.B., and Bolling, S.F. 2008. Altered hyperlipidemia, hepatic steatosis, and hepatic peroxisome proliferator-activated receptors in rats with intake of tart cherry. *J Med Food 11*: 252–259.

Seymour, E.M., Tanone, II, Urcuyo-Llanes, D.E., Lewis, S.K., Kirakosyan, A., Kondoleon, M.G., Kaufman, P.B., and Bolling, S.F. 2011. Blueberry intake alters skeletal muscle and adipose tissue peroxisome proliferator-activated receptor activity and reduces insulin resistance in obese rats. *J Med Food 14*: 1511–1518.

Shaughnessy, K.S., Boswall, I.A., Scanlan, A.P., Gottschall-Pass, K.T., and Sweeney, M.I. 2009. Diets containing blueberry extract lower blood pressure in spontaneously hypertensive stroke-prone rats. *Nutr Res 29*: 130–138.

Sikora, J., Broncel, M., Markowicz, M., Chalubinski, M., Wojdan, K., and Mikiciuk-Olasik, E. 2012. Short-term supplementation with *Aronia melanocarpa* extract improves platelet aggregation, clotting, and fibrinolysis in patients with metabolic syndrome. *Eur J Nutr 51*: 549–556.

Simeonov, S.B., Botushanov, N.P., Karahanian, E.B., Pavlova, M.B., Husianitis, H.K., and Troev, D.M. 2002. Effects of *Aronia melanocarpa* juice as part of the dietary regimen in patients with diabetes mellitus. *Folia Med (Plovdiv) 44*: 20–23.

Skoczynska, A., Jedrychowska, I., Poreba, R., Affelska-Jercha, A., Turczyn, B., Wojakowska, A., and Andrzejak, R. 2007. Influence of chokeberry juice on arterial blood pressure and lipid parameters in men with mild hypercholesterolemia. *Pharm Rep 59*: 177–182.

Slimestad, R., and Solheim, H. 2002. Anthocyanins from black currants (*Ribes nigrum* L.). *J Agric Food Chem 50*: 3228–3231.

Spalding, K.L., Arner, E., Westermark, P.O. et al. 2008. Dynamics of fat cell turnover in humans. *Nature 453*: 783–787.

Stull, A.J., Cash, K.C., Johnson, W.D., Champagne, C.M., and Cefalu, W.T. 2010. Bioactives in blueberries improve insulin sensitivity in obese, insulin-resistant men and women. *J Nutr 140*: 1764–1768.

Sun, S., Ji, Y., Kersten, S., and Qi, L. 2012. Mechanisms of inflammatory responses in obese adipose tissue. *Annu Rev Nutr 32*: 261–286.

Sun, T., Simon, P.W., and Tanumihardjo, S.A. 2009. Antioxidant phytochemicals and antioxidant capacity of biofortified carrots (*Daucus carota* L.) of various colors. *J Agric Food Chem 57*: 4142–4147.

Surwit, R.S., Kuhn, C.M., Cochrane, C., McCubbin, J.A., and Feinglos, M.N. 1988. Diet-induced type II diabetes in C57BL/6J mice. *Diabetes 37*: 1163–1167.

Takikawa, M., Inoue, S., Horio, F., and Tsuda, T. 2010. Dietary anthocyanin-rich bilberry extract ameliorates hyperglycemia and insulin sensitivity via activation of AMP-activated protein kinase in diabetic mice. *J Nutr 140*: 527–533.

Thole, J.M., Kraft, T.F., Sueiro, L.A., Kang, Y.H., Gills, J.J., Cuendet, M., Pezzuto, J.M., Seigler, D.S., and Lila, M.A. 2006. A comparative evaluation of the anticancer properties of European and American elderberry fruits. *J Med Food 9*: 498–504.

Tinahones, F.J., Murri-Pierri, M., Garrido-Sanchez, L., Garcia-Almeida, J.M., Garcia-Serrano, S., Garcia-Arnes, J., and Garcia-Fuentes, E. 2009. Oxidative stress in severely obese persons is greater in those with insulin resistance. *Obesity (Silver Spring) 17*: 240–246.

Truong, V.D., Deighton, N., Thompson, R.T., McFeeters, R.F., Dean, L.O., Pecota, K.V., and Yencho, G.C. 2010. Characterization of anthocyanins and anthocyanidins in purple-fleshed sweetpotatoes by HPLC-DAD/ESI-MS/MS. *J Agric Food Chem 58*: 404–410.

Tsuda, T., Horio, F., Uchida, K., Aoki, H., and Osawa, T. 2003. Dietary cyanidin 3-O-beta-D-glucoside-rich purple corn color prevents obesity and ameliorates hyperglycemia in mice. *Journal of Nutrition 133*: 2125–2130.

Valcheva-Kuzmanova, S., Kuzmanov, K., Mihova, V., Krasnaliev, I., Borisova, P., and Belcheva, A. 2007a. Antihyperlipidemic effect of *Aronia melanocarpa* fruit juice in rats fed a high-cholesterol diet. *Plant Foods Hum Nutr 62*: 19–24.

Valcheva-Kuzmanova, S., Kuzmanov, K., Tancheva, S., and Belcheva, A. 2007b. Hypoglycemic and hypolipidemic effects of *Aronia melanocarpa* fruit juice in streptozotocin-induced diabetic rats. *Methods Find Exp Clin Pharmacol 29*: 101–105.

Viollet, B., Foretz, M., Guigas, B., Horman, S., Dentin, R., Bertrand, L., Hue, L., and Andreelli, F. 2006. Activation of AMP-activated protein kinase in the liver: A new strategy for the management of metabolic hepatic disorders. *J Physiol 574*: 41–53.

Vuong, T., Benhaddou-Andaloussi, A., Brault, A., Harbilas, D., Martineau, L.C., Vallerand, D., Ramassamy, C., Matar, C., and Haddad, P.S. 2009. Antiobesity and antidiabetic effects of biotransformed blueberry juice in KKA(y) mice. *Int J Obes (Lond) 33*: 1166–1173.

Wang, Q., Han, P., Zhang, M., Xia, M., Zhu, H., Ma, J., Hou, M., Tang, Z., and Ling, W. 2007. Supplementation of black rice pigment fraction improves antioxidant and anti-inflammatory status in patients with coronary heart disease. *Asia Pac J Clin Nutr 16 Suppl 1*: 295–301.

Wang, H., and Peng, D.Q. 2011. New insights into the mechanism of low high-density lipoprotein cholesterol in obesity. *Lipids Health Dis 10*: 176.

Wiseman, W., Egan, J.M., Slemmer, J.E., Shaughnessy, K.S., Ballem, K., Gottschall-Pass, K.T., and Sweeney, M.I. 2011. Feeding blueberry diets inhibits angiotensin II-converting enzyme (ACE) activity in spontaneously hypertensive stroke-prone rats. *Can J Physiol Pharmacol 89*: 67–71.

Wu, X., Gu, L., Prior, R.L., and McKay, S. 2004. Characterization of anthocyanins and proanthocyanidins in some cultivars of *Ribes*, *Aronia*, and *Sambucus* and their antioxidant capacity. *J Agric Food Chem 52*: 7846–7856.

Wu, X., Kang, J., Xie, C., Burris, R., Ferguson, M.E., Badger, T.M., and Nagarajan, S. 2010. Dietary blueberries attenuate atherosclerosis in apolipoprotein E-deficient mice by upregulating antioxidant enzyme expression. *J Nutr 140*: 1628–1632.

Wu, X.L., and Prior, R.L. 2005. Systematic identification and characterization of anthocyanins by HPLC-ESI-MS/MS in common foods in the United States: Fruits and berries. *J Agr Food Chem 53*: 2589–2599.

Xia, M., Ling, W.H., Ma, J., Kitts, D.D., and Zawistowski, J. 2003. Supplementation of diets with the black rice pigment fraction attenuates atherosclerotic plaque formation in apolipoprotein e deficient mice. *J Nutr 133*: 744–751.

Xia, X., Ling, W., Ma, J., Xia, M., Hou, M., Wang, Q., Zhu, H., and Tang, Z. 2006. An anthocyanin-rich extract from black rice enhances atherosclerotic plaque stabilization in apolipoprotein E-deficient mice. *J Nutr 136*: 2220–2225.

Zeyda, M., Farmer, D., Todoric, J., Aszmann, O., Speiser, M., Gyori, G., Zlabinger, G.J., and Stulnig, T.M. 2007. Human adipose tissue macrophages are of an anti-inflammatory phenotype but capable of excessive pro-inflammatory mediator production. *Int J Obes (Lond) 31*: 1420–1428.

Zeyda, M., and Stulnig, T.M. 2009. Obesity, inflammation, and insulin resistance—A mini-review. *Gerontology 55*: 379–386.

Zhang, S.H., Reddick, R.L., Piedrahita, J.A., and Maeda, N. 1992. Spontaneous hypercholesterolemia and arterial lesions in mice lacking apolipoprotein E. *Science 258*: 468–471.

Zhang, M.W., Zhang, R.F., Zhang, F.X., and Liu, R.H. 2010. Phenolic profiles and antioxidant activity of black rice bran of different commercially available varieties. *J Agric Food Chem 58*: 7580–7587.

Zhang, B.B., Zhou, G., and Li, C. 2009. AMPK: An emerging drug target for diabetes and the metabolic syndrome. *Cell Metab 9*: 407–416.

Zunino, S.J., Parelman, M.A., Freytag, T.L., Stephensen, C.B., Kelley, D.S., Mackey, B.E., Woodhouse, L.R., and Bonnel, E.L. 2011. Effects of dietary strawberry powder on blood lipids and inflammatory markers in obese human subjects. *Br J Nutr 108*: 900–909.

8 Anthocyanins, Anthocyanin Derivatives, and Colorectal Cancer

Li-Shu Wang, Chieh-Ti Kuo, Dan Peiffer,
Claire Seguin, Kristen Stoner, Yi-Wen Huang,
Tim H.-M. Huang, Nita Salzman, Zhongfa Liu,
Daniel Rosenberg, Guang-Yu Yang, Wencai Yang,
Xiuli Bi, Steven Carmella, Stephen Hecht, and
Gary Stoner

CONTENTS

8.1   INTRODUCTION

Anthocyanins are a ubiquitous group of water-soluble plant metabolites of the flavonoid family. They are "nature's colors," responsible for the blue, purple, red, and
intermediate colors of leaves, flowers, vegetables, and fruits, especially berries. The
daily intake of anthocyanins in the U.S. diet is estimated to be as much as 180–
215 mg, or about ninefold higher than that of other flavonoids such as quercetin,
genistein, and apigenin [1]. Epidemiological studies suggest that the consumption of
anthocyanins lowers the risk for diabetes, arthritis, cardiovascular disease, and cancer due, at least in part, to their anti-inflammatory and antioxidant activities [2]. The
chemistry, bioavailability, and chemopreventive effects of anthocyanins in multiple
tissues have been summarized in our earlier publication [3] and will not be addressed
here. This chapter focuses exclusively on mechanistic studies of the prevention of
colorectal cancer with anthocyanins, including results from *in vitro* cell culture systems, *in vivo* animal model systems, and human studies. In addition, recent studies
from our laboratory and those of our collaborators of the effects of anthocyanins on
inflammation-associated colorectal cancer, that is ulcerative colitis, are discussed.
Finally, this chapter summarizes the known chemopreventive effects of protocatechuic acid (PCA), a major metabolite of anthocyanins. Additional *in vitro* and *in
vivo* studies of the potential preventative effects of PCA in animal model and human
systems are recommended.

8.2   MECHANISMS OF COLORECTAL CANCER CHEMOPROTECTION BY ANTHOCYANINS

8.2.1   *In Vitro* Studies

Anthocyanins elicit multiple effects on cultured colon cancer cells including reduction of cell proliferation, induction of apoptosis, anti-inflammation, anti-invasion,
inhibition of drug resistance, and promoter demethylation (Table 8.1).

Using the metastatic colorectal cancer cell line, LoVo/ADR, delphinidin from
blueberries and cyanidin from black raspberries were found to increase the accumulation of cellular reactive oxygen species (ROS), inhibit glutathione reductase, and
deplete cellular glutathione [4]. These results suggest that delphinidin may be used
as a sensitizing agent in the therapy of metastatic colorectal cancer [4]. Mechanisms
for the induction of apoptosis and cell cycle arrest by delphinidin were studied in
detail by Yun et al. [5]. Treatment of HCT116 cells with delphinidin resulted in
decreased cell viability, induction of apoptosis, cleavage of PARP, activation of cas-
pases-3, -8, and -9, an increase in Bax with a concomitant decrease in Bcl-2 protein,
and G2/M phase cell cycle arrest. This study also examined the effect of delphinidin on NFκB signaling and showed that the compound inhibited the expression of
IKKα, phosphorylation and degradation of IκBα, phosphorylation of NFκB/p65 at
Ser(536), nuclear translocation of NFκB/p65, NFκB/p65 DNA binding activity, and
transcriptional activation of NFκB. Clearly, these studies demonstrate the ability of
delphinidin to influence a major signaling pathway (NFκB) associated with cellular
proliferation, apoptosis, and inflammation.

TABLE 8.1

Anticarcinogenic Effects of Anthocyanin and Anthocyanin-Rich Extracts in Colon Cancer Cells

Anticancer Effect	Anthocyanins/ Metabolites	Concentration	Cells	Ref.
	Inhibition of Drug Resistance			
↑ ROS accumulation ↓ Glutathione reductase ↓ Glutathione	Cyanidin, delphinidin	25–100 μM	LoVo/ADR	[4]
	Anticell Proliferation and Induction of Apoptosis			
↑ Cleavage of PARP, caspases-3, -8, and -9, Bax to Bcl-2 ratio, G2/M cell cycle arrest ↓ IKKα, IκBα, p-NFκB/p65 (Ser 536), nuclear translocation, DNA binding, and transcriptional activation of NFκB/p65	Delphinidin	30–240 μM	HCT116	[5]
↓ mTOR, Akt, ↑ AMPKα1	Anthocyanin extract from Vitis coignetiae Pulliat (Meoru in Korea)	25–400 μg/ mL	HT-29	[6]
↑ p38-MAPK, ↓ Akt	Anthocyanin extract from Vitis coignetiae Pulliat (Meoru in Korea)	5–60 μg/mL	HCT116	[7]
↑ Cell-cycle arrest at G1/G0 and G2/M phase ↑ p21, p27 ↓ Cyclin A, cyclin B ↓ COX-2	Aronia meloncarpa	50 mg/mL	HT-29 but did not inhibit proliferation of normal line NCM460	[9]
	Anti-Inflammation			
↓ COX-1 and COX-2	Cyanidin-3-O-glucoside	12.5–200 μg/ mL	HCT116	[10]
↓ IL-12	Hull blackberry extract	40 μg/mL	HT-29	[11]
	Anti-Invasiveness			
↓ MMP-2, MMP-9, claudin, and NFκB	Anthocyanin extract from Vitis coignetiae Pulliat	15–75 μg/mL	HT-29	[8]

continued

TABLE 8.1 (continued)
Anticarcinogenic Effects of Anthocyanin and Anthocyanin-Rich Extracts in Colon Cancer Cells

Anticancer Effect	Anthocyanins/ Metabolites	Concentration	Cells	Ref.
Promoter Demethylation				
↓ Promoter demethylation of tumor suppressor genes, for example, p16, sfrp2, sfrp5, wif1 ↓ DNMT1 and DNMT3B activities	Black raspberry extract	5–25 μg/mL	HCT116, SW480, Caco2	[19]

Note: ROS, reactive oxygen species; AMPK1α, AMP-activated protein kinase 1 α; IKK, IkappaB kinase; NFκB, nuclear factor kappa B; COX, cyclooxygenase; IL, interleukin; MMP, matrix metalloproteinases; sfrp, secreted Frizzled-related protein; wif, Wnt inhibitory factor; DNMT, DNA methyltransferase.

Anthocyanins from *Meoru* in Korea were shown to suppress mTOR and Akt pathways through the activation of AMP-activated protein kinase (AMPK) 1α in HT29 colon cancer cells [6,7]. These same anthocyanins also inhibited genes associated with tissue invasiveness, for example, MMP-2, MMP-9, claudin, and NFκB [8]. Another study reported that blockage of the cell cycle by an anthocyanin-rich extract from *Aronia meloncarpa E* was associated with increased expression of p21WAF1 and p27KIP, and decreased expression of cyclin A, cyclin B, and COX-2 [9]. The anti-inflammatory effects of cyanidin-3-glucoside and hull blackberry extract were associated with their ability to decrease COX-1 and COX-2 [10], and IL-12 [11], respectively. Anti-invasiveness of anthocyanins from *Meoru* in Korea for colorectal cancer cells was also reported [8].

DNA methylation, particularly at promoter regions of genes that regulate important cellular functions, is the best characterized epigenetically mediated transcriptional silencing [12]. DNA methylation in mammalian cells is regulated by a family of highly related DNA methyltransferase enzymes (DNMT1, DNMT3a, and DNMT3b), which mediate the transfer of methyl groups from *S*-adenosylmethionine to the 5′ position of cytosine bases in the dinucleotide sequence CpG. DNMT1 functions as the maintenance DNA methyltransferase in mammalian cells, and is therefore responsible for accurately replicating genomic DNA methylation patterns during the S phase of the cell cycle [12]. In contrast, *de novo* methylation of DNA is believed to be performed by the DNMT3a and DNMT3b enzymes, which possess both maintenance and *de novo* DNA methylation activities [13]. Both groups of enzymes however, have been shown to exhibit some level of both maintenance and *de novo* methylation *in vitro*, suggesting that this classification of the DNMTs may be oversimplified [14]. Confirming the importance of DNA methylation in tumorigenesis, studies in several tumor types, including those of colon, bladder, and kidney, have shown all three DNMTs to be overexpressed [15]. When DNMT1 and DNMT3b are knocked out in colon cancer cell lines, methylation of tumor suppressor genes such as p16 is almost entirely eliminated and the gene is reexpressed [16]. Inhibition of DNMTs, therefore, may lead to demethylation and reactivation of the silenced genes. Aberrant methylation of tumor suppressor genes by DNMTs may be a promising target for chemoprevention [17,18].

We recently demonstrated that black raspberry (BRB)-derived anthocyanins are capable of causing promoter demethylation of tumor suppressor genes, for example, p16, in human colon cancer cell lines through suppressing the activities of DNA methyltransferases 1 and 3B (DNMT1 and 3B) (Figure 8.1). The colocalization of black raspberry-derived anthocyanins with DNMT3B suggests the possible direct binding of anthocyanins to these enzymes (Figure 8.2). A similar staining pattern was observed with DNMT1 and black raspberry anthocyanins suggesting that DNMT1 colocalizes with these anthocyanins [19].

A study compared the antiproliferative effects of anthocyanins from *Aronia meloncarpa* on normal versus colon cancer cells *in vitro* and found that these compounds are more effective in inhibiting the growth of cancer cells [9]. The reasons for this observation were not investigated. Our laboratory reported that an ethanol extract from black raspberries selectively inhibited the growth and induced apoptosis in a highly tumorigenic rat esophageal epithelial cell line (RE-149 DHD) but not in a

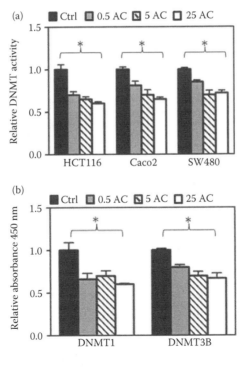

FIGURE 8.1 Black raspberry-derived anthocyanins (AC) suppressed DNMT1 and DNMT3B. (a) Total DNMT activity in nuclear extracts from human colon cancer cells, HCT116, Caco2, and SW480, treated with AC at 0.5–25 μg/mL for 3 days was decreased. In a cell-free *in vitro* inhibition assay, AC inhibited DNMT1 and DNMT3B (b). * $P < 0.05$.

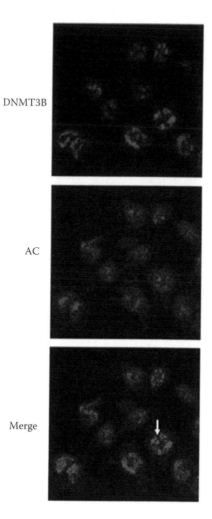

FIGURE 8.2 **(See color insert.)** Cellular uptake of black raspberry-derived anthocyanins (AC) and colocalization of AC with DNMT3B in HCT116 cells. AC uptake was observed in HCT116 cells treated with AC at 25 μg/mL for 1 day. AC in red presented in both cytoplasm and nuclei. Same cells were stained with DNMT3B shown in green. Colocalization of AC with DNMT3B appeared in yellowish as indicated by arrowhead.

weakly tumorigenic line (RE-149) [20]. The uptake of anthocyanins from the extract into RE-149 DHD cells far exceeded their uptake into RE-149 cells which may have accounted for the selective effects of the extract. This might explain the results in human normal colon versus colon cancer cells.

The concentrations of anthocyanins or extracts required to elicit anticarcinogenic effects *in vitro* are much higher (~10–200 μM) than the levels of these compounds observed in the blood of animals administered anthocyanin-containing diets or juices (~2–25 nM). This may be due, in part, to the instability of anthocyanins at physiological pH. The half-life of most anthocyanins in culture medium at pH 7.0 is less than 5 h [2]. Thus, perhaps the most appropriate use of the *in vitro* data in Table 8.1 is that of identifying possible mechanistic biomarkers for clinical investigations with the anthocyanins or anthocyanin-containing foodstuffs.

8.2.2 ANIMAL STUDIES

Because relatively few animal investigations have evaluated anthocyanin-rich extracts for colon cancer prevention, we will also discuss studies using anthocyanin-rich foods, for example, berries, to prevent colon cancer in rodents. Although anthocyanins are responsible for some of the effects observed, other components likely contribute to the preventive effects as well. In this section, effects of anthocyanin extracts and anthocyanin-rich foods in rodent models of spontaneous, chemically induced, and ulcerative colitis-associated colon cancer are discussed (Table 8.2).

8.2.2.1 Spontaneous Intestinal Carcinogenesis

The Apc(Min) mouse has one nonfunctional allele of the Apc gene leading to dys-regulated signaling of the Wnt/β-catenin pathway and the spontaneous development of intestinal adenomas [22]. Muc2 knockout (KO) mice develop intestinal adenomas and adenocarcinomas in response to chronic inflammation [22]. We and our collaborators reported that BRBs significantly inhibit intestinal tumor formation in both models; reducing tumor incidence by 45% and tumor multiplicity by 60% in Apc(Min) mice and tumor incidence and multiplicity by 50% in Muc2 KO mice [22]. Mechanistic studies revealed that BRBs inhibit tumor development in Apc(Min) mice by suppressing β-catenin signaling and in Muc2 KO mice by reducing chronic inflammation. Intestinal cell proliferation was inhibited by BRBs in both animal models; however, the extent of mucus cell differentiation was not changed in either model [22]. Collectively, our data suggest that BRBs are highly effective in preventing intestinal tumor development in both Apc(Min) and Muc2 KO mice through targeting multiple signaling pathways.

An anthocyanin-rich red grape extract reduced overall adenoma burden and marginally reduced adenoma multiplicity in Apc(Min) mice when compared with mice on control diet [23]. The proliferation index (Ki-67) in colonic adenomatous crypts was significantly decreased in control mice versus mice receiving the grape extract, and this correlated with reduced expression of Akt. Further, Min mice fed on anthocyanin-rich tart cherry extract had 74% fewer cecal tumors, but the percent reduction in overall colon tumors (17%) and small intestinal tumors (30%) was not significant [24]. In a subsequent study using a similar protocol, Min mice fed the anthocyanin-rich tart cherry extract plus the nonsteroidal anti-inflammatory drug, sulindac, had significantly fewer tumors in the proximal and medial thirds of the small intestine, but not in the distal third, relative to mice given sulindac alone [25]. Lastly, both cyanidin-3-glucoside and an anthocyanin mixture from bilberry at the same dietary concentration decreased intestinal adenoma numbers in Min mice by 30–45% [26]. In this same study, anthocyanins were detected in plasma, and both glucuronide and methylated metabolites of the anthocyanins were detectable in the intestinal mucosa and urine.

TABLE 8.2

Anticarcinogenic Effects of Anthocyanin-Rich Extracts and Anthocyanin-Rich Foods in Animal Models of Colon Cancer

Animal Model	Anthocyanins/ Metabolites	Concentration	Mechanisms	Ref.
Spontaneous Intestinal Carcinogenesis				
Apc(Min) and Muc2 knockout	BRBs	10% in diet	↓ Tumor formation ↓ Cell proliferation ↓ β-catenin	[22]
Apc(Min)	Red grape extract	0.3% in diet	↓ Adenoma burden ↓ Proliferation ↓ Ki-67, Akt,	[23]
Apc(Min)	Tart cherry extract	375–3000 mg/ kg diet	↓ Adenoma burden	[24,25]
Apc(Min)	Bilberry	0.03–0.3% in diet	↓ Adenoma number	[26]
Chemical-Induced Colon Cancer				
Azoxymethane-induced colon cancer in F-344 rats	BRBs	2.5–10% in diet	↓ ACF and tumor multiplicity ↓ urinary 8-hydroxy-2′-deoxyguanosine (8-OHdG)	[27]
Azoxymethane-induced colon cancer in F-344 rats	Anthocyanin-rich extract from bilberry, chokeberry, grape	3.85 g/kg diet	↓ Aberrant crypt foci ↓ Proliferation ↓ COX-2	[28]
1,2-dimethylhydrazine (DMH) and 2-amino-1-methyl-6-phenylimidazo[4,5-b] pyridine (PhIP)-induced colon cancer in F344 rats	Purple sweet potatoes, red cabbage, purple corn color	5% in diet	↓ Incidence and multiplicity of adenoma and adenocarcinoma	[29,30]
Ulcerative Colitis-Associated Colon Cancer				
Dextran sodium sulfate (DSS)-induced ulcerative colitis	BRBs	5% and 10% in diet	↓ Colonic injury ↓ TNFα, IL-1β ↓ p-IkBα, COX-2, PGE$_2$	[31]
Dextran sodium sulfate (DSS)-induced ulcerative colitis	BRBs	5% in diet	↓ Promoter methylation of dkk3 ↓ HDAC2	[32]

TABLE 8.2 (continued)
Anticarcinogenic Effects of Anthocyanin-Rich Extracts and Anthocyanin-Rich Foods in Animal Models of Colon Cancer

Animal Model	Anthocyanins/ Metabolites	Concentration	Mechanisms	Ref.
IL-10 knockout mice	BRBs	5% and 10% in diet	↓ Tumor incidence ↓ Extent of inflammation, ulcer formation, and epithelial hyperplasia ↓ Myeloperoxidase-labeled inflammatory cells, nitro-oxidative stress, and cell proliferation Modulate genes in inflammatory processes and T-cell activation	[33,34]
IL-10 knockout mice	BRBs	5% in diet	↓ Promoter methylation of dkk2, dkk3, sfrp1 ↓ HDAC1, HDAC2, MBD2, DNMT3B Protectively modulate genes in TLR4 and the Wnt pathways	[35]

Note: BRBs, black raspberries; HDAC, histone deacetylase; MBD, methyl-binding domain protein; DNMT, DNA methyltransferase; TLR, toll-like receptor.

8.2.2.2　Chemically-Induced Colon Cancer

In the azoxymethane (AOM)-induced model of colon cancer in F344 rats, diets containing 2.5%, 5%, and 10% BRBs significantly decreased total tumors (adenomas and adenocarcinomas) by up to 71%, and urinary 8-hydroxy-2′-deoxyguanosine (8-OHdG) levels by up to 83%, respectively [27]. The reduction in urinary 8-OHdG levels indicated that berries reduce ROS-induced DNA damage in animals. Lala et al. [28], using the AOM-induced rat colon cancer model, reported that an anthocyanin-rich extract from bilberry, chokeberry, and grape significantly reduced AOM-induced aberrant crypt foci by 26–29%. This reduction was associated with decreased cell proliferation and COX-2 gene expression, however, the levels of urinary 8-OHdG were similar among rats fed different diets. Anthocyanins from purple sweet potatoes, red cabbage, and purple corn (at 5% in the diet) significantly reduced colorectal carcinogenesis by 48%, 63%, and 89%, respectively, in rats treated with 1,2-dimethylhydrazine, but the mechanism(s) of tumor inhibition was not investigated [29,30].

8.2.2.3　Ulcerative Colitis-Associated Colon Cancer

Ulcerative colitis (UC) is a chronic inflammatory disease of the colonic mucosa that can dramatically increase the risk of colon cancer. In dextran sodium sulfate

(DSS)-induced UC in C57BL/6J mice, dietary BRBs markedly reduced DSS-induced acute injury to the colonic epithelium [31]. This protection included better maintenance of body mass, and reductions in colonic shortening and ulceration. BRB treatment, however, did not affect the levels of either plasma nitric oxide or colon malondialdehyde, biomarkers of oxidative stress that are otherwise increased by DSS-induced colonic injury. BRB treatment suppressed several key proinflammatory cytokines, including tumor necrosis factor α and interleukin 1β in the colon. Further examination of the inflammatory response revealed that BRB treatment reduced the levels of phospho-IκBα within the colonic tissue. Colonic COX-2 levels were also dramatically suppressed by BRB treatment, with a concomitant decrease in plasma prostaglandin E2. These findings suggest a potent anti-inflammatory effect of BRBs during DSS-induced colonic injury and support a possible therapeutic or preventative role in the pathogenesis of UC and related neoplastic events. In addition, promoter methylation of dickkopf 3 (dkk3), a Wnt pathway antagonist, is increased in the colon of DSS-treated mice and BRBs were found to decrease dkk3 methylation. This was associated with decreased histone deacetylase 2 (HDAC2) protein expression [32].

Interleukin 10 (IL-10) is an important regulatory anti-inflammatory cytokine associated with the development of colonic colitis [33]. Essentially all IL-10 KO mice develop colitis after 3 months of age; histopathology reveals transmural inflammation, predominantly in the cecum. The administration of a diet containing twice the recommended amount of iron leads to the development of colonic adenocarcinoma in more than 80% of IL-10 KO mice within 5 months. The colorectal tumor incidence in IL-10 KO mice fed control diet was 71.4% (10/14 mice with tumors). Significant reductions in colon tumor incidence [38% (5/13 mice) and 30.8% (4/13 mice)] were observed in mice receiving 5% and 10% BRB diets, respectively [33]. Histopathologic analysis of chronic inflammatory activity in the colon showed that the berries significantly inhibited the overall inflammatory index, including the extent of inflammation, ulcer formation, and epithelial hyperplasia [33]. Mechanistic studies suggest that inhibition of colitis-associated carcinogenesis by dietary BRBs might relate to their protective effects on modulation of myeloperoxidase-labeled inflammatory cells, nitro-oxidative stress, and cell proliferation [33]. Later studies using transcriptional gene array analysis with 24,611 gene probes revealed that 797 genes were differentially expressed in IL-10 KO mice when compared to IL-10 wild-type mice and the BRB diets altered the expression of 168 of the 797 genes toward normal levels of expression. Further pathway analysis of the 168 genes demonstrated that many of them are involved in inflammatory processes and T-cell activation [34]. More recently, we demonstrated increased promoter methylation of dkk2, dkk3, and secreted frizzled-related protein 1(sfrp1), all Wnt pathway antagonists, in the colon of IL-10 KO mice when compared with colonic tissues from IL-10 wild-type mice suggesting that the Wnt pathway is dysregulated and this may contribute to the development of ulcerative colitis in the IL-10 KO mice [35]. The methylation levels of these genes in bone marrow and spleen were lower than those in colon. BRBs significantly decreased promoter methylation of dkk2 in colon and dkk3 in spleen and these effects were associated with decreased protein expression of HDAC1, HDAC2, MBD2, and DNMT3B. The expression of 84 genes in the Wnt and toll-like receptor pathways was different in the colon of IL-10 KO mice when compared to the colon of wild-type mice. BRBs modulated 80% (67/84) and 95% (80/84) of differentially expressed genes toward normal levels of expression in the Wnt and toll-like receptor pathways, respectively. Protein expression of nuclear β-catenin was also decreased by BRBs. We conclude that BRBs are capable of reversing aberrant promoter methylation of genes in the Wnt signaling pathway presumably through inhibition of proteins regulating DNA methylation. These changes could result in protective modulation of the Wnt pathway and/or its interaction with the toll-like receptor pathway which, in turn, would reduce ulcerative colitis-associated inflammation [35]. These results suggest that BRBs, as a food-based prevention agent, have potential for the prevention of inflammation-associated carcinogenesis in the colon.

8.2.3 HUMAN STUDIES

As stated earlier, naturally occurring anthocyanins and anthocyanin-rich foods possess colorectal cancer chemopreventive properties in rodent models. Two intervention studies stated later used anthocyanin-rich foods in patients with colorectal cancer. Thomasset et al. 2009 [36] investigated whether mirtocyan, an anthocyanin-rich standardized bilberry extract, exhibits chemotherapeutic efficacy in patients with colorectal cancer [36]. Twenty-five colorectal cancer patients scheduled to undergo resection of primary tumor or liver metastases received 0.4, 2.8, or 5.6 g mirtocyan (containing 0.5–2.0 g anthocyanins) daily for 7 days before surgery. Bilberry anthocyanins were analyzed by high performance liquid chromatography (HPLC) with visible or mass spectrometric detection. Proliferation was measured by immunohistochemistry of Ki-67 in colorectal tumors, and insulin-like growth factor (IGF)-I was measured in plasma. Mirtocyan anthocyanins and their methyl and glucuronide metabolites were identified in plasma, colorectal tissue, and urine, but not in liver. Anthocyanin concentrations in plasma and urine were roughly dose-dependent and reached ~179 ng/g in tumor tissue at the highest dose. In tumor tissue from all patients treated with mirtocyan, proliferation was decreased by 7% compared with preintervention values. The low dose of mirtocyan caused a small but nonsignificant reduction in circulating IGF-I. These results suggest that repeated administration of bilberry anthocyanins exerts pharmacodynamic effects in colorectal cancer patients.

Recently, our group evaluated the effects of BRBs on biomarkers of cell proliferation, apoptosis, angiogenesis, and methylation of relevant tumor suppressor genes in the Wnt signaling pathway in tumors from colorectal cancer patients [37]. Biopsies of adjacent normal tissues and colorectal adenocarcinomas were taken from 20 patients before and after oral consumption of BRB powder (60 g/day, containing around 1.8 g anthocyanins) for 1–9 weeks. Methylation status of promoter regions of five tumor suppressor genes was quantified. Protein expression of DNMT1 and genes associated with cell proliferation, apoptosis, angiogenesis, and Wnt signaling were measured. The methylation of three Wnt inhibitors, sfrp2, sfrp5, and wif1, upstream genes in the Wnt pathway, and pax6a, a developmental regulator, was modulated in a protective direction by BRBs in normal tissues and in colorectal tumors only in patients who received BRB treatment for an average of 4 weeks, but not in all 20 patients with 1–9 weeks of BRB treatment. The methylation effects were associated with decreased expression of the enzyme, DNMT1. BRBs modulated expression of genes associated with Wnt signaling, dysregulated in 85% of sporadic colorectal cancers [38], proliferation, apoptosis, and angiogenesis in a protective direction. These data provide evidence of the ability of BRBs to demethylate tumor suppressor genes and to modulate other biomarkers of tumor development in the human colon and rectum. While demethylation of genes did not occur in colorectal tissues from all treated patients, the positive results with the secondary endpoints suggest that additional studies of BRBs for the prevention of colorectal cancer in humans now appear warranted.

8.3 MECHANISMS OF COLORECTAL CANCER CHEMOPROTECTION BY PCA

8.3.1 PCA Is Detected in Plasma and Intestinal Tissues from Rodents Fed Diets Containing Anthocyanins or Anthocyanin-Rich Foods

Pharmacokinetic studies have shown that <1% of the administered dose of anthocyanins from various berry types is absorbed into blood [39]. In addition, the recovery of known anthocyanin metabolites (glucuronidated and methylated compounds) in urine is very low relative to the administered dose [40]. Therefore, pharmacologists have been interested in determining if anthocyanins are metabolized to other compounds with chemopreventive potential. Seeram et al. [41] demonstrated that cyanidin glucosides spontaneously convert to PCA and other hydroxybenzoic acids *in vitro* at pH 7.4. Another study reported that the glucosides almost completely disappeared within 60 min and formed different dimerization products (via the quinoid anhydrobase), PCA and an aldehyde (via an α-diketone intermediate) [42]. Tsuda et al. [43] reported the presence of PCA in the plasma of rats fed cyanidin glucosides. The plasma concentration of PCA exceeded that of the cyanidin glucosides by a factor of 8 whereas cyanidin itself (the aglycone) was not detectable. Moreover, they detected PCA and cyanidin in the intestine, and demonstrated *in vitro* that PCA is formed within 15 min after the addition of cyanidin to plasma [43]. Collectively, these findings suggest that anthocyanins are rapidly metabolized to PCA *in vitro* and *in vivo*, and PCA itself may be responsible for the protective effects of anthocyanins. Our group fed F344 rats either control AIN-76A diet or AIN-76A diet containing 5% freeze-dried BRBs for 6 weeks and measured anthocyanins and PCA in the colon, urine, and feces. As expected, anthocyanins and PCA were not detectable in the colon, urine, and feces from the control group (Figures 8.3a and 8.3b). In BRB-fed rats, the amounts of anthocyanins detected in urine were >100-fold higher than those in the colon and feces suggesting that urine is the major excretion route of anthocyanins (Figure 8.3a). PCA content in the feces was significantly higher than that in urine, and there were no differences in PCA content between colon and feces (Figure 8.3b). Our data agree with previous findings that anthocyanins are metabolized to PCA. However, the detection of both anthocyanins and PCA in the colon suggests that both constitutents might be protective against colorectal cancer.

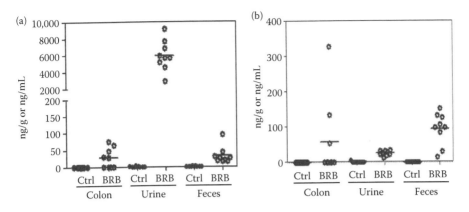

FIGURE 8.3 Anthocyanins (a) and PCA (b) in biospecimens from rats fed control (Ctrl) and whole BRB (BRB) diet.

8.3.2 Effects of BRBs and BRB Anthocyanins on the Composition and Diversity of Colon Microbiota in F344 Rats

Feces were collected from F344 rats prior to introduction of either 5% BRBs or an anthocyanin-enriched fraction of BRBs into the diet. They were collected again at 3 and 6 weeks after BRB or anthocyanin supplementation. The V3–V5 region of the 16S rRNA gene was amplified from fecal genomic DNA for each specimen using barcoded primers, pooled, and sequenced using the Roche 454 sequencer as described [44]. Sequence counts for each bacterial OTU and Order were exported from Qiime. The dataset was read into R (x64, 2.13.0) for analysis. The R package Vegan 1.17-9 was used to calculate Shannon diversity, principal component analysis, hierarchical clustering, and adonis statistical testing. A total of 8995 OTUs were found in the BRB-treated samples. The number of sequences per sample ranged from 5058 to 18,180 and were classified into 26 distinct Orders. Heatmap visualization at the Order level suggests that BRB treatment increased the abundance of Lactobacillus, Coriobacteriales, Bifidobacteriales, Bacteroides, and Clostridiales (Figure 8.4a). Bray–Curtis analysis showed a strong time-dependent change in bacterial diversity of the microbiota in response to dietary treatment with BRBs (Figure 8.4b) ($P < 0.001$) (unpublished data). Similar to the results with BRBs, dietary treatment with the anthocyanin-rich fraction of BRBs resulted in significant shifts in the intestinal microbiota (data not shown). These results suggest that BRBs and their component anthocyanins can influence the diversity of gut microflora which may affect the metabolism of the anthocyanins to PCA and/or other metabolites.

8.3.3 Colorectal Cancer Prevention Studies with PCA in In Vitro and In Vivo Models, and in Humans

To date, only a few studies have investigated the chemopreventive effects of PCA in colorectal cancer. Table 8.3 summarizes results from *in vitro* and *in vivo* studies.

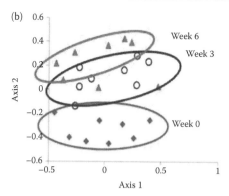

FIGURE 8.4   **(See color insert.)** (a) Heatmap analyses of the fecal microbiota. Heatmaps were generated for the BRB-treated rats by normalizing the percent abundance of each bacterial order within each sample to the average number of sequences for all 24 samples in each treatment group. Green-to-red shading indicates relatively lower-to-higher number of sequences. (b) Bray–Curtis analysis of the intestinal microbiota in response to diet. Nonmetric multidimensional scaling (NMDS) ordination of the Bray–Curtis distances showed that the samples at 3 and 6 weeks were progressively distant (i.e., different in biodiversity) from the pretreatment samples for BRB diet supplementation. The R package Ecodist 1.2.3 was used to calculate Bray–Curtis distances and perform NMDS.

8.3.3.1   *In Vitro* Studies

With respect to PCA's antioxidant activities, in human colon cancer cell lines HT-29 and HCT-116, 50 μM PCA did not cause significant changes in COX-1 and COX-2 activities but reduced liposome oxidation [41].

8.3.3.2   Animal Studies

PCA at 500 and 1000 ppm in the diet significantly inhibited colon tumor incidence and multiplicity in AOM-treated F344 rats [45]. This was associated with decreases in cell proliferation as measured by bromodeoxyuridine (BrdU) labeling, counts of silver-stained nuclear organizer regions (AgNORs), and ornithine decarboxylase (ODC) activity. PCA did not produce toxic effects at the doses administered. Using the same model, dietary PCA at 1000 and 2000 ppm significantly decreased aberrant crypt foci (ACF) in the colon compared with that in rats treated with AOM alone [46].

TABLE 8.3
Anticolon Cancer Effects of Protocatechuic Acid

Models	Protocatechuic Acid Source	Concentration	Effects	Reference
		In vitro		
HT-29 and HCT-116	Degradation product from tart cherry	50 μM	↓ Liposome oxidation	[41]
		In vivo		
Azoxymethane-induced colon cancer in F-344 rats	Synthetic	500–1000 ppm	↓ Tumor incidence, multiplicity ↓ Cell proliferation 45	
Azoxymethane-induced colon cancer in F-344 rats	Synthetic	1000–2000 ppm	↓ Aberrant crypt foci	[46]

8.3.3.3 Human Studies

In a recent intervention study, six healthy volunteers each consumed 1 L of a commercial Sicilian red orange juice containing 71 mg of total cyanidin glucosides as the major anthocyanins [47]. PCA was the main metabolite of the cyanidin glucosides. Less than 1% of the administered cyanidin glucosides was present as methylated and glucuronidated metabolites in blood and urine which is consistent with previous studies in the literature. For the first time, PCA was identified as the major metabolite of cyanidin glucosides, accounting for about 72% of the ingested anthocyanins with 44% in serum and 28% in the feces [47]. There was no detectable PCA in the urine. Detecting PCA in human fecal samples demonstrated for the first time the ability of gut microflora to degrade cyanidin glucosides to PCA [47]. In another intervention study, lymphocyte DNA resistance to oxidative stress was improved in 16 healthy female volunteers who took 600 mL/day of blood orange juice, produced by Oranfrizer (Scordia), containing 21 mg cyanidin-3-glucoside for 21 days [48]. Together, these results suggest that PCA can be absorbed through the colon and might be released slowly and continuously through the gut into the bloodstream where it represents a defense against oxidative damage in the body [49].

8.4 CONCLUSIONS

Although anthocyanins are poorly absorbed systemically, they can interact directly with the tissues lining the colon by direct uptake and can have a beneficial effect on the colon environment. Anthocyanin-rich materials have been found to be protective in the colon. Because of the presumed systemic low bioavailability of the anthocyanins, researchers are now beginning to focus more effort on studies of anthocyanin metabolites such as PCA. However, the reader should be aware of results from a recent study comparing the uptake of quercetin and resveratrol into plasma and whole blood of humans. This study reported that up to 76% of the analytes from these compounds were unaccounted for when only plasma was examined due to the binding of the analytes to the cellular fraction of whole blood [50]. If this is also true for the anthocyanins, pharmacokinetic studies to date may have underestimated their actual uptake into blood. It seems important, therefore, to analyze whole blood rather than plasma to avoid underestimating the absorption of anthocyanins and perhaps other polyphenols in bioavailability studies. Investigations to improve techniques for the measurement of protein-bound anthocyanins or other polyphenols should be encouraged.

Even if pharmacokinetic studies may have underestimated the actual uptake of anthocyanins into blood and tissues, the utility of these compounds for chemoprevention is compromised by the fact that they are difficult to synthesize and to isolate in pure form from natural sources. Accordingly, they are likely to be too expensive for routine chemoprevention. In this regard, PCA is attractive because it is commercially available and relatively inexpensive. In addition, it does not appear to be toxic at doses where it elicits chemopreventive effects. We suggest, therefore, that additional studies be undertaken to further evaluate PCA and/or its metabolites for chemopreventive efficacy in both preclinical and clinical studies.

It is important to note that a recent publication [51] showed that berry anthocyanidins synergistically suppress growth and invasive potential of human non-small-cell lung cancer cells. The anthocyanidin mixture from bilberry is more effective than individual anthocyanidins in the induction of cell-cycle arrest, apoptosis, invasion, and migration. The effective dose of delphinidin, the most potent anthocyanidin in this tested system, in the anthocyanidin mixture was eightfold lower than delphinidin alone. Therefore, the results clearly demonstrate synergism. Applying the same idea, the consumption of whole foods such as berries but not individual berry component could be more beneficial.

REFERENCES

1. Hertog MG, Hollman PC, Katan MB, Kromhout D. Intake of potentially anticarcinogenic flavonoids and their determinants in adults in The Netherlands. *Nutr. Cancer.* 1993;20:21–29.
2. Prior RL, Wu X. Anthocyanins: Structural characteristics that result in unique metabolic patterns and biological activities. *Free Radic. Res.* 2006;40:1014–28.
3. Wang L-S, Stoner GD. Anthocyanins and their role in cancer prevention. *Cancer Lett.* 2008;269:281–90.
4. Cvorovic J, Tramer F, Granzotto M, Candussio L, Decorti G, Passamonti S. Oxidative stress-based cytotoxicity of delphinidin and cyanidin in colon cancer cells. *Arch. Biochem. Biophys.* 2010;501:151–7.
5. Yun JM, Afaq F, Khan N, Mukhtar H. Delphinidin, an anthocyanidin in pigmented fruits and vegetables, induces apoptosis and cell cycle arrest in human colon cancer HCT116 cells. *Mol. Carcinog.* 2009;48:260–70.
6. Lee YK, Lee WS, Kim GS, Park OJ. Anthocyanins are novel AMPKα1 stimulators that suppress tumor growth by inhibiting mTOR phosphorylation. *Oncol. Rep.* 2010;24:1471–77.

7. Shin DY, Lee WS, Lu JN, Kang MH, Ryu CH, Kim GY, Kang HS, Shin SC, Choi YH. Induction of apoptosis in human colon cancer HCT-116 cells by anthocyanins through suppression of Akt and activation of p38-MAPK. *Int. J. Oncol.* 2009;35:1499–504.

8. Yun JW, Lee WS, Kim MJ, et al., Characterization of a profile of the anthocyanins isolated from Vitis coignetiae Pulliat and their anti-invasive activity on HT-29 human colon cancer cells. *Food Chem. Toxicol.* 2010;48:903–9.

9. Malik M, Zhao C, Schoene N, Guisti MM, Moyer MP, Magnuson BA. Anthocyanin-rich extract from Aronia meloncarpa E. induces a cell cycle block in colon cancer but not normal colonic cells. *Nutr. Cancer* 2003;46:186–96.

10. Reddy MK, Alexander-Lindo RL, Nair MG. Relative inhibition of lipid peroxidation, cyclooxygenase enzymes, and human tumor cell proliferation by natural food colors. *J. Agric. Food Chem.* 2005;53:9268–73.

11. Dai J, Patel JD, Mumper RJ. Characterization of blackberry extract and its antiproliferative and anti-inflammatory properties. *J. Med. Food* 2007;10:258–65.

12. Bestor TH. The DNA methyltransferases of mammals. *Hum. Mol. Genet.* 2000; 9:2395–402.

13. Okano M, Xie S, Li E. Cloning and characterization of a family of novel mammalian DNA (cytosine-5) methyltransferases. *Nat. Genet.* 1998;19:219–20.

14. Pradhan S, Bacolla A, Wells R, Roberts R. Recombinant human DNA (cytosine-5) methyltransferase. I. Expression, purification, and comparison of de novo and maintenance methylation. *J. Biol. Chem.* 1999;274:33002–10.

15. Robertson KD, Uzvolgyi E, Liang G, Talmadge C, Sumegi J, Gonzales FA, Jones PA. The human DNA methyltransferases (DNMTs) 1, 3a and 3b: Coordinate mRNA expression in normal tissues and overexpression in tumors. *Nucl. Acids Res.* 1999;27:2291–98.

16. Rhee I, Bachman KE, Park BH, et al., DNMT1 and DNMT3b cooperate to silence genes in human cancer cells. *Nature* 2002;416:552–56.

17. Issa J-PJ. DNA methylation as a therapeutic target in cancer. *Clin Cancer Res.* 2007;13:1634–47.

18. Yoo CB, Jones PA. Epigenetic therapy of cancer: Past, present and future. *Nat Rev Drug Discov.* 2006;5:37–50.

19. Kuo CT, Seguin C, Stoner K, Cho S, Weng Y, Tichelaar J, Huang YW, Huang TH, Stoner GD, Wang LS. Tumor suppressor gene demethylation and reactivation in human colon cancer cells by black raspberry anthocyanins. *103rd American Association for Cancer Research (AACR), 2012, AACR Proceedings* 2012;53:1628.

20. Zikri NN, Riedl KM, Wang L-S, Lechner J, Schwartz SJ, Stoner GD. Black raspberry components inhibit proliferation, induce apoptosis, and modulate gene expression in rat esophageal epithelial cells. *Nutr. Cancer* 2009;61:816–26.

21. Duthie S, Jenkinson A, Crozier A, Mullen W, Pirie L, Kyle J, Yap LS, Christen P, Duthie GG. The effects of cranberry juice consumption on antioxidant status and biomarkers relating to heart disease and cancer in healthy human volunteers. *Eur. J. Nutr.* 2006;45:113–22.

22. Bi X, Fang W, Wang L-S, Stoner GD, Yang W. Black raspberries inhibit intestinal tumorigenesis in Apc1638 + /– and Muc2 – /– mouse models of colorectal cancer. *Cancer Prev. Res. (Phila).* 2010;3:1443–50.

23. Cai H, Marczylo TH, Teller N, Brown K, Steward WP, Marko D, Gescher AJ. Anthocyanin-rich red grape extract impedes adenoma development in the ApcMin mouse: Pharmacodynamic changes and anthocyanin levels in the murine biophase. *Eur. J. Cancer.* 2010;46: 811–17.

24. Bobe G, Wang B, Seeram NP, Nair MG, Bourquin LD. Dietary anthocyanin-rich tart cherry extract inhibits intestinal tumorigenesis in APCMin mice fed suboptimal levels of sulindac. *J. Agric. Food Chem.* 2006;54:9322–28.

25. Kang S-Y, Seeram NP, Nair MG, Bourquin LD. Tart cherry anthocyanins inhibit tumor development in ApcMin mice and reduce proliferation of human colon cancer cells. *Cancer Lett.* 2003;194:13–19.

26. Cooke D, Schwarz M, Boocock D, Winterhalter P, Steward WP, Gescher AJ, Marczylo TH. Effect of cyanidin-3-glucoside and an anthocyanin mixture from bilberry on adenoma development in the ApcMin mouse model of intestinal carcinogenesis—Relationship with tissue anthocyanin levels. *Int. J. Cancer* 2006;119:2213–20.

27. Harris GK, Gupta A, Nines RG, Kresty LA, Habib SG, Frankel WL, LaPerle K, Gallaher DD, Schwartz SJ, Stoner GD. Effects of lyophilized black raspberries on azoxymethane-induced colon cancer and 8-hydroxy-2'-deoxyguanosine levels in the Fischer 344 rat. *Nutr. Cancer* 2001;40:125–33.

28. Lala G, Malik M, Zhao C, He J, Kwon Y, Giusti MM, Magnuson BA. Anthocyanin-rich extracts inhibit multiple biomarkers of colon cancer in rats. *Nutr. Cancer* 2006;54:84–93.

29. Hagiwara A, Miyashita K, Nakanishi T et al. Pronounced inhibition by a natural anthocyanin, purple corn color, of 2-amino-1-methyl-6-phenylimidazo[4,5-b]pyridine (PhIP)-associated colorectal carcinogenesis in male F344 rats pretreated with 1,2-dimethylhydrazine. *Cancer Lett.* 2001;171:17–25.

30. Hagiwara A, Yoshino H, Ichihara T et al. Prevention by natural food anthocyanins, purple sweet potato color and red cabbage color, of 2-amino-1-methyl-6-phenylimidazo[4,5-b] pyridine (PhIP)-associated colorectal carcinogenesis in rats initiated with 1,2-dimethylhydrazine. *J. Toxicol. Sci.* 2002;27:57–68.

31. Montrose DC, Horelik NA, Madigan JP, Stoner GD, Wang L-S, Bruno RS, Park HJ, Giardina C, Rosenberg DW. Anti-inflammatory effects of freeze-dried black raspberry powder in ulcerative colitis. *Carcinogenesis* 2011;32:343–50.

32. Stoner K, Kuo CT, Huang YW et al. Dietary black raspberries-modulated DNA methylation in dextran sodium sulfate (DSS)-induced ulcerative colitis. *Inflamm. Bowel Dis.* 2010;17:S25.

33. Liao J, Chung Y, Wang LS, Yang A, Stoner G, Yang G. Inhibition of chronic colitis-induced carcinogenesis in IL-10 knockout mice by dietary supplementation of black raspberries. *Frontiers Cancer Pre. Res.* 2008:A132.

34. Liao J, Chung Y, Wang LS, Li H, Stoner G, Yang G. Modulation of inflammatory gene expression profile and inhibition of chronic colitis-induced carcinogenesis by dietary supplementation of black raspberries in IL-10 knockout mice. *100th American Association for Cancer Research (AACR), 2009, AACR Proceedings.* 2009;50:56.

35. Kuo CT, Stoner K, Huang Y, Yu J, Huang TH, Yearsley M, Yang GY, Wang LS. Anti-inflammatory effects of black raspberries in ulcerative colitis are associated with demethylation of genes in the Wnt signaling and protective modulation of toll-like receptor pathway. *102th American Association for Cancer Research (AACR), 2011, AACR Proceedings.* 2011;52:816.

36. Thomasset S, Berry DP, Cai H, et al., study of oral anthocyanins for colorectal cancer chemoprevention. *Cancer Prev. Res. (Phila).* 2009;2:625–33.

37. Wang LS, Arnold M, Huang Y-W et al. Modulation of genetic and epigenetic biomarkers of colorectal cancer in humans by black raspberries: A phase I pilot study. *Clin. Cancer Res.* 2011;17:598–610.

38. Klaus A, Birchmeier W. Wnt signalling and its impact on development and cancer. *Nat. Rev. Cancer.* 2008;8:387–98.

39. Mallery S, Stoner G, Larsen P, Fields H, Rodrigo K, Schwartz SJ, Tian Q, Dai J, Mumper RJ. Formulation and in-vitro and in-vivo evaluation of a mucoadhesive gel containing freeze dried black raspberries: Implications for oral cancer chemoprevention. *Pharm. Res.* 2007;24:728–37.

40. McGhie TK, Walton MC. The bioavailability and absorption of anthocyanins: Towards a better understanding. *Mol. Nutr. Food Res.* 2007;51:702–13.

41. Seeram NP, Bourquin LD, Nair MG. Degradation products of cyanidin glycosides from tart cherries and their bioactivities. *J. Agric. Food Chem.* 2001;49:4924–9.

42. Fleschhut J, Kratzer F, Rechkemmer G, Kulling S. Stability and biotransformation of various dietary anthocyanins in vitro. *Eur. J. Nutr.* 2006;45:7–18.

43. Tsuda T, Horio F, Osawa T. Absorption and metabolism of cyanidin 3-O-β-D-glucoside in rats. *FEBS Lett.* 1999;449:179–82.

44. McKenna P, Hoffmann C, Minkah N, Aye PP, Lackner A, Liu Z, Lozupone CA, Hamady M, Knight R, Bushman FD. The macaque gut microbiome in health, lentiviral infection, and chronic enterocolitis. *PLoS Pathog.* 2008;4:e20.

45. Tanaka T, Kojima T, Suzui M, Mori H. Chemoprevention of colon carcinogenesis by the natural product of a simple phenolic compound protocatechuic acid: Suppressing effects on tumor development and biomarkers expression of colon tumorigenesis. *Cancer Res.* 1993;53:3908–13.

46. Kawamori T, Tanaka T, Kojima T, Suzui M, Ohnishi M, Mori H. Suppression of azoxymethane-induced rat colon aberrant crypt foci by dietary protocatechuic acid. *Jpn. J. Cancer Res.* 1994;85:686–91.

47. Vitaglione P, Donnarumma G, Napolitano A, Galvano F, Gallo A, Scalfi L, Fogliano V. Protocatechuic acid is the major human metabolite of cyanidin-glucosides. *J. Nutr.* 2007;137:2043–48.

48. Riso P, Visioli F, Gardana C, Grande S, Brusamolino A, Galvano F, Galvano G, Porrini M. Effects of blood orange juice intake on antioxidant bioavailability and on different markers related to oxidative stress. *J. Agric. Food Chem.* 2005;53:941–47.

49. Galvano F, Salamone F, Nicolosi A, Vitaglione P. Anthocyanins-based drugs for colon cancer treatment: The nutritionist's point of view. *Cancer Chemother. Pharmacol.* 2009;64:431–32.

50. Biasutto L, Marotta E, Garbisa S, Zoratti M, Paradisi C. Determination of quercetin and resveratrol in whole blood—implications for bioavailability studies. *Molecules* 2010;15:6570–79.

51. Kausar H, Jeyabalan J, Aqil F, Chabba D, Sidana J, Singh IP, Gupta RC. Berry anthocyanidins synergistically suppress growth and invasive potential of human non-small-cell lung cancer cells. *Cancer Lett.* 2012;325:54–62.

9 Anthocyanins in Visual Performance and Ocular Diseases

Francois Tremblay and Wilhelmina Kalt

CONTENTS

9.1 INTRODUCTION

Traditional pharmacopeia dating back centuries ago has been promoting the use of medicinal plant and animal products for the treatment of a plethora of health conditions, including eye-related diseases. For instance, bear bile has been used in Asia for more than 3000 years to treat visual disorders, and it is only recently that tauroursodeoxycholic acid was found to be a bile constituent that promoted cell survival in models of retinal degeneration (Boatright et al. 2006). As early as the twelfth century, the German herbalist and composer Hildegarde von Bingen proposed to use bilberries as healing plants (reported in Morazzoni and Bombardelli 1996). In more recent history, the interest on bilberries for vision improvement related to their use by World War II Royal Air Force pilots who were said to ingest bilberries before their flight mission to improve their night vision (reported in Ulbricht et al. 2009). During the early 1960s, numerous studies were published in European literature supporting the claim of the beneficial effects of anthocyanins (ACNs) on the various aspects of vision such as visual acuity and dark adaptation. These claims were challenged by more rigorous, randomized, and placebo-controlled clinical studies performed more recently in normal individuals (reviewed extensively in Canter and Ernst 2004; Kalt et al. 2010). Furthermore, over the past 20 years, research involving patients with various ocular disorders has suggested that ACNs may play a significant role in the evolution of several visual dysfunctions. Other studies involving animal models of human disease have provided the supporting evidence that ACNs not only have antioxidative properties but may also have more direct involvement in the metabolic pathways, possibly acting as allosteric modulators (Yanamala et al. 2012).

ACNs are the pigments responsible for the blue, red, and purple coloration of berries. The commercial blueberry species such as wild blueberries (*Vaccinium angustifolium* Aiton), various cultivated blueberries (*V. corymbosum* sp.), and bilberries (*V. myrtillus* L.) are very rich in ACNs. ACNs belong to the category of phytochemicals called flavonoids. ACNs and other flavonoids are under active investigation for their effects as anticarcinogenic, antiproliferative and antimutagenic agents, as antiviral and antimicrobial mediators, and for their involvement in anti-inflammatory processes, cardiovascular protection, microcirculation improvement, prevention of diabetic disorders, as well as in neuroprotection (reviewed in He and Giusti 2010; Pascual-Teresa and Sanchez-Ballesta 2007; Zafra-Stone et al. 2007).

In the following sections, we will attempt to summarize and appraise the evidence that ACNs play a role in the normal vision and in the prevention of ocular pathologies.

9.2   A BRIEF REVIEW OF THE VISUAL SYSTEM

Light, the natural stimulus for the visual system, reaches the retina, the neural structure lying at the back of the eye, through a series of transparent media such as the cornea, the aqueous humour that fills the cavity between the cornea and the lens, and the vitreous humour that fills most of the eyeball cavity (Figure 9.1b). The retina is responsible for the transduction of light into an electrical signal that can be conveyed by the optic nerve to areas of the brain where visual perception occurs (Figure 9.1d). The retina has an extremely high rate of metabolic activity. When normalized for its mass, the retina is the most oxygen-consuming tissue in the body, with a consumption level of ~50% higher than the brain or kidneys (Anderson 1968). Photoreceptors within the retina, rods for night vision, and cones for chromatic day vision, are responsible for capturing photons and, through a series of biochemical events referred to as phototransduction cascade, they transform light into variations in the cellular membrane potential. Changes in the membrane potential are transmitted to second-order neurons, the horizontal and bipolar cells, before being integrated by the ganglion cells (RGCs), whose axons extend to form the optic nerves and to reach various brain structures devoted to visual and multisensorial perception. For more in-depth description of vision processes, good reviews have been published by Latek et al. (2012), Lamb and Pugh (2004), and Crouch et al. (1996).

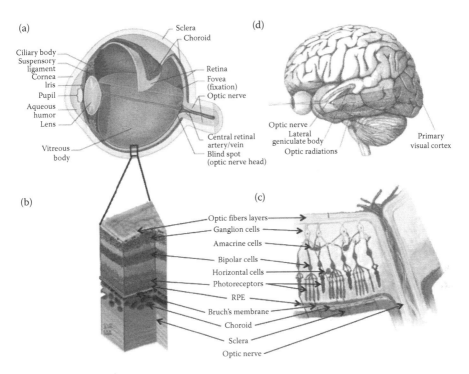

FIGURE 9.1 (**See color insert.**) Anatomy of the primate visual system. (a) Macroanatomy of the eyeball, which includes the transparent media (cornea, lens, aqueous humor filling the space between the lens and the cornea, and the vitreous filling the posterior part of the eye), the ciliary body and suspensory ligaments responsible for accommodation, and the retina lying at the back of the eye. (http://visionandeyecare.wordpress.com/tag/retina/). (b) The retina is composed of several layers of cells, with the photoreceptors facing the RPE that is separated from the choroidal vascular plexus by Bruch's membrane. (Modified from http://www.mcponline.org/content/9/6/1031.full.pdf + html). (c) The retina is composed of various types of cells. Photoreceptors, bipolar, and ganglion cells are glutamatergic excitatory cells, whereas horizontal (GABAergic) and amacrine (mostly GABAergic and glycinergic) are providing inhibitory inputs. Axons from the ganglion cells gather in the nerve fiber layer (NFL) to reunite at the optic nerve. (Modified from http://www.cidpusa.org/senses.htm). (d) Ganglion cell axons in the optic nerve synapse mainly in the lateral geniculate body, a small relay thalamic structure that conveys the information to the primary visual cortex via the optic radiations.

Interestingly, and almost counterintuitively, photoreceptors are located in close apposition to the retinal pigmentary epithelium (RPE) that acts as the barrier between the retina and the circulating blood in the choroid. This is counterintuitive because the light has to cross all layers of the retina before reaching the photoreceptors; however, this arrangement may be dictated by the photoreceptors' strong oxygen requirement; the choroidal circulation is under autonomic regulation, which means that photoreceptors are submitted to oxygen concentration that depends on the body's general demand. The central retinal artery that emerges from the optic nerve head supplies the rest of the retina; as for most brain circulation and unlike the

(a)

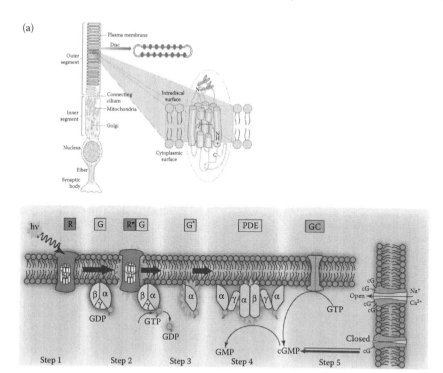

FIGURE 9.2 (**See color insert.**) Molecular aspects of vision. (a) Photoreceptor's outer segment is rich in phospholipid membrane that is continuously renewed and phagocytized by RPE cells. Embedded in these membranes is the whole machinery for phototransduction. The isomerization of 11-*cis* retinal into all-*trans*-retinal activates catalytic sites on rhodopsin(R) that is now free to activate the G-protein transducin(G) to induce phosphodiesterase (PDE) activation, which in turn reduces the intracellular level of cGMP. This will close the sodium channels on the cellular membrane and will provoke its hyperpolarization that will be transmitted through synaptic contact to other retinal cells. The activated rhodopsin will be deactivated by the binding of rhodopsin kinase, recoverin, and arrestin. (b) The all-*trans*-retinal cannot be reisomerized within the photoreceptor; so, it is transported out of the discs by the ATP-binding cassette transporter where it is reduced in the form of all-*trans* retinol, to be then carried to the RPE by binding to interphotoreceptor retinoid binding protein (IRBP). Once in the RPE cell, it binds to cellular retinol binding protein (CRBP) and can be either esterified by lecithin–retinol acyltransferase (LRAT) or isomerized back to 11-*cis* form (RPE65-isomerase hydrolase complex) to be stored by binding to cellular retinal binding protein (CRABL) or reoxidized to 11-*cis*-retinal where it is returned to the photoreceptor. Retinol binding proteins (RBP4) act as the carrier protein for retinol (vitamin A alcohol) from the liver supply. (Adapted from Hargrave, PA 2001, Rhodopsin. In: eLS. John Wiley & Sons Ltd, Chichester. http://www.els.net [doi: 10.1038/npg.els.0000072].)

choroid circulation, these blood vessels are under autonomic regulation, that is, they are regulated by local demand.

In a nutshell, phototransduction (Figure 9.2a) involves the photoactive molecule in the outer segment of the photoreceptor built around a covalent association between a seven transmembrane protein, the opsin, and a chromophore, the 11-cis-retinal, which

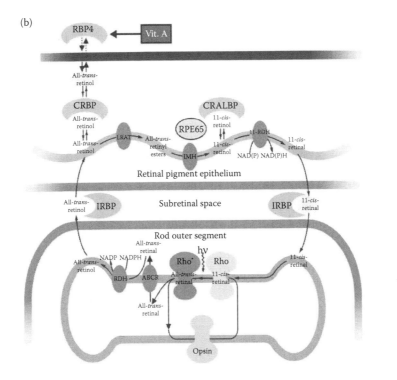

(b)

FIGURE 9.2   (continued)

is a derivative of vitamin A. The phototransduction cascade is initiated by the absorption of light by the chromophore, 11-*cis*-retinal, which results in its isomerization to all-*trans*-retinal. The isomerization to the all-*trans*-retinal causes the molecule to straighten, changes the distribution of electrical forces within the opsin molecule, freeing catalytic sites that allow the interaction with the G-protein transducin that, in turn, activates a cyclic guanosine monophosphate (GMP) phosphodiesterase, the activation of which causes the hydrolysis of cyclic GMP and the consecutive closure of sodium membrane channels. The net result of this process is a hyperpolarization of the photoreceptor that causes a change in the synaptic activity that is processed by the second-order retinal cells.

Photoreceptors cannot reisomerize the all-*trans*-retinal to its former 11-*cis* state; this process is taken care of by the RPE cells through the intervention of chaperone molecules such as retinoid binding proteins (RBP) that carry the all-*trans* retinol to the RPE where it can be enzymatically reisomerized and sent back to the photoreceptor; this constitutes the visual cycle summarized in Figure 9.2c. Furthermore, RPE cells are also implicated in phagocytosis of continuously shedding the photoreceptor's outer segments. Thus, RPE cells are instrumental to the completion of the phototransduction process and are essential to the efficient elimination of membrane phospholipids rich in long-chain polyunsaturated fatty acids sensitive to photooxidation. Any dysregulation of RPE function may thus result in severe visual threat. Furthermore, all-*trans*-retinal cannot be reisomerized locally and, if left free within the cell, could enter the bisretinoid pathway that ends with the production of *N*-retinylidene-*N*-retinylethanolamine (A2E), a highly toxic by-product (Sparrow et al. 2010). A2E is a constituent of lipofuscin, a substance that accumulates in RPE cells in many retinal disorders; it can mediate a detergent-like perturbation of cell membranes when photooxidized and becomes toxic to the RPE cell through the generation of singlet oxygen and the oxidation of A2E at carbon–carbon double bonds.

9.3　ACN BIOAVAILABILITY TO OCULAR STRUCTURES

To determine the role of ACNs in the visual function, it is necessary to understand what ACN forms and concentrations occur in ocular cells and tissues where their effects may occur. The cumulative evidence obtained from bioavailability studies on ACNs and ACN metabolites in the plasma and urine indicate that ACNs are not well absorbed by the body and are rapidly eliminated from circulation after a single dose (Williamson and Clifford 2010). On the basis of elimination profiles of ACN from the plasma and urine, <0.01% of an administered dose can be accounted for in these fluids (Wu et al. 2002). Native anthocyanidin glycosides have been identified in the human plasma and urine where they can be found at nanomolar concentrations, several orders of magnitude lower than other phenolic compounds (Manach et al. 2005). In human blood serum, the intact glycosylated forms and also mono- or polyacylated forms were found (Mazza et al. 2002). In rats fed with blackberries, the native ACNs as well as ACN metabolites such as methylated and glucuronidated ACNs were identified in various tissues (Talavéra et al. 2005; Felgines et al. 2009). Phenolic metabolites resulting from the cleavage of dietary ACNs by colonic microflora may also have to play a role *in vivo* (Rechner et al. 2002; Ichiyanagi et al. 2005; Walton et al. 2006). For instance, in rats fed with cyanidin-3-*O*-glucoside (C3G), the plasma concentration of protocatechuic acid, a catabolite of C3G, rose significantly and exceeded the plasma C3G concentration by 8–10 fold (Tsuda et al. 1999).

With respect to *in vitro* studies, ACNs have been tested in various cellular and noncellular models in their native forms that is, as they occur in the berries or in foods. Since ACNs undergo phase 2 metabolism *in vivo* (Dulebohn et al. 2008), the *in vitro* models may not accurately represent the suite of ACN-based compounds that may occur *in vivo*. For example, through phase 2 metabolism, ACNs may become glucuronidated, to affect their polarity, molecular weight, and possibly their physiological effects. Also, it is worth noting that *in vitro* studies often employ ACN concentrations far in excess of their estimated *in vivo* concentration.

Orally administered ACNs have been found in the brain and eye tissues after long-term feeding, indicating that these compounds have crossed the highly selective blood–brain barrier (BBB) and can become localized in the brain areas where blueberry supplementation has been reported to confer protection or improved function (reviewed in Youdim et al. 2004; Andres–Lacueva et al. 2005; Passamonti et al. 2005; Matsumoto et al. 2006; Kalt et al. 2008; Talavéra et al. 2005). In acute studies using rats, ACNs and ACN metabolites were found in brain structures as early as 10 min after intragastric administration (Passamonti et al. 2005). The mechanism by which ACNs cross the BBB is unknown although there is some evidence from *in situ* research demonstrating that P-glycoprotein transporters may play a role in the movement of flavonoids into the brain (Milbury and Kalt 2010). Passive diffusion of ACNs across the BBB is not favored since anthocyanidin glycosides are highly polar molecules of relatively high-molecular weight (>400 MW).

ACNs appear to have a longer residence time in tissues than in the plasma (Kalt et al. 2008; Milbury and Kalt 2010). In pigs fed with diets supplemented with 0%, 1%, 2%, or 4% w/w blueberries for 4 weeks and euthanized after 18–21 h fasting, no ACNs were detected in the plasma or in the urine but intact ACNs were identified in the brain and the eye, suggesting that ACNs can accumulate in tissues, rather than being in a rapid equilibrium with ACNs in blood circulation. The concentrations of various ACNs also varied between tissues (Kalt et al. 2008) with the eye having the highest concentration compared to two other brain regions and the liver. Differential tissue concentration was also documented in the rat (Ichiyanagi et al. 2006; Matsumoto et al. 2006). It has also been proposed that the simultaneous intake of other food or simply the presence of other flavonoids may change/delay the absorption profiles of ACNs in the plasma (Walton et al. 2006; Matsumoto et al. 2007; Serafini et al. 2009).

However, there is almost no information on how the long-term consumption may modify ACN absorption, metabolism, and residence time in the body. But this question is very relevant to *in vivo* studies that feed ACNs over periods of weeks and even months. It is also noteworthy that studies cited in this chapter have administered ACNs in a variety of ways, including by intravenous and intraperitoneal injection, by injection into the eye, and where specifically stated, into the vitreous. While some of these studies document the beneficial effects after ACN treatment, none of them studies test for the presence of ACN in ocular tissues following treatment. The methods for the accurate liquid chromatography–mass spectrometry (LC–MS) detection of ACN and their metabolites are challenging due to their instability at physiological pH, very limited capacity for digestive absorption, and the dampening effects on LC–MS detection due to the tissue and fluid matrices where they are found. To address the key questions regarding the role of ACNs in vision physiology, it will be essential to determine better the ACN concentration and to profile the cells and tissues that are relevant to their benefits.

9.4 *IN VITRO* ASSAY ON VISUAL CELLS

9.4.1 STUDIES INVOLVING RPE CELLS

The RPE is a monolayer of pigmented cells strategically situated between the photoreceptor cells and the choroidal vascular supply. Together with the retina, it constitutes a functional unit involved in the recycling of 11-*cis*-retinal, absorption of light and protection against photooxidation, selective transepithelial transport, spatial buffering of ions in the subretinal space, and phagocytosis of shed photoreceptor membranes rich in photosensitive long-chain polyunsaturated fatty acids.

In vitro studies using a variety of immortalized RPE cells revealed that these cells benefit from the antioxidative properties of ACNs and other flavonoids. Milbury et al. (2007) examined ARPE-19 cells treated with a purified bilberry (*Vaccinium mytillus*

L.) extract that contained either ACNs or non-ACN phenolics before they received an oxidative challenge with H_2O_2. Both the ACNs and non-ACN extracts were found to reduce the abundance of reactive oxygen species arising from H_2O_2, thus providing evidence for the modulation of oxidative stress defense enzymes heme oxygenase-1 and glutathione S-transferase-pi (GSH) by bilberry ACNs in RPE *in vitro*. However, neither extracts were providing a sufficient level of protection to affect cell survival. Both bilberry extracts increased the expression of the phase 2 proteins hemeoxygenase-1 and GSH transferase.

Hanneken et al. (2006) showed that, unlike other flavonoids, the ACN aglycones cyanidin, malvidin, and peonidin were poorly effective in protecting 1-day ARPE-19 cells from 250 µM H_2O_2. They also found that other flavonoids, including quercetin, fisetin, and galangin, protect RPE cells by the stimulation of antioxidant response element, an upstream regulatory element that is shared by all the genes encoding phase-2 proteins, thus demonstrating an alternative mechanism to explain the *in vitro* antioxidative effects of flavonoids. The authors proposed that the difference in efficacy might be partly explained by the molecule's hydrophobicity, indicating that it should readily pass through cell membranes and accumulate in intracellular compartments in this particular *in vitro* assay.

The senescence-related changes in RPE can lead to abnormal physiological functions and are closely associated to the development of retinal dystrophies. In a series of *in vitro* experiments using the established models of replicative senescence and light-induced damage, Liu et al. (2012) demonstrated an increase of the survival rate from 14.6% after the 10th replication in control conditions, to 76.9% when the RPE cells were treated with 0.1 mg/mL of blueberry ACN extracts. They also showed a decrease of senescent cells from 57.2% after light exposure at 2500 lux for 12 h, to 35.1% and 32.5% when RPE cells were treated with 1.0 and 10 µg/mL blueberry ACN extracts, respectively, with no effect when the dose was 0.1 µg/mL. Even though no putative mechanisms were proposed, this study clearly suggests that blueberry ACNs can be effective in preventing RPE cells from senescence and incurring light-induced damage. It must be noted that these *in vitro* studies used ACNs concentrations that were higher than what would occur *in vivo* and did not include any of the ACN metabolites that would be present after *in vivo* digestive absorption and metabolism.

Bestrophin is a 68-kDa transmembrane protein localized to the basolateral plasma membrane of the RPE cells that functions as Ca^{2+}-regulated chloride channels. Bestrophin has been associated with several macular dysfunctions (Marmorstein et al. 2009). Docking studies with ACNs and bestrophin from Priya et al. (2012) suggest that all ACNs and in particular malvidin 3,5-diglucoside, cyanidin 3,5-diglucoside, and petunidin 3,7-diglucoside have favorable binding potential with the involvement of active sites. This research illustrates the potential for ACNs to modify metabolic pathways by allosteric interactions with proteins involved with vision physiology.

9.4.2 Studies Involving RGCs

RGCs are third-order neurons, the first ones to generate the action potentials. The axons of RGCs form the optic nerve that terminates in the brain subcortical structures involved in visual perception, eye movement, coordination, and circadian rhythms. Their long axon is fragile and involved in many pathological processes such as glaucoma and inflammatory optic neuropathies. RGCs have also been cultured *in vitro* and used as experimental models.

Tanaka et al. (2012) evaluated the protective effects of purple rice (*Oryza sativa* L.) bran extract against H_2O_2 and tunicamycin-induced damage in cultured RGC (RGC-5). The purple pigmentation of this rice bran is due to C3G, peonidin-3-*O*-glucoside, and cyanidin-3-*O*-gentiobioside. Tunicamycin is a glucosamine-containing nucleoside antibiotic that inhibits protein folding in the endoplasmic reticulum, ultimately leading to cell death. The treatment of RGC-5 cells with 100 µg/mL bran extract (2.9% C3G and 0.2% peonidin-3-*O*-glucoside) significantly inhibited cell death induced by both H_2O_2 and tunicamycin in a concentration-dependent manner. In addition, they investigated the differences of potency among various ACNs in their tunicamycin model. Cyanidin and delphinidin inhibited tunicamycin-induced cell death, but peonidin and malvidin had no effect. These data suggest that at least one pair of vicinal hydroxyl groups on the β-ring in anthocyanidin is essential to efficiently reduce the endoplasmic reticulum stress. Interestingly, the authors suggested that the *in vitro* protective effects against H_2O_2 were through the traditional antioxidative properties. However, they suggest that the endoplasmic reticulum effects were exerted not through the endoplasmic reticulum stress-related proteins but rather via mechanisms involving the caspase-3 pathway, which is downstream in the apoptosis pathway. *In vivo* experiments were also conducted after the intraocular injection of tunicamycin that produced a reduction in the ganglion cell number, and thinning of the inner plexiform and nuclear layers of the retina. The bran extract injected simultaneously showed the protection of RGCs but was ineffective in reducing the other retinal damage.

Using the same line of immortalized RGC-5 cells, Matsunaga et al. (2009) showed that the pretreatment with bilberry anthocyanosides and/or their main ACN aglycone constituents (cyanidin, delphinidin, and malvidin) significantly inhibited SIN-1-induced neurotoxicity (3-(4-morpholinyl) sydnonimine hydrochloride, a peroxynitrite donor) and radical activation. Furthermore, the morphological retinal damage induced by *N*-methyl-D-aspartate (NMDA) was inhibited in *in vivo* mice models. This is in opposition to the results in cortical neuron cultures, where black soybeans ACN did not protect against NMDA-induced neuronal cell death but were effective against oxygen–glucose deprivation (Bhuiyan et al. 2012). On the other hand, Maher and Hanneken (2005) observed no protection by cyanidin against glutathione depletion in RGC-5 cell line, whereas other flavonols such as quercetin provided significant protection. They hypothesized that the specific location of hydroxyl groups, unsaturation of the C ring, and a greater hydrophobicity are important to the flavonoid efficacy against oxidative stress-induced cell death in this model.

The antioxidative properties of ACNs from *in vitro* models of retinal cells and others are quite compelling. However, the tissue concentration of ACNs reported *in vivo* is not high enough to significantly contribute to the protection already provided by endogenous antioxidants. However, the contribution to antioxidant defense by ACN metabolites, that are formed after digestion, is currently unknown. it is exciting that even at low concentrations, ACN may influence the various cell-signaling pathways involved in the normal and pathophysiological visual processes.

9.5 ACN INTERACTIONS WITH THE PHOTOTRANSDUCTION CASCADE

Phototransduction, which is the biochemical cascade that transduces light into a neuronal signal in retinal photoreceptors, involves the heterotrimeric guanosine triphosphate (GTP)-binding protein (G protein) signaling pathway (see Figure 9.1b).

Studies in the French literature speculated in the early 1970s that ACNs accelerate the synthesis of rhodopsin and modulate the retinal enzymatic activity (Cluzel et al. 1970). Alfieri and Sole (1966) reported that the intravenous administration of ACNs (160 mg/kg) improved the time taken by the rod electroretinograms (ERGs) to return to normal values after a short retinal bleaching in rabbits. Tronche and Bastide (1967) proposed that this improvement in the visual function was probably related to an increase in the regeneration rate of rhodopsin. The effects of ACNs in rhodopsin regeneration have been examined in more detail using various *in vitro* approaches. In 1979, Ruckstuhl et al. (1979) demonstrated some inhibitory action of flavonoids on cyclic GMP phosphodiesterase a molecule that controls the time course of light response in vertebrate rods (Capovilla et al. 1983). In 1990, Virmaux et al. (1990) reported that the activity of cyclic GMP–phosphodiesterase extracted from illuminated rod outer segment was strongly increased by synthetic cyanidin chloride in the range 2.5–10 μM (up to 180% of basal activity at 5 μM level). Interestingly, these authors found a biphasic activity curve, with the activation by cyanidin chloride up to 5–10 μM and inhibition with higher concentration. Peonidin and peonin chloride exhibited different maximum activation at 25 μM and above 500 μM, respectively, whereas amentoflavon, a biflavonoid similar to cyanidin but without the flavylium ion, showed a very strong inhibition of cyclic GMP– phosphodiesterase but no activation within the 0.01–50 μM levels. Thus, it appears that ACNs could either activate or inhibit cyclic GMP-phosphodiesterase, and that various ACNs have different half-maximal inhibitory concentrations. On the other hand, Matsumoto et al. (2003) reported the insignificant effects of 10 and 50 μM of black currant (*Ribes nigrum*) ACNs on cyclic GMP phosphodiesterase activity when tested at various stages of phosphodiesterase activation in the light and the dark.

Matsumoto and Yoshizawa (2008) also reported that specific black currant ACNs stimulate the regeneration of rhodopsin. The kinetics of the incorporation of purified 11-*cis*-retinal into an opsin membrane fraction prepared from frog rod outer segments was studied in the presence and absence of 20 μM purified black currant ACNs. Of the four ACNs contained in black currant, namely, the glucoside and rutinoside of cyanidin and delphinidin, only the two cyanidin glycosides increased the rate of rhodopsin regeneration. Kinetic analysis revealed that the K_m for rhodopsin formation was lowered by the presence of cyanidin glycosides although it was not possible to determine which specific reaction was affected. Wahid et al. (2011) also looked at rhodopsin regeneration in a frog preparation, analyzing ERGs after the administration of Korean black raspberry (*Rubus coreanus*) extract that contains a large proportion of flavonoids. In this model, after the exposure to intense background light, the threshold of ERG takes time to recover in the dark and this measure is an indirect reflection of rhodopsin regeneration. In the presence of *R. coreanus* extract, this time was greatly shortened.

Over the recent years, evidence has been accumulating on the possible modulation of the structure and function of G-protein-coupled receptors through the binding of allosteric ligands (De Amici et al. 2010; Latek et al. 2012; Yanamala et al. 2012). The inherent flexibility of G-protein-coupled receptors permits the dynamic and conformational changes that can be triggered by only a fraction of the energy derived from ligand binding. The high-resolution structure of rhodopsin is well known and can be used to investigate these interactions. Yanamala et al. (2009) showed that C3G can bind directly to rhodopsin in the dark and upon light activation. This binding probably occurs in the cytoplasmic domain of rhodopsin. They report that dark-adapted and light-activated states of rhodopsin show preferences for different C3G species. The same group (Tirupula et al. 2009) also documented that the regeneration of purified rhodopsin in detergent micelles is accelerated in the presence of C3G. They suggested a destabilizing effect on rhodopsin structure with modest alteration of G-protein activation and the rates at which the light-activated rhodopsin (in its metarhodopsin II state) decays to opsin and free retinal. These results indicate that the mechanism of C3G-enhanced regeneration may be based on changes in opsin structure, promoting access to the retinal binding pocket. Computational docking studies such as the one previously reported for bestophin protein in the RPE (Priya et al. 2012) will help our understanding of ACN interaction with molecules involved in visual function. In that same vein, the flavanone eriodictyol has recently been reported to act as a retinoid surrogate *in vitro* and to affect opsin-mediated changes in transducin activation (Hanneken et al. 2009). In computational docking experiments, eriodictyol had a very favorable binding energy in the opsin-binding pocket, thus representing another possible docking site for ACNs.

Yanamala et al. (2009) also documented the pH-dependent nature of the docking interactions between opsin and ACN. The pH of the extracellular environment of the photoreceptor is increased by light exposure (Yamamoto et al. 1992) and vertebrate retinae undergo diurnal pH changes (Dmitriev and Mangel 2001) such that retinal pH is more alkaline during the day. The differences in photoreceptor pH may thus be another modifying factor that influences the results of ACN studies in vision.

9.6 ACTIONS OF ACNS ON VARIOUS OCULAR DYSFUNCTIONS

9.6.1 MYOPIA

Myopia is a common ocular condition that is characterized by the eye being too long for its optical power, which is determined by the cornea and the crystalline lens, or that is optically too powerful for its axial length. Myopia is thought to be caused by both genetic and environmental factors and is particularly prominent in the Asian population. Myopia can produce difficulties of seeing in dim light because of increased aberration and can also be associated with the fatigue arising from sustained near-visual tasks (asthenopia). Experimental models for myopia with axial-length elongation have been developed in many animal species where wearing negative-power lenses during development result in increased myopia.

In a prospective, randomized, placebo-controlled, and crossover study, Kamiya et al. (2012) examined 30 eyes of 30 middle-aged healthy volunteers with myopia who were randomly assigned to one of two oral regimens: fermented bilberry extract (400 mg/day) or placebo (4-week treatment, 4-week washout design). The fermented bilberry extract was effective in improving the accommodation from 4.62 ± 1.88 D before treatment to 5.33 ± 2.03 D after treatment and improving mesopic (dim light) contrast sensitivity from 1.04 ± 0.16 before, to 1.13 ± 0.17 after treatment although no possible mechanism of this effect could be offered. Unfortunately, they did not quantify the individual ACN concentration in the fermented bilberry extracts; it is only known to contain 15 different ACNs, derived from five aglycones (cyanidin, delphinidin, malvidin, peonidin, and petunidin).

The effects of the oral intake of a black currant anthocyanoside concentrate on work-induced transient refractive alteration and subjective asthenopia symptoms (visual fatigue) were examined in a double-blind, placebo-controlled, crossover study with 21 healthy human subjects (Nakaishi et al. 2000). The refractive error after the video task increased by 0.12 ± 0.28 dipoters toward myopia in the placebo group while that change was limited to 0.03 ± 0.25 in the experimental group. The authors also reported significant improvement in subjective asthenopia (by way of questionnaire) and some mild improvement in dark-adapted threshold (lowest intensity detectable after a prolonged period spent in total darkness) by 0.1 log abs (equivalent to 4 cd/m^2). All these improvements were statistically significant although the changes observed would hardly be noticeable in everyday life under nonexperimental conditions.

In a randomized, double-blind, placebo-controlled trial in 60 low-to-moderate myopia subjects presenting with symptoms of asthenopia and reduced night vision, Lee et al. (2007) reported an improvement in contrast sensitivity function at all spatial frequencies tested in subjects fed for 4 weeks with 170 mg anthocyanide oligomers (Eyezone®, Hanmi Pharmaceuticals, Seoul, Korea), as well as an improvement in the subjective complaints of asthenopia. However, the experimental and placebo groups differed in age and sex, factors known to affect the contrast sensitivity function; so, conclusions are guarded as no crossover design was used.

One theory for the development of refractive myopia during continuous visual strain is that the ciliary muscle becomes spastic as a result of excessive contraction during close-up work, leading to spasmodic refractive power of the lens. As a consequence, the myopic ciliary muscle cannot relax sufficiently to allow the lens to focus on distant images. Matsumoto et al. (2005) found in bovine ciliary muscle preparations that delphinidin-3-rutinoside provided both relaxation activity during the contraction caused by endothelin-1 and an inhibitory effect on the magnitude of the endothelin-1-induced contraction. This effect was mediated through endothelin-B receptors and involved the production/release of nitric oxide, to inhibit myosin regulatory light-chain phosphorylation and/or acceleration of dephosphorylation. In this way, delphinidin-3-rutinoside caused ciliary muscle relaxation and produced an inhibitory effect on the endothelin-1-induced contraction.

Finally, Iida et al. (2010) used negative lenses (−8 D) placed on the right eye of 8-day-old chicks (left eye used as control), for three consecutive days to induce axial eye length increase in a control group, in keeping with the previous experiments that established that model. Black currant extract orally administered at a daily dose of 400 mg/kg significantly inhibited the enlargement of axial length (0.22 ± 0.03

compared to 0.41 ± 0.04 mm for the control group). No difference was found between the left eyes of either control or black currant extract groups, showing that the beneficial effect was restricted to the induced myopia and had no effect on the normal development of the eye.

Thus, it appears that myopic development can be mitigated by ACN intervention at the level of the ciliary muscle, to allow more relaxation of the lens and regulation of the eye axial length during development. The associated symptoms such as asthenopia and decreased contrast sensitivity may also be alleviated by the introduction of ACN in the diet. The endothelin pathway is the only one that has been studied so far but it is worth noting that the interaction of ACN with the endothelin pathway has also received support from studies in glaucoma (see Section 9.6.5).

9.6.2 CATARACT

Cataracts are developmental or degenerative opacities of the lens of the eye, generally characterized by a gradual painless loss of vision. Cataracts are the leading cause of impaired vision in the Occidental countries, with a large percentage of the geriatric population exhibiting some signs of lens opacity. The treatment consists of surgical removal of the opacities. Cataracts are characterized by electrolyte disturbances and a decrease in proteolytic enzymes resulting in osmotic imbalances, and the development of aggregates of insoluble material including oxidized proteins. The mechanisms involved are nonenzymatic glycation, oxidative stress, and the polyol pathway (Gupta et al. 2009). More than 40 flavone derivatives were tested *in vitro* and were found to be potent inhibitors of aldose reductase, the enzyme that initiates cataract formation in diabetes (Laurens et al. 1983).

The lens of the eye is devoid of antioxidant enzymes, superoxide dismutase (SOD), catalase and glutathione peroxidase, and is completely dependent on the nutritional antioxidants, including vitamin E, vitamin C, selenium, and carotenes for its antioxidant defenses (Gupta et al. 2009). Most large-scale interventional studies such as Age-Related Eye Disease Study (AREDS), Beaver Dam Eye Study, Nurses Health Study, and Blue Mountains Eye Studies found a weak or no relationship between antioxidant (such as vitamin A, E, or lutein) intake and cataractogenesis in population studied in developed countries (Head 2001; West et al. 2006; reviewed in Seddon 2007). However, other compounds with strong antioxidant activity such as rhamnocitrin (Bodakhe et al. 2012), ebselen (2-phenyl-1,2-benziselenazol-3 [2H]-one) (Aydemir et al. 2012), as well as many flavonoids (Varma et al. 1977; Stefek 2011) have been shown to inhibit cataractogenesis in *in vitro* and *in vivo* rat models. Another large epidemiological study, this time focusing on flavonoid intakes, failed to demonstrate any link with cataractogenesis prevention; so, the topic remains controversial (Knekt et al. 2002).

Morimitsu et al. (2002) used a rat lens organ culture system and 15 mM xylose to mimic the cataract formation due to diabetes. Five ACN monomers from grape skins showed inhibitory activities for lens opacity when present in the *in vitro* media at a concentration higher than $10 \mu M$, with malvidin and delphinidin-3-glucoside showing stronger inhibitory activities than cyanidin, petunidin, and peonidin-3-glucoside in that assay system. They also looked at ACN (mainly C3G) and polyphenol extracted from colored rice (asamurasaki-2 [Japanese black] and chikushi-akamochi-2 [Japanese red]), to find that ACNs were effective in that assay model but their concentration in the extract was too low to exclude any other polyphenol actions.

The technology to produce artificial corneal grafts is developing rapidly (Tan et al. 2012) and the effects of ACNs on these processes have been examined. Song et al. (2010) investigated the effects of different concentrations of bilberry extracts on the cell viability, cell cycle, and the expression of hyaluronic acid and glycosaminoglycans of cultured human corneal limbal epithelial cells and showed that the extract (10^{-5} M C3G) promoted cell growth to about 120% compared with the control group after 24 h incubation. Also, three concentrations (10^{-6}, 10^{-5}, and 10^{-4} M C3G) were effective in increasing cell viability to 112.9%, 130.1%, and 113.8%, respectively, after 48-h incubation. Cell differentiation also increased compared to control, so that the proportion of cells in generation 0/generation 1 phase decreased from 84.88% to 60.15%, whereas the ratio of generation 2 to mitotic cell phases increased from 8.95% to 21.20%. As the production of artificial corneal grafts progresses, these *in vitro* beneficial effects of ACNs are promising. Glycosaminoglycans are also the major constituents of Bruch's membrane, which is part of the BBB in the eye and hyaluronic acid is the major constituent of the vitreous body; so, a possible interaction of ACN in these two systems might warrant further investigation.

Interestingly, Willis et al. (2005) found that hippocampal embryonic cells proliferate and differentiate when implanted in an intraocular substrate in mature rats when blueberry extract supplementation is provided. The success of grafts occurred in a manner that would be expected in immature rats and was in contrast to the mature rats on the control diet. This provides further evidence that cell proliferation and differentiation are favored by blueberry-enriched diet *in vivo*, which supports the findings *in vitro*. Together, the results may open new avenues in the fields of corneal grafts and possibly of populating retinal grafts using stem cells.

Fursova et al. (2005) found that the supplementation with bilberry extract (20 mg/kg of body weight [BW] including 4.5 mg of anthocyanidin) decreased serum and retinal lipid peroxides and slowed the development of cataract and macular degeneration in senescence-accelerated OXYS rats, a hypertensive strain that presents with a shortened life span and early phenotypes of age-related disorders.

To our knowledge, the clinical evidence for the effect of ACNs on cataract development is insufficient. In one report of 50 patients with senile cataracts, Bravetti (1989) reported that a combination of bilberry, standardized to contain 25% anthocyanosides (180 mg twice daily), and vitamin E in the form of DL-tocopheryl acetate (100 mg twice daily) for 4 months stopped the progression of cataracts in 96% of the subjects treated compared to 76% in the control group. These results are quite substantial; however, the study methods for the cataract assessment and the analysis are not described with sufficient details to reproduce the study. In addition, the effects of bilberry cannot be separated from vitamin E; so, more clinical studies are needed to determine the real implication of ACN as well as the magnitude of the effect that can be achieved.

In vitro evidence of cataractogenesis inhibition and graft stabilization are now available and promising, but clinical evidence is needed. Clinical trials are impeded by the time frame of such studies. In the meantime, studies using animal models of senile cataract are available and may provide unique opportunities to investigate these processes within a shorter time period.

9.6.3 Vascular Pathologies/Inflammation

Abnormal angiogenesis causes many ocular diseases, such as diabetic retinopathy, age-related macular degeneration, and retinopathy of prematurity. While there is evidence of the various ACN effects on angiogenesis, it is beyond the scope of this chapter to review this work beyond some key studies related to ocular disorders.

In a multicentric study involving 88 patients with nonproliferative diabetic retinopathy and associated macular edema, Kim et al. (2008) found an improvement in contrast sensitivity but no change in visual acuity, macroanatomy of the retina (numbers of hard exudates, microaneurysms, and leaking points on angiograms), or retinal thickness as measured by optical coherence tomography after treatment with one capsule a day of bilberry *Vaccinium myrtillus* extract (170 mg/capsule, Tagen-F®, Kukje pharmaceutical) for 1 year. In contrast, Huismans (1988) reported a decrease in the number of microaneurysms per fundus in 25% of the diabetic retinopathy cases that were followed, as well as an increase in capillary resistance from 307 to 437 mm Hg in 40 patients with diabetic retinopathy who were treated with extracts of bilberry.

ACNs purified from purple corn added to the diet (+0.2%) for 5 weeks in the KK-A^y mice, a murine model of obesity and diabetes, reduce blood glucose concentration and enhanced insulin sensitivity; the adiponectin and its receptors expression were not found responsible for this amelioration (Sasaki et al. 2007). Rather, ACNs upregulated the glucose transporter 4 and downregulated the retinoid binding protein 4 in the white adipose tissue, which is accompanied by downregulation of the inflammatory adipocytokines. The retinoid binding protein 4 is not found in the eye but its downregulation may reduce the supply of retinoid to the RPE, where RBP1 is acting as the main transporter system to the photoreceptors. Recently, Dobri et al. (2013) used the new nonretinoid RBP4 antagonist A1120 to significantly reduce the accumulation of lipofuscin bisretinoids A2E in the retina of Abca4$^{-/-}$ mice, thus now linking the reduction of RBP4 in adipocytes due to an enriched ACN diet to a reduction in the accumulation of lipofuscin A2E in the retina. This may have an implication in many retinal dystrophies.

Anecdotal evidence from non-English literature (Coutinho and Barcaui 1975; Scharrer and Ober 1981; Repossi et al. 1987; Perossini et al. 1987) also provides some support for the positive effects of ACNs in diabetic retinopathy and other vascular diseases.

High-fat diet associated with obesity and hyperglycemia has induced the differential expression of several stress-related genes in the mouse retina (Mykkänen et al. 2012). Despite minor effects on this mouse phenotype, a 12-week diet rich in bilberries (5%) mitigated the upregulation of crystallins and reduced the expression of genes in the mitogen-activated protein kinase (MAPK) pathway and increased these in the glutathione metabolism pathway. The biological process of eye development and molecular function of the structural constituents of the eye lens also seemed to be modulated by bilberries.

The role of ACNs in inflammatory and ischemic processes is discussed in the other chapters of this book; so, this section will be restricted to inflammation and ischemia pertaining to eye conditions. Inflammatory processes are involved in pathogenic mechanisms in vision-threatening retinal diseases such as diabetic retinopathy and age-related macular degeneration. Experimental models of uveitis have shown to be particularly useful in examining the effects of ACNs in this condition.

ACN effects were studied in a mouse model of endotoxin-induced uveitis where retinal inflammation was monitored after a single intraperitoneal injection of lipopolysaccharide (LPS) (Miyake et al. 2012). Prior to the LPS treatment bilberry extract (commercially available by Wakasa Seikatsu (Kyoto, Japan) was orally administered in a dose of 500 mg/kg BW (39% ACN content) once daily for 4 days prior to the insult. The study included a placebo and a control group. ERGs to monitor the retinal function were performed 1.5 and 24 h after the LPS injection. Compared to controls, the bilberry-fed group had significantly better ERG signals, higher concentration of rhodopsin in retina samples, and less shortening of rod photoreceptor outer segments. Also, the activation of the inflammatory transcription factor STAT3 was almost totally prevented. In addition to its anti-inflammatory effects, the ACN-rich bilberry extract ameliorated the intracellular elevation of reactive oxygen species and activated nuclear factor kappa-light-chain-enhancer of activated B cells (NF-κB), a redox-sensitive transcription factor, in the inflamed retina.

Using a similar mouse model, Yao et al. (2010) showed the oral administration of bilberry extract (contained 42% ACNs) at dosages of 50, 100, and 200 mg/kg/day for 5 days before the LPS injection. After treatment, whole-eye homogenates in the bilberry-treated group had smaller elevation in nitric oxide, lower malondialdehyde concentration and a higher oxygen-radical absorbance capacity, a greater glutathione and vitamin C concentration, and higher total SOD and glutathione peroxidase activity.

An experimental model of the ischemic retina can be created by increasing the intraocular pressure (IOP) above the perfusion pressure of the central retinal artery or by ligating the central retinal artery. Our group is currently investigating the effect of an acute increase in IOP and a transient blockage of the central retinal artery by examining ERG of rats fed by gavage with either concentrated blueberry juice for 6 weeks (2.8 mg C3G eq per day) or a placebo prior to the insult. The preliminary results are supportive of neuroprotection in these two models of ischemia (unpublished results). Our results in the acute IOP are supported by recent findings from Vin et al. (2012) who used resveratrol (30 mg/kg) administered daily for 5 days via intraperitoneal injection and retinal ischemic injury induced by the elevation of IOP for 45 min on the 3rd day of treatment to show that ERGs and retinal histology were significantly preserved after that short treatment. These two models are also widely used in experimental glaucoma models.

9.6.4 RETINOPATHIES

Retinal dystrophies embrace a wide diversity of genetic and drug/environment-induced causes. The age-related macular dystrophy (ARMD), a disease leading to severe loss in vision and legal blindness in the elderly population, has benefited from antioxidant research. In ARMD, the dysfunction in four functionally interrelated tissues is observed, that is, photoreceptors, RPE, Bruch's membrane, and choriocap-illaries. However, the impairment of RPE cell functions is an early and crucial event in the molecular pathways leading to clinically relevant ARMD changes. During the impairment of RPE function, the accumulation of lipofuscin and the formation of drusens (both showing the accumulation of oxidized membrane products), as well as inflammation and neovascularization, contribute to the development of ARMD.

Large-scale studies and meta-analyses including more than 60,000 subjects with ARMD, regarding the use of antioxidant therapies using β-carotene, vitamin C, and E and zinc supplementation do not prevent ARMD but may have a mild effect in slowing down the progression to advanced ARMD and visual acuity loss in people with signs of the disease (reviewed in Evans 2008). No data are available for the use of ACNs in these epidemiological studies.

On the contrary, there are several studies available on experimental models of ARMD. Tremblay et al. (2012) used a light-induced retinopathy (LIR), a commonly used model of photoreceptor dysfunction as found in ARMD and in retinitis pigmen-tosa, a common inherited pan-retinal degenerative disorder. LIR is induced by expos-ing nocturnal animals such as mice and rats to a moderate amount of light for several days or brighter lights for several hours; such exposure induces a photoreceptor degen-eration that is proportional to the exposure to light. In that particular study, Wistar and Brown–Norway rats were fed by gavage with long (7 weeks) and short (2 weeks) intervention with fortified blueberry juice (1 mL; 11.2 mg/kg C3Geq) or with a placebo solution (7 weeks) that contained the abundant non-ACN blueberry phenolic chloro-genic acid. The authors found both long- and short-term intervention with blueberry-enriched diet to provide neuroprotection in the Wistar rats. Specifically, there was a 60% reduction in ERG signal with the placebo solution, and <15% reduction with either blueberry treatment. However, no effect was seen in the Brown–Norway rats where there was an approximately 78% reduction in ERG signal in both the groups. With a milder insult (shorter exposition to light) in the Brown–Norway rats, some neu-roprotection was observed, but with less efficacy than with the Wistar rats. They also looked at the timing of the ACN intervention and found that if the dietary treatment commenced after the light insult, there was no protection of the retina. Altogether, these results support the role of ACN in protecting the retina from light insult. Several mechanisms were proposed for this protection including the classical antioxidative properties of ACN, and also a possible allosteric binding to rhodopsin's cytoplas-mic domains. The study could not elucidate why the Wistar and Brown–Norway rats responded differently, either to the susceptibility of light damage or to the neuroprotec-tion offered by ACNs. However, it alerts the scientific community that genetic back-ground is important to consider when using animal models.

Liu et al. (2011) also demonstrated ACN neuroprotection in rabbit LIR, using 6–24 mg/kg daily doses of ACN obtained from Chinese blueberries (*Vaccinium* spe-cies not specified, possibly several species) for 4 weeks prior to the insult. In that study, neuroprotection was observed only in the rod, but not in the cone photorecep-tors. Histological evidence indicated less damage after LIR in mice when purple rice extract (2.9% ACN) was intravitreously injected at a dose of 10 μg per eye 2 h before the light exposure (Tanaka et al. 2011). These authors also looked at visible light injury in cultured murine photoreceptor cells (661 W) and found that the purple rice extract was effective in preventing photoinduced photoreceptor death, and more specifically that cyanidin and peonidin were significantly contributing to that effect. They also confirmed that antioxidative properties of the purple rice extract, cyanidin, peonidin, and Trolox played a major role in the neuroprotection of the 661 W cells because the larger increase of the reactive oxygen species level 24 h after light exposure in the vehicle group was not observed in the treatment group.

Albino rats fed with Caucasian bilberry (*V. arctostaphylos*) aqueous extract at the dose of 20 mg/kg BW daily for 10 days prior to and 10 days subsequent to light insult (2100 lux for 24 h) also showed some preservation of ERG function (1032 µV in control, 486 µV in placebo group, and 671 µV in blueberry-fed group) as well as histological evidence of neuroprotection by ACNs (Stepanyan and Topchyan 2007).

Photoreceptors can also be damaged through various pharmacological interventions. For instance, Paik et al. (2012) used intraperitoneal injection of *N*-methyl-*N*-nitrosourea, a deoxyribonucleic acid (DNA)-methylating agent to induce photoreceptor degeneration. Black soybean seed coat ACN extract given orally (50 mg/kg BW) to rats for 4 weeks had 31% and 37% greater amplitude of the a- and b waves of the ERG, respectively, in comparison with rats that did not receive the ACN extract. Also, although not quantified, the expression of glial fibrillary acidic protein, a marker of glial activation associated with neuronal degeneration, was downregulated in the ACN-treated group.

Matsunaga et al. (2009) induced a retinal degeneration in a murine model by intravitreal injection of *N*-methyl-D-aspartic acid. This model induces the formation of nitrite, nitrotyrosine, and lipid peroxidation, leading to the death of RGCs. They showed that the concomitant injection of bilberry ACN (at 10 or 100 µg per eye, reported corresponding to 84 mg/kg if administrated by oral route made by the calculation based on Morazzoni et al. 1991) reduced by 25% of the formation of terminal deoxynucleotidyl transferase nick end labeling (TUNEL)-positive cells 24 h after injury in ACN injection, resulting in an increase of close to 40% in the RGC population and 32% in the thickness of the inner plexiform layer of the retina. This study is unique in the use of the intravitreal administration of ACN that would likely have a significant impact on the *in vivo* ACN concentration and their metabolites. Notably, effects were observed as early as 24 h after the ACN administration.

Yu et al. (2012) used freeze-dried grape powder containing a mixture of resveratrol, flavans, flavonols, ACN, simple phenolics, and 1% FloraGlo, a patented marigold extract containing 5% lutein and 0.2% zeaxanthin xanthophylls to study the age-related retinal changes in the β-5 integrin knockout mice. This mouse model is deficient in phagocytosis of the outer segment of rod photoreceptors that results in the progressive retinal degeneration and relates to human retinal diseases such as retinitis pigmentosa. Both enriched diets, when given early in life, significantly delayed age-related blindness by improving cone and rod photoreceptor functionalities independent of detectable changes in the outer segment structure or photoreceptor survival. Grape- and lutein diets dramatically decreased the by-products generated from the oxidation of polyunsaturated fatty acids (from the photoreceptor's membrane-continuous phagocytosis by RPE) by 75% and 68%, respectively. Other mice models of retinitis pigmentosa such as the rd1 and rd10 mice have benefited from antioxidant treatment (Komeima et al. 2007; Yoshida et al. 2012) but no data are available for the specific use of ACN. A2E is a photooxidative by-products from abnormal accumulation of all-trans retinoid in photoreceptors. It is found associated with lipofuscin accumulation in several retinal dystrophies such as ARMD. ACNs appear to prevent the photooxidative damage from A2E in ARPE19 *in vitro* cell culture, as well as some protection against membrane permeabilization. (Jang et al. 2005) and their role in downregulation of retinoid binding proteins prevents its accumulation in RPE tissues.

Clinical trials assessing the potential benefits of ACNs in retinopathies have not yet been conducted. The heterogeneity of these disorders and their slow progression in humans are significant challenges in assessing ACN effects in human retinopathies. Fortunately, the evidence of ACN benefits is available from animal models of retinal dystrophies, especially in light-induced retinal degeneration. This positive evidence should foster expanded research.

9.6.5 GLAUCOMA

Glaucoma is a multifactorial group of eye diseases in which the optic nerve is damaged, resulting in the death of RGCs. The morphological and functional changes in affected humans include a cupping of the optic nerve head, a loss of retinal nerve fiber layer, and visual field defects. The risk factors for glaucoma include elevated IOP, heredity, high myopia, and race. It is likely that the microenvironment in the optic nerve head reacts to stressors such as IOP and ischemia, to ultimately cause RGC axonal damage but the exact etiology is not known. Currently, the only treatment that has been well established for glaucoma is the reduction of IOP although other therapeutic avenues such as neuroprotection and the use of vasodilators have been explored.

The group of Dr. Ohguro has reported several clinical studies in patients with various etiologies of glaucoma. In a small placebo-controlled, double-blind crossover study of 12 healthy normal subjects (Ohguro et al. 2012a), black currant ACNs (50 mg/day) or placebo was orally administrated once daily for 4 weeks to healthy subjects. A statistically significant decrease from the baseline in the mean IOP, averaging 1.89 ± 1.58 mmHg at 2 weeks and 1.19 ± 1.77 mmHg at 4 weeks was observed in the black currant ACN-treated healthy subjects whereas no statistical difference in IOP was observed in the placebo group. They also followed for 2 years a group of 21 glaucoma patients currently treated with a single antiglaucoma medication. They showed that, during the 24 months of treatment with 50 mg ACN per day, the mean IOP did not significantly change in the patient treated with only antiglaucoma medications, whereas it was significantly lower in the other group in patients who also took the black currant ACN, adding a further reduction of 0.58–0.79 mmHg (2–3% further reduction) over the last year of the follow-up. The mean visual field defect was also shown to have deteriorated in the placebo group whereas it remained stable in the black currant ACN group.

Another study from the same group (Ohguro et al. 2012b) included 21 patients of the above study who were on one antiglaucoma medication and added 17 others who were on multitherapy for the treatment of elevated IOP. Because of the heterogeneity of the group, the effect of ACN on IOP was not anymore significant after 24 months of treatment. However, Humfrey mean visual field defects and ocular blood circulation by laser-speckled flowgraphy were significantly improved.

Ohguro et al. (2007) also report on a group of 31 normotensive glaucoma patients who received black currant ACNs (25 mg/day) for 6 months. This was a pilot study with only a pre–post design. An increase in blood flow and no progression in visual field defects were observed. No significant changes in the mean IOP were observed but the level of endothelin-1 was increased to reach the normal levels. Endothelin-1 is a potent vasoactive peptide whose role in glaucoma etiology is still controversial (see Chauhan 2008); this study, as well as the one of Matsumoto et al. (2005) on ciliary muscle contraction, supports some interaction of ACNs with the endothelin pathway.

IOP can also be modulated through the endocannabinoid system, where the stimulation of cannabinoid receptor-1 lowered IOP (reviewed in Tomida et al. 2004). Korte et al. (2009) studied ACN effects using a competitive radioligand-binding assay and identified cyanidin and delphinidin as ligands with moderate affinity to human cannabinoid receptor-1. Russo et al. (2004) anecdotally documented an improvement in the dark sensitivity of 0.1 log in one patient, using a graduated administration of Δ9-tetrahydocannabinol (0–20 mg, as Marinol®) versus placebo in a double-blind trial. Three other subjects were tested pre–post smoking kir, again with an improvement in night vision.

The positive influence of ACN on IOP was also supported by Steigerwalt et al. (2010), who studied the IOP evolution in a group of 79 patients with ocular hypertension but no evidence for glaucomatous defect. The group was divided into three, each of whom receiving for 2 months a supplementation with 80 mg Mirtogenol®, a tablet of bilberry extract standardized to contain 36% ACNs combined with Pycnogenol®, an extract from maritime pine bark that is rich in various phenolics, a standard glaucoma treatment in the form of Latanoprost® eye drops, or a combination of the two. The Myrtogenol supplementation in itself was a sufficient treatment to lower the mean IOP from a baseline 38.1 ± 2.0 to 29.0 mmHg and to improve the ocular blood flow as measured by Color Doppler imaging. The Latanoprost group lowered IOP from baseline 37.7 ± 2.0 to 27.2 mmHg, therefore as effectively as the ACN treatment but with results achieved in only 4 weeks. The combination of Myrtogenol with Latanoprost further decreased the IOP from 38.0 ± 3.1 to 23 mmHg, thus suggesting a synergetic effect. It is interesting to note here that the pine bark extract in itself appears to be a potent vasoactive substance (see review in Schönlau and Rohdewald 2001) so that the respective effects of ACN and pine bark extract cannot be differentiated.

In an electrophysiological study of patients with a variety of eye conditions, Caselli (1985) documented eight glaucoma cases that showed improvement in critical fusion frequency as measured by ERG from 13.08 ± 1.31 to 14.58 ± 1.24 Hz, 90 min after a single oral dose of 200 mg of a dry extract of ACN. An increase in critical fusion frequency is associated with faster retinal processing. This is one of the few studies that looked at short-term effects of ACN in glaucoma patients.

Experimental models of glaucoma involve an acute increase in IOP or transient ischemia of the central retinal artery. These two models have been reviewed in

Section 9.3 of this chapter. The models of chronic mild IOP have also been developed and using such a model, Ko et al. (2005) documented that free radical balance changes are part of the pathophysiology of this condition. Therefore, the general antioxidative properties of ACN could also come into play in glaucoma.

Thus, it appears that ACN per se can decrease the IOP to a level equivalent to the standard antiglaucomatous medication. ACN action may also act synergistically to potentiate the effects of the primary medication. Endothelin- and cannabinoid-signaling pathways, both known for their impact on IOP, appear to be modulated by the introduction of ACNs in the diet.

9.7 CLINICAL IMPACT OF ACNS ON VISUAL PERFORMANCE

Many ACN clinical studies using psychophysical and ERG approaches have been published. These are particularly abundant in the European literature and often have been published in journals not listed in computerized databases and are in languages other than English. A survey of bilberry ACN effects on vision in dark conditions can be found in the excellent review by Canter and Ernst (2004) and the reader could refer to this review for extensive description of the earlier European studies. In the next two sections, we will concentrate on the most relevant and most recent psychophysical reports. Some other appreciation of the early ACN literature in vision can also be found in Morazzoni and Bombardelli (1996), Kalt et al. (2010), and Milbury (2012).

9.7.1 NONRANDOMIZED PLACEBO-CONTROLLED AND PRE–POST DESIGNS

The late 1960s European literature boasted the fact that World War II Royal Air Force pilots were eating blueberry jam before flying at night to improve their vision. Many studies from that period looking at the various aspects of night vision performances are thus available. In this section, we review the studies that were validated by at least a control arm, then we briefly mention the studies with a less-convincing designs, such as pre–post designs with no control group.

Lee et al. (2007) randomized 60 subjects with asthenopia and mild myopic refractive errors to a treatment using purified high-dose anthocyanoside oligomer (100 mg tablet comprising 85% anthocyanoside oligomer) or a placebo for 4 weeks. Before and after treatment, the subjects completed a questionnaire to determine their clinical symptoms and were also assessed for nocturnal visual function using contrast sensitivity testing. Questionnaire data analysis showed that, following treatment, 22 (73.3%) anthocyanoside subjects reported improved symptoms, whereas only one placebo subject reported an improvement. Contrast sensitivity significantly improved in the anthocyanoside group and remained stable in the placebo group. The mean contrast sensitivity change in the anthocyanoside group was 2.41 ± 1.91 dB, compared to -0.66 ± 2.66 dB for the placebo group. ACN oligomers arise during processing due to the various condensation reactions and are structurally very diverse. The digestive absorption and metabolism of ACN oligomers are unknown and make comparison with other studies more challenging.

In a study of 31 patients with deficits of night vision who were randomized in two 4-week treatments of either cyaninoside chloride or heleninene (xanthophyll dipalmitate), Sole et al. (1984) found improvements in ERG's alpha point. The alpha point refers to the time point where the cone and rod potentials reach equal amplitude and improvement in that time means rod photoreceptors were recovering their prebleach potential faster. Also, a 10% increase in electro-oculographic potentials (that refers to the *trans*-epithelial potential of the RPE) was reported in the cyaninoside chloride group, whereas no differences were noted with the patients randomized to the xanthophyll palmitate group (Sole et al. 1984). Visual acuity was also significantly improved in both the groups by around 0.8 log units, a significant amount. Using similar alpha-point criteria and using a double-blind placebo-controlled parallel design, Alfieri and Sole (1966) documented the dark adaptation in 12 normal, fasted subjects after a single dose of almost 3 g of ACN. The time to reach the alpha point decreased from 9 to 6.5 min in the ACN-treated group, 1 and 2 h after ACN ingestion, suggesting an increase in the rod photoreceptor sensitivity. Also, using a double-blind, placebo-controlled design, Sbrozzi et al. (1983) showed a decrease of 3.75 min in the time of point alpha after treatment of 60 mg ACN per day for 6 days and a reduction of 1.3 min 2 h after a single 60 mg ACN dose in a group of 16 patients. Ponte and Lauricella (1977) also used ERG to show that a single dose of 600 mg ACN, with and without β-carotene, produced a faster dark adaptation 2 and 4 h after treatment. On the other hand, in a double-blind, placebo-controlled study on 46 normal adults who had taken 300 mg ACN per day for 3 and 7 days, only a mild decrease in ERG threshold and no change in the ERG alpha point were reported by Sala et al. 1979. The recovery of visual acuity after visual "dazzling" with bright light was not hastened by the ACN intervention.

The improvements in dark adaptation were also reported by Urso (1967) in 25 patients presenting the pathological decrease in night vision after a single or a 2-day treatment with 25 mg of ACNs and β-carotene. Also, Fiorini et al. (1965) reported the improvement in dark vision after treatment with 100 mg ACN in all patients enrolled, 11 of who were patients with normal vision, 11 with refractive errors, and 11 with vision pathologies. Zavarise (1968) documented an improvement in dark adaptation threshold in 14 patients with hemeralopia (poor or no vision in bright light) 2 days and up to 3 months after treatment with 150 ACN daily. The authors also report that dark adaptation values returned to pretreatment values after the cessation of the ACN treatment but the quantitative data cannot be found in this manuscript.

An observational study documented that subjects with poor night vision showed some decrease in dark adaptation threshold after a single dose of 400 mg ACN, whereas those with a dark adaptation threshold already in the lower range showed no change (Belleoud et al. 1966). A few subjects with poor initial dark adaptation threshold in the placebo group did not improve. It should be noted that this study enrolled only aircraft pilots who were tested before and after night-flying missions; so, the timing of vision testing after the initial dose of ACN varied. Also, the number of subjects in the experimental and control groups was not balanced.

More recently, Pescosolido et al. (2012) used the pattern visual-evoked potentials, a technique collecting brain activity at the occipital cortex and thus encompassing both the retinal and cortical processing, to study responses of the visual system to dimlight stimuli in 23 normal volunteers. The nutritional supplement Myoops® (SOOFT Italia Spa, Montegiorgio, Italy) was taken three times a day for 45 days.

Each tablet contained 123 mg *Sambucus nigra* extract containing 6.5% anthocyanosides, 30 mg *Vitis vinifera* extract containing 20% procyanidin oligomers, 6 mg lutein (FloraGlo Lutein®, Kemin, Des Moines, IA), 800 µg vitamin A, and 30 mg vitamin E. They observed an increase in amplitude in foveal responses and a decrease in implicit time in parafoveal responses. It is noteworthy that no control group or crossover design that would include a washout period was used. The authors explained their result by the different proportion of cone and rods in foveal and parafoveal areas but in reality, small changes in evoked potentials are difficult to interpret as they can be easily induced by the subject's focusing or attention.

The improvement in the mean sensitivity of visual fields was also reported. Gandolfo (1990) used the standard Octopus perimetry and observed an improvement in the mean field sensitivity from 26.31 ± 1.16 to 26.64 ± 0.95 db following the daily consumption of a product that included 530 mg ACN, β-carotene, and vitamin E for 7 days in a group of 30 patients with myopia. Using a noncommercial perimetric equipment, a marked increase in retinal sensitivity was also observed in 32 of 42 eyes of medium-to-high-myopia patients treated with 320 mg ACN per day for 90 days (Virno et al. 1986). Some patients were reported to maintain this improved sensitivity 1 month after cessation of the treatment.

9.7.2 RANDOMIZED CROSSOVER STUDIES

Levy and Glovinsky (1998) used a single oral dose of 12, 24, or 36 mg ACN (given as Strix®) (Halsoprodukter, Forserum, Sweden tablets containing 12 mg ACNs as blueberry extract and 2 mg β-carotene) in a double-masked, placebo-controlled, crossover study involving 16 young normal volunteers and failed to show any effect on full-field scotopic retinal threshold, dark adaptation rate, or mesopic contrast sensitivity at 4, 8, and 24 h after treatment. The same study design was applied but with longer treatment and follow-up time (1 day before and daily during a 4-day treatment period 3–5 h following the morning drug application) (Zadok et al. 1999), again with negative results.

A double-blind, placebo-controlled, crossover design examining night vision in 15 young normal volunteers to examine the effects of 160 mg bilberry extract (25% ACN) was taken three times a day for 21 days, followed by a 4-week washout period. The protocol included multiple measurements during each period including three measurements of night visual acuity and contrast sensitivity before treatment, then at days 1, 4–6, 12–14, and 19–21 during treatment, and once weekly during the washout period. No difference in night visual acuity or contrast sensitivity during any of the measurement periods was observed for either the bilberry ACN or the placebo (Muth et al. 2000).

Finally, an abstract report of a double-blind, crossover-randomized controlled trial of 119 healthy individuals who consumed 160 mg bilberry extracts (25% ACN) daily for 28 days indicated that there was no improvement in visual acuity, contrast sensitivity, and dark-adapted threshold tests (Mayser and Wilhelm 2001). Further analysis of groups of subjects that had similar dark-adapted test results at the first visit was carried out to reduce the intrinsic variability in these data, but still no difference between the two groups was detected.

In two double-blind placebo-controlled, crossover studies that included measures of dark-adapted sensitivity, no improvement was found due to ACN treatment (Levy and Glovinsky 1998; Zadok et al. 1999) that contradicts reports from earlier literature (Alfieri and Sole 1966; Belleoud et al. 1966). However, a study with 37 normal subjects found an improvement in dark adaptation 4 h after a single dose of 400 mg ACN although no significant effect was found after 24 h (Jayle and Aubert 1964). Jayle and Aubert also documented a reduction in the central scotoma ("blind spot") and its density was followed by the absorption of ACNs. In a subsequent study, Jayle et al. (1965) confirmed the improvement in dark-adapted threshold in 60 young subjects 1 and 2 h after a single dose of 600 mg ACN.

A single ingestion of black currant ACN in the form of powdered capsule concentrate at doses of 12.5, 20, or 50 mg was found to decrease the dark adaptation threshold by 0.17 log asb with the highest dose only 2 h after treatment, in a group of 12 young and healthy volunteers (Nakaishi et al. 2000). This experiment was reported as a double-blind, placebo-controlled, crossover study. Another experiment, this time using 540 mg of the black currant powder diluted in an artificial juice (corresponding to a dose of 50 mg ANC), looked at the effect of intense reading task on the refractive power of the eye and associated visual fatigue. The experiment revealed less-severe transient refractive alteration after intensive video-terminal use in the ACN-treated arm. The subjective assessment of asthenopia symptoms (visual fatigue) indicated significant fewer symptoms when subjects were treated with ACN.

Similarly, in an unpublished study reporting on two double-blind, placebo-controlled, crossover design clinical trials involving 60 and 72 healthy volunteers aged 35–65-years old, our group showed that dark adaptation, visual acuity, and contrast sensitivity in the dark were not affected by blueberry ACN doses ranging between 270 and 350 mg/d after 3, 8, or 12 weeks. However, the time required to recover the normal visual acuity after a light-dazzling stimulus (i.e., photostress test), which heavily relies on rhodopsin availability, was found to be significantly improved in both trials. This is consistent with *in vitro* studies showing improved phototransduction turnover (Matsumoto and Yoshizawa 2008; Tirupula et al. 2009).

9.8 CONCLUSIONS AND PERSPECTIVES

The specific actions of ACN that are relevant to vision have been demonstrated *in vitro*. ACN are reported to influence photo-transduction processes via several mechanisms. ACN effects in photo-transduction are difficult to interpret holistically, and how these *in vitro* results may relate to normal human vision in the light or dark is unknown. ACN can act as antioxidants directly *in vitro* by their redox action *in situ* and also indirectly by upregulating antioxidant defense mechanisms. The very high respiratory demand and light- and ultraviolet (UV)-absorbing activity of the eye means that ocular tissues are subjected to a high level of oxidative stress. ACN have been shown *in vitro* to affect cell-signaling processes via allosteric mechanisms in ways that may be beneficial to vision. Indeed, the anti-inflammation activity of ACN is due in part to allosteric effects, which is noteworthy because inflammation underlies many major diseases and stress conditions, including various retinopathies, ischemia, and so on. The significance of ACN as allosteric effectors, antioxidants, or as effectors of photo-transduction will significantly depend on how well the various *in vitro* models simulate the *in vivo* milieu, especially with respect to the concentration and forms of ACN that are present.

In studying animals in relation to ACN and vision physiology, often, a very high oral ACN dose is used when compared to amounts typically found in food. These animal studies may be short (hours) or longer term (several months). Other more invasive ACN administration methods have been used (intravenous, intraperitoneal, and intraocular). Animal studies discussed here typically involve the imposition of a stress and/or employ various genetic disease models. The various nonsurgical and surgical procedures intended to simulate various retinopathies and ocular diseases or their damage have been employed. Inflammation has also been examined in relation to ACN and vision. ACN have been shown to have effects that may mitigate inflammation in other systems such as neuronal function, diabetes, and in vascular tissue. Acute and chronic inflammation are the major elements in several ocular diseases and damage to the eye, for example, in retinal ischemia and glaucoma. While the evidence cited here of ACN effects in vision physiology is promising, it is not known if any ACN benefits may accrue to free-living human populations who consume ACN over a long period, with respect to their risk of ocular disease or vision loss during aging.

Despite evidence obtained from *in vitro* and *in vivo* studies, the possible significance of ACN in vision and eye health will rest on the evidence obtained from clinical studies. However, these types of clinical studies are challenging for various reasons. Testing psychophysical parameters such as visual perception demand a great deal of rigor in the design of the study and a host of other factors. For example, the composition of ACN products should be considered with respect to the amounts of native monomeric ACN present versus the larger ACN oligomers that arise during processing because the bioavailability of these various ACN forms will be different. The survey by Canter and Ernst (2004) of clinical evidence for bilberry ACN effects on scotopic vision strongly emphasizes the need for well-designed placebo-controlled crossover designs using modern and ideally objective methods to test vision in humans. These approaches will be essential to unravel what may be relatively small ACN effects arising from a normally sighted and diverse human populations. Clinical studies with populations at risk of vision impairment or people with existing vision pathologies will also be important; however, these patient populations are difficult to identify and are often already using other therapies that complicate the outcomes of the study. An opportunity may exist to determine ACN effects via epidemiological studies. A recent report found that high ACN intake was associated with a reduced risk of myocardial infarction in women (Cassidy et al. 2013). Using similar epidemiological approaches, it may be possible to determine if there is an association between ACN intake and vision health outcomes over a long period (Knekt et al. 2002).

The possible beneficial link between ACN consumption and vision and eye health has received great scientific and commercial attention. Part of this recent interest is due to public awareness of the possible health benefits of fruit and vegetables in general, as reports accumulate of their benefits as antioxidants and as protective agents against degenerative processes, disease, and aging. Notably, the link between ACN and vision benefits has one of the longest histories in the area of fruit and vegetable phytochemicals and health. Currently, the aging and affluent Western demographic is strongly motivated to protect their vision health during aging through a variety of means. Scientific research and commercial activity are being fueled by the desire to maintain personal freedom, and mitigate the personal and public costs attributable to vision loss. It is hoped that this chapter serves as an up-to-date and balanced overview of the more significant research in the field of ACN and vision.

REFERENCES

Alfieri, R and Sole, P 1966, Influence of anthocyanosides, in oral–perlingual administration, on the adapto-electroretinogram (AERG) in red light in humans, *Comptes Rendus Seances Society of Biology Filiales*, 160(8), 1590–1593.

Anderson, B 1968, Ocular effects of changes in oxygen and carbon dioxide tension, *Transactions of the American Ophthalmological Society*, 66, 423–474.

Andres-Lacueva, C, Shukitt-Hale, B, Galli, RL, Jauregui, O, Lamuela-Raventos, RM, and Joseph, JA 2005, Anthocyanins in aged blueberry-fed rats are found centrally and may enhance memory, *Nutritional Neuroscience*, 8(2), 111–120.

Aydemir, O, Güler, M, Kaya, MK, Deniz, N, and Ustundağ, B 2012, Protective effects of ebselen on sodium–selenite-induced experimental cataract in rats, *Journal of Cataract and Refractive Surgery*, 38(12), 2160–2166.

Belleoud, L, Leluan, D, and Boyer, Y 1966, Etude des effets des glucosides d' anthocyan sur la vision nocturne des controleurs d' aerodrome, *Revue Metallurgie Aeronautique Spatiale*, 18, 3–7.

Besch, DD, Jägle, HH, Scholl, HPNH, Seeliger, MWM, and Zrenner, EE 2003, Inherited multifocal RPE-diseases: Mechanisms for local dysfunction in global retinoid cycle gene defects, *Vision Research*, 43(28), 14–14. doi:10.1016/j.visres.2003.09.020.

Bhuiyan, MIH, Kim, JY, Ha, TJ, Kim, SY, and Cho, K-O 2012, Anthocyanins extracted from black soybean seed coat protect primary cortical neurons against *in vitro* ischemia, *Biological and Pharmaceutical Bulletin*, 35(7), 999–1008.

Boatright, JH, Moring, AG, McElroy, C, Phillips, MJ, Do, VT, Chang, B, Hawes, NL et al. 2006, Tool from ancient pharmacopoeia prevents vision loss, *Molecular Vision*, 12, 1706–1714.

Bodakhe, SH, Ram, A, Verma, S, and Pandey, DP 2012, Anticataract activity of rhamnocitrin isolated from *Bauhinia variegata* stem bark, *Oriental Pharmacy and Experimental Medicine*, 12(3), 227–232.

Bravetti, G 1989, Preventive medical treatment of senile cataract with vitamin E and anthocyanosides: Clinical evaluation, *Annual Ottalmological Clinical Ocular*, 115, 109.

Canter, PH and Ernst, E 2004, Anthocyanosides of *Vaccinium myrtillus* (bilberry) for night vision—A systematic review of placebo-controlled trials, *Survey of Ophthalmology*, 49(1), 38–50.

Capovilla, M, Cervetto, L, and Torre, V 1983, The effect of phosphodiesterase inhibitors on the electrical activity of toad rods, *The Journal of Physiology*, 343, 277–294.

Caselli, L 1985, Studio clinico ed elettroretinografico sull'attivita degli antocianosidi, *Archives of Medicine International*, 37, 29–35.

Cassidy, A, Mukamal, KJ, Liu, L, Franz, M, Eliassen, AH, and Rimm, EB 2013, High anthocyanin intake is associated with a reduced risk of myocardial infarction in young and middle-aged women, *Circulation*, 127(2), 188–196.

Chauhan, BC 2008, Endothelin and its potential role in glaucoma, *Canadian Journal of Ophthalmology*, 43(3), 356–360.

Cluzel, C, Bastide, P, Wegman, R, and Tronche, P 1970, Enzymatic activities of retina and anthocyanoside extracts of *Vaccinium myrtillus* (lactate dehydrogenase, alpha-hydroxy-butyrate dehydrogenase, 6-phosphogluconate dehydrogenase, glucose-6-phosphate dehydrogenase, alpha-glycerophosphate dehydrogenase, 5-nucleotidase, phosphoglucose isomerase), *Biochemical Pharmacology*, 19(7), 2295–2302.

Coutinho, D and Barcaui, PP 1975, Os antocianosideos em oftalmologica, *Review of Brasileiro Oftalmologia*, 34, 47–62.

Crouch, RK, Chader, GJ, Wiggert, B, and Pepperberg, DR 1996, Retinoids and the visual process, *Photochemistry and Photobiology*, 64(4), 613–621.

De Amici, M, Dallanoce, C, Holzgrabe, U, Tränkle, C, and Mohr, K 2010, Allosteric ligands for G protein-coupled receptors: A novel strategy with attractive therapeutic opportunities, *Medicinal Research Reviews*, 30(3), 463–549.

Dmitriev, AV and Mangel, SC 2001, Circadian clock regulation of pH in the rabbit retina, *The Journal of Neuroscience: The Official Journal of the Society for Neuroscience*, 21(8), 2897–2902.

Dobri, N, Qin, Q, Kong, J, Yamamoto, K, Liu, Z, Moiseyev, G, Ma, J-X, Allikmets, R, Sparrow, JR, and Petrukhin, K 2013, A1120, a nonretinoid RBP4 antagonist, inhibits formation of cytotoxic bisretinoids in the animal model of enhanced retinal lipofuscinogenesis, *Investigative Ophthalmology and Visual Science*, 54(1), 85–95.

Dulebohn, RV, Yi, W, Srivastava, A, Akoh, CC, Krewer, G, and Fischer, JG 2008, Effects of blueberry (*Vaccinium ashei*) on DNA damage, lipid peroxidation, and phase II enzyme activities in rats, *Journal of Agricultural and Food Chemistry*, 56(24), 11700–11706.

Evans, J 2008, Antioxidant supplements to prevent or slow down the progression of AMD: A systematic review and meta-analysis, *Eye (London)*, 22(6), 751–760.

Felgines, C, Texier, O, Garcin, P, Besson, C, Lamaison, J-L, and Scalbert, A 2009, Tissue distribution of anthocyanins in rats fed a blackberry anthocyanin-enriched diet, *Molecular Nutrition and Food Research*, 53(9), 1098–1103.

Fiorini, G, Biancacci, A, and Graziano, FM 1965, Modificazione perimetriche ed adattometriche dopo ingestion di mirtillina associata a betacarotene, *Annual Ottalmological Clinical Ocular*, 91(6), 371–386.

Fursova, AZA, Gesarevich, OGO, Gonchar, AMA, Trofimova, NAN, and Kolosova, NGN 2005, Dietary supplementation with bilberry extract prevents macular degeneration and cataracts in senesce-accelerated OXYS rats, *Advances in Gerontology=Uspekhi Gerontologii/Rossiiskaia Akademiia Nauk, Gerontologicheskoe Obshchestvo*, 16, 76–79.

Gandolfo, E 1990, Monitoraggio perimetrico di soggetti miopi in trattamento farmacologico a longo termine con un'associazone tra antocianosidi E vitamine, *Bollettino Ocular*, 69, 57–71.

Gupta, SK, Selvan, V, Agrawal, SS and Saxena, R 2009, Advances in pharmacological strategies for the prevention of cataract development, *Indian Journal of Ophthalmology*, 57(3), 175.

Hanneken, A, Lin, F-F, Johnson, J, and Maher, P 2006, Flavonoids protect human retinal pigment epithelial cells from oxidative-stress-induced death, *Investigative Ophthalmology and Visual Science*, 47(7), 3164–3177.

Hanneken, AM, Harris, R, Olson, A, Crouch, RK, Cornwall, C, and Kono, M 2009, Modulation of opsin signaling by flavonoids, *ARVO Meeting Abstracts*, 50(5), 2722.

Hargrave, PA 2001, Rhodopsin. In: eLS. John Wiley & Sons Ltd, Chichester. http://www.els.net [doi: 10.1038/npg.els.0000072].

He, J and Giusti, MM 2010, Anthocyanins: Natural colorants with health-promoting properties, *Annual Review of Food Science and Technology*, 1(1),163.

Head, KAK 2001, Natural therapies for ocular disorders, part two: Cataracts and glaucoma, *Alternative Medicine Review: A Journal of Clinical Therapeutic*, 6(2), 141–166.

Huismans, H 1988, Traitement medicamenteux de la retinopathie diabetique en pratique ophtalmologique, *Der Augenspiegel*, 34(6), 16–21.

Ichiyanagi, T, Shida, Y, Rahman, MM, Hatano, Y, Matsumoto, H, Hirayama, M, and Konishi, T 2005, Metabolic pathway of cyanidin 3-*O*-β-D-glucopyranoside in rats, *Journal of Agricultural and Food Chemistry*, 53(1), 145–150.

Ichiyanagi, T, Shida, Y, Rahman, MM, Hatano, Y, and Konishi, T 2006, Bioavailability and tissue distribution of anthocyanins in bilberry (*Vaccinium myrtillus* L.) extract in rats, *Journal of Agricultural and Food Chemistry*, 54(18), 6578–6587.

Iida, H, Nakamura, Y, Matsumoto, H, Takeuchi, Y, Harano, S, Ishihara, M, and Katsumi, O 2010, Effect of black-currant extract on negative lens-induced ocular growth in chicks, *Ophthalmic Research*, 44(4), 242–250.

Jang, YP, Zhou, J, Nakanishi, K, and Sparrow, JR 2005, Anthocyanins protect against A2E photooxidation and membrane permeabilization in retinal pigment epithelial cells, *Photochemistry and Photobiology*, 81(3), 529–536.

Jayle, GE, Aubry, M, Gavini, H, Braccini, G, and la Baume De, C 1965, Etude concernant l'action sur la vision nocture des anthocyanosides extraits de *Vaccinium myrtillus*, *Annual Ocular*, 198, 556–562.

Jayle, GEG and Aubert, LL 1964, Action of anthocyanin glycosides on the scotopic and mesopic vision of the normal subject, *Therapie*, 19, 171–185.

Kalt, W, Blumberg, JB, McDonald, JE, Vinqvist-Tymchuk, MR, Fillmore, SAE, Graf, BA, O'Leary, JM, and Milbury, PE 2008, Identification of anthocyanins in the liver, eye, and brain of blueberry-fed pigs, *Journal of Agricultural and Food Chemistry*, 56(3), United States, 705–712.

Kalt, W, Hanneken, A, Milbury, P, and Tremblay, F 2010, Recent research on polyphenolics in vision and eye health, *Journal of Agricultural and Food Chemistry*, 58(7), 4001–4007.

Kamiya, K, Kobashi, H, Fujiwara, K, Ando, W, and Shimizu, K 2012, Effect of fermented bilberry extracts on visual outcomes in eyes with myopia: A prospective, randomized, placebo-controlled study, *Journal of Ocular Pharmacological Therapy*, Mary Ann Liebert, Inc. 140 Huguenot Street, 3rd floor, New Rochelle, NY 10801, USA, p. 12103108010 9000.

Kim, ES, Yu, SY, Kwon, SJ, Kwon, OW, Kim, SY, Kim, TW, Ahn, JK et al. 2008, Clinical evaluation of patients with nonproliferative diabetic retinopathy following medication of anthocyanoside: Multicenter study, *Journal of Korean Ophthalmology*, 49(10), 1629–1633.

Knekt, P, Kumpulainen, J, Järvinen, R, Rissanen, H, Heliövaara, M, Reunanen, A, Hakulinen, T, and Aromaa, A 2002, Flavonoid intake and risk of chronic diseases, *The American Journal of Clinical Nutrition,* 76, 560–568.

Ko, M-L, Peng, P-H, Ma, M-C, Ritch, R, and Chen, C-F 2005, Dynamic changes in reactive oxygen species and antioxidant levels in retinas in experimental glaucoma, *Free Radical Biology and Medicine*, 39(3), 365–373.

Komeima, K, Rogers, BS, and Campochiaro, PA 2007, Antioxidants slow photoreceptor cell death in mouse models of retinitis pigmentosa, *Journal of Cellular Physiology*, 213(3), 809–815.

Korte, G, Dreiseitel, A, Schreier, P, Oehme, A, Locher, S, Hajak, G, and Sand, PG 2009, An examination of "anthocyanins and anthocyanidins" affinity for cannabinoid receptors, *Journal of Medicinal Food*, 12(6), 1407–1410.

Lamb, TD and Pugh, EN 2004, Dark adaptation and the retinoid cycle of vision, *Progress in Retinal and Eye Research*, 23(3), England, 307–380.

Latek, D, Modzelewska, A, Trzaskowski, B, Palczewski, K, and Filipek, S 2012, G protein-coupled receptors—Recent advances, *Acta Biochimica Polonica,* 59(4), 515–529.

Laurens, A, Giono-Barber, P, and Silla, O 1983, Aldose reductase inhibitors: A new therapeutic method for certain complications of diabetes? *Therapie*, 38, 659–663.

Lee, J, Lee, HK, Kim, CY, Hong, YJ, Choe, CM, You, TW, and Seong, GJ 2007, Purified high-dose anthocyanoside oligomer administration improves nocturnal vision and clinical symptoms in myopia subjects, *The British Journal of Nutrition*, 93(06), 895.

Levy, Y and Glovinsky, Y 1998, The effect of anthocyanosides on night vision, *Eye*, 12(6), Nature Publishing Group, 967–969.

Liu, Y, Song, X, Han, Y, Zhou, F, Zhang, D, Ji, B, Hu, J et al. 2011, Identification of anthocyanin components of wild Chinese blueberries and amelioration of light-induced retinal damage in pigmented rabbit using whole berries, *Journal of Agricultural and Food Chemistry*, 59(1), United States, 356–363.

Liu, Y, Song, X, Zhang, D, Zhou, F, Wang, D, Wei, Y, Gao, F et al. 2012, Blueberry anthocyanins: Protection against ageing and light-induced damage in retinal pigment epithelial cells, *The British Journal of Nutrition*, 108(1), 16–27.

Maher, P and Hanneken, A 2005, Flavonoids protect retinal ganglion cells from oxidative stress-induced death, *Investigative Ophthalmology and Visual Science*, 46(12), 4796–4803.

Manach, C, Williamson, G, Morand, C, Scalbert, A, and Rémésy, C 2005, Bioavailability and bioefficacy of polyphenols in humans. I. Review of 97 bioavailability studies, *The American Journal of Clinical Nutrition*, 81(1 suppl), 230S–242S.

Marmorstein, AD, Cross, HE, and Peachey, NS 2009, Functional roles of bestrophins in ocular epithelia, *Progress in Retinal and Eye Research*, 28(3), 206–226.

Matsumoto, H, Ito, K, Yonekura, K, Tsuda, T, Ichiyanagi, T, Hirayama, M, and Konishi, T 2007, Enhanced absorption of anthocyanins after oral administration of phytic acid in rats and humans, *Journal of Agricultural and Food Chemistry*, 55(6), 2489–2496.

Matsumoto, H, Kamm, KE, Stull, JT, and Azuma, H 2005, Delphinidin-3-rutinoside relaxes the bovine ciliary smooth muscle through activation of ETB receptor and NO/cGMP pathway, *Experimental Eye Research*, 80(3), 313–322.

Matsumoto, H, Nakamura, Y, Iida, H, Ito, K, and Ohguro, H 2006, Comparative assessment of distribution of blackcurrant anthocyanins in rabbit and rat ocular tissues, *Experimental Eye Research*, 83(2), 348–356.

Matsumoto, H, Nakamura, Y, Tachibanaki, S, Kawamura, S, and Hirayama, M 2003, Stimulatory effect of cyanidin 3-glycosides on the regeneration of rhodopsin, *Journal of Agricultural and Food Chemistry*, 51(12), 3560–3563.

Matsumoto, HH and Yoshizawa, TT 2008, Rhodopsin regeneration is accelerated via noncovalent 11-*cis* retinal–opsin complex—A role of retinal binding pocket of opsin, *Photochemistry and Photobiology*, 84(4), 985–989.

Matsunaga, N, Imai, S, Inokuchi, Y, Shimazawa, M, Yokota, S, Araki, Y, and Hara, H 2009, Bilberry and its main constituents have neuroprotective effects against retinal neuronal damage *in vitro* and *in vivo*, *Molecular Nutrition and Food Research*, 53(7), 869–877.

Mayser, HM and Wilhelm, H 2001, Effects of anthocyanosides on contrast vision (abstract), *Investigative Ophthalmology and Visual Science*, 42(suppl.), 63.

Mazza, G, Kay, CD, Cottrell, T, and Holub, BJ 2002, Absorption of anthocyanins from blueberries and serum antioxidant status in human subjects, *Journal of Agricultural and Food Chemistry*, 50(26), United States, 7731–7737.

Milbury, PE 2012, Flavonoid intake and eye health, *Journal of Nutrition in Gerontology and Geriatrics*, 31(3), 254–268.

Milbury, PE and Kalt, W 2010, Xenobiotic metabolism and berry flavonoid transport across the blood–brain barrier, *Journal of Agricultural and Food Chemistry*, 58(7), 3950–3956.

Milbury, PE, Graf, B, Curran-Celentano, JM, and Blumberg, JB 2007, Bilberry (*Vaccinium myrtillus*) anthocyanins modulate heme oxygenase-1 and glutathione S-transferase-pi expression in ARPE-19 cells, *Investigative Ophthalmology and Visual Science*, 48(5), 2343–2349.

Miyake, S, Takahashi, N, Sasaki, M, Kobayashi, S, Tsubota, K, and Ozawa, Y 2012, Vision preservation during retinal inflammation by anthocyanin-rich bilberry extract: Cellular and molecular mechanism, *Laboratory Investigation; a Journal of Technical Methods and Pathology*, 92(1), 102–109.

Morazzoni, P and Bombardelli, E 1996, *Vacciium myrtillus* L, *Fototherapia*, 67, 3–29.

Morazzoni, P, Livio, S, Scilingo, A, and Malandrino, S 1991, *Vaccinium myrtillus* anthocyanosides pharmacokinetics in rats, *Arzneimittelforschung*, 41, 128–131.

Morimitsu, Y, Kubota, K, Tashiro, T, Hashizume, E, Kamiya, T, and Osawa, T 2002, Inhibitory effect of anthocyanins and colored rice on diabetic cataract formation in the rat lenses, *International Congress Series*, 1245, 503–508.

Muth, ER, Laurent, JM, and Jasper, P 2000, The effect of bilberry nutritional supplementation on night visual acuity and contrast sensitivity, *Alternative Medicine Review: A Journal of Clinical Therapeutic*, 5(2), Thorne Research Inc., 164–173.

Mykkänen, OT, Kalesnykas, G, Adriaens, M, Evelo, CT, Törrönen, R, and Kaarniranta, K 2012, Bilberries potentially alleviate stress-related retinal gene expression induced by a high-fat diet in mice, *Molecular Vision*, 18, 2338–2351.

Nakaishi, H, Matsumoto, H, Tominaga, S, and Hirayama, M 2000, Effects of black currant anthocyanoside intake on dark adaptation and VDT work-induced transient refractive alteration in healthy humans, *Alternative Medicine Review: A Journal of Clinical Therapeutic*, 5(6), 553–562.

Ohguro, I., Ohguro, H., and Nakazawa, M 2007, Hirosaki University Repository for Academic Resources: Effects of Anthocyanins in Black Currant on Retinal Blood Flow Circulation of Patients with Normal Tension Glaucoma. A Pilot Study. *Hirosaki Medical Journal*, 59, 23–32.

Ohguro, H, Ohguro, I and Yagi, S 2012a, Effects of black currant anthocyanins on intraocular pressures in healthy volunteers and patients with glaucoma, *Journal of Ocular Pharmacology*, 29(1), 61–67.

Ohguro, H, Ohguro, I, Katai, M, and Tanaka, S 2012b, Two-year randomized, placebo-controlled study of black currant anthocyanins on visual field in glaucoma, *Ophthalmologica Journal International d'Ophthalmologie International Journal of Ophthalmology Zeitschrift für Augenheilkunde*, 228(1), 26–35.

Paik, S-S, Jeong, E, Jung, SW, Ha, TJ, Kang, S, Sim, S, Jeon, JH, Chun, M-H, and Kim, I-B 2012, Anthocyanins from the seed coat of black soybean reduce retinal degeneration induced by *N*-methyl-*N*-nitrosourea, *Experimental Eye Research*, 97(1), 55–62.

Pascual-Teresa, S and Sanchez-Ballesta, MT 2007, Anthocyanins: From plant to health, *Phytochemistry Reviews*, 7(2), 281–299.

Passamonti, S, Vrhovsek, U, Vanzo, A, and Mattivi, F 2005, Fast access of some grape pigments to the brain, *Journal of Agricultural and Food Chemistry*, 53(18), 7029–7034.

Perossini, M, Guidi, G, Chiellini, S, and Siravo, D 1987, Studio clinico sull'impiego degli antocianosidi del mirtillo (tegens) nel trattamento delle microangiopatie retiniche di topo diabetico ed ipertensivo, *Annual Ottalmological Clinical Ocular*, 113, 1173–1190.

Pescosolido, N, Di Blasio, D, Rusciano, D, Belcaro, G, and Nebbioso, M 2012, The effect of night vision goggles on the retinocortical bioelectrical activity and its improvement by food supplement, *Panminerva Medica*.

Ponte, F and Lauricella, M 1977, Effect of *Vaccinium myrtillus* total extract on the recovery in the dark of the human electroretinogram, *Atti VII Simposio ISCERG Istambul*, 107, 47–55.

Priya, SSL, Devi, PR, Eganathan, P, and Topno, NS 2012, Structure prediction of bestrophin for the induced-fit docking of anthocyanins, *Bioinformation*, 8(16), 742–748.

Rechner, AR, Kuhnle, G, Bremner, P, Hubbard, GP, Moore, KP, and Rice-Evans, CA 2002, The metabolic fate of dietary polyphenols in humans, *Free Radical Biology and Medicine*, 33(2), 220–235.

Repossi, P, Malagola, R, and De Cadillhac, C 1987, Influenza degli antocianosidi sulle mallatie vasali da alterata permeabilitica, *Annual Ottalmological Clinical Ocular*, 113, 357–361.

Ruckstuhl, M, Beretz, A, Anton, R and Landry, Y 1979, Flavonoids are selective cyclic GMP phosphodiesterase inhibitors, *Biochemical Pharmacology*, 28(4), 535–538.

Russo, EB, Merzouki, A, Mesa, JM, Frey, KA, and Bach, PJ 2004, Cannabis improves night vision: A case study of dark adaptometry and scotopic sensitivity in kif smokers of the Rif mountains of northern Morocco, *Journal of Ethnopharmacology*, 93(1), 99–104.

Sala, D, Rossi, PL, and Rolando, SD 1979, Effetto degli antocianosidi sulle "performances" visive alle basse luminanze, *Miverva Oftalmology*, 21, 283–285.

Sasaki, R, Nishimura, N, Hoshino, H, Isa, Y, Kadowaki, M, Ichi, T, Tanaka, A et al. 2007, Cyanidin 3-glucoside ameliorates hyperglycemia and insulin sensitivity due to down-regulation of retinol binding protein 4 expression in diabetic mice, *Biochemical Pharmacology*, 74 (11), 1619–1627.

Sbrozzi, F, Landini, J, and Zago, M 1983, Night vision affected by anthocyanosides. An electroretinographic test, *Minerva Oftalmology*, 24, 189–193.

Scharrer, A and Ober, M 1981, Anthocyanoside in der behandlung von retinopathien, *Klin Monatsbl Augenheilkd*, 178, 386–389.

Schönlau, F and Rohdewald, P 2001, Pycnogenol for diabetic retinopathy. A review, *International Ophthalmology*, 24(3), 161–171.

Seddon, JMJ 2007, Multivitamin–multimineral supplements and eye disease: Age-related macular degeneration and cataract, *The American Journal of Clinical Nutrition*, 85(1), 304S–307S.

Serafini, M, Testa, MF, Villaño, D, Pecorari, M, van Wieren, K, Azzini, E, Brambilla, A, and Maiani, G 2009, Antioxidant activity of blueberry fruit is impaired by association with milk, *Free Radical Biology and Medicine*, 46(6), 769–774.

Sole, P, Rigal, D, and Peyresblanques, J 1984, Effects of cyaninoside chloride and heleniene on mesopic and scotopic vision in myopia and night blindness, *Journal of Français d'Ophthalmologie*, 7(1), 35–39.

Song, J, Li, Y, Ge, J, Duan, Y, Sze, SC-W, Tong, Y, Shaw, P-C et al. KY 2010, Protective effect of bilberry (*Vaccinium myrtillus* L.) extracts on cultured human corneal limbal epithelial cells (HCLEC), *Phytotherapy Research*, 24(4), 520–524.

Sparrow, JR, Wu, Y, Kim, CY, and Zhou, J 2010, Phospholipid meets all-*trans*-retinal: The making of RPE bisretinoids, *The Journal of Lipid Research*, 51(2), 247–261.

Stefek, M 2011, Natural flavonoids as potential multifunctional agents in prevention of diabetic cataract, *Interdisciplinary Toxicology*, 4(2), 69–77.

Steigerwalt, RDR, Belcaro, GG, Morazzoni, PP, Bombardelli, EE, Burki, CC, and Schönlau, FF 2010, Mirtogenol potentiates latanoprost in lowering intraocular pressure and improves ocular blood flow in asymptomatic subjects, *Clinical Ophthalmology (Auckland, N.Z.)*, 4, 471–476.

Stepanyan, RV and Topchyan, HV 2007, Processing drug dosage forms from Caucasian bilberry and their influence on the eye retina, *The New Armenial Medical Journal*, 1, 65–74.

Talavéra, S, Felgines, C, Texier, O, Besson, C, Gil-Izquierdo, A, Lamaison, J-L, and Rémésy, C 2005, Anthocyanin metabolism in rats and their distribution to digestive area, kidney, and brain, *Journal of Agricultural and Food Chemistry*, 53(10), 3902–3908.

Tan, DTHD, Dart, JKGJ, Holland, EJE, and Kinoshita, SS 2012, Corneal transplantation, *Lancet*, 379(9827), 1749–1761.

Tanaka, J, Nakanishi, T, Ogawa, K, Tsuruma, K, Shimazawa, M, Shimoda, H, and Hara, H 2011, Purple rice extract and anthocyanidins of the constituents protect against light-induced retinal damage *in vitro* and *in vivo*, *Journal of Agricultural and Food Chemistry*, 59(2), 528–536.

Tanaka, J, Nakanishi, T, Shimoda, H, Nakamura, S, Tsuruma, K, Shimazawa, M, Matsuda, H, Yoshikawa, M, and Hara, H 2012, Purple rice extract and its constituents suppress endoplasmic reticulum stress-induced retinal damage *in vitro* and *in vivo*, *Life Sciences*, 92(1), 17–25.

Tirupula, KC, Balem, F, Yanamala, N, and Klein Seetharaman, J 2009, pH-dependent interaction of rhodopsin with cyanidin–3-glucoside. 2. Functional aspects, *Photochemistry and Photobiology*, 85(2), Wiley Online Library, 463–470.

Tomida, I, Pertwee, RG, and Azuara-Blanco, A 2004, Cannabinoids and glaucoma, *The British Journal of Ophthalmology*, 88(5), 708–713.

Tremblay, F, Waterhouse, J, Nason, J, and Kalt, W 2012, Prophylactic neuroprotection by blueberry-enriched diet in a rat model of light-induced retinopathy, *The Journal of Nutritional Biochemistry*, 24, 647–655.

Tronche, P and Bastide, P 1967, Effet des glycosides 'anthocyanes sur la cinetique de regeneration du pourpre retinien chez le lapin, *C.R. Society of Biology*, 161, 2473–2476.

Tsuda, T, Horio, F, and Osawa, T 1999, Absorption and metabolism of cyanidin 3-*O*-beta-D-glucoside in rats, *FEBS Letters*, 449, (2–3), 179–182.

Ulbricht, C, Basch, E, Basch, S, Bent, S, Boon, H, Burke, D, Costa, D, Falkson, C, Giese, N, Goble, M, Hashmi, S et al. J 2009, An evidence-based systematic review of bilberry (*Vaccinium myrtillus*) by the Natural Standard Research Collaboration, *Journal of Dietary Supplements*, 6(2), Informa UK Ltd, UK, 162–200.

Urso, G 1967, Azione degli antocianosidi del "*Vaccinium myrtillus*" associati a betacarotene sulls sensibilita' lumina, *Annual Ottalmological Clinical Ocular*, 93, 931–938.

Varma, SD, Mizuno, A, and Kinoshita, JH 1977, Diabetic cataracts and flavonoids, *Science (New York, NY)*, 195(4274), 205–206.

Vin, AP, Hu, H, Zhai, Y, Zee Von, CL, Logeman, A, Stubbs, EB, Perlman, JI, and Bu, P 2012, Neuroprotective effect of resveratrol prophylaxis on experimental retinal ischemic injury, *Experimental Eye Research*, 108, 72–75.

Virmaux, N, Bizec, JC, Nullans, G, Ehret, S, and Mandel, P 1990, Modulation of rod cyclic GMP–phosphodiesterase activity by anthocyanidin derivatives, *Biochemical Society Transplant*, 18(4), 686–687.

Virno, M, Motolese, E, Garofalo, G, and Giraldi, JP 1986, Effeti degli antocianosidi di mirtillo sulla sensibilita retinica di pazienti miopi all'seame perimetrico computerizzato, *Bollettino Ocular*, 65, 315–326.

Wahid, FF, Jung, HH, Khan, TT, Hwang, K-HK, Park, JSJ, Chang, S-CS, Khan, MAM, and Kim, YYY 2011, Effects of *Rubus coreanus* extract on visual processes in bullfrog's eye, *Journal of Ethnopharmacology*, 138 (2), 7–7.

Walton, MC, Lentle, RG, Reynolds, GW, Kruger, MC, and McGhie, TK 2006, Anthocyanin absorption and antioxidant status in pigs, *Journal of Agricultural and Food Chemistry*, 54(20), 7940–7946.

West, AL, Oren, GA, and Moroi, SE 2006, Evidence for the use of nutritional supplements and herbal medicines in common eye diseases, *American Journal of Ophthalmology*, 141(1), 157–166.

Williamson, G and Clifford, MN 2010, Colonic metabolites of berry polyphenols: The missing link to biological activity?, *The British Journal of Nutrition*, 104(suppl 3), S48–S66.

Willis, L, Bickford, P, Zaman, V, Moore, A, and Granholm, A-C 2005, Blueberry extract enhances survival of intraocular hippocampal transplants, *Cell Transplantation*, 14(4), 213–223.

Wu, XX, Cao, GG, and Prior, RLR 2002, Absorption and metabolism of anthocyanins in elderly women after consumption of elderberry or blueberry, *Journal of Nutrition*, 132(7), 1865–1871.

Yamamoto, F, Borgula, GA, and Steinberg, RH 1992, Effects of light and darkness on pH outside rod photoreceptors in the cat retina, *Experimental Eye Research*, 54(5), 685–697.

Yanamala, N, Gardner, E, Riciutti, A, and Klein-Seetharaman, J 2012, The cytoplasmic rhodopsin–protein interface: Potential for drug discovery, *Current Drug Targets*, 13(1), 3–14.

Yanamala, N, Tirupula, KC, Balem, F, and Klein Seetharaman, J 2009, pH-dependent interaction of rhodopsin with cyanidin-3-glucoside. 1. Structural aspects, *Photochemistry and Photobiology*, 85(2), Wiley Online Library, 454–462.

Yao, NN, Lan, FF, He, R-RR, and Kurihara, HH 2010, Protective effects of bilberry (*Vaccinium myrtillus* L.) extract against endotoxin-induced uveitis in mice, *Transactions of the IRE Professional Group on Audio*, 58(8), 4731–4736.

Yoshida, N, Ikeda, Y, Notomi, S, Ishikawa, K, Murakami, Y, Hisatomi, T, Enaida, H, and Ishibashi, T 2012, Laboratory evidence of sustained chronic inflammatory reaction in retinitis pigmentosa, *Ophthalmology*, 120(1), e5–e12.

Youdim, KA, Shukitt-Hale, B, and Joseph, JA 2004, Flavonoids and the brain: Interactions at the blood–brain barrier and their physiological effects on the central nervous system, *Free Radical Biology and Medicine*, 37(11), 1683–1693.

Yu, C-C, Nandrot, EF, Dun, Y, and Finnemann, SC 2012, Dietary antioxidants prevent age-related retinal pigment epithelium actin damage and blindness in mice lacking αvβ5 integrin, *Free Radical Biology and Medicine*, 52(3), 660–670.

Zadok, D, Levy, Y, and Glovinsky, Y 1999, The effect of anthocyanosides in a multiple oral dose on night vision, *Eye (London)*, 13(6), 734–736.

Zafra-Stone, S, Yasmin, T, Bagchi, M, Chatterjee, A, Vinson, JA, and Bagchi, D 2007, Berry anthocyanins as novel antioxidants in human health and disease prevention, *Molecular Nutrition and Food Research*, 51(60), 675–683.

Zavarise, G 1968, Sull'effetto del trattamento prolungato con antocianusidi sul senso luminoso, *Annual Ottalmological Clinical Ocular*, 94, 209–214.

10 Effects of Anthocyanins on Neuronal and Cognitive Brain Functions

Erika K. Ross, Aimee N. Winter,
and Daniel A. Linseman

CONTENTS

10.1 THE PATHOLOGICAL BASIS OF NEURODEGENERATION

The advent of rapid medical advance has brought with it a myriad of positive impacts, not the least of which is a dramatic increase in average life expectancy, primarily among those living in developed countries. However, as the average age of the population increases, new problems have arisen such as a corresponding increase in the incidence of neurodegenerative diseases (Melo et al., 2011). Neurodegenerative diseases include common disorders such as Alzheimer's disease and Parkinson's disease, as well as other less prevalent fatal diseases like amyotrophic lateral sclerosis (ALS, Lou Gehrig's disease) and Huntington's disease. All neurodegenerative disorders are distinguished by the chronic and progressive loss of neurons in specific regions of the brain and/or spinal cord, resulting in characteristic impairment of motor and/or cognitive function (Jellinger, 2009). Also of equal clinical relevance is ischemia–reperfusion injury, which is most commonly caused by stroke (Chen et al., 2011). Neuronal injury due to stroke does not fit the classical definition of a neurodegenerative disease; however, it is responsible for causing damage to localized brain regions, and results in neuronal death through many of the same pathways as neurodegenerative diseases (Chen et al., 2011; Dong et al., 2009; Jellinger, 2009).

While the individual pathologies of each disorder vary, several factors, including oxidative stress, excitotoxicity, neuroinflammation, and cellular apoptosis, have been identified as common elements of most neurodegenerative diseases. These factors have been shown to contribute significantly to the neuronal cell death underlying these disorders and consequently, to the overall onset and progression of neurodegeneration (Jellinger, 2009). Given their contributions to cell death, these pathways represent attractive and in some cases, untapped, targets for treatment. Thus, understanding the role of these factors in neuronal cell death is a prerequisite to discovering new and more effective therapeutic agents to mitigate the progression of neurodegeneration.

10.1.1 ROLE OF OXIDATIVE STRESS IN NEURODEGENERATION

For at least the last decade, the role of oxidative stress in neurodegenerative disease has been the topic of intense scientific investigation. This condition, in which reactive oxygen species (ROS) accumulate within the cell and overwhelm endogenous antioxidant defenses, ultimately results in the activation of apoptotic (programmed) and other cell death pathways. This typically occurs when proper detoxification and disposal of intracellular ROS is interrupted which leads to oxidative damage of vital cellular macromolecules, including DNA, proteins, and lipid membranes (Lin and Beal, 2006). Evidence of a role for oxidative stress in neurodegeneration is plentiful and varied, but can be categorized based on two primary factors that lead to the generation or accumulation of ROS. These include mitochondrial dysfunction and mutations in the proteins responsible for disposing of ROS.

The process of cellular respiration and the associated reactions that take place within the electron transport chain in the mitochondria are some of the greatest sources of intracellular ROS (Lin and Beal, 2006). Normally, these ROS are scavenged and detoxified before significant damage can be done to cellular machinery; however, in many examples of neurodegenerative disease, the ability of the cell to scavenge these reactive species is significantly reduced (Kelsey et al., 2010; Lin and Beal, 2006). As a result, ROS produced by the mitochondria are free to cause extensive oxidative damage to mitochondrial components, further disrupting organelle function and prompting additional production of ROS (Lin and Beal, 2006). Neurons are particularly susceptible to this aberrant generation of ROS owing to their elevated need for energy production by the mitochondria (Krantic et al., 2005; Murata et al., 2008). Indeed, oxidative stress and mitochondrial dysfunction have been implicated in Alzheimer's disease as one of the earliest noted events in overall disease pathology, preceding even the deposition of characteristic amyloid beta (Aβ)-plaques (Nunomura et al., 2001; Pratico et al., 2001). Oxidative stress and mitochondrial dysfunction also play notable roles in Parkinson's disease. For instance, mutated α-synuclein, which is involved in the formation of Lewy bodies, has been shown to impair mitochondrial function in a transgenic mouse model of Parkinson's disease (Song et al., 2004). In addition, deficits in complex I of the electron transport chain are a pathological hallmark of Parkinson's disease (Schapira et al., 1989). Indeed, chemical inhibitors of complex I, including MPTP (1-methyl-4-phenyl-1,2,3,6-tetrahydropyridine) and rotenone, have been extensively utilized as neurotoxin-based models of Parkinson's disease (Jackson-Lewis et al., 2012). These factors play a key pathological role in ischemia–reperfusion injury as well, in which the flood of oxygen into the brain during reperfusion results in increased production of superoxide radicals by mitochondria, oxidative damage, and mitochondrial dysfunction. In addition, induction of neuronal nitric oxide synthase (nNOS) by ischemia causes increased production of nitric oxide (NO) radicals. Superoxide and NO can also interact to form peroxynitrite which is capable of nitrosylating important cellular proteins (Eliasson et al., 1999).

In addition to the feed-forward reaction of increased ROS leading to mitochondrial dysfunction leading to increased ROS, mutations in a number of key antioxidant proteins have also been observed in several neurodegenerative diseases. This, in turn, may lead to either the loss of protein function or the acquisition of new and deleterious functions as is the case with gain-of-toxic function mutations in Cu/Zn superoxide dismutase (SOD1) (Liu et al., 2004). This enzyme normally scavenges damaging superoxide radicals; however, when mutated it has been shown to selectively associate with mitochondrial membranes within spinal motor neurons where it causes significant damage to proteins, mitochondrial lipids, and DNA (Liu et al., 2004). Gain-of-toxic function mutations in SOD1 are causative in approximately 20% of cases of familial ALS (Rosen et al., 1993). On the other hand, loss of function mutations in DJ-1, a protein involved in sensing oxidative stress and promoting antioxidant response element-dependent gene transcription via activation of Nrf2, have been implicated in development of early onset Parkinson's disease (Gan et al., 2010). Loss of function mutations in DJ-1 reduce cellular capacity to cope with ROS (Bonifati et al., 2003). Thus, either gain-of-toxic function or loss of function mutations in key antioxidant enzymes or transcriptional regulators result in the failure of the cell to remove intracellular ROS, allowing harmful free radicals and other reactive species to amass.

Additionally, mutations in these and many other proteins associated with neurodegeneration can magnify mitochondrial dysfunction. For instance, at least six nuclear gene mutations associated with familial Parkinson's disease occur in genes coding for proteins that are linked to the mitochondria (Lin and Beal, 2006). In the G93A mutant SOD1 mouse model of ALS, mice expressing this mutant protein suffer significant mitochondrial abnormalities, displaying increased degeneration of the mitochondria and impaired energy metabolism in spinal cord motor neurons (Kong and Xu, 1998; Mattiazzi et al., 2002). These and other similar findings establish a definitive link between gene mutations, mitochondrial dysfunction, and oxidative stress.

It is important to note, however, that oxidative stress is not always preceded by mitochondrial dysfunction or gene mutations. In some cases, oxidative stress from extrinsic sources, such as environmental toxins or pollutants, may play a causative role in the expression of other facets of disease pathology (Nunomura et al., 2001; Pratico et al., 2001). As mentioned before, this is the case in Alzheimer's disease, in which oxidative stress heralds the processing of amyloid precursor protein (APP) by β- and γ-secretase to form Aβ (Nunomura et al., 2001; Pratico et al., 2001; Tamagno et al., 2008). Aggregation of Aβ fragments follows, resulting in the classic disease pathology observed in Alzheimer's brains (Ross and Poirier, 2004). This may also be true for other neurodegenerative disorders, such as sporadic ALS, in which oxidative stress due to environmental factors occurs in the apparent absence of genetic mutations. These extrinsic sources of ROS may precipitate mitochondrial dysfunction and cell death; however, it is currently unknown if this is truly the case, or if oxidative stress is merely a secondary consequence of some other underlying disease pathology (Barber et al., 2006).

10.1.2 Role of Excitotoxicity in Neurodegenerative Disease

Excitotoxicity is a process specific to neurons through which overstimulation of cell surface receptors by excitatory neurotransmitters causes calcium overload (Dong et al., 2009). This, in turn, triggers a series of signaling cascades that culminate in a number of downstream processes, such as mitochondrial membrane depolarization, calpain activation, increased production of ROS and NO, and degradation of membranes, proteins, and DNA, which eventually lead to cell death (Jung et al., 2009; Ward et al., 2000). Glutamate is the most common neurotransmitter involved in excitotoxicity, binding primarily to N-methyl-D-aspartate (NMDA) and α-amino-3-hydroxy-5-methyl-4-isoxazolepropionic acid (AMPA) receptors, and causing influx of Ca^{2+} ions from the extracellular space (Dong et al., 2009). High cell surface expression of NMDA or AMPA receptors, or constant activation of these receptors by increased levels of receptor agonists over time, are shown to be associated with the development of neurodegenerative disease (Dong et al., 2009).

Excitotoxicity plays a role in a variety of injurious events within the central nervous system (CNS), such as trauma, ischemia or stroke, and the neurodegenerative diseases discussed here (Dong et al., 2009; Sun et al.. 2006; Szydlowska and Tymianski, 2010; Van Den Bosch et al., 2000). Evidence for its role in ischemia and neurodegeneration is vast, and shows that excitotoxicity plays a key pathogenic role in a variety of neurodegenerative disorders. In Alzheimer's disease, for example, atypical upregulation of NMDA receptors is shown to favor the production of Aβ fragments, disrupting synaptic transmission and promoting glutamate accumulation, which subsequently leads to excitotoxic cell death. *In vivo* studies confirm that increased stimulation of NMDA receptors in the presence of Aβ fragments can indeed cause degeneration characteristic of this disease (Miguel-Hidalgo et al.,

2002; Parameshwaran et al., 2008). In some forms of familial Parkinson's disease, there is a notable proliferation of excitatory synapses within the substantia nigra due to mutations in the *parkin* gene. Parkin plays a significant role in the regulation of excitatory synaptic transmission, thus its loss of function sensitizes cells in the substantia nigra to excitotoxic stimuli (Helton et al., 2008). Ischemia causes an increased release of glutamate, overstimulating NMDA receptors in post-synaptic cells (Szydlowska and Tymianski, 2010). Strong evidence for the role of excitotoxicity in ALS is also substantial. Long-term activation of AMPA receptors in rats by infusion with kainic acid, an AMPA receptor agonist, results in the development of progressive deficiencies in motor function and selective loss of motor neurons (Sun et al., 2006). This pathology mimics that of ALS, suggesting a role for excitotoxicity in this disease (Sun et al., 2006). Further, the intrinsic properties of motor neurons, which possess a substantial number of calcium-permeable AMPA receptors, also suggest that motor neurons may be exceptionally prone to excitotoxic events, accounting for their selective loss in ALS (Van Den Bosch et al., 2000).

Furthermore, there is evidence to suggest that excitotoxicity and oxidative stress may share an intimate relationship. A number of events triggered by an increase in cytosolic calcium that occurs with excitotoxicity have been shown to increase the production of ROS within the cell. In Alzheimer's, for example, it was found that excitation of NMDA receptors was linked strongly with the production of ROS by NADPH oxidase (Shelat et al., 2008). Oxidative stress caused by excitotoxicity compounds the effects of massive calcium influx, making neuronal degeneration a multi-faceted process.

10.1.3   ROLE OF NEUROINFLAMMATION IN NEURODEGENERATION

Neuroinflammation, as the name implies, refers to an inflammatory response within the nervous system. While this process can be advantageous, acting as the nervous system's immune response to harmful antigens, inappropriate activation of an inflammatory response, particularly in the CNS, can also have devastating consequences (Di Filippo et al., 2010). In the latter case, prolonged inflammation in particular regions of the CNS often contributes to neuronal death, thereby facilitating the progression of neurodegenerative disease (Wyss-Coray and Mucke, 2002).

Unlike other organs and tissues, the CNS is equipped with its own innate immune system composed of glial cells; neuronal support cells, which, like the body's immune system, have two primary states (Di Filippo et al., 2010). These include a resting state and an active state in which glial cells respond to aversive stimuli, such as infection or traumatic injury (Di Filippo et al., 2010). While most glial cells participate in this response, microglia are the primary cell type responsible for neuroinflammation (Di Filippo et al., 2010). These cells, which are normally maintained in a resting state, become activated in response to physical or chemical disruptions of their microenvironment to mediate the protection and repair of nearby neurons (Block et al., 2007; Nimmerjahn et al., 2005; Wyss-Coray and Mucke, 2002). One factor that may stimulate the transition of resting microglia to activated microglia is the aggregation of abnormal or misfolded proteins, such as senile plaques composed of Aβ in Alzheimer's disease, the α-synuclein containing Lewy bodies of Parkinson's disease, and accumulation of aggregated TDP-43 protein in ALS and frontotemporal lobar degeneration (Langenhove et al., 2012; Wyss-Coray and Mucke, 2002). Additionally, chemical signals such as the release of signaling molecules or glial activating factors by injured neurons, such as in ischemia, can induce this transition of microglia from a resting state to an activated state (Block et al., 2007; Wyss-Coray and Mucke, 2002).

Once activated, microglia release a host of cytokines and chemokines, small molecules that modulate immune responses in neighboring cells, altering their gene expression and stimulating the synthesis of inducible nitric oxide synthase (iNOS), which is responsible for the production of NO (Di Filippo et al., 2010). NO, a highly reactive and damaging free radical, is abundantly secreted by activated microglia, prompting the onset of nitrosative stress which is a condition similar to oxidative stress in which reactive nitrogen species (RNS) overcome cellular antioxidant defenses (Block et al., 2007). While the ultimate goal of releasing cytokines and NO is the removal of pathogens, if prolonged, these aspects of the inflammatory response will eventually culminate in the release of apoptogenic factors and cell death (Wyss-Coray and Mucke, 2002).

Neuroinflammation appears to play a significant pathological role across a spectrum of neurodegenerative disorders, and postmortem examination has demonstrated substantial evidence in this regard. For example, brains of Parkinson's patients show elevated levels of activated microglia in the substantia nigra. Moreover, signs of nitrosative stress are apparent in the brains of both Parkinson's and Alzheimer's patients (Di Filippo et al., 2010; Heneka and O'Banion, 2007; Hirsch and Hunot, 2009; Hunot et al., 1996). This is also true of ALS, in which the spinal cord and motor cortex demonstrate increased numbers of activated microglia in close proximity to damaged motor neurons (Henkel et al., 2004).

10.1.4 ROLE OF APOPTOSIS IN NEURODEGENERATION

Activation of apoptotic pathways is the ultimate culmination of the events discussed above, and represents the final point at which cell death can be prevented by therapeutic intervention. While there are countless causes of apoptotic cell death, most activate one of two major pathways: the intrinsic pathway mediated by the release of apoptogenic factors from the mitochondria, and the extrinsic pathway mediated by death receptors on the cell surface (Kajta, 2004; Krantic et al., 2005). These pathways, though initiated independently, converge with the activation of caspase-3, also known as the executioner caspase, which is ultimately responsible for cellular death (Kajta, 2004; Krantic et al., 2005). Owing to the vast array of factors that may induce cells to undergo apoptosis, both pathways are observed in neurodegenerative disease (Kajta, 2004).

The substantial role of mitochondrial dysfunction in neurodegeneration provides an excellent example for activation of the intrinsic apoptotic pathway in neuronal death. In this signaling cascade, intracellular factors, such as ROS, RNS, and calcium influx, cause the release of cytochrome c from the mitochondria and activation of caspase-9. The intrinsic initiator caspase-9 and cytochrome c interact with APAF-1 to form a large oligomeric complex, the apoptosome, which in turn activates caspase-3 (Luetjens et al., 2000; Ott et al., 2007; Slee et al., 1999). Once activated, caspase-3 cleaves many critical cell proteins and induces the fragmentation of DNA, resulting in cell death (Sakahira et al., 1998). Because activation of the intrinsic apoptotic pathway is dependent upon fluctuations in intracellular levels of calcium and mitochondrial permeability, both excitotoxicity and oxidative stress induce cell death principally via this mechanism (Luetjens et al., 2000; Ott et al., 2007). In the case of neuroinflammation, however, the extrinsic pathway is favored (Kajta, 2004).

The extrinsic apoptotic pathway is dependent upon the extracellular release of death receptor ligands which bind their cognate cell surface death receptors and initiate caspase activation (Kajta, 2004; Krantic et al., 2005). These ligands include several cytokines which are released during inflammation, such as Fas ligand and tumor necrosis factor α, or TNFα (Kajta, 2004). Binding of these ligands to the death domain of a receptor initiates an alternative signaling cascade that depends upon activation of the extrinsic initiator caspase-8 and the subsequent activation of caspase-3 (Varfolomeev et al., 1998). The rest of the death process is then identical to that observed in the intrinsic apoptotic pathway, concluding with DNA fragmentation (Kajta, 2004; Krantic et al., 2005). It is also important to note that there is crosstalk between the extrinsic and intrinsic pathways. The extrinsic initiator caspase-8 can cleave the Bcl-2 family member Bid to a truncated form known as tBid. In turn, tBid can stimulate Bax-dependent permeabilization of the outer mitochondrial membrane and release cytochrome c, resulting in the activation of the intrinsic cascade. This is known as intrinsic amplification of the extrinsic cascade (Niizuma et al., 2009).

Given the multifactorial nature of neurodegeneration, it becomes clear that identifying therapeutic agents capable of providing significant neuroprotection is challenging. While many substances have been identified as potential treatment options for a variety of neurodegenerative disorders, most are specifically tailored to mitigate only one of the causative factors discussed here (Figure 10.1). As an illustration, the antiglutamatergic agent, Riluzole, is currently the only drug approved by the FDA for the treatment of ALS, yielding a modest increase in lifespan of only two to three months (Miller et al., 2007). However, this drug does not address the role of either oxidative stress or neuroinflammation in this disease, severely limiting its therapeutic efficacy. Examples such as this clearly demonstrate the need for treatments that attack multiple components of disease pathology, including oxidative stress, excitotoxicity, neuroinflammation, and by extension, apoptosis.

Anthocyanins are flavonoid compounds that give berries and flowers their distinctive red, purple, and blue coloring. These unique chemical entities possess intrinsic antioxidant activity as well as anti-inflammatory properties and the capacity to modulate key pro-apoptotic and antiapoptotic signaling pathways. Here, we discuss the use of anthocyanins as potential preventative and therapeutic agents for the treatment of neurodegenerative disorders. We will explore their neuroprotective mechanisms, current applications as neuroprotective agents, bioavailability, and metabolism. With their powerful antioxidant, anti-inflammatory, and antiapoptotic properties, these compounds may provide a natural solution to the growing problem of neurodegeneration in our aging society.

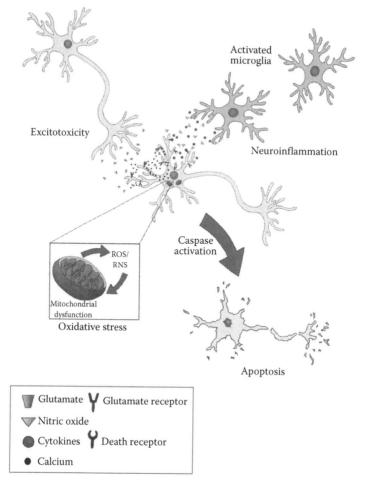

FIGURE 10.1 **(See color insert.)** The pathology of neurodegeneration. The causes of neurodegeneration are diverse and multifaceted, often operating in concert. Thus, at any given time, neurons in affected areas of the brain or the spinal cord may be subjected to a cyclic progression of neurotoxic events. Glutamatergic excitoxicity, the major form of excitotoxicity observed in neurodegenerative disease, is caused by excessive release of glutamate from presynaptic cells or increased expression of NMDA and AMPA receptors on postsynaptic cells. This results in a rapid influx of extracellular calcium, inducing several damaging cellular processes, the most detrimental of which may be the loss of mitochondrial membrane potential and subsequent mitochondrial dysfunction. The aberrant generation and accumulation of ROS by dysfunctional mitochondria then result in a state of oxidative stress in which free radicals and other reactive species are free to do significant damage to vital cellular components. Such damage can lead to the release of inflammatory signals by injured cells, provoking a response from surrounding glial cells. These in turn, release cytokines and high levels of NO, further exacerbating existing oxidative damage and mitochondrial dysfunction with additional injury by RNS. These events ultimately culminate in the release or activation of apoptogenic factors, caspase-3 activation, and apoptotic cell death.

10.2   ANTHOCYANIN ABSORPTION AND BLOOD–BRAIN BARRIER PERMEABILITY

The ability of anthocyanins to traverse biological membranes is a major part of their appeal. Intact anthocyanins taken orally are absorbed rapidly into many relevant tissues (El Mohsen et al., 2006; Milbury and Kalt, 2010; Vanzo et al., 2011). There is considerable interest into the mechanism behind this rapid absorption, as revealing this process could lead to further development of these nutraceuticals for treatment of many pertinent diseases. It is likely that absorption initially occurs in the stomach, which accounts for the rapid kinetics of anthocyanins appearing in plasma (Passamonti et al., 2003). There is accumulating data suggesting that anthocyanin transport occurs via a bilitranslocase transporter, primarily into the vascular endothelium, and subsequently into target tissue (Maestro et al., 2010; Vanzo et al., 2008). Despite their rapid uptake into the systemic circulation, the overall bioavailability of anthocyanins is very low due to their rapid metabolism and excretion principally as glucuronidated, methylated, and glycosylated species (McGhie and Walton, 2007).

Anthocyanins are present in brain endothelial cells and in brain parenchymal tissue of rodents after one oral dose, suggesting that they are capable of crossing the blood–brain barrier (BBB) (Mohsen et al., 2006; Youdim et al., 2003). They are rapidly transported to the brain after ingestion with detection occurring within minutes of a single dose, and once there they may accumulate to concentrations up to 0.21 nmol/g of tissue (Passamonti et al., 2005; Talavera et al., 2005). It is posited that anthocyanins enter the brain in a similar fashion to their absorption in the stomach and GI tract. There is a bilirubin-binding motif in bilitranslocase in the CNS, which is likely the transporter involved in transporting anthocyanins into the brain (Battiston et al., 1999). This interaction occurs between the anthocyanin and the bilitranslocase due to hydrogen bonds and not charge, which provides further selectivity for this transport mechanism (Karawajczyk et al., 2007). There is also evidence that suggests P-glycoprotein transporters play a role in flavonoid transport into the brain (Youdim et al., 2004). Furthermore, it has been documented that the ability of specific anthocyanins to be transported across the basolateral membrane is inversely correlated with the number of hydroxyl groups possessed by the compound, and the presence of lipophilic moieties such as methoxy groups on the B ring of the anthocyanin increases transport across the basolateral membrane (Yi et al., 2006). This indicates that the relative lipophilicity of the different anthocyanins gives rise to further selectivity for their ability to be transported across biological membranes, including the BBB, such that some anthocyanins are likely to be more readily transported than others (Yi et al., 2006).

Although they do penetrate the BBB, the concentrations of parent anthocyanins detected in brain tissue after oral administration fall well below the concentrations which display significant antioxidant and neuroprotective effects *in vitro*. This finding suggests that metabolites of anthocyanins may also play significant neuroprotective roles *in vivo*. Moreover, the ability of chronic anthocyanin administration to induce accumulation of anthocyanins in brain has not been studied. Ultimately, there is very little known about anthocyanin transport, specifically for BBB permeation, or the possible accumulation of anthocyanins or their metabolites in the brain after chronic oral dosing. Both of these areas warrant further investigation; however, due to their intrinsic neuroprotective properties, anthocyanins remain an appealing option for the treatment of neuro-degenerative disease, nonetheless.

10.3 ANTHOCYANINS AS NOVEL NEUROPROTECTIVE AGENTS

10.3.1 INTRINSIC ANTIOXIDANT ACTIVITY

Anthocyanins have a polyphenolic cationic structure that activates an array of cel-lular responses (Figure 10.2). Anthocyanins have a robust antioxidant capacity, which appears to be highly effective in several models of neurodegenerative disease (Paik et al., 2012; Shih et al., 2011; Stintzing et al., 2002; Traustadottir et al., 2009). They have a high oxygen radical absorption capacity (ORAC) value, which is par-tially responsible for this neuroprotective effect (Zafra-Stone et al., 2007; Zhu et al., 2010). More specifically, anthocyanins act as antioxidants because of their ability to directly scavenge free radicals and to prevent ROS formation in affected cells. For example, anthocyanins decrease the formation of ROS in *in vitro* models of Aβ peptide-induced toxicity, as well as in H_2O_2 injury (Hwang et al., 2012; Shih et al., 2011). Additionally, using the highly accurate method of electron spin resonance (ESR) spectroscopy, it has been shown that anthocyanins have a high affinity to scavenge 1,1-diphenyl-2-picrylhydrazyl (DPPH), alkyl, and hydroxyl free radicals, and that this increases dose dependently (Hwang et al., 2012).

Anthocyanins activate the antioxidant response element (ARE), which controls expression of a large array of endogenous antioxidant and phase II detoxification genes (Shih et al., 2007). This effect appears to be via activation of the Nrf2 tran-scription factor (Hwang et al., 2011). In addition, anthocyanins have been shown to enhance the activities of free radical scavenging enzymes, including superoxide

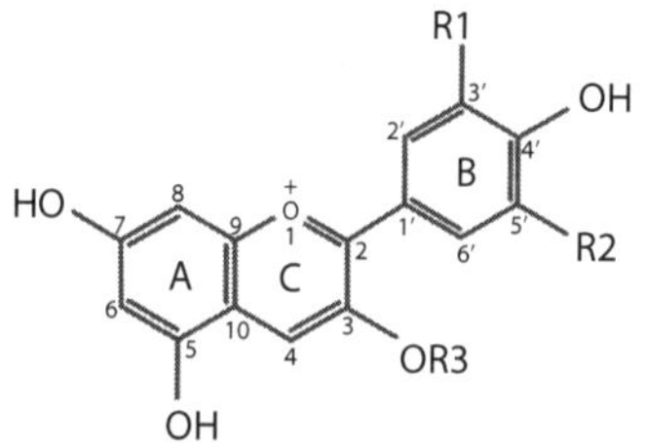

	R_1	R_2
Cyanidin	OH	H
Delphinidin	OH	OH
Malvidin	OCH_3	OCH_3
Pelargonidin	H	H
Petunidin	OCH_3	OH
Peonidin	OCH_3	H
Rosinidin	OCH_3	H

R_3 = Sugar moiety

FIGURE 10.2 Structures of common anthocyanins.

dismutase, catalase, and glutathione peroxidase (Hwang et al., 2012; Kelsey et al., 2011; Lu et al., 2010). Finally, levels of the endogenous antioxidant glutathione have been shown to increase with anthocyanin treatment, suggesting that anthocyanins exert their protective effects in part, via modulation of endogenous antioxidant defenses, and not exclusively due to their intrinsic antioxidant activity (Toufektsian et al., 2008).

10.3.2   EFFECTS ON Ca²⁺ HOMEOSTASIS

Calcium homeostasis is an important factor in suppressing excitotoxicity and preventing neuronal apoptosis in relevant neurodegenerative diseases. Anthocyanins have been shown to effectively prevent disturbances in intracellular Ca^{2+} levels during neurodegeneration and during normal aging (Joseph et al., 2009). Aβ peptide treatment of many cell types increases Ca^{2+} levels and subsequently causes apoptosis. In one study using Neuro-2a neuroblastoma cells, treatment with malvidin-3-O-glucoside significantly decreased intracellular Ca^{2+} concentrations compared to Aβ peptide alone (Shih et al., 2011). Moreover, delphinidin, which is an anthocyanidin base, returned Ca^{2+} levels to normal physiological levels in a model of actinomycin D-induced toxicity in PC12 cells (Martin et al., 2003). Additionally, purple sweet potato anthocyanins have been shown to attenuate membrane potential loss and subsequent apoptosis in PC12 cells induced by Aβ peptide (Ye et al., 2010). Though there has been little investigation into the mechanism underlying this effect, these studies suggest a significant role for anthocyanin-mediated Ca^{2+} regulation in relevant neurodegenerative diseases.

10.3.3   ANTI-INFLAMMATORY PROPERTIES

Induction of inflammatory gene expression and subsequent production of pro-inflammatory cytokines is a common feature in neurodegeneration. Methods that target these inflammatory processes may be beneficial in limiting neuronal apoptosis associated with disease. Anthocyanins exhibit significant anti-inflammatory properties, as they have been shown to inhibit various inflammatory biomarkers, including interleukin-8 (IL-8) (Zafra-Stone et al., 2007). In addition to decreased IL-8 production, pomegranate anthocyanins suppress activation of nuclear transcription factor kappaB (NFκB), which is responsible for the expression of several pro-inflammatory genes. They were also shown to inhibit mitogen-activated protein kinases (MAPKs), c-Jun N-terminal kinase (JNK) and extracellular signal-regulated kinase (ERK), which are responsible for the expression of a number of pro-inflammatory cytokines (Rasheed et al., 2009). Furthermore, cherry and blackberry anthocyanins act as potent cyclooxygenase-2 (COX-2) inhibitors; COX-2 is an important pro-inflammatory enzyme involved in the synthesis of prostacyclin (Cuevas-Rodriguez et al., 2010; Saric et al., 2009; Zdarilova et al., 2010). Anthocyanins have been shown to inhibit up to 95% of COX activity at high concentrations (250 μg/mL) (Mulabagal et al., 2009). These cumulative data suggest that anthocyanins may be useful in attenuating inflammatory processes that are associated with neurodegenerative disease.

10.3.4 ANTIAPOPTOTIC PROPERTIES

Anthocyanins modulate specific antiapoptotic pathways to increase survival in neuronal cells. They have been demonstrated to regulate both caspase-dependent and caspase-independent cell death pathways (Reddivari et al., 2007). It is mainly by acting on a variety of signaling proteins that they promote cell survival. Anthocyanins from purple sweet potato were shown to suppress JNK activity, release of cytochrome c from mitochondria, and cell apoptosis in D-galactose-treated mice. Additionally, it was demonstrated that this protective effect was dependent on phosphatidylinositol 3-kinase (PI3K) an upstream activator of pro-survival Akt (Lu et al., 2010; Ye et al., 2010). Anthocyanins also block the p53 and JNK signaling pathways in a model of cerebral artery ischemia in rats (Shin et al., 2006). Anthocyanins also prevent the release of apoptosis-inducing factor (AIF) from mitochondria, which may be a principle mechanism by which they inhibit caspase-independent cell death (Min et al., 2011). Anthocyanins have also been shown to increase the expression of pro-survival proteins such as Bcl-2 and to decrease levels of the pro-apoptotic protein Bax, effects which play a significant role in the prevention of 6-hydroxydopamine (6-OHDA)-induced toxicity, a model of Parkinson's disease (Kim et al., 2010).

10.4 EVIDENCE FOR THE NEUROPROTECTIVE EFFECTS OF ANTHOCYANINS

The recent application of anthocyanins as neuroprotective agents has thus far yielded promising results both *in vitro* and *in vivo*. The efficacy of anthocyanins against various types of neurotoxicity and degeneration stems primarily from their anti-oxidant, anti-inflammatory, and antiapoptotic activities, discussed above. Given the central roles of oxidative damage, inflammation, and apoptotic cell death in several neurodegenerative disorders, it is perhaps not surprising that these nutraceuticals have demonstrated their neuroprotective potential within a diverse array of model systems. The following section highlights the beneficial effects of anthocyanins in diverse cell and animal models of Parkinson's disease, Alzheimer's disease, ischemia, and several sources of oxidative stress.

10.4.1 ANTHOCYANINS AND PARKINSON'S DISEASE

Parkinson's disease is caused by the progressive loss of dopaminergic neurons in the substantia nigra due to a culmination of oxidative damage, inflammation, and excitotoxicity. The resulting deficiency in dopamine release in the striatum causes the characteristic impairment of motor function observed in this disease. Though the use of anthocyanins as a potential treatment for Parkinson's disease in patients is presently untested, a number of studies have been conducted *in vitro* and in animal models of Parkinson's disease which show promise for this approach.

Dopamine-induced oxidative damage plays a major role in the pathology of Parkinson's disease, and is often used as an *in vitro* model for the study of this disease. A study in COS-7 fibroblast-like cells showed that dopamine treatment disrupts Ca^{2+} homeostasis and induces cell death. Treatment with anthocyanin-enriched extracts from boysenberry and blackcurrants significantly protected COS-7 cells from dopamine toxicity (Ghosh et al., 2007). Additional evidence for protection against a similar oxidative insult, 6-OHDA, as well as 1-methyl-4-phenylpyridinium (MPP+), both well-known models of toxin-induced Parkinson's disease, was obtained using the SH-SY5Y dopaminergic cell line and primary dopaminergic neurons isolated from embryonic rats (Kim et al., 2010). Utilizing an anthocyanin-enriched extract from mulberry significantly increased cell viability in both SH-SY5Y cells and primary dopaminergic neurons, while reducing ROS generation and extracellular NO concentrations (Kim et al., 2010). Moreover, anthocyanin treatment reduced expression of pro-apoptotic factors such as Bax and increased expression of the pro-survival Bcl-2 protein ultimately preventing caspase-3 activation and apoptosis in SH-SY5Y cells (Kim et al., 2010). These findings strongly implicate the antioxidant and anti-apoptotic properties of anthocyanins in their neuroprotective effects observed in these *in vitro* models of Parkinson's disease.

The therapeutic potential of anthocyanins in Parkinson's disease was also explored by Kim et al. using the MPTP neurotoxin mouse model of this disease (2010). Oral administration of anthocyanin extract from mulberry significantly reduced the loss of dopaminergic neurons in the nigrostriatal pathway of MPTP-treated mice, while also reducing expression of pro-apoptotic Bax (Kim et al., 2010). Similarly, 6-OHDA unilaterally lesioned rats, a model of hemi-parkinsonism, also benefitted from receiving oral treatment with the anthocyanidin pelargonidin, showing significantly enhanced survival of neurons within the lesioned substantia nigra and lower levels of lipid peroxidation in comparison to lesioned rats that did not receive pelargonidin (Roghani et al., 2010). Neuroprotection in both models of Parkinson's disease can likely be attributed to the antioxidant properties of anthocyanins that attenuate oxidative damage and suppress the expression of pro-apoptotic factors.

10.4.2 ANTHOCYANINS AND ALZHEIMER'S DISEASE, AGING, AND MEMORY

Loss of cognitive function and memory is a problem affecting vast portions of the elderly and aging population. In many instances, these deficits may be the result of Alzheimer's disease, currently the most common cause of dementia in aging populations (Sompol et al., 2008). As aging is commonly regarded as the primary risk factor for Alzheimer's disease, studies to identify natural agents that preserve memory and cognition over time are extensive and involve a series of diverse solutions to this problem. Among these, anthocyanins have garnered attention as being highly effective at slowing or halting age-associated cognitive decline, in addition to showing great potential as a viable therapeutic option for Alzheimer's disease.

In vitro studies using anthocyanins to counteract the neurotoxicity induced by Aβ peptides are numerous. For instance, treatment of PC12 cells with Aβ showed a marked increase in apoptosis accompanied by Ca^{2+} influx, increases in intracellular ROS, loss of mitochondrial membrane potential, and subsequent increases in caspase-3 activity. Each of these toxic effects was significantly attenuated by treatment with an anthocyanin-rich extract from purple sweet potatoes (Ye et al., 2010). A similar study on Aβ toxicity in this cell line demonstrated that treatment with delphinidin also maintained intracellular levels of Ca^{2+} at basal levels, in addition to preventing the accumulation of hyperphosphorylated tau proteins which typically form neurofibrillary tangles that contribute to cell death (Kim et al., 2010). In the murine neuroblastoma Neuro-2A cell line, treatment with the anthocyanins malvidin and oenin prevented Aβ neurotoxicity by mitigating oxidative damage by ROS, maintaining Ca^{2+} homeostasis, upregulating numerous pro-survival genes, and lowering expression of β- and γ-secretase (Shih et al., 2011). Interestingly, Aβ neurotoxicity was attenuated in SH-SY5Y cells treated with cyanidin-O-3-glucoside by preventing aggregation and binding of Aβ oligomers to cell membranes, results that suggest an alternative neuroprotective mechanism for anthocyanins involving blockage of protein aggregate formation (Tarozzi et al., 2010).

Even within this small sampling of studies, it is clear that anthocyanins possess a multiplicity of properties that allow them to efficiently mitigate the numerous damaging effects of Aβ exposure. These effects hinge upon not only their powerful antioxidant properties, but also their ability to influence levels of intracellular Ca^{2+}, and thereby preserve proper mitochondrial function and prevent release of apoptogenic factors. Further, induction of pro-survival proteins and suppression of pro-apoptotic protein expression appears to play a significant role in the neuroprotection afforded by these compounds, as does their capacity to inhibit protein aggregation.

Similar positive findings have been observed in *in vivo* studies using the SAMP8 senescence accelerated mouse model of Alzheimer's disease, which shows the characteristic deposition of Aβ proteins associated with Alzheimer's plaque pathology (Shih et al., 2010). SAMP8 mice showed significant decreases in levels of antioxidant proteins in both the brain and liver which were restored by diet supplementation with mulberry extract (Shih et al., 2010). Furthermore, treatment with the extract significantly reduced the expression of Aβ and stress-activated protein kinases, including *c*-Jun *N*-terminal kinase (JNK) and p38 MAP kinase (Shih et al., 2010).

Other *in vivo* work has concentrated heavily on the effects of anthocyanin supplementation on memory and cognition in aged mice and rats, with less emphasis on Alzheimer's disease *per se*. Improvements in memory and cognition have been attributed to the ability of anthocyanins to cross the BBB and enter the cerebral cortex (Andres-Lacueva et al., 2005; Williams et al., 2008). A direct correlation between increasing concentrations of anthocyanins in the cortex and enhanced memory has been shown in blueberry fed rats (Andres-Lacueva et al., 2005). Since this initial discovery, results have shown consistently that the introduction of anthocyanin-rich foods or extracts into the diets of aged animals significantly improves learning and memory. For instance, in a D-galactose-induced mouse model of aging, it was found that mice receiving anthocyanin-rich purple sweet potato extract demonstrated enhanced memory retention as compared to their counterparts who did not receive the extract, resulting in maintenance of normal field behavior (Lu et al., 2010; Shan et al., 2009). This improvement in cognitive and behavioral performance was accompanied by preservation of antioxidant enzyme activity, decreases in indices of oxidative stress and inflammation, including lipid peroxidation and iNOS activity, respectively, and significantly lower levels of cytosolic apoptogenic factors and active caspase-3 (Lu et al., 2010; Shan et al., 2009). Another study using the same purple sweet potato extract demonstrated that treatment with anthocyanins promoted increased levels of the estrogen receptor α protein in the hippocampus of rats treated with the oxidative stressor, domoic acid, and that protection by anthocyanins was dependent upon this increase (Lu et al., 2012). Corresponding decreases in the levels of endoplasmic reticulum-related stress, oxidative stress, and concentrations of apoptogenic proteins were also observed, resulting in the reversal of cognitive deficits caused by domoic acid exposure (Lu et al., 2012). In a natural model of aging, 18-month-old rats fed a diet supplemented with blueberries showed rapid improvement in working-spatial memory over a period of weeks. This functional enhancement corresponded to increased levels of cAMP-response element binding protein (CREB) phosphorylation and brain-derived neurotrophic factor (BDNF) in the hippocampus, both of which are involved in cellular signaling and transcription during long-term memory formation (Williams et al., 2008). Increases in these proteins were attributed to enhanced phosphorylation and activation of extracellular signal-regulated kinase 1/2 (ERK1/2), indicating that blueberry anthocyanins may play a role in direct modulation of this signaling pathway (Williams et al., 2008). These findings are in agreement with several studies showing that flavonoids, such as anthocyanins, display diverse effects in the modulation of several MAPK cascades. These include those pathways that activate CREB, BDNF, and their downstream effectors, which are responsible for the plasticity of synaptic morphology and strength and are shown to be upregulated in response to blueberry anthocyanin supplementation (Spencer, 2008, 2010; Williams et al., 2008). This evidence is suggestive of a direct role for anthocyanins in mediating the formation and retention of memories in the long term.

Recent studies have taken these findings one step further, introducing anthocyanin-rich foods and drinks into the diets of older adults at risk for dementia. One clinical trial using blueberries and another using Concord grape juice indicated that a diet supplemented with these anthocyanin-rich substances significantly improved mild memory impairment in these individuals (Krikorian et al., 2010a, 2010b). While such clinical findings are currently limited in scope and number, these studies allude to a potential role for anthocyanins as therapeutic agents in age-related impairment of memory and cognition.

As a whole, this evidence emphasizes the beneficial effects of the antioxidant and anti-inflammatory properties of anthocyanins on the aging brain, while highlighting a novel role for these compounds in cell signaling and memory formation and retention. Additionally, studies on aging have yielded insight into the unique ability of anthocyanins to cross the BBB, a factor which may prove limiting for the efficacy of other nutraceuticals and therapeutic drugs in the treatment of neurodegenerative disorders. One area of investigation which is notably lacking regarding the potential therapeutic effects of anthocyanins on cognitive decline is the influence these compounds might have on neurite outgrowth and synaptic plasticity.

10.4.3 ANTHOCYANINS AND ISCHEMIA

Ischemic injury, particularly stroke, is one of the leading causes of death worldwide (Chen et al., 2011). Because of its pervasiveness, the need to identify agents that can effectively mitigate or reverse the cognitive and motor deficiencies that occur due to ischemic brain damage is the topic of intense research. Anthocyanins have thus far proven to be effective in this capacity as they demonstrate the capacity to minimize damage associated with ischemia–reperfusion injury in multiple *in vitro* and *in vivo* animal models.

While *in vitro* studies examining anthocyanin effects on ischemic injury are few in number, so far they have produced exciting results with regard to the potential of anthocyanins as neuroprotective agents against this type of insult. For example, extracts from mulberry fruit, and cyanidin-3-*O*-β-D-glucopyranoside, one of the primary anthocyanin constituents of this extract, significantly reduce the effects of ischemic damage in PC12 cells. Specifically, these anthocyanins increased viability of cells exposed to oxygen glucose deprivation (OGD), an *in vitro* model of stroke, or hydrogen peroxide-induced oxidative stress (Kang et al., 2006). Similar results were found in a study by Bhuiyan et al. (2011) in which the primary rat cortical neurons exposed to OGD and "reperfusion" were protected by treatment with cyanidin-3-*O*-glucoside extracted from mulberries. In addition to reducing apoptosis, anthocyanin treatment reduced indices of oxidative stress and also maintained mitochondrial membrane potential. These studies highlight both the antioxidant activity of anthocyanins and their capacity to limit mitochondrial dysfunction and suppress activation of intrinsic apoptotic pathways.

In vivo experiments testing anthocyanins and anthocyanin-containing extracts in models of ischemic damage are more common and have established the neuroprotective actions of these compounds both during occlusion and after reperfusion-induced injury. This activity involves the mitigation of oxidative stress and excitotoxicity. For instance, rats treated with anthocyanins or polyphenolic compounds from red wine, which include high concentrations of anthocyanins, prior to middle cerebral artery occlusion and reperfusion, displayed significantly diminished infarct volume in comparison to rats not receiving anthocyanins (Ritz et al., 2008a, 2008b; Shin et al., 2006). Additionally, anthocyanin treatment significantly reduced DNA fragmentation and prevented activation of the apoptogenic factors JNK and p53, suggesting that anthocyanins or their metabolites actively protect neurons against injury both during the ischemic episode and immediately after reperfusion (Shin et al., 2006). Studies using polyphenolic compounds from red wine have shown that neuroprotection during occlusion may be due to a reduction in the levels of excitatory amino acids such as glutamate. This, in turn, prevents ischemia-induced excitotoxic events (Ritz et al., 2008a, 2008b). Treatment with red wine polyphenols also increased blood flow to the brain upon reperfusion by causing arterial vasodilatation, thus reducing ischemic injury (Ritz et al., 2008a, 2008b). *In vivo* studies of anthocyanin containing extracts also demonstrate that treatment with these extracts promotes increased expression of free radical scavengers such as ascorbic and uric acids during and after occlusion, mitigating the effects of oxidative damage (Nade et al., 2010; Ritz et al., 2008a,b). Anthocyanin extracts also display an ability to restore cognitive function over a longer term of treatment during the recovery phase from ischemic injury. Rats receiving *Hibiscus rosa sinensis* root extract, for instance, demonstrated a preserved capacity for learning and memory in the days following cerebral artery occlusion and reperfusion (Nade et al., 2010).

Collectively, the results of these studies demonstrate that consumption of anthocyanins prior to and after ischemic events can ultimately preserve and restore healthy brain function. This suggests that a diet rich in anthocyanin-containing fruits and vegetables may be a beneficial strategy for patients at high risk for stroke.

10.4.4 ANTHOCYANINS, OXIDATIVE STRESS, AND OTHER MODELS OF NEURONAL INJURY

Since oxidative stress is a prominent feature of most neurodegenerative diseases, research into nutraceutical antioxidants as potential treatment agents has intensified in recent years, revealing anthocyanins to be remarkably potent mitigators of damage caused by ROS. While not every study conducted has focused specifically on neurodegenerative disease, many pose exciting implications for the future treatment of these disorders. Anthocyanins are shown to be effective in alleviating oxidative damage induced by a variety of insults, which include models of mitochondrial oxidative stress and mitochondrial dysfunction, excitotoxicity, neuroinflammation, and apoptotic signaling cascades. Moreover, the effects of anthocyanins in models of oxidative stress and neuronal injury not typically associated with neurodegeneration have also been examined.

In vitro research testing the neuroprotective effects of anthocyanins on mitochondrial dysfunction has yielded encouraging results. A series of experiments conducted by Kelsey et al. (2011) utilized HA14-1, an inhibitor of the pro-survival Bcl-2 protein which is found in mitochondrial membranes, to induce mitochondrial oxidative stress in primary rat cerebellar granule neurons. Treatment with HA14-1 depleted mitochondrial pools of glutathione (GSH), a powerful endogenous antioxidant, caused extensive oxidation of the mitochondrial lipid, cardiolipin, and promoted mitochondrial fragmentation; however, these effects were thoroughly negated by treatment with either cyanidin-3-*O*-glucoside (kuromanin) or pelargonidin-3-*O*-glucoside (callistephin) (2011). Both anthocyanins displayed a significant capacity to maintain GSH concentrations within the mitochondria, and prevented oxidation of mitochondrial lipids and mitochondrial fragmentation, significantly improving mitochondrial health and cell viability (Kelsey et al., 2011). Additionally, in a cell-free assay, treatment with kuromanin and callistephin significantly increased glutathione peroxidase activity, which uses GSH to detoxify harmful organic hydroperoxides (Kelsey et al., 2011). In accord with these findings, an earlier study demonstrated that the major anthocyanin constituents of plums are highly efficient superoxide radical scavengers, showing activity even above and beyond other powerful antioxidants such as quercetin (Chun et al., 2003). Due to the dependency on oxygen for cellular respiration, superoxide is one of the major free radicals generated within the mitochondria, making these essential organelles highly susceptible to oxidative damage by this particular species (Ott et al., 2007). With this in mind, anthocyanins could play a major role in promoting mitochondrial health and function through a multitude of direct and indirect pathways like those discussed here.

Other *in vitro* studies have put a strong emphasis on the ability of anthocyanins to limit oxidative damage caused by hydrogen peroxide (H_2O_2) in different neuronal cell lines. H_2O_2 is a product of many cellular reactions and possesses the potential to do significant oxidative damage if left unchecked, making it a physiologically relevant insult for *in vitro* studies. These studies indicate that treatment with anthocyanin fractions from açai berries, strawberries, and boysenberries, pure isolates of the conjugate, cyanidin 3-*O*-glucopyranoside, and protocatechuic acid, a major metabolite of anthocyanins, provides a host of protective effects against H_2O_2-induced toxicity in SH-SY5Y and PC12 cells (Ghosh et al., 2006; Heo and Lee, 2005; Spada et al., 2009; Tarozzi et al., 2007). These include improvements in mitochondrial function, decreases in DNA fragmentation, reductions in multiple indices of oxidative stress, protein damage, and lipid peroxidation, increases in the activity of the antioxidant enzymes SOD and catalase, and a general increase in cell viability (Ghosh et al., 2006; Heo and Lee 2005; Spada et al., 2009; Tarozzi et al., 2007). Interestingly, comparisons between the neuroprotective potential of strawberry, banana, and orange extracts in Heo and Lee's research (2005) indicated that strawberry extracts protected PC12 cells to a higher degree than either banana or orange extracts, which do not contain anthocyanins, suggesting that fruits containing high concentrations of anthocyanins may be a more effective option for treatment of neurodegenerative disorders.

Various models of oxidative stress *in vivo* have given similarly positive results and confirm that many of the defensive mechanisms promoted by anthocyanins *in vitro* are also active *in vivo*. Injection of carbon tetrachloride (CCl_4) promotes the generation of free radical species within the brains of rats, resulting in significant oxidative damage to proteins and lipids, which are prevented by treatment with anthocyanin-rich grape juice (Dani et al., 2008). Additionally, in accordance with the *in vitro* studies discussed above, elevated levels of both catalase and SOD were found in the striatum and substantia nigra of anthocyanin-treated rats (Dani et al., 2008). In a model of psychological stress induced by removing the whiskers of rats, similar results were obtained (Rahman et al., 2008). Whisker removal significantly increased the production of ROS in the brains of rats, leading to significant neuronal injury. This consisted of a rise in lipid peroxidation products, oxidized proteins, and changes in levels of dopamine and its metabolites, resulting in motor dysfunction and psychological distress (Rahman et al., 2008). Introduction of anthocyanins into the diet of the rats prior to having their whiskers cut, however, maintained dopamine concentrations at their normal levels and ameliorated oxidative damage in several brain regions, showing the potential of these compounds to prevent injury resulting from psychological distress associated with neurodegenerative disease (Rahman et al., 2008). Anthocyanins have also been investigated for their ability to attenuate several forms of injury associated with ethanol neurotoxicity in the developing brain, including oxidative stress (Ke et al., 2011). In this model, treatment of infant mice with anthocyanins significantly reduced indices of oxidative stress and prevented activation of caspase-3 and apoptosis in the brain (Ke et al., 2011). Most notably, treatment with anthocyanins prevented activation of glycogen synthase kinase 3β (GSK3β) and p47phox, which plays a role in regulating NADPH oxidase, and in turn, stimulating superoxide production (Ke et al., 2011).

Although oxidative stress is a major focus of anthocyanin research, current studies have also demonstrated that these compounds are effective against a more diverse array of insults associated with neurodegeneration. Bilberry extract was recently used as a neuroprotective agent in a model of retinal ganglion degeneration caused by the injection of NMDA into the eyes of mice (Matsunaga et al., 2009). Increases in NMDA are associated with excitotoxicity and caused significant damage to retinal ganglion cells in injected mice, including a decrease in cell number and increases in DNA fragmentation. Both of these effects were significantly mitigated by simultaneous injection with bilberry extract (Matsunaga et al., 2009). This suggests that anthocyanins possess the ability to minimize excitotoxic damage in addition to their powerful antioxidant effects. Evidence also exists for the ability of anthocyanins to substantially attenuate neuroinflammatory damage. An *in vitro* study of neuroinflammation using BV-2 microglial cells demonstrated that anthocyanins significantly reduced the release of cytokines and limited the induction of pro-inflammatory proteins such as iNOS and cyclooxygenase 2 (COX2), as well as several others in activated microglia (Poulose et al., 2012). These findings are mirrored in an *in vivo* experiment by Wang et al. (2010). Lipopolysaccharide, a compound known to induce an inflammatory response in the brain, was shown to cause significant damage to the brains of mice by stimulating the release of numerous cytokines and upregulating pro-inflammatory proteins, including iNOS and COX2 (Wang et al., 2010). This damage was accompanied by a corresponding increase in apoptogenic factors, and subsequent decline in cognition and memory; however, administration of anthocyanins from purple sweet potatoes substantially diminished these effects and restored cognitive function to the mice (Wang et al., 2010).

Taken together, the results of a relatively large number of studies clearly establish the ability of anthocyanins to function not only as potent antioxidants, but also as moderators of both excitotoxicity and neuroinflammation. As all of these sources of neuronal injury are involved to varying degrees in most forms of neurodegeneration, the capability of anthocyanins to assuage damage generated by these three mechanisms makes them incredibly appealing as a novel treatment strategy for neurodegenerative disorders. Moreover, prevention of these causes of neuronal injury obviously precludes the onset of apoptotic cell death, which is a major underlying mechanism of neuronal cell death in neurodegenerative disease. Indeed, the research conducted so far to elucidate the neuroprotective effects of anthocyanins has demonstrated that these impressive pleiotropic properties make these compounds a viable option for future development as therapeutic treatments for a host of neurodegenerative disorders.

10.5 ANTHOCYANIN METABOLITES AND NEURODEGENERATION

There are several anthocyanin metabolites that show promise for prevention and treatment in neurodegenerative disease. Anthocyanin metabolites may be responsible for the high bioactivity of anthocyanin compounds, despite their apparently low bioavailability. The various identified metabolites are more stable than the parent anthocyanins in a variety of conditions, which make them more likely candidates as the bioactive moieties in target tissues (Keppler and Humpf, 2005). Indeed, recent research suggests that different anthocyanin conjugates are found in high concentrations in circulation, including sulfated, methylated, glucuronidated, and glycosylated conjugates (Gonthier et al., 2003; Manach et al., 2005). In one study, 10 separate metabolites from anthocyanins were detected in human blood after consumption of anthocyanin-rich chokeberry extract, including methylated, dimethylated, and glucuronidated forms of cyanidin-based anthocyanins, as well as one metabolite

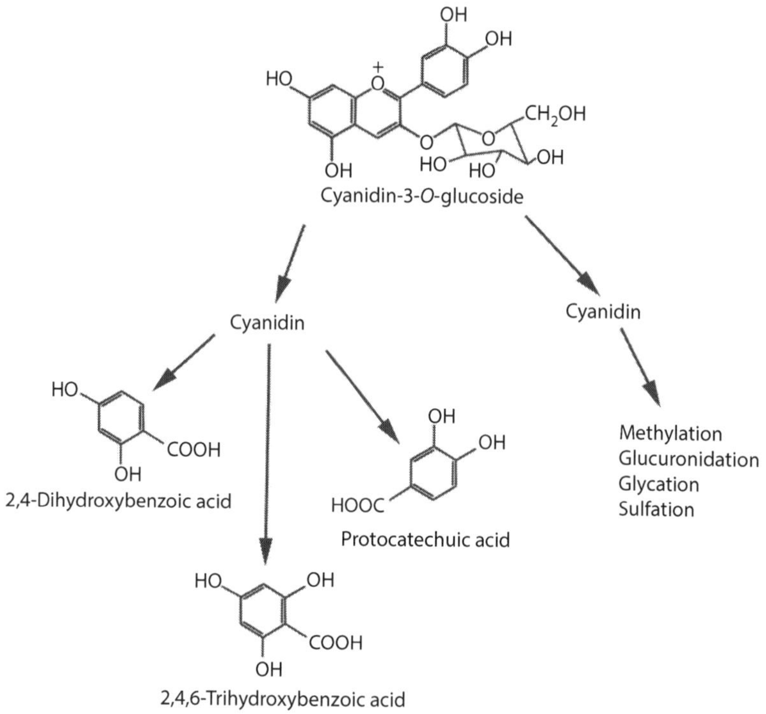

FIGURE 10.3   Common anthocyanin metabolites and conjugates from cyanidin-3-*O*-glucoside.

displaying oxidative modification (Kay et al., 2004). In addition, several studies have indicated that the most common forms of anthocyanin metabolites are phenolic acids and aldehydes that are produced largely by gut microflora (Fleschhut et al., 2006; Forester and Waterhouse, 2010; Woodward et al., 2009). Phenolic acids are the product of hydroxylation of the B ring of the parent anthocyanin, and thus vary in their structure and chemical properties; however, all anthocyanins produce a universal aldehyde metabolite, 2,4,6-trihydroxybenzaldehyde as another product of this reaction (Fleschhut et al., 2006; Forester and Waterhouse, 2010; Woodward et al., 2009) (Figure 10.3; Table 10.1). As physiologically active concentrations of parent anthocyanins are not readily detected in the CNS, it is important to investigate the efficacy of several of these prominent anthocyanin metabolites (Del Rio et al., 2010; Woodward et al., 2009). Protocatechuic acid (PCA) and the anthocyanidin base are common anthocyanin metabolites. During metabolism, the anthocyanidin in its aglycone form has a tendency to stay intact and retain its basic function. From there, conjugates of the anthocyanin are readily formed (Aura et al., 2005; Kay et al., 2004). PCA concentrations have been detected up to eight times higher than the parent anthocyanin in rodent plasma. PCA has also been shown to remain in target tissues for longer periods of time than the parent anthocyanins (Azzini et al., 2010; Tsuda et al., 1999). Significantly, both PCA and the anthocyanidin base provided significant protection and decreased ROS formation in SH-SY5Y neuronal cells treated

TABLE 10.1
Common Anthocyanin Metabolites

Anthocyanidin Base	Phenolic Acid	Aldehyde Metabolite
Pelargonidin	4-Hydroxybenzoic acid	2,4,6-Trihydroxybenzaldehyde
Cyanidin	Protocatechuic acid	
Delphinidin	Gallic acid	
Peonidin	Vanillic acid	
Petunidin	3-O-Methoxygallic acid	
Malvidin	Syringic acid	

with H_2O_2 (Tarozzi et al., 2007). Additionally, oral anthocyanidin treatment in rats decreased oxidative stress and toxicity associated with an array of cellular insults (reviewed in Domitrovic, 2011). Finally, PCA induced the Nrf2 transcription factor in murine macrophages, though whether similar effects would be observed in neuronal or glial cells has not been investigated (Vari et al., 2011).

Although studies have shown that anthocyanin metabolites accumulate within relevant tissues within the brain, their ability to interact with and cross the BBB remains poorly characterized. To date, there have been no studies showing directly the interaction between the various anthocyanin metabolites and the BBB; however, *in vitro* evidence has demonstrated that the anthocyanidin base and phenolic acid derivatives of various anthocyanins display differential abilities to interact with lipid-rich environments. The ability of anthocyanins, their anthocyanidin bases, and their phenolic acid derivatives to reduce lipid peroxidation was examined by Brown and Kelly (2007). In the case of cyanidin-3-*O*-glucoside and its metabolites, it was found that PCA displayed the greatest capacity to extend the time it took for peroxidation to occur, and significantly decreased the overall lipid peroxidation in the system. PCA was followed in potency by the anthocyanidin, cyanidin, and finally by cyanidin-3-*O*-glucoside itself, which showed the least efficiency in this regard (Brown and Kelly, 2007). These results suggest that the removal of hydrophilic moieties, particularly hydroxyl groups, increases the capacity of these compounds to interact with a lipid-rich environment to prevent damage (Brown and Kelly, 2007). Significantly, it was also observed in this study that delphinidin and its phenolic acid metabolite, gallic acid, which possess an additional hydroxyl group in comparison to cyanidin and PCA respectively, showed a diminished capacity to interact with the lipophilic environment to reduce lipid peroxidation (Brown and Kelly, 2007). From this, it can be assumed that some of the phenolic acid or aglycon derivatives of anthocyanins may interact with biological membranes such as the BBB more efficiently than parent anthocyanins. Further support of this concept can be obtained from the published partition coefficient (log P) values for parent anthocyanins and their metabolites. For example, the parent anthocyanin, cyanidin-3-*O*-glucoside (kuromanin), has a log P value of 0.51, while its metabolites PCA, cyanidin, and 2,4,6-trihydroxybenzaldehyde are more lipophilic, displaying higher log P values of 1.32, 2.41, and 1.18, respectively (ALOGPS).

There are relatively few studies into the neuroprotective properties of isolated anthocyanin metabolites and their ability to cross the BBB; however, with the pool of data suggesting their relevant concentrations and accumulation in tissue, it is important to further investigate their neuroprotective effects both *in vitro* and *in vivo* (Prior and Wu, 2006).

10.6 CONCLUSIONS AND FUTURE DIRECTIONS

In summary, anthocyanins are a unique group of nutraceutical polyphenols that display intrinsic antioxidant activity, as well as effects on calcium regulation, antiinflammatory properties, and antiapoptotic activity via modulation of cell signaling pathways and gene expression. Because of their pleiotropic activities, anthocyanins represent viable therapeutics for various neurodegenerative disorders, the etiology of which is multi-faceted and involves oxidative stress, excitotoxicity, neuroinflammation, and neuronal apoptosis. Several important areas require further study including the effects of anthocyanins on neurite outgrowth and synaptic plasticity and the role of anthocyanin metabolites in the neuroprotective mechanism of action of these compounds. Finally, well-controlled studies assessing the possible therapeutic effects of these nutraceuticals in patients suffering from Alzheimer's, Parkinson's,

ALS, or stroke may be warranted given their promising effects *in vitro* and in animal models of disease coupled with their low inherent toxicity.

ACKNOWLEDGMENTS

The authors are funded by a Merit Review grant from the Department of Veterans Affairs and NIH R01NS062766.

REFERENCES

Andres-Lacueva, C., Shukitt-Hale, B., Galli, R.L., Jauregui, O., Lamuela-Raventos, R.M., and Joseph, J.A. 2005. Anthocyanins in aged blueberry-fed rats are found centrally and may enhance memory. *Nutritional Neuroscience*, 8(2), 111–120.

Atamna, H., Mackey, J., and Dhahbi, J.M. 2012. Mitochondrial pharmacology: Electron transport chain bypass as strategies to treat mitochondrial dysfunction. *Biofactors*, 38(2), 158–166.

Aura, A.M., Martin-Lopez, P., O'Leary, K.A., Williamson, G., Oksman-Caldentey, K.M., Poutanen, K., and Santos-Buelga, C. 2004. *In vitro* metabolism of anthocyanins by human gut microflora. *European Journal of Nutrition*, 44(3), 133–142.

Azzini, E., Vitaglione, P., Intorre, F. et al. 2010. Bioavailability of strawberry antioxidants in human subjects. *British Journal of Nutrition*, 104(8), 1165–1173.

Barber, S.C., Mead, R.J., and Shaw, P.J. 2006. Oxidative stress in ALS: A mechanism of neuro-degeneration and a therapeutic target. *Biochimica et Biophysica Acta*, 1762, 1051–1067.

Battiston, L., Macagno, A., Passamonti, S., Micali, F., and Sottocasa, G.L. 1999. Specific sequence-directed anti-bilitranslocase antibodies as a tool to detect potentially bilirubin-binding proteins in different tissues of the rat. *FEBS Letters*, 453(3), 351–355.

Bhuiyan, M.I.H., Kim, H., Kim, S.Y., and Cho, K. 2011. The neuroprotective potential of cyanidin-3-glucoside fraction extracted from mulberry following oxygen-glucose depri-vation. *The Korean Journal of Physiology and Pharmacology*, 15, 353–361.

Block, M.L., Zecca, L., and Hong, J. 2007. Microglia-mediated neurotoxicity: Uncovering the molecular mechanisms. *Nature Reviews*, 8, 57–69.

Bonifati, V., Rizzu, P., van Baren, M.J. et al. 2003. Mutations in the DJ-1 gene associated with autosomal recessive early-onset Parkinsonism. *Science*, 299, 256–259.

Brown, J.E. and Kelly, M.F. 2007. Inhibition of lipid peroxidation by anthocyanins, anthocy-anidins, and their phenolic degradation products. *European Journal of Lipid Science and Technology*, 109, 66–71.

Chen, H., Yoshioka, H., Kim, G.S. et al. 2011. Oxidative stress in ischemic brain dam-age: Mechanisms of cell death and potential molecular targets for neuroprotection. *Anitoxidants and Redox Signaling*, 14(8), 1505–1517.

Chun, O.K., Kim, D., and Lee, C.Y. 2003. Superoxide radical scavenging activity of the major polyphenols in fresh plums. *Journal of Agricultural and Food Chemistry*, 51, 8067–8072.

Cuevas-Rodriguez, E.O., Dia, V.P., Yousef, G.G., Garcia-Saucedo, P.A., Lopez-Medina, J., Paredes-Lopez, O., Gonzalez de Mejia, E., and Lila, M.A. 2010. Inhibition of pro-inflam-matory responses and antioxidant capacity of Mexican blackberry. *Journal of Agricultural and Food Chemistry*, 58(17), 9542–9548.

Dani, C., Pasquali, M.A.B., Oliveira, M.R., Umezu, F.M., Salvador, M., Henriques, J.A.P., and Moreira, J.C.F. 2008. Protective effects of purple grade juice on carbon tetrachloride-induced oxidative stress in brains of adult Wistar rats. *Journal of Medicinal Food*, 11(1), 55–61.

Del Rio, D., Borges, G., and Crozier, A. 2010. Berry flavonoids and phenolics: Bioavailability and evidence of protective effects. *British Journal of Nutrition*, 104, S67–S90.

Di Filippo, M., Chisserini, D., Tozzi, A., Picconi, B., and Calabresi, P. 2010. Mitochondria and the link between neuroinflammation and neurodegeneration. *Journal of Alzheimer's Disease*, 20, S369–S379.

Domitrovic, R. 2011. The molecular basis for the pharmacological activity of anthocyans. *Curent Medicinal Chemistry*, 18(29), 4454–4469.

Dong, X., Wang, Y., and Qin, Z. 2009. Molecular mechanisms of excitotoxicity and their relevance to pathogenesis of neurodegenerative diseases. *Acta Pharmacologica Sinica*, 30(4), 379–387.

Eliasson, M.J.L., Huang, Z., Ferrante, R.J., Sasamata, M., Molliver, M.E., Snyder, S.H., and Moskowitz, M.A. 1999. Neuronal nitric oxide synthase activation and peroxynitrite formation in ischemic stroke linked to neural damage. *The Journal of Neuroscience*, 19(4), 5910–5918.

El Mohsen, M.A., Marks, J., Kuhnle, G., Moore, K., Debnam, E., Kaila Srai, S., Rice-Evans, C., and Spencer, J.P. 2006. Absorption, tissue distribution and excretion of pelargonidin and its metabolites following oral administration to rats. *British Journal of Nutrition*, 95(1), 51–58.

Fleschhut, J., Kratzer, F., Rechkemmer, G., and Kulling, S.E. 2006. Stability and biotransformation of various dietary anthocyanins *in vitro*. *European Journal of Nutrition*, 45(1), 7–18.

Forester, S.C. and Waterhouse, A.L. 2010. Gut metabolites of anthocyanins, gallic acid, 3-*O*-methylgallic acid, and 2,4,6-trihydroxybenzaldehyde, inhibit cell proliferation of Caco-2 cells. *Journal of Agricultural and Food Chemistry*, 58(9), 5320–5327.

Gan, L., Johnson, D.A., and Johnson, J.A. 2010. Keap1-Nrf2 activation in the presence and absence of DJ-1. *The European Journal of Neuroscience*, 31(6), 967–977.

Ghosh, D., McGhie, T.K., Fisher, D.R., and Joseph, J.A. 2007. Cytoprotective effects of anthocyanins and other phenolic fractions of Boysenberry and blackcurrant on dopamine and amyloid β-induced oxidative stress in transfected COS-7 cells. *Journal of the Science of Food and Agriculture*, 87, 2061–2067.

Ghosh, D., McGhie, T.K., Zhang, J., Adaim, A., and Skinner, M. 2006. Effects of anthocyanins and other phenolics of boysenberry and blackcurrant as inhibitors of oxidative stress and damage to cellular DNA in SH-SY5Y and HL-60 cells. *Journal of the Science of Food and Agriculture*, 86, 678–686.

Gonthier, M.P., Cheynier, V., Donovan, J.L., Manach, C., Morand, C., Mila, I., Lapierre, C., Remesy, C., and Scalbert, A. 2003. Microbial aromatic acid metabolites formed in the gut account for a major fraction of the polyphenols excreted in urine of rats fed red wine polyphenols. *Journal of Nutrition*, 133(2), 461–467.

Helton, T.D., Otsuka, T., Lee, M., Mu, Y., and Ehlers, M.D. 2008. Running and loss of excitatory synapses by the parkin ubiquitin ligase. *Proceedings of the National Academy of Sciences of the United States of America*, 105(49), 19492–19497.

Heneka, M.T. and O'Banion, M.K. 2007. Inflammatory processes in Alzheimer's disease. *Journal of Neuroimmunology*, 184, 69–91.

Henkel, J.S., Englehardt, J.I., Siklos, L., Simpson, E.P., Kim, S.H., Pan, T., Goodman, C., Siddique, T., Beers, D.R., and Appel, S.H. 2004. Presence of dendritic cells, MCP-1, and activated microglia/macrophages in amyotrophic lateral sclerosis spinal cord tissue. *Annals of Neurology*, 55(2), 221–235.

Heo, H.J. and Lee, C.Y. 2005. Strawberry and its anthocyanins reduce oxidative stress-induced apoptosis in PC12 cells. *Journal of Agricultural and Food Chemistry*, 53, 1984–1989.

Hirsch, E.C. and Hunot, S. 2009. Neuroinflammation in Parkinson's disease: A target for neuroprotection? *Lancet Neurology*, 8, 382–397.

Hunot, S., Boissiere, F., Faucheux, B., Brugg, B., Mouatt-Prigent, A., Agid, Y., and Hirsch, E.C. 1996. Nitric oxide synthase and neuronal vulnerability in Parkinson's disease. *Neuroscience*, 72(2), 355–363.

Hwang, J.W., Kim, E.K., Lee, S.J., Kim, Y.S., Moon, S.H., Jeon, B.T., and Kim, E.T. 2012. Antioxidant activity and protective effect of anthocyanin oligomers on H_2O_2-triggered G2/M arrest in retinal cells. *Journal of Agricultural and Food Chemistry*, *60*(17), 4282–4288.

Hwang, Y.P., Choi, J.H., Yun, H.J., et al. 2011. Anthocyanins from purple sweet potato attenuate dimethylnitrosamine-induced liver injury in rats by inducing Nrf2-mediated antioxidant enzymes and reducing COX-2 and iNOS expression. *Food and Chemical Toxicology*, *49*(1), 93–99.

Jackson-Lewis, V., Blesa, J., and Przedborski, S. 2012. Animal models of Parikinson' disease. *Parkinsonism and Related Disorders*, *18*(Suppl. 1), S183–S185.

Jellinger, K. 2009. Recent advances in our understanding of neurodegeneration. *Journal of Neural Transmission*, *116*, 1111–1162.

Joseph, J.A., Shukitt-Hale, B., and Willis, L.M. 2009. Grape juice, berries, and walnuts affect brain aging and behavior. *Journal of Nutrition*, *139*(9), 1813S–1817S.

Jung, K., Chu, K., Lee, S., Park, H., Kim, J., Kang, K., Kim, M., Lee, S.K., and Roh, J. 2009. Augmentation of nitrite therapy in cerebral ischemia by NMDA receptor inhibition. *Biochemical and Biophysical Research Communications*, *378*(3), 507–512.

Kajta, M. 2004. Apoptosis in the central nervous system: Mechanisms and protective strategies. *Polish Journal of Pharmacology*, *56*, 689–700.

Kang, T.H., Hur, J.Y., Kim, H.B., Ryu, J.H., and Kim, S.Y. 2006. Neuroprotective effects of the cyanidin-3-O-β-D-glucopyranoside isolated from mulberry fruit against cerebral ischemia. *Neuroscience Letters*, *391*, 168–172.

Karawajczyk, A., Drgan, V., Medic, N., Oboh, G., Passamonti, S., and Novic, M. 2007. Properties of flavonoids influencing the binding to bilitranslocase investigated by neural network modelling. *Biochemical Pharmacology*, *73*(2), 308–320.

Kay, C.D., Mazza, G., Holub, B.J., and Wang, J. 2004. Anthocyanin metabolites in human urine and serum. *British Journal of Nutrition*, *91*(6), 933–942.

Ke, Z., Liu, Y., Wang, X., et al. 2011. Cyanidin-3-glucoside ameliorates ethanol neurotoxicity in the developing brain. *Journal of Neuroscience Research*, *89*(10), 1676–1684.

Kelsey, N., Hulick, W., Winter, A., Ross, E., and Linseman, D. 2011. Neuroprotective effects of anthocyanins on apoptosis induced by mitochondrial oxidative stress. *Nutritional Neuroscience*, *14*(6), 249–259.

Kelsey, N.A., Wilkins, H.M., and Linseman, D.A. 2010. Nutraceutical antioxidants as novel neuroprotective agents. *Molecules*, *15*(11), 7792–7814.

Keppler, K., and Humpf, H. 2005. Metabolism of anthocyanins and their phenolic degradation products by the intestinal microflora. *Bioorganic & Medicinal Chemistry*, *13*(17), 5195–5205.

Kim, H.G., Ju, M.S., Shim, J.S., Kim, M.C., Lee, S., Huh, Y., Kim, S.Y., and Oh, M.S. 2010. Mulberry fruit protects dopaminergic neurons in toxin-induced Parkinson's disease models. *British Journal of Nutrition*, *104*, 8–16.

Kong, J. and Xu, Z. 1998. Massive mitochondrial degeneration in motor neurons triggers the onset of amyotrophic lateral sclerosis in mice expressing a mutant SOD1. *The Journal of Neuroscience*, *18*(9), 3241–3250.

Krantic, S., Mechawar, N., Reix, S., and Quirion, R. 2005. Molecular basis of programmed cell death in neurodegeneration. *TRENDS in Neurosciences*, *28*(12), 670–676.

Krikorian, R., Nash, T.A., Shidler, M.D., Shukitt-Hale, B., and Joseph, J.A. 2010a. Concord grape juice supplementation improves memory function in older adults with mild cognitive impairment. *British Journal of Nutrition*, *103*, 730–734.

Krikorian, R., Shidler, M.D., Nash, T., Kalt, W., Vinqvist-Tymchuk, M.R., Shukitt-Hale, B., and Joseph, J.A. 2010b. Blueberry supplementation improves memory in older adults. *Journal of Agricultural and Food Chemistry*, *58*, 3996–4000.

Langenhove, T.V., Zee, J.V., and van Broeckhoven, C. 2012. The molecular basis of the fron-totemporal lobar degeneration-amyotrophic lateral sclerosis spectrum. *Ann Med*, *44*(8), 817–828.

Lin, M.T. and Beal. M.F. 2006. Mitochondrial dysfunction and oxidative stress in neurodegen-erative diseases. *Nature*, *443*, 787–795.

Liu, J., Lillo, C., Jonsson, P.A., Vande Velde, C. et al. 2004. Toxicity of familial ALS-linked SOD1 mutants from selective recruitment to spinal mitochondria. *Neuron*, *43*, 5–17.

Lu, J., Wu, D., Zheng, Y., Hu, B., and Zhang, Z. 2010. Purple sweet potato color alleviates D-galactose-induced brain aging in old mice by promoting survival of neurons via PI3K pathway and inhibiting cytochrome c-mediated apoptosis. *Brain Pathology*, *20*, 598–612.

Lu, J., Wu, D., Zheng, Y., Hu, B., Cheng, W., and Zhang, Z. 2012. Purple sweet potato color attenuates domoic acid-induced cognitive deficits by promoting estrogen receptor-α-mediated mitochondrial biogenesis signaling in mice. *Free Radical Biology and Medicine*, *52*, 646–659.

Luetjens, C.M., Bui, N.T., Sengpiel, B., Munstermann, G., Poppe, M., Krohn, A.J., Bauerbach, E., Krieglstein, J., and Prehn, J.H. M. 2000. Delayed mitochondrial dysfunction in exci-totxic neuron death: Cytochrome c release and a secondary increase in superoxide pro-duction. *The Journal of Neuroscience*, *20*(15), 5715–5723.

Maestro, A., Terdoslavich, M., Vanzo, A., Kuku, A., Tramer, F., Nicolin, V., Micali, F., Decorti, G., and Passamonti, S. 2010. Expression of bilitranslocase in the vascular endothelium and its function as a flavonoid transporter. *Cardiovascular Research*, *85*(1), 175–183.

Manach, C., Williamson, G., Morand, C., Scalbert, A., and Remesy, C. 2005. Bioavailability and bioefficacy of polyphenols in humans. I. Review of 97 bioavailability studies. *The American Journal of Clinical Nutrition*, *81*(1 Suppl.), 230S–242S.

Martin, S., Giannone, G., Andriantsitohaina, R., and Martinez, M.C. 2003. Delphinidin, an active compound of red wine, inhibits endothelial cell apoptosis via nitric oxide path-way and regulation of calcium homeostasis. *British Journal of Pharmacology*, *139*(6), 1095–1102.

Matsunaga, N., Imai, S., Inokuchi, Y., Shimazawa, M., Yokota, S., Araki, Y., and Hara, H. 2009. Bilberry and its main constituents have neuroprotective effects against retinal neu-ronal damage *in vitro* and in vivo. *Molecular Nutrition & Food Research*, *53*, 869–877.

Mattiazzi, M., D'Aurelio, M., Gajewski et al. 2002. Human SOD1 causes dysfunction of oxi-dative phosphorylation in mitochondria of transgenic mice. *The Journal of Biological Chemistry*, *277*(23), 29626–29633.

McGhie, T.K. and Walton, M.C. 2007. The bioavailability and absorption of anthocyanins: Towards a better understanding. *Molecular Nutrition and Food Research*, *51*(6), 702–713.

Melo, A., Montiero, L., Lima, R.M.F., de Oliviera, D.M., de Cerqueira, M.D., and El-Bachá, R.S. 2011. Oxidative stress in neurodegenerative diseases: Mechanisms and therapeu-tic perspectives. *Oxidative Medicine and Cellular Longevity*, *2011*. doi:10.1155/2011/467180.

Miguel-Hidalgo, J.J., Alvarez, X.A., Cacbelos, R., and Quack, G. 2002. Neuroprotection by memantine against neurodegeneration induced by β-amyloid(1–40). *Brain Research*, *958*(1), 210–221.

Milbury, P.E. and Kalt, W. 2010. Xenobiotic metabolism and berry flavonoid transport across the blood–brain barrier. *Journal of Agricultural and Food Chemistry*, *58*(7), 3950–3956.

Miller, R.G., Mitchell, J.D., Lyon, M., and Moore, D.H. 2007. Riluzole for amyotrophic lat-eral sclerosis (ALS)/motor neuron disease (MND). *Cochrane Database of Systematic Reviews*, Issue 1. Art. No.: CD001447. DOI: 10.1002/14651858.CD001447.pub2.

Min, J., Yu, S.W., Baek, S.H., Nair, K.M., Bae, O.N., Bhatt, A., Kassab, M., Nair, M.G., and Majid, A. 2011. Neuroprotective effect of cyanidin-3-O-glucoside anthocyanin in mice with focal cerebral ischemia. *Neuroscience Letters*, *500*(3), 157–161.

Mulabagal, V., Lang, G.A., DeWitt, D.L., Dalavoy, S.S., and Nair, M.G. 2009. Anthocyanin content, lipid peroxidation and cyclooxygenase enzyme inhibitory activities of sweet and sour cherries. *Journal of Agricultural and Food Chemistry*, 57(4), 1239–1246.

Murata, T., Ohtsuka, C., and Terayama, Y. 2008. Increased mitochondrial oxidative damage and oxidative DNA damage contributes to the neurodegenerative process in sporadic amyotrophic lateral sclerosis. *Free Radical Research*, 42(3), 221–225.

Nade, V.S., Kawale, L.A., Dwivedi, S., and Yadav, A.V. 2010. Neuroprotective effect of *Hibiscus rosa sinensis* in an oxidative stress model of cerebral post-ischemic reperfusion injury in rats. *Pharmaceutical Biology*, 48(7), 822–827.

Niizuma, K., Endo, H., and Chan, P.H. 2009. Oxidative stress and mitochondrial dysfunction as determinants of ischemic neuronal death and survival. *Journal of Neurochemistry* 109(Suppl. 1): 133–138.

Nimmerjahn, A., Kirchhoff, F., and Helmchen, F. 2005. Resting microglial cells are highly dynamic surveillants of brain parenchyma in vivo. *Science*, 308, 1314–1318.

Nunomura, A., Perry, G., Aliev, G. et al. 2001. Oxidative damage is the earliest event in Alzheimer's disease. *Journal of Neuropathology and Experimental Neurology*, 60(8), 759–767.

Ott, M., Gogvadze, V., Orrenius, S., and Zhivotovsky, B. 2007. Mitochondria, oxidative stress and cell death. *Apoptosis*, 12, 913–922.

Paik, S.S., Jeong, E., Jung, S.W., Ha, T.J., Kang, 2012. Anthocyanins from the seed coat of black soybean reduce retinal degeneration induced by *N*-methyl-*N*-nitrosourea. *Experimental Eye Research*, 97(1), 55–62. doi:http://dx.doi.org/10.1016/j.exer.2012.02.010.

Parameshwaran, K., Dhanasekaran, M., and Suppiramaniam, V. 2008. Amyloid beta peptides and glutamatergic synaptic dysregulation. *Experimental Neurology*, 210(1), 7–13.

Passamonti, S., Vrhovsek, U., Vanzo, A., and Mattivi, F. 2003. The stomach as a site for anthocyanins absorption from food. *FEBS Letters*, 544(1–3), 210–213.

Passamonti, S., Vrhovsek, U., Vanzo, A., and Mattivi, F. 2005. Fast access of some grape pigments to the brain. *Journal of Agrucultural and Food Chemistry*, 53(18), 7029–7034.

Poulose, S.M., Fisher, D.R., Larson, J., Bielinski, D.F., Rimando, A.M., Carey, A.N., Schauss, A.G., and Shukitt-Hale, B. 2012. Anthocyanin-rich acai (*Euterpe oleracea* Mart.) fruit pulp fractions attenuate inflammatory stress signaling in mouse brain BV-2 microglial cells. *Journal of Agricultural and Food Chemistry*, 60(4), 1084–1093.

Pratico, D., Uryu, K., Sleight, S., Trojanowski, J., and Lee, V.M. Y. 2001. Increased lipid peroxidation precedes amyloid plaque formation in an animal model of Alzheimer's amyloidosis. *The Journal of Neuroscience*, 21(12), 4183–4187.

Prior, R.L. and Wu, X. 2006. Anthocyanins: Structural characteristics that result in unique metabolic patterns and biological activities. *Free Radical Research*, 40(10), 1014–1028.

Rahman, M.M., Ichiyanagi, T., Komiyama, T., Sato, S., and Konishi, T. 2008. Effects of anthocyanins on psychological stress-induced oxidative stress and neurotransmitter status. *Journal of Agricultural and Food Chemistry*, 56, 7545–7550.

Rasheed, Z., Akhtar, N., Anbazhagan, A.N., Ramamurthy, S., Shukla, M., and Haggi, T.M. 2009. Polyphenol-rich pomegranate fruit extract (POMx) suppresses PMACI-induced expression of pro-inflammatory cytokines by inhibiting the activation of MAP kinases and NF-kappaB in human KU812 cells. *Journal of Inflammation (London)*, 6. doi:10.1186/1476-9255-6-1.

Reddivari, L., Vanamala, J., Chinthariapalli, S., Safe, S.H., and Miller, J.C., Jr. 2007. Anthocyanin fraction from potato extracts is cytotoxic to prostate cancer cells through activation of caspase-dependent and caspase-independent pathways. *Carcinogenesis*, 28(10), 2227–2235.

Ritz, M., Curin, Y., Mendelowitsch, A., and Andriantsitohaina, R. 2008a. Acute treatment with red wine polyphenols protects from ischemia-induced excitotoxicity, energy failure and oxidative stress in rats. *Brain Research*, 1239, 226–234.

Ritz, M., Ratajczak, P., Curin, Y., Cam, E., Mendelowitsch, A., Pinet, F., and Andriantsitohaina, R. 2008b. Chronic treatment with red wine polyphenol compounds mediates neuroprotection in a rat model of ischemic cerebral stroke. *The Journal of Nutrition*, *138*, 519–525.

Roghani, M., Niknam, A., Jalali-Nadoushan, M., Kisalari, Z., Khalili, M., and Baluchnejad-mojarad, T. 2010. Oral pelargonidin exerts dose-dependent neuroprotection in 6-hydroxy-dopamine rat model of hemi-parkinsonism. *Brain Research Bulletin*, *82*, 279–283.

Rosen, D.R., Siddique, T., Patterson, D. et al. 1993. Mutations in Cu/Zn superoxide dismutase gene are associated with familial amyotrophic lateral sclerosis. *Nature, 362*, 59–62.

Ross, C.A. and Poirier, M.A. 2004. Protein aggregation and neurodegenerative disease. *Nature Medicine*, *10*, S10–S17.

Sakahira, H., Enari, M., and Nagata, S. 1998. Cleavage of CAD inhibitor in CAD activation and DNA degradation during apoptosis. *Nature*, *39*, 96–99.

Saric, A., Sobocanec, S., Balog, T., Kusic, B., Sverko, V., Drgovic-Uzelac, V., and Levaj, B. 2009. Improved antioxidant and anti-inflammatory potential in mice consuming sour cherry juice (*Prunus cerasus* cv. *Maraska*). *Plant Foods for Human Nutrition*, *64*(4), 231–237.

Schapira, A.H., Cooper, J.M., Dexter, D., Jenner, P., Clark, J.B., and Marsden, C.D. 1989. Mitochondrial complex I deficiency in Parkinson's disease. *Lancet*, *1*(8649), 1269.

Shan, Q., Lu, J., Zheng, Y., Li, J., Zhou, Z., Hu, B., Zhang, Z., Fan, S., Mao, Z., Wang, Y., and Ma, D. 2009. Purple sweet potato color ameliorates cognition deficits and attenuates oxidative damage and inflammation in aging mouse brain induced by D-galactose. *Journal of Biomedicine and Biotechnology*, *2009*. doi:10.1155/2009/564737.

Shelat, P.B., Chalimoniuk, M.H., Wang, J.A., Strosznajder, J.B., Lee, J.C., Sun, A.Y., Simonyi, A., and Sun, G.Y. 2008. Amyloid beta peptide and NMDA induce ROS from NADPH oxidase and AA release from cytosolic phospholipase A2 in cortical neurons. *Journal of Neurochemistry*, *106*, 45–55.

Shih, P., Chan, Y., Liao, J., Wang, M., and Yen, G. 2010. Antioxidant and cognitive promotion effects of anthocyanin-rich mulberry (*Morus atropurpurea* L.) on senescence-accelerated mice and prevention of Alzheimer's disease. *Journal of Nutritional Biochemistry*, *21*, 598–605.

Shih, P., Wu, C., Yeh, C., and Yen, G. 2011. Protective effects of anthocyanins against amyloid β-peptide-induced damage in Neuro-2A cells. *Journal of Agricultural and Food Chemistry*, *59*, 1683–1689.

Shih, P.H., Yeh, C.T., and Yen, G.C. 2007. Anthocyanins induce the activation of phase II enzymes through the antioxidant response element pathway against oxidative stress-induced apoptosis. *Journal of Agricultural and Food Chemistry*, *55*(23), 9427–9435.

Shin, W., Park, S., and Kim, E. 2006. Protective effect of anthocyanins in middle cerebral artery occlusion and reperfusion model of cerebral ischemia in rats. *Life Sciences*, *79*, 130–137.

Slee, E.A., Harte, M.T., Kluck, R.M., et al. 1999. Ordering the cytochrome c-initiated caspase cascade: Hierarchical activation of caspases-2, -3, -6, -7, -8, and -10 in a caspase-9 dependent manner. *Journal of Cell Biology, 144*(2): 281–292.

Sompol, P., Ittarat, W., Tangpong, J., Chen, Y., Doubinskaia, I., Batinic-Haberle, I., Abdul, H.M., Butterfield, D.A., and St. Clair, D.K. 2008. A neuronal model of Alzheimer's disease: An insight into the mechanisms of oxidative stress-mediated mitochondrial injury. *Neuroscience*, *153*(1), 120–130.

Song, D.D., Shults, C.W., Sisk, A., Rockenstein, E., and Masliah, E. 2004. Enhanced substantia nigra mitochondrial pathology in human α-synuclein transgenic mice after treatment with MPTP. *Experimental Neurology*, *186*(2),158–172.

Spada, P.D.S., Dani, C., Bortolini, G.V., Funchal, C., Henriques, J.A.P., and Salvador, M. 2009. Frozen fruit pulp of *Euterpe oleracea* Mart. (acai) prevents hydrogen peroxide-induced damage in the cerebral cortex, cerebellum, and hippocampus of rats. *Journal of Medicinal Food*, *12*(5), 1084–1088.

Spencer, J.P.E. 2008. Flavonoids: Modulators of brain function? *British Journal of Nutrition,* *99*(E-Suppl. 1), ES60–ES77.

Spencer, J.P.E. 2010. Impact of fruit flavonoids on memory and cognition. *British Journal of* *Nutrition, 104,* S40–S47.

Stintzing, F.C., Stintzing, A.S., Carle, R., Frei, B., and Wrolstad, R.E. 2002. Color and anti-oxidant properties of cyanidin-based anthocyanin pigments. *Journal of Agricultural and* *Food Chemistry, 50*(21), 6171–6182.

Sun, H., Kawahara, Y., Ito, K., Kanazawa, I., and Kwak, S. 2006. Slow and selective death of spinal motor neurons *in vivo* by intrathecal infusion of kainic acid: Implications for AMPA receptor-mediated excitotoxicity in ALS. *Journal of Neurochemistry, 98,* 782–791.

Szydlowska, K. and Tymianski, M. 2010. Calcium, ischemia, and excitotoxicity. *Cell Calcium,* *47,* 122–129.

Talavera, S., Felgines, C., Texier, O., Besson, C., Gil-Izquierdo, A., Lamaison, J.L., and Remesy, C. 2005. Anthocyanin metabolism in rats and their distribution to digestive area, kidney, and brain. *Journal of Agricultural and Food Chemistry, 53*(10), 3902–3908.

Tamagno, E., Guglielmotto, M., Aragno, M. et al. 2008. Oxidative stress activates a posi-tive feedback between the γ- and β-secretase cleavages of β-amyloid precursor protein. *Journal of Neurochemistry, 104,* 683–695.

Tarozzi, A., Morroni, F., Hrelia, S., Angeloni, C., Marchesi, A., Cantelli-Forti, G., and Hrelia, P. 2007. Neuroprotective effects of anthocyanins and their *in vivo* metabolites in SH-SY5Y cells. *Neuroscience Letters, 424,* 36–40.

Tarozzi, A., Morroni, F., Merlicco, A., Bolondi, C., Teti, G., Falconi, M., Cantelli-Forti, G., and Hrelia, P. 2010. Neuroprotective effects of cyanidin3-*O*-glucopyranoside on amy-loid beta (25–35) oligomer-induced toxicity. *Neuroscience Letters, 473,* 72–76.

Toufektsian, M.C., de Lorgeril, M., Nagy, N. et al. 2008. Chronic dietary intake of plant-derived anthocyanins protects the rat heart against ischemia-reperfusion injury. *Journal* *of Nutrition, 138*(4), 747–752.

Traustadottir, T., Davies, S.S., Stock, A.A., Su, Y., Heward, C.B., Roberts, L.J., 2nd, and Harman, S.M. 2009. Tart cherry juice decreases oxidative stress in healthy older men and women. *Journal of Nutrition, 139*(10), 1896–1900.

Tsuda, T., Horio, F., and Osawa, T. 1999. Absorption and metabolism of cyanidin 3-*O*-beta-D-glucoside in rats. *FEBS Letters, 449*(2–3), 179–182.

Van Den Bosch, L., Vandenberghe, W., Klaassen, H., Van Houtte, E., and Robberecht, W. 2000. Ca^{2+}-permeable AMPA receptors and selective vulnerability of motor neurons. *Journal of the Neurological Sciences, 180,* 29–34.

Vanzo, A., Terdoslavicj, M., Brandoni, A., Torres, A.M., Vrhovsek, U., and Passamonti, S. 2008. Uptake of grape anthocyanins into the rat kidney and the involvement of bilitrans-locase. *Molecular Nutrition & Food Research, 52*(10), 1106–1116.

Vanzo, A., Vrhovsek, U., Tramer, F., Mattivi, F., and Passamonti, S. 2011. Exceptionally fast uptake and metabolism of cyanidin 3-glucoside by rat kidneys and liver. *Journal of* *Natural Products, 74*(5), 1049–1054.

Varfolomeev, E., Schuchmann, M., Luria, V. et al. 1998. Trageted disruption of the mouse caspase-8 gene ablates cell death induction by the TNF receptors, Fas/Apo1, and DR3 and is lethal prenatally. *Immunity, 9,* 267–276.

Vari, R., D'Archivio, M., Filesi, C., Carotenuto, S., Scazzocchio B., Santangelo, C., Giovannini, C., and Masella, R. 2011. Protocatechuic acid induces antioxidant/detoxifying enzyme expression through JNK-mediated Nrf2 activation in murine macrophages. *Journal of* *Nutritional Biochemistry 22*(5): 409–417.

Wang, Y., Zheng, Y., Lu, J., Chen, G., Wang, X., Feng, J., Ruan, J., Sun, X., Li, C., and Sun, Q. 2010. Purple sweet potato color suppresses lipopolysaccharide-induced acute inflammatory response in mouse brain. *Neurochemistry International, 56,* 424–430.

Ward, M.W., Rego, A.C., Frenguelli, B.G., and Nicholls, D.G. 2000. Mitochondrial membrane potential and glutamate excitotoxicity in cultured cerebellar granule neurons. *The Journal of Neuroscience*, *20*(19), 7208–7219.

Williams, C.M., Mohsen, M.A.E., Vauzour, D., Rendeiro, C., Butler, L.T., Ellis, J.A., Whiteman, M., and Spencer, J.P.E. 2008. Blueberry-induced changes in spatial working memory correlate with changes in hippocampal CREB phosphorylation and brain-derived neurotrophic factor (BDNF) level. *Free Radical Biology & Medicine*, *45*, 295–305.

Woodward, G., Kroon, P., Cassidy, A., and Kay, C. 2009. Anthocyanin stability and recovery: Implications for the analysis of clinical and experimental samples. *Journal of Agricultural and Food Chemistry*, *57*(12), 5271–5278.

Wyss-Coray, T. and Mucke, L., 2002. Inflammation in neurodegenerative disease: A double-edged sword. *Neuron*, *35*, 419–432.

Ye, J., Meng, X., Yan, C., and Wang, C. 2010. Effect of purple sweet potato anthocyanins on β-amyloid-mediated PC-12 cells death by inhibition of oxidative stress. *Neurochemical Research*, *35*, 357–365.

Yi, W., Akoh, C.C., Fischer, J., and Krewer, G. 2006. Absorption of anthocyanins from blueberry extracts by Caco-2 human intestinal cell monolayers. *Journal of Agricultural and Food Chemistry*, *54*, 5651–5658,

Youdim, K.A., Dobbie, M.S., Kuhnie, G., Proteggente, A.R., Abbott, N.J., and Rice-Evans, C. 2003. Interaction between flavonoids and the blood–brain barrier: *In vitro* studies. *Journal of Neurochemistry*, *85*(1), 180–192.

Youdim, K.A., Shukitt-Hale, B., and Joseph, J.A. 2004. Flavonoids and the brain: Interactions at the blood–brain barrier and their physiological effects on the central nervous system. *Free Radical Biology and Medicine*, *37*(11), 1683–1693.

Zafra-Stone, S., Yasmin, T., Bagchi, M., Chatterjee, A., Vinson, J.A., and Bagchi, D. 2007. Berry anthocyanins as novel antioxidants in human health and disease prevention. *Molecular Nutrition & Food Research*, *51*(6), 675–683.

Zdarilova, A., Rajnochova Svobodova, A., Chytilova, K., Simanek, V., and Ulrichova, J. 2010. Polyphenolic fraction of *Lonicera caerulea* L. fruits reduces oxidative stress and inflammatory markers induced by lipopolysaccharide in gingival fibroblasts. *Food and Chemical Toxicology*, *48*(6), 1555–1561.

Zhu, F., Cai, Y.Z., Yang, X., Ke, J., and Corke, H. 2010. Anthocyanins, hydroxycinnamic acid derivatives, and antioxidant activity in roots of different chinese purple-fleshed sweetpotato genotypes. *Journal of Agricultural and Food Chemistry*, *58*(13), 7588–7596.

11 Role of Anthocyanins in Skin Aging and UV-Induced Skin Damage

Leonel E. Rojo, Diana E. Roopchand,
Brittany Graf, Diana M. Cheng, David Ribnicky,
Bertold Fridlender, and Ilya Raskin

CONTENTS

11.1 INTRODUCTION

The visible changes associated with chronological aging and chronic sun exposure, especially to the face, head, and neck areas, are particularly concerning for a significant percentage of the general population. This fact, along with the powerful influence of advertisement and the popular press, has led to an increasing demand for natural and efficient cosmetic ingredients that claim to reduce manifestations of skin aging (Baumann et al., 2009). More importantly, while skin cancers account for up to 40% of the newly diagnosed cancers in the United States (Afaq et al., 2005a), there are no natural preventive methods to avoid cutaneous malignancies associated with chronic sunlight exposure for individuals with pigmentary traits associated with high cancer risk (Zanetti et al., 1996). Consequently, new effective antiaging and chemopreventive agents are in high demand. Although many of the skin-protective claims attributed to botanical products still lack sufficient scientific evidence, the use of natural bioactives with potential antiaging and/or skin-protective properties continues to receive attention from consumers. During the last decade, a substantial body of knowledge has been produced in this area (Chiu and Kimball, 2003; Afaq et al., 2002, 2005a; Afaq, 2011).

Polyphenols (Afaq et al., 2005a; Afaq and Katiyar, 2011; Kao et al., 2007; Kim et al., 1998) and, most recently, anthocyanins (Afaq et al., 2009, 2011; Lila, 2004;

Schmidt et al., 2008; Schreckinger et al., 2010; Tsoyi et al., 2008d) have been reported as potentially effective agents to prevent signs of skin aging and protect the skin from external injuries caused by ultraviolet (UV) radiation (Afaq et al., 2011, 2010; Schreckinger et al., 2010; Tsoyi et al., 2008c). A better understanding of the role of UV radiation, reactive oxygen species (ROS), inflammation, and extracellular matrix (ECM) remodeling in skin pathophysiology has allowed researchers to propose the specific molecular targets for anthocyanins and/or anthocyanin-rich extracts. Although some of the current research describes promising skin-protective effects for anthocyanins, most of the proposed dermatological applications still await clinical validation. This chapter reviews the current scientific literature on the potential of anthocyanins in preserving skin health and preventing skin aging.

11.2 SKIN AGING

Skin aging affects the dermis, epidermis, and hypodermis of the skin (Gomez and Berman, 1985; Giangreco et al., 2008). It not only makes the skin look different but also makes it more vulnerable to external injuries (Giangreco et al., 2008). The epidermis, the most external layer of the skin, is mainly composed of keratinocytes and is directly exposed to environmental aggressions (Figure 11.1). The dermis, rich in connective tissues (structural proteins), such as collagen and elastin, is under the epidermis and gives the young skin its characteristic strength, extensibility, and elasticity (Figure 11.1). Skin aging is an intrinsic biological process, which inevitably starts once a person reaches puberty (Farage et al., 2009) and is manifested by the appearance of skin wrinkles, dryness, thinning of the skin, loss of subcutaneous fat, and uneven pigmentation (Giacomoni, 2008). Each individual's genetic background

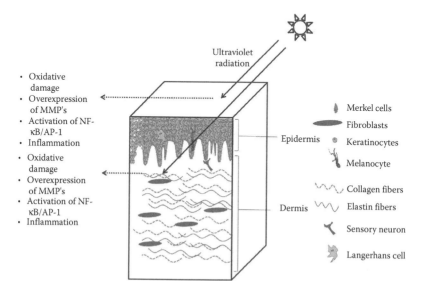

FIGURE 11.1 (**See color insert.**) Schematic representation of the skin architecture and mechanisms of skin damage.

dictates when and how quickly the so-called "intrinsic aging process" unfolds (Koehler et al., 2008). A number of factors can accelerate the intrinsic skin-aging process, which includes weakened deoxyribonucleic acid (DNA)-repairing mechanisms, alterations of the mitochondrial function, slower repair of the ECM, and alterations in cell cycle regulation. The most important extrinsic (accelerating) factor of skin aging is exposure to UV radiation, although diet and smoking can also play a key role in enhancing the appearance of signs of skin aging (Fazio et al., 1989; Fisher et al., 2002; Sakuraoka et al., 1996). It is well known that the acute exposure to high doses of UV radiation triggers the various inflammatory pathways and oxidative damage in the epidermis, dermis, and adnexal organs of the skin, especially UVB radiation (290–320 nm) (Afaq et al., 2005a; Fisher et al., 2002; Ting et al., 2003). The UVA radiation (320–400 nm) is less powerful than UVB, but can penetrate deeper into the skin. In addition, the chronic exposure to high levels of UV radiation can lead to accelerated skin aging (photoaging), hyperkeratosis or atrophy, and precancerous lesions, such as squamous cell carcinomas (Afaq et al., 2005a; Farage et al., 2009). One of the key molecular alterations associated with UV-induced skin damage is the overexpression of metalloproteinases (MMPs), a family of zinc-dependent endopeptidases capable of degrading proteins of the ECM, primarily collagen and elastin. MMPs play an important physiological role in skin regeneration and cell migration (adhesion/dispersion). However, repeated exposure to UV radiation induces the overexpression of specific MMPs (e.g., MMP-9 and MMP-2) leading to the degradation of skin collagen and elastin, incomplete repair of the ECM, loss of skin elasticity and resilience, and the appearance of skin wrinkles. UV radiation also triggers the increase in redox-sensitive transcription factors, including nuclear factor kappa-B, (NF-κB) and activator protein-1 (AP-1). Consequently, researchers are actively looking for natural compounds or mixtures capable of blocking UV radiation, suppressing UV-mediated oxidative damage, inhibiting UV-induced overexpression of MMPs, modulating NF-κB/AP-1 pathways, and decreasing skin inflammation (Figure 11.1).

11.3 ANTHOCYANINS AND SKIN PROTECTION

The known antioxidant power of anthocyanins has led researchers to study their potential in preventing noncommunicable chronic diseases (Cao et al., 2000; Chirinos et al., 2006; Grace et al., 2009; Prior and Wu, 2006; Rojo et al., 2012; Schreckinger et al., 2010). However, the potential of anthocyanins in preventing oxidative skin damage such as UV-induced erythema, skin cancer, and photoaging has received less attention and only a relatively limited number of reports has addressed this question (Afaq et al., 2005b, 2007, 2010, 2011). As the overexposure to UVB radiation is among the most relevant risk factors for oxidative damage to the skin, researchers have used various chemical and biological models to explore the potential of anthocyanins in preventing UVB-induced skin damage. A recent *in vitro* chemical study showed that a cosmetic formulation containing anthocyanins from TNG73 purple sweet potato, at a concentration of 0.61 mg/100 g of cream, could absorb up to 46% of the incident UV radiation (Chan et al., 2010). Although this study was not performed using cellular or animal models of skin damage and has not been clinically confirmed, the results suggest that the topical application of anthocyanins from TNG73 purple sweet potato at very low doses may prevent UV-induced skin damage by decreasing the amount of UVB radiation reaching the epidermis. This mechanism of skin protection is not unexpected, considering that anthocyanins also attenuate UV damage in plants (Woodall and Stewart, 1998; Harvaux and Kloppstech, 2001). Anthocyanins absorb strongly in the visible and UV spectrums, with maximum absorbances in the ranges of 500–550 and 280–320 nm (Harborne, 1958). The UV absorption capacity of anthocyanins varies depending on their specific aglycones, sugar conjugation, and acylation patterns. Consequently, in some colored plant species, other C6–C3–C6 flavonoids, but not anthocyanins, are responsible for the UV protection (Woodall and Stewart, 1998). More importantly, for anthocyanin-rich UV-blocking formulations, it has been reported that acylated anthocyanins containing coumaric acid, caffeic acid, and ferulic acid display the enhanced adsorption of UVB radiation (Harborne, 1958). Chan et al. (2010) also concluded that acidic ethanol-extracted anthocyanins have better radical scavenging ability, higher total phenolic content, and stronger reducing ability than acidic water-extracted anthocyanins from TNG73 purple sweet potato.

A variety of cellular and animal models have been used to elucidate the pharmacological mechanism by which anthocyanins prevent UV-induced damage to the skin (Table 11.1). A recent study using the reconstituted human skin (EpiD5erm(TM) FT-200) showed that pomegranate-derived extracts and juices rich in anthocyanins prevented UVB-induced damage to the dermal structures (Afaq et al., 2009). In this study, the pomegranate-derived products were applied to reconstituted human skin 1 h prior to a 12-h UVB (60 mJ/cm^2) irradiation period. The pomegranate-derived products significantly inhibited protein oxidation, elevation of cyclobutane pyrimidine dimers (CPD), and 8-dihydro-2′-deoxyguanosine (8-OHdG), suggesting the protective effects against the oxidative damage to proteins and DNA. According to the authors, anthocyanin-rich products from pomegranate also protected the ECM of the skin by ameliorating the UVB-induced overexpression of various MMPs, such as collagenase (MMP-1), gelatinase (MMP-2, MMP-9), stromelysin (MMP-3), marilysin (MMP-7), and elastase (MMP-12). Similarly, another study showed that an extract from the blueberry (*Vaccinium uliginosum* L.), rich in cyanidin-3-glucoside, petunidin-3-glucoside, malvidin-3-glucoside, and delphinidin-3-glucoside, prevented UVB-induced overexpression of MMPs and upregulated the UVB-induced suppression of collagen synthesis in human fibroblasts (Bae et al., 2009). These results suggest that anthocyanins from the blueberry may offer protection against photoaging.

Another report by Cimino et al. (2006) showed that the anthocyanin cyanidin-3-*O*-glucoside (C3G) inhibited UV-induced translocation of the transcription factors NF-κB and AP-1 and other inflammatory responses in keratinocytes. According to these data, C3G could provide multifaceted protection against skin damage since NF-κB and AP-1 are the key modulators of several cellular survival programs of skin cells, including the synthesis of inflammatory mediators, and effectors of both innate and adaptive immunity. C3G was also found to prevent the UV-induced overexpression of IL-8, caspase-3 activation, and DNA fragmentation in human keratinocytes (Cimino et al., 2006). This evidence points to a potential protective role of C3G-rich extracts, not only against UVB accumulative skin damage, but also against

TABLE 11.1
Skin-Protective Effects Reported for Anthocyanins

Anthocyanin(s) Tested	Reported Mode of Action	Type of Study	Associated Skin Disease	Reference
Anthocyanins (+ reduced glutathione)	Reduction of erythema after radiation therapy in patients with breast cancer	Clinical/ human	Radiation of dermatitis, discomfort associated with breast irradiation	Enomoto et al. (2005)
Cyanidin-3-O-β-glucopyranoside	Protection against UVA-induced oxidative stress in human keratinocytes	*In vitro*	Photoaging, hyperkeratosis, skin atrophy, precancerous lesions, and skin cancer	Tarozzi et al. (2005)
cyanidin-3-O-glucoside	Reduction of UVB-induced translocation of NF-κ B and AP-1, overexpression of the cytokines IL-8, apoptosis, and DNA fragmentation in cultured human keratinocytes.	*In vitro*	Photoaging, UV-induced erythema	Cimino et al. (2006)
Anthocyanins (+ proanthocyanidin) from Jacquez grapes	Reduction of IL-1α and PGE2, malondialdehyde/4-hydroxynonenal, protein carbonyl groups, and oxidized glutathione, in human reconstructed dermis	*In vitro*	Photoaging, UV-induced erythema	Tomaino et al. (2006)
Delphinidin	Protection of human HaCaT keratinocytes and mouse skin against UVB-mediated oxidative stress and apoptosis	*In vitro* and *in vivo*	Photoaging and skin cancer	Afaq et al. (2007)
Anthocyanin-rich extract from red orange	Reduction of UVB-induced translocation of NF-κ B and AP-1, anti-inflammation in cultured human keratinocytes.	*In vitro*	Photoaging, UV-induced erythema	Cimino et al. (2007)
Black soybean seed anthocyanins	Prevention of UVB-induced apoptotic cell death, inflammation, COX-2, and PGE2. Decreased production of NF-κ B and inhibition of phosphatidylinositol 3-kinase/ Akt pathway	*In vitro* and *in vivo*	Photoaging, hyperkeratosis, skin atrophy, precancerous lesions, and skin cancer	Tsoyi et al. (2008)
Blueberry anthocyanins	Amelioration of UVB-induced damage to human dermal fibroblasts	*In vitro*	Photoaging, precancerous lesions, and skin cancer	Bae et al. (2009)

continued

TABLE 11.1 (continued)
Skin-Protective Effects Reported for Anthocyanins

Anthocyanin(s) Tested	Reported Mode of Action	Type of Study	Associated Skin Disease	Reference
Bilberry anthocyanins	Reduction of UVA-stimulated oxidative damage to keratinocytes	*In vitro*	Photoaging, hyperkeratosis, skin atrophy, precancerous lesions, and skin cancer	Bae et al. (2009)
Anthocyanins from TNG73 purple potato	Absorption of 46% incident UV radiation (0.61 mg/100 g of cream)	*In vitro*	Sun burns, photoaging, and UV-induced erythema	Chan et al. (2010)

psoriasis, characterized by hyperactive NF-κB in keratinocytes. A similar effect was documented using anthocyanins from bilberry and human keratinocytes as a model of dermal UV-induced damage (Svobodova et al., 2008). This latter study showed that anthocyanins from bilberry reduce UVA-stimulated ROS formation and lipid peroxidation.

Analogous skin-protective mechanisms were documented in two separate publications from the same research group (Tsoyi et al., 2008a,b). According to these studies, anthocyanins from black soybean coats may offer protection against UV-induced damage not only to cultured keratinocytes, but also *in vivo* to hairless mice skin. At least two different modes of action were identified in these reports: (i) a reduction of UVB-induced elevation of cyclooxygenase-2 (COX-2) and prostaglandin E2 (PGE(2)) through an NF-κB-dependent pathways (Tsoyi et al., 2008b) and (ii) the prevention of apoptotic cell death by inhibiting caspase-3 activation and reduction of proapoptotic Bax protein levels (Tsoyi et al., 2008a). Delphinidin, an ubiquitous anthocyanin, commonly found in edible berries (Escribano-Bailon et al., 2006; Rojo et al., 2012; Schreckinger et al., 2010), has also shown the protective effect to human HaCaT keratinocytes and mouse skin against UVB-mediated oxidative stress and apoptosis (Afaq et al., 2009). Similarly, another anthocyanin, cyanidin-3-*O*-β-glucopyranoside, was found to prevent UVA-induced damage to human keratinocytes (Tarozzi et al., 2005).

11.4   CURRENT AND FUTURE WORK

It is well known that oxidative damage, inflammation, apoptotic cell death, and overexpression of MMPs play a key role in skin aging and certain forms of UV-induced skin damage. The accumulated scientific evidence suggests that anthocyanins may offer the protection against UV-induced precancerous lesions and possibly delay the appearance of signs of skin aging (Table 11.1). The protective effect of anthocyanins and mode of action have been partially described in several *in vitro* and *in vivo* models of skin damage. However, the current preclinical evidence is seemingly insufficient to conclude that anthocyanins are solely responsible for the skin-protective properties observed *in vitro* and *in vivo* (Table 11.1) because various polyphenols that are different from anthocyanins may be present in the test materials used for these studies.

Additional work is needed to address another important "innovation gap"; the development of chemically stable and clinically effective anthocyanin-rich formulations for dermatological applications. Only one clinical study was available at the time this chapter was written. It reported that a multicomponent formulation containing anthocyanins and glutathione significantly reduced skin erythema after radiation therapy in patients with breast cancer (Enomoto et al., 2005). Unfortunately, the report provided very scarce information regarding the specific group of anthocyanins and doses used for topical applications.

Our research group has recently reported that anthocyanins, along with other polyphenols, can be efficiently separated from highly polar carbohydrates, bound, concentrated, and stabilized into protein-rich, food matrixes, such as defatted soybean flour (DSF) and soy protein isolate (SPI), while preserving their pharmacological effects (Roopchand et al., 2012). The stability of anthocyanins captured in this type of protein-rich matrices was verified up to 50 weeks (Figure 11.2) at 37°C. This form of stabilized anthocyanins opens challenging avenues to develop stable

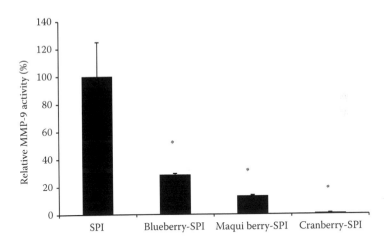

FIGURE 11.2   Effect of anthocyanin-rich protein matrix on MMP-9 activity *in vitro*. The polyphenols including anthocyanins from different sources were concentrated and stabilized in SPI. The concentrations of anthocyanins bound to SPI are shown in Table 11.2. The SPI-enriched matrices were suspended in water mixed with recombinant human MMP-9 (1 μg/mL in PBS) and dye-quenched (DQ) gelatin (1 mg/mL in PBS), a fluorescent quenched substrate of MMP-9. Anthocyanin-enriched matrices (1.5 mg/mL) were incubated with 0.4 μg/mL MMP-9, and 50 μg/mL DQ gelatin at 37°C for 30 min, centrifuged to precipitate solids, and the supernatant was transferred to a 96-well plate to measure MMP-9 activity. The data are reported as the percentage of the inhibition of MMP-9 activity relative to control (SPI). The values correspond to the mean of the three replicates ± SD (*), $P < 0.05$, and *t*-test.

dermatological and cosmetic anthocyanin-rich formulations with the application in cosmetics and food products. We also evaluated whether the SPI with electrostatically bound and concentrated anthocyanins and other polyphenols from maqui berry, blueberry, and cranberry retains the antioxidant capacity and human MMP-9 inhibitory activity of its components. According to our results, these anthocyanin-rich matrices not only displayed a powerful antioxidant capacity (Table 11.2), but also inhibited collagen degradation by human MMP-9 (Figure 11.3), a collagenase known to participate in UV-induced skin damage. The molecular mechanisms by

TABLE 11.2
ORAC Antioxidant Capacity of Different Anthocyanin-Rich Soy Protein Matrices from Fruits

	Anthocyanins (mg/g)	Phenolics (mg/g)	ORAC[a] (Trolox Equivalents (μmol/g)	Representative Kinetic Curve of AAPH[b]-Induced Fluorescence Decay
Maqui berry–SPI	32	80	1090 ± 130	
Blueberry–SPI	29	251	1380 ± 360	
Cranberry–SPI	16	290	1550 ± 200	
SPI	—	—	230 ± 110	

Note: SPI, Soy protein isolate.

[a] ORAC, Oxygen radical absorbance capacity.

[b] AAPH, 2,2′-Azobis (2-amidinopropane) hydrochloride.

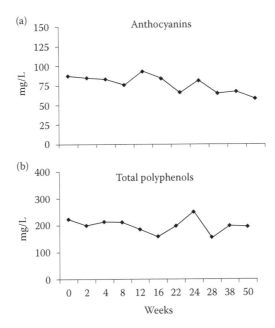

FIGURE 11.3 Stability of blueberry anthocyanins and polyphenols bound to DSF. The concentration of (a) monomeric anthocyanins and (b) total polyphenols are eluted from blueberry polyphenol-enriched DSF after the indicated number of weeks postincubation at 37°C. Polyphenolic compounds were eluted from DSF with 75% methanol, 20% water, 5% acetic acid solution, and the quantifications were done as described elsewhere. (From Roopchand, D. et al. 2012. *Food Chemistry*, 131, 1193–1200.)

which polyphenols bound to SPI inhibit MMP-9 activity remain to be elucidated; our hypothesis is that the specific and nonspecific mechanisms may explain this inhibition and are worthy of further investigations.

11.5 CONCLUSIONS

The previous publications and the presented data suggest that anthocyanins from plants can prevent skin aging and UV-induced skin damage, particularly in formulations that enhance their stability to temperature, pH, and light (Gironés-Vilaplana et al., 2012; Roopchand et al., 2012; Schreckinger et al., 2010). For example, acylated anthocyanins have shown increased stability to pH offering a natural and safer alternative to synthetic dyes for food and cosmetics (Giusti and Wrolstadb, 2003). However, the skin-protective properties of acylated anthocyanins from plants are scarcely studied (Schreckinger et al., 2010). Other authors have addressed this problem by stabilizing anthocyanins from *Aristotelia chilensis* in beverages using lemon juice (Gironés-Vilaplana et al., 2012). Stabilizing anthocyanins by electrostatically binding them to protein matrices may provide another strategy for protecting their structural integrity, function, and color (Roopchand et al., 2012). We hope that in the future, the intense and beautiful colors of free or stabilized anthocyanins may offer a possibility for developing naturally colored cosmetics with skin-protecting and antiaging properties.

ACKNOWLEDGMENTS

This work was supported in part by the NIH training grant T32 AT004094 (supporting DER and DMC) and by P50AT002776-01 and 2P50AT002776-06 grants from the National Center for Complementary and Alternative Medicine (NCCAM) and the Office of Dietary Supplements (ODS) that funds the Botanical Research Center of Pennington Biomedical Research Center and the Department of Plant Biology and Pathology at Rutgers University.

REFERENCES

Afaq, F. 2011. Natural agents: Cellular and molecular mechanisms of photoprotection. *Archives of Biochemistry and Biophysics,* 508, 144–151.

Afaq, F., Adhami, V.M., Ahmad, N., and Mukhtar, H. 2002. Botanical antioxidants for chemoprevention of photocarcinogenesis. *Frontiers in Bioscience,* 7, D784–D792.

Afaq, F., Adhami, V.M., and Mukhtar, H. 2005a. Photochemoprevention of ultraviolet B signaling and photocarcinogenesis. *Mutation Research-Fundamental and Molecular Mechanisms of Mutagenesis,* 571, 153–173.

Afaq, F. and Katiyar, S.K. 2011. Polyphenols: Skin photoprotection and inhibition of photocarcinogenesis. *Mini-Reviews in Medicinal Chemistry,* 11, 1200–1215.

Afaq, F., Khan, N., Syed, D.N., and Mukhtar, H. 2010. Oral feeding of pomegranate fruit extract inhibits early biomarkers of UVB radiation-induced carcinogenesis in Skh-1 hairless mouse epidermis. *Photochemistry and Photobiology,* 86, 1318–1326.

Afaq, F., Saleem, M., Krueger, C.G., Reed, J.D., and Mukhtar, H. 2005b. Anthocyanin- and hydrolyzable tannin-rich pomegranate fruit extract modulates Mapk and NF-kappa B pathways and inhibits skin tumorigenesis in CD-1 mice. *International Journal of Cancer,* 113, 423–433.

Afaq, F., Syed, D.N., Khan, N., and Mukhtar, H. 2011. A dietary anthocyanidin delphinidin promotes epidermal differentiation and activation of caspase-14 in human epidermal keratinocytes and reconstituted skin. *Journal of Investigative Dermatology,* 131, S48–S48.

Afaq, F., Syed, D.N., Malik, A., Hadi, N., Sarfaraz, S., Kweon, M.H., Khan, N., Abu Zaid, M., and Mukhtar, H. 2007. Delphinidin, an anthocyanidin in pigmented fruits and vegetables, protects human HaCaT keratinocytes and mouse skin against UVB-mediated oxidative stress and apoptosis. *Journal of Investigative Dermatology,* 127, 222–232.

Afaq, F., Zaid, M.A., Khan, N., Dreher, M., and Mukhtar, H. 2009. Protective effect of pomegranate-derived products on UVB-mediated damage in human reconstituted skin. *Experimental Dermatology,* 18, 553–561.

Bae, J.-Y., Lim, S.S., Kim, S.J., Choi, J.-S., Park, J., Ju, S.M., Han, S.J., Kang, I.-J., and Kang, Y.-H. 2009. Bog blueberry anthocyanins alleviate photoaging in ultraviolet-B irradiation-induced human dermal fibroblasts. *Molecular Nutrition and Food Research,* 53, 726–738.

Baumann, L., Woolery-Lloyd, H., and Friedman, A. 2009. "Natural" ingredients in cosmetic dermatology. *Journal of Drugs in Dermatology: Jdd,* 8, S5–S9.

Cao, G., Sanchez-Moreno, C., and Prior, R.L. 2000. Procyanidins, anthocyanins and antioxidant capacity in wines. *Faseb Journal,* 14, A564–A564.

Chan, C.F., Lien, C.Y., Lai, Y.C., Huang, C.L., and Liao, W.C. 2010. Influence of purple sweet potato extracts on the UV absorption properties of a cosmetic cream. *Journal of Cosmetic Science,* 61, 333–341.

Chirinos, R., Campos, D., Betalleluz, I., Giusti, M.M., Schwartz, S.J., Tian, Q., Pedreschi, R., and Larondelle, Y. 2006. High-performance liquid chromatography with photodiode array detection (HPLC–DAD)/HPLC–mass spectrometry (MS) profiling of anthocyanins from Andean mashua tubers (*Tropaeolum tuberosum Ruiz and Pavon*) and their contribution to the overall antioxidant activity. *Journal of Agricultural and Food Chemistry,* 54, 7089–7097.

Chiu, A. and Kimball, A.B. 2003. Topical vitamins, minerals and botanical ingredients as modulators of environmental and chronological skin damage. *British Journal of Dermatology,* 149, 681–691.

Cimino, F., Ambra, R., Canali, R., Saija, A., and Virgili, F. 2006. Effect of cyanidin-3-*O*-glucoside on UVB-induced response in human keratinocytes. *Journal of Agricultural and Food Chemistry,* 54, 4041–4047.

Enomoto, T.M., Johnson, T., Peterson, N., Homer, L., Walts, D., and Johnson, N. 2005. Combination glutathione and anthocyanins as an alternative for skin care during external-beam radiation. *American Journal of Surgery,* 189, 627–630.

Escribano-Bailon, M.T., Alcalde-Eon, C., Munoz, O., Rivas-Gonzalo, J.C., and Santos-Buelga, C. 2006. Anthocyanins in berries of maqui (*Aristotelia chilensis (Mol.) Stuntz*). *Phytochemical Analysis,* 17, 8–14.

Farage, M.A., Miller, K.W., Berardesca, E., and Maibach, H.I. 2009. Clinical implications of aging skin: Cutaneous disorders in the elderly. *American Journal of Clinical Dermatology,* 10, 73–86.

Fazio, M.J., Olsen, D.R., and Uitto, J.J. 1989. Skin aging: Lessons from cutis laxa and elastoderma. *Cutis,* 43, 437–44.

Fisher, G.J., Kang, S., Varani, J., Bata-Csorgo, Z., Wan, Y., Datta, S., and Voorhees, J.J. 2002. Mechanisms of photoaging and chronological skin aging. *Archives of Dermatology,* 138, 1462–1470.

Giacomoni, P.U. 2008. Advancement in skin aging: The future cosmeceuticals. *Clinics in Dermatology,* 26, 364–366.

Giangreco, A., Qin, M., Pintar, J.E., and Watt, F.M. 2008. Epidermal stem cells are retained *in vivo* throughout skin aging. *Aging Cell,* 7, 250–259.

Gironés-Vilaplana, A., Pedro, M., Christina, G.-V., and Diego, A.M. 2012. A novel beverage rich in antioxidant phenolics: Maqui berry (*Aristotelia chilensis*) and lemon juice. *Lwt-Food Science and Technology*, 47, 1–8.

Giusti, M.M. and Wrolstadb, R.E. 2003. Acylated anthocyanins from edible sources and their applications in food systems. *Biochemical Engineering Journal,* 14, 217–225.

Gomez, E.C. and Berman, B. 1985. The aging skin. *Clinics in Geriatric Medicine,* 1, 285–305.

Grace, M.H., Ribnicky, D.M., Kuhn, P., Poulev, A., Logendra, S., Yousef, G.G., Raskin, I., and Lila, M.A. 2009. Hypoglycemic activity of a novel anthocyanin-rich formulation from lowbush blueberry, *Vaccinium angustifolium Aiton. Phytomedicine,* 16, 406–415.

Harborne, J.B. 1958. Spectral methods of characterizing anthocyanins. *Biochemical Journal,* 70, 22–28.

Harvaux, M. and Kloppstech, K. 2001. The protective functions of carotenoid and flavonoid pigments against excess visible radiation at chilling temperature investigated in Arabidopsis npq and tt mutants. *Planta,* 213, 953–966.

Kao, E.S., Wang, C.J., Lin, W.L., Chu, C.Y., and Tseng, T.H. 2007. Effects of polyphenols derived from fruit of *Crataegus pinnatifida* on cell transformation, dermal edema and skin tumor formation by phorbol ester application. *Food and Chemical Toxicology,* 45, 1795–1804.

Kim, J., Hwang, J.S., Cho, Y.K., Han, Y.K., Jeon, J.L., and Yang, K.H. 1998. Protective effects of green tea polyphenols on the ultraviolet-induced dermal extracellular damage. *Journal of Investigative Dermatology,* 110, 599–599.

Koehler, M.J., Hahn, S., Preller, A., Elsner, P., Ziemer, M., Bauer, A., Konig, K., Buckle, R., Fluhr, J. W., and Kaatz, M. 2008. Morphological skin ageing criteria by multiphoton laser scanning tomography: Non-invasive *in vivo* scoring of the dermal fibre network. *Experimental Dermatology,* 17, 519–523.

Lila, M.A. 2004. Anthocyanins and human health: An *in vitro* investigative approach. *Journal of Biomedicine and Biotechnology,* 2004, 306–313.

Prior, R.L. and Wu, X.L. 2006. Anthocyanins: Structural characteristics that result in unique metabolic patterns and biological activities. *Free Radical Research,* 40, 1014–1028.

Rojo, L.E., Ribnicky, D., Logendra, S., Poulev, A., Rojas-Silva, P., Kuhn, P., Dorn, R., Grace, M.H., Lila, M.A., and Raskin, I. 2012. *In vitro* and *in vivo* anti-diabetic effects of anthocyanins from maqui berry (*Aristotelia chilensis*). *Food Chemistry,* 131, 387–396.

Roopchand, D., Grace, M.H., Kuhn, P., Cheng, D., Plundrich, N., Pouleva, A., Howell, A., Fridlender, B., Lila, M., and Raskin, I. 2012. Efficient sorption of polyphenols to soybean flour enables natural fortification of foods. *Food Chemistry,* 131, 1193–1200.

Sakuraoka, K., Tajima, S., Seyama, Y., Teramoto, K., and Ishibashi, M. 1996. Analysis of connective tissue macromolecular components in Ishibashi rat skin: Role of collagen and elastin in cutaneous aging. *Journal of Dermatological Science,* 12, 232–237.

Schmidt, B., Ribnicky, D.M., Poulev, A., Logendra, S., Cefalu, W.T., and Raskin, I. 2008. A natural history of botanical therapeutics. *Metabolism,* 57, S3–S9.

Schreckinger, M.E., Lotton, J., Lila, M.A., and De Mejia, E.G. 2010. Berries from South America: A comprehensive review on chemistry, health potential, and commercialization. *Journal of Medicinal Food,* 13, 233–246.

Svobodova, A., Rambouskova, J., Walterova, D., and Vostalova, J. 2008. Bilberry extract reduces UVA-induced oxidative stress in HaCaT keratinocytes: A pilot study. *BioFactors (Oxford, England),* 33, 249–266.

Tarozzi, A., Marchesi, A., Hrelia, S., Angeloni, C., Andrisano, V., Fiori, J., Cantelli-Forti, G., and Hrelia, P. 2005. Protective effects of cyanidin-3-*O*-beta-glucopyranoside against UVA-induced oxidative stress in human keratinocytes. *Photochemistry and Photobiology,* 81, 623–629.

Ting, W.W., Vest, C.D., and Sontheimer, R. 2003. Practical and experimental consideration of sun protection in dermatology. *International Journal of Dermatology,* 42, 505–513.

Tsoyi, K., Bin Park, H., Kim, Y.M., Chung, J.L., Shin, S.C., Shim, H.J., Lee, W.S. et al. 2008a. Protective effect of anthocyanins from black soybean seed coats on UVB-induced apoptotic cell death *in vitro* and *in vivo*. *Journal of Agricultural and Food Chemistry,* 56, 10600–10605.

Tsoyi, K., Bin Park, H., Kim, Y.M., Il Chung, J., Shin, S.C., Lee, W.S., Seo, H.G., Lee, J.H., Chang, K.C., and Kim, H.J. 2008b. Anthocyanins from black soybean seed coats inhibit UVB-induced inflammatory cyclooxygenase-2 gene expression and PGE(2) production through regulation of the nuclear factor-kappa B and phosphatidylinositol 3-kinase/Akt pathway. *Journal of Agricultural and Food Chemistry,* 56, 8969–8974.

Tsoyi, K., Park, H.B., Kim, Y.M., Chung, J.I., Shin, S.C., Lee, W.S., Seo, H.G., Lee, J.H., Chang, K. C., and Kim, H. J. 2008c. Anthocyanins from black soybean seed coats inhibit UVB-induced inflammatory cyclooxygenase-2 gene expression and PGE2 production through regulation of the nuclear factor-kappa B and phosphatidylinositol 3-kinase/Akt pathway. *Journal of Agricultural Food Chemistry,* 56, 8969–8974.

Tsoyi, K., Park, H.B., Kim, Y.M., Chung, J.I., Shin, S.C., Shim, H.J., Lee, W.S. et al. 2008d. Protective effect of anthocyanins from black soybean seed coats on UVB-induced apoptotic cell death *in vitro* and *in vivo*. *Journal of Agricultural Food Chemistry,* 56, 10600–10605.

Woodall, G.S. and Stewart, G.R. 1998. Do anthocyanins play a role in UV protection of the red juvenile leaves of Syzygium? *Journal of Experimental Botany,* 49, 1447–1450.

Zanetti, R., Rosso, S., Martinez, C., Navarro, C., Schraub, S., Sancho-Garnier, H., Franceschi, S. et al. The multicentre South European study "Helios". I: Skin characteristics and sunburns in basal cell and squamous cell carcinomas of the skin. *British Journal of Cancer,* 73, 1440–1446.

12 Anthocyanins, Innate Immunity, and Exercise

Suzanne Maria Hurst and Roger Donald Hurst

CONTENTS

12.1 INTRODUCTION

Fruits (and some vegetables) contain a broad array of phytochemicals, including anthocyanins, which occur as glycosides bound to various sugar molecules. Anthocyanins are reactive compounds with at least eight distinct molecular structures, which occur in an aqueous environment in different structural forms dependent upon pH, time, and temperature (McGhie and Walton, 2007, He and Giusti, 2010). Anthocyanins within fruits and vegetables are thought to serve to either attract pollinating insects and/or seed-distributing animals or act as part of the plant defense mechanism against ultraviolet (UV) exposure and/or microbe/insect infestation (Lev-yadun and Gould, 2009). Fruit and vegetables that contain anthocyanins possess strong inherent antioxidant properties, which have led to the derived foods being marketed as "foods with high anti-oxidant capacity" and suggested health benefits (Zafra-Stone et al. 2007, He and Giusti, 2010, Miguel 2011). However, recent studies reveal that fruit- and vegetable-derived anthocyanins possess other biological actions that are independent of antioxidant capability with the upregulation of adaptive signaling mechanisms within mammalian cells that directly modulate the innate immunity (Cho et al. 2001, Shih et al. 2007, Miguel 2011). The concentration of anthocyanins required to initiate these adaptive mammalian cellular events is below that required to exhibit an enhanced antioxidant capability, and recent studies (Birringer 2011) show that low concentrations of anthocyanins are able to activate hormetic adaptive cellular signaling events, which may underlie both their antioxidant-dependent and antioxidant-independent efficacy on the immune system.

Exercise is known to initiate an increase in reactive oxygen and nitrogen species (ROS, RNS) activating an acute inflammation and resulting in an oxidative stress (Radak et al. 2007, Walsh et al. part I, 2011). The intensity, duration, and adequate rest periods between each exercise session determine whether the outcome is beneficial or detrimental to health. Dietary intervention has been applied in various exercise strategies in an effort for an individual to maintain health while striving to improve the physical performance. Moderate exercise-induced ROS have been shown to exhibit a hormetic action to initiate adaptive cellular events that upregulate both antioxidant and immune capability (Radak et al. 2007), suggesting that the consumption of foods rich in anthocyanins may facilitate the hormetic action of exercise. However, these positive adaptive events may be suppressed or delayed in exercise regimes that are intense, prolonged, or unaccustomed, indicating that beneficial dietary intervention requires the ability to ameliorate excessive oxidative stress/inflammation as well as to facilitate long-term cellular adaptation. In this chapter, we overview how anthocyanins within foods modulate the various aspects of innate immunity that may have important implications for the maintenance of immune and antioxidant efficacy required for gaining the optimum benefits from different exercise regimes.

12.1.1 THE INNATE IMMUNE RESPONSE

The mammalian immune system serves to coordinate the appropriate defense mechanisms with an aim to maintain health and well-being. It basically consists of two main components, innate (or natural) immunity or acquired (or adaptive) immunity, and its overall effectiveness is dependent upon the coordinated interplay between the two components (Getz 2005). While acquired immunity is involved in the recognition (memory) and activation of the appropriate response to specific antigens, innate immunity is nonspecific (no memory) and recognizes a broad array of signals (e.g., Toll-like receptors, Beutler 2009) expressed on microbes as well as signals from damaged or infected cells. Innate immunity is essentially the "first line of defense," and consists of an intricate network of biochemical and cellular events that limits, neutralizes, and facilitates the removal of foreign material (Medzhitov and Janeway, 2000). Both complement and inflammatory factors are interlinked and serve to regulate the appropriate cellular response. Inflammation is an important physiological response to infection, irritation or damage, and serves to create a physical barrier against the spread of infection or tissue damage, facilitating the removal of any foreign or damaged cells and promoting the appropriate repair, recovery, and adaptation processes required to restore the tissue function (Ryan and Majno, 1977). Leukocytes of the innate immune response are circulating within the blood, mobilized, and synthesized from hemopoietic organs such as the bone marrow. Briefly, natural killer (NK) cells are important in combating viral infection and recognize cells either infected or transformed with a virus and initiate cytotoxicity to clear the compromised cells. Circulating neutrophils, monocytes, or tissue-resident macrophages (differentiated monocytes) are the major phagocytes involved in the removal of invading microbes or the damaged tissue. Once phagocytosed, microbes are killed by an oxidative burst and release a number of ROS. This is pro-inflammatory and further attracts other immune cells to the site of infection or tissue damage (Kohchi et al. 2009). Another type of immune cells primarily associated with innate immunity are dendritic cells, which are resident in mucosal tissues and the skin (Langerhans cells), where they are involved in the presentation of antigens to the immune cells of the acquired immune system.

12.1.2 ANTHOCYANINS AND INNATE IMMUNITY: HUMAN INTERVENTION STUDIES

Human intervention studies involving fruit extracts rich in anthocyanins provide an insight into how anthocyanins may modulate the innate immunity and their possible therapeutic or prophylactic applications for the amelioration of chronic inflammatory conditions and/or facilitating the efficacy of the innate immunity. Recent studies by our group have explored how consumption of a New Zealand black currant fruit extract rich in anthocyanins (~240 mg per person dose) influences the innate immune efficacy in humans. The incubation of blood with lipopolysaccharide (LPS) collected 2 h post-ingestion of the extract revealed a differential time-dependent expression in the key inflammatory cytokines (tumor necrosis factor α [TNF]) and interleukin-6 [IL-6]) that correlated with the presence of plasma of black currant-derived anthocyanins (Figure 12.1). Our findings suggest that the ingestion of fruit-derived anthocyanins may improve the effectiveness of the innate immune response against the invading microbes or tissue damage, leading to a speedier immune resolution and tissue recovery. Other human studies involving the ingestion of whole fruit juices or extracts rich in anthocyanins for longer periods have shown an improvement in both the antioxidant capability and efficacy of the innate immunity. The daily consumption of 100% Concord grape juice (240 mL—containing a total of 1894 mg/L gallic acid polyphenolic compounds, of which 223 mg/L malvidin were anthocyanins) for 9 weeks in middle-aged male and female individuals resulted in not only an increase in antioxidant but also in immune surveillance capability, due to an increase in circulating gamma–delta ($\gamma\delta$) T lymphocytes observed after 7 weeks (Rowe et al. 2011). Whereas the consumption of a bilberry/black currant extract (300 mg/day) for 3 weeks in individuals aged between 40 and 74 years showed a reduction in pro-inflammatory cytokines associated with NF-kB (nuclear factor kappa-light-chain-enhancer of activated B cells) activation (Karlsen et al. 2007). While it is unclear whether anthocyanins accumulate in the tissue and have significant long-term health benefits, these studies indicate that the sustained ingestion of foods rich in anthocyanins may improve the overall body antioxidant and innate immune capability, which may be applied to alleviate chronic inflammatory conditions, such as coronary vascular disease, diabetes type II, colitis, and multiple sclerosis. The comprehension of the underlying mechanisms involved in the facilitation of the overall effectiveness of the innate immunity is however still in its infancy. The current consensus from cell studies indicates that while anthocyanins exert a hormetic action on cellular antioxidant systems, the antioxidant-independent effects on innate immunity are mediated

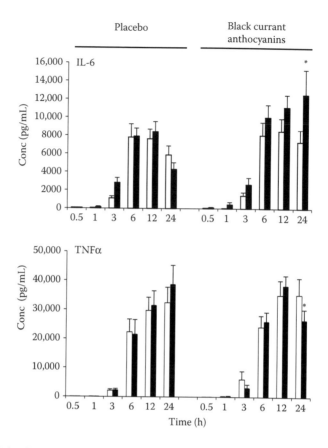

FIGURE 12.1 Consumption of black currant anthocyanins has a variable effect on LPS-stimulated inflammatory cytokine production. The blood collected prior (open bars) to and 2 h (filled bars) after the ingestion of a placebo or an anthocyanin-rich New Zealand black currant extract (total 240 mg anthocyanins) was diluted (1:2) with culture media and incubated with 10 ng/mL purified LPS for 0.5–24 h. The plasma was collected and assessed for either IL-6 (top panel) or TNFα (bottom panel). The results are expressed as picogram per milliliters and are the mean of at least eight separate volunteers. *Represents the statistical significance ($P < 0.05$) at specified times pre- and 2 h postingestion.

by many processes that may be immune, cell specific, and/or dependent upon the type of anthocyanin.

12.1.3 ANTHOCYANINS AND INNATE IMMUNITY: *IN VITRO* STUDIES

Anthocyanins are thought to modulate innate immunity via antioxidant-dependent and antioxidant-independent events. The ability of anthocyanins to scavenge oxygen- and nitrogen-reactive species has been shown in a number of cells and thought to be due to the inherent antioxidant capacity of the anthocyanin compounds themselves (Bianchi et al. 2001, Zafra-Stone et al. 2007, Miguel 2011). However, anthocyanins have also been shown to upregulate the activity and expression of selective antioxidant enzymes possessed by cells (Fiander and Schneider 2000, Turner 2009) indicating that anthocyanins may mediate a dual effect on ROS and RNS. Moreover, human feeding studies (Bitsch et al. 2004, McGhie and Walton 2007, Zafra-Stone et al. 2007, Yang et al. 2011) show that the bioavailability of anthocyanins after the ingestion of fruit and/or vegetables high in anthocyanins (>300 mg/100 g) is transient. The concentration of anthocyanins detected in the plasma post-consumption is also considerably lower than the antioxidant capacity of the original dietary source and yet, an enhanced cellular antioxidant capacity may still be observed. These findings suggest that the ability of anthocyanins to upregulate cellular antioxidant capacity may be more important than their ability to scavenge oxygen and nitrogen species. Fruit-derived anthocyanins (and their primary metabolite—protocatechuic acid (Vari et al. 2011)) activate the redox NF-E2-related factor 2 (nrf2)/antioxidant response element (ARE) signaling network to upregulate the expression of ARE proteins that regulate both antioxidant and innate immune processes (Cho et al. 2001, Shih et al. 2007). The underlying mechanism of the activation is unclear and may involve a chemical conversion of diphenols into quinines, involving an autooxidation to an electrophilic quinine-type molecule (Miguel 2011) or a transient increase in intracellular ROS, a known modulator of nrf2/ARE signaling in the innate immunity as observed in anthocyanin-induced modulation of nicotinamide adenine dinucleotide phosphate–oxidase (NADPH)–quinone oxidoreductase (NQO) 1 in neutrophils (Thimmulappa et al. 2006). Additionally, cyandin-3-O-β-glucoside has been shown to increase the antioxidant cellular capacity independently of nrf2/ARE signaling by upregulating the antioxidant-reduced glutathione (GSH) synthesis. This is reported to occur through a novel antioxidant mechanism involving phosphorylation and activation of cyclic adenosine monophosphate (cAMP) response element binding protein (Zhu et al. 2012).

Anthocyanins have also been shown to modulate innate immunity via processes independent of their antioxidant capability. Anthocyanin glycosides have been shown to modulate LPS- and TNFα-stimulated inflammatory responses in several cell types. The evaluation of different anthocyanin glycosides revealed that delphinidin glycosides were potent at inhibiting LPS-stimulated cyclooxygenase (COX)-2 expression in RAW macrophage 264 cells, involving the inhibition of the degradation of inhibitory kappa beta–alpha (IkB-α) in the NF-kB signaling pathway to reduce the activation of NF-kB (Hou et al. 2005). In contrast, purified anthocyanins have been shown to increase LPS-induced TNFα production in a macrophage cell line, enhancing the pro-inflammatory response (Wang and Mazza, 2002). Furthermore, anthocyanins increase the expression of vascular cell adhesion molecule (VCAM)-1 and decrease the expression of intercellular adhesion molecule (ICAM)-1 (adhesion molecules that regulate the attachment and transmigration of leukocytes from the circulation to the site of tissue infection/damage) induced by TNFα on endothelial cells (Nizamutdinova et al. 2009).

12.2 EXERCISE AND INNATE IMMUNITY

Although most studies to date explore anthocyanin's antioxidant-dependent and antioxidant-independent mechanisms separately, it is becoming evident from both human intervention and *in vitro* cellular studies that the interplay between the two may underlie the ability of foods rich in anthocyanins to improve immune effectiveness. This may be important in many conditions of body stress and/or pathology and therefore be of interest and discussed here are the benefits associated with exercise.

It is generally accepted that regular exercise is essential in the maintenance of health and well-being. However, the mechanisms underlying this health benefit are dependent upon a variety of factors such as the type, intensity and duration of exercise, and factors intrinsic to the exercising individual such as age, genetics, gender, and immune status. Studies show that individuals who take part in regular exercise are not only less susceptible to common infections (such as viruses that cause the common cold, Nieman 1994, Walsh et al. 2011, part I, Romeo et al. 2010), but also demonstrate an improvement in the efficacy of immunization (Kohut et al. 2004) and a reduced recovery time from tissue damage/trauma (Woods et al. 2006) compared to those individuals who remain sedentary. However, excessive intensive exercise is known to lead to compromised immunity, increased susceptibility to infection, and increased recovery time from minor ailments and tissue traumas (Bandyopadhyay et al. 2012). The relationship between exercise and susceptibility to infection (i.e., innate immune efficacy) has been described as a "J" curve. In terms of hormesis, a low or moderate level of physical activity or exercise has a beneficial immunity effect and intense, prolonged, or unaccustomed exercise causes an inhibitory and/or detrimental effect on immunity (Nieman et al. 1993). In the following sections, we focus our discussion on the potential application of foods rich in anthocyanins and their potential roles in modulating the innate immune efficacy applicable for different exercise regimes.

12.2.1 ANTHOCYANINS, EXERCISE, AND INNATE IMMUNITY: REDUCED SUSCEPTIBILITY TO INFECTION

The intensity and duration of exercise and its influence on the innate immunity may be pivotal in activating the appropriate downstream-adaptive cellular processes that influence the overall immune capability. In general, exercise of high intensity or long duration can cause immunosuppression and increased susceptibility to infection (Nieman 1994, Walsh et al. 2011, part II, Romeo et al. 2010). Prolonged moderate-to-high-intensity exercise has been associated with impaired immune function with respect to neutrophil chemotaxis, phagocytosis (degranulation and oxidative burst), and NK (cell cytotoxicity) activity. Furthermore, this level of exercise is associated with glutamine deficit, which may precipitate an anti-inflammatory action and general immunosuppression via hypothalamic–pituitary–adrenal (HPA) axis-stimulated glucocorticoid release (Wood et al. 2006). It is recognized that following strenuous exercise a window of 3–72 h exists in which the immune function is compromised and the susceptibility to opportunistic bacteria/viruses pathogens is evident; this is referred to as the "window of infection risk" (Nieman 1994, Kakanis et al. 2010, Walsh et al. 2011, part II). Therefore, the period of immune recovery is an important consideration for athletes trying to create a balance between maintaining optimal immunity and achieving maximal performance. Dietary interventions have been explored both to narrow the "window of infection risk" and to facilitate the adaptive health benefits of regular moderate exercise. Since the immune system is a complicated network of biochemical and cellular events, selecting foods and/or supplements that target and act synergistically on various aspects of the innate immune system may be more beneficial than those that focus solely on suppressing or controlling oxidative stress. Some antioxidant supplements have been shown to counteract the beneficial aspects of moderate exercise (Teixeira et al. 2009, Peternelj et al. 2011) and may even delay the recovery from strenuous exercise bouts and increase the susceptibility to opportunistic infectious agents.

Foods rich in anthocyanins have been applied to counteract the oxidative stress effects of prolonged intense exercise, especially in training athletes. There have been no studies to date exploring how the consumption of anthocyanins benefits the innate immunity after intense prolonged exercise. Instead, research has focused on the antioxidant properties of anthocyanins for exercise. The consumption of chokeberry juice (23 mg anthocyanins/100 mL) by athletes during a 1-month training camp was shown to cause an increase in their antioxidant capacity and a reduction in exercise-induced oxidative damage to erythrocytes (Pilaczynska-Szczesniak et al. 2005). Similarly, cyclists who ingested a polyphenolic extract (2.3 g/trial) containing 21% black grapes, raspberry, and redcurrants juice (758.6 mg/L anthocyanins) demonstrated an overall reduction of oxidative stress parameters in response to intense exercise bouts in a controlled physical training program (Morillas-Ruiz et al. 2006). The selection of these anthocyanin-rich fruit products in the above studies was reported to be based on their inherent antioxidant capacity. However, it is more likely that the reduction in the overall oxidative stress observed in these studies may be due to fruit-derived polyphenols (including anthocyanins), modulating different aspects of the innate immunity via both antioxidant-dependent and antioxidant-independent cellular events. As indicated, the bioavailability of anthocyanins is transient and it is inconclusive whether long-term ingestion of foods rich in anthocyanins leads to tissue accumulation to levels able to influence tissue function. McAnulty et al. (2011) explored the effect of blueberry consumption in trained athletes on the innate immunity in trained athletes after completing a single bout of intense exercise. They found that the ingestion of a single dose (375 g) of blueberries (containing a variety of anthocyanins) 2 h prior to exercise not only attenuated oxidative stress biomarkers, but also elevated plasma anti-inflammatory cytokines such as IL-10. In the same report, an increase in circulating NK cells was also observed after a 6-week consumption of blueberries (250 g/day). This finding is similar to another human feeding study (Rowe et al. 2011) where an increase in circulating γδ T lymphocytes was observed after drinking grape juice for 9 weeks. Taken together, these unrelated observations tentatively suggest that the consumption of foods rich in anthocyanins may improve the innate immune surveillance and may serve to reduce the "window of infection risk" following intense exercise.

12.2.2 Anthocyanins, Exercise, and Innate Immunity: Complement Regular Moderate Exercise

Nonathletes taking part in moderate exercise most likely do not require nutritional supplements, and the inclusion of antioxidant/anti-inflammatory dietary supplements probably exhibits no health benefits or may even counteract the beneficial adaptive processes stimulated by the exercise. However, the "appropriate" incorporation of certain nutrients in an individuals' exercise regime may "complement" the health benefits of regular exercise of moderate intensity. Physical activity or moderate exercise combined with sufficient recovery times has a beneficial effect on immune capability and, if performed regularly, it may reduce the susceptibility to infection (Nieman 1994, Walsh et al. 2011, parts I and II). These health benefits are dependent upon the activation of appropriate adaptive pathways, which will be dictated by the intensity and duration of the exercise. This is especially important during periods of training where these cellular adaptations may serve to maintain immune capability while maintaining and/or improving performance. The production of ROS or RNS during exercise is important for the activation of downstream protective systems, including the upregulation of adaptive mechanisms, and their redox activity underlies their hormetic action during exercise (Radak et al. 2007, Li et al. 2006). Eukaryotic cells have evolved an intricate adaptive network to counteract the damaging effect of pro-oxidants. ROS and RNS have been shown to upregulate antioxidant enzyme systems, such as superoxide dismutase (SOD), inducible nitric oxide synthase (iNOS) via the NF-kB family of transcription factors (Li 2007). For example, myotubules exposed to hydrogen peroxidase exhibit the upregulation of transcripts for the antioxidant enzymes, catalase, glutathione peroxidase (GPX), zinc–SOD, and manganese–SOD (Franco et al. 1999). In addition, exercise-induced ROS is capable of upregulating a variety of cytoprotective proteins that will facilitate the innate immunity. The expression of these proteins is coordinated through the ability of different pro-oxidants to activate members of the kelch-like ECH-associated protein 1(Keap1)/Nrf2/ARE signaling pathway, involving single or multiple copies of the ARE gene. As discussed earlier in this chapter, anthocyanins (either directly or indirectly) activate the nrf2/ARE signaling pathway to upregulate selective ARE-regulated proteins involved in both antioxidant-dependent and antioxidant-independent mechanisms that serve to regulate the innate immunity (Kensler et al. 2007). Therefore, it is feasible that foods rich in anthocyanins may complement the efficacious actions of moderate exercise on the innate immunity via the upregulation of Keap1/nrf2/ARE signaling pathway.

It has been proposed that foods and/or supplements should target or "prime" the innate immunity thereby generically enhancing immunosurveillance against opportunistic pathogens and reducing the risk of infection (Walsh et al. 2011, part II). Exercise is also known to "prime" the circulating leukocytes, such as neutrophils, which may improve the effectiveness of an innate immune response. Indeed, both moderate and intense exercise are shown to increase the number of circulating neutrophils (Peake et al. 2004); however, the intensity and duration of the exercise have a differential action on their effectiveness. Moderate exercise is shown to improve neutrophil function (Peake 2002), whereas intense exercise transiently suppresses oxidative burst capability. Neutrophil priming includes a variety of cellular events, including selective granule mobilization, calcium mobilization, and serine/tyrosine phosphorylation of various target molecules (Swain et al. 2002). Moderate and intense exercise have been shown to improve neutrophil phagocytic capability of microbial pathogens, thereby supporting the notion that ROS generated during moderate exercise primes neutrophils and serves to facilitate its ability to remove the invading microbe (Ortega et al. 1993, García et al. 2011). With regard to other leukocytes, low-intensity exercise has been shown to prime the circulating monocytes promoting the differentiation toward an M2 macrophage phenotype via peroxisome proliferator-activated receptor γ (PPARγ) constituting a novel anti-inflammatory benefit (Yakeu et al. 2010). However, although exercise has been shown to modulate the effectiveness of NK cytotoxicity (Nieman et al. 1993), it is not known whether it involves a priming event, such as that observed with microbes (Boysen et al. 2011) or dendritic cells (Long 2007). Since the ingestion of foods rich in anthocyanins have been shown to improve the efficacy of the innate immunity, it is possible that the underlying mechanism may also involve "priming" the circulating leukocytes. The preliminary results from a recent study conducted by our group showed that the consumption of a New Zealand black currant extract rich in anthocyanins prior to a moderate exercise of 30 min maintained neutrophil phagocytosis capability. The underlying events involved an upregulation in the expression of selective complementary and antibody receptors that "primed" the neutrophils for the recognition of opsonized *Escherichia coli* bacteria (unpublished observations). Furthermore, a study by Chang et al. (2007) showed that basketball players consuming a diet containing polyphenolic-rich purple sweet potato leaves (PSPL, 900 mg polyphenols/day) during a 2-week training period showed an improvement in the cytotoxic capability of the circulating NK cells (no change in the overall leukocyte numbers was observed). The authors conclude that the daily consumption of PSPL during the training period had an additive effect on the innate immune efficacy and may reduce the susceptibility to infection, although this was not explored in that study. A prior study showed that anthocyanins were the major polyphenolic compounds present in the PSPL extract (Miyazaki et al. 2008).

12.2.3 Anthocyanins, Exercise, and Innate Immunity: Modulate Inflammatory Responses

Acute inflammation is the physiological response to exercise and is an essential trigger in the activation of downstream anti-inflammatory and repair mechanisms that lead to the resolution and recovery of the damaged tissue. Regular moderate exercise is considered to promote an "anti-inflammatory" environment by a combination of exercise-induced events, including the upregulation of "anti-inflammatory" cytokines and/or through the reduction of visceral fat (Petersen and Pedersen, 2005, Woods et al. 2006, Gleeson et al. 2011). Both cross-sectional comparisons and longitudinal exercise-training studies demonstrate a long-term anti-inflammatory effect (Kasapis and Thompson, 2005). A study with professional cyclists has revealed that exercise enhanced the anti-inflammatory response after the injection of bacterial endotoxin (Starkie et al. 2003). The ingestion of bilberry/black currant extract rich in anthocyanins for 3 weeks in middle-aged individuals was shown to reduce pro-inflammatory plasma cytokines (Karlsen et al. 2007) suggesting that the consumption of foods rich in anthocyanins may complement the anti-inflammatory actions of regulating moderate exercise. Furthermore, a recent study by our group (Lyall et al. 2009) revealed that the ingestion of a New Zealand black currant extract rich in anthocyanins prior to and immediately after a moderate exercise not only attenuated oxidative stress, but also facilitated the acute inflammatory response to LPS. The outcome of exercise-induced inflammation has been linked to the intensity of the exercise (Nieman et al. 2012). If the exercise is particularly strenuous or unaccustomed, then it may result in delayed onset muscle soreness (DOMS, Armstrong 1984), and if there is insufficient recovery between exercise bouts, then adaptive events (Tibal l 2004, Jackson 2005) may be delayed, inflammation may not resolve, and tissues may not recover quickly. If this continues over a long period, overtraining and an inability to maintain pro- and anti-inflammatory homeostasis may lead to a subclinical chronic inflammation within vulnerable tissues such as muscle. Delayed and sometimes inappropriate tissue recovery (i.e., scarring) may result in impaired function. Both prophalytic and therapeutic strategies have been applied to counteract the symptoms of DOMS and to speedup the recovery (Connolly et al. 2003). Recently, there has been a focus on identifying dietary foods or supplements that target and control the underlying oxidative stress and inflammation. While to date, there has been mixed results using the therapeutic action of antioxidant dietary supplements to suppress oxidative stress, foods containing a cocktail of nutrients that sequentially target various stages of the inflammation, repair, and recovery/adaptation process may be more successful (Walsh et al. 2011, part II).

The consumption of cherries containing a variety of polyphenols, including anthocyanins, has ameliorated various aspects of the skeletal muscle damage induced by damaging eccentric exercise leading to a speedier recovery of muscle

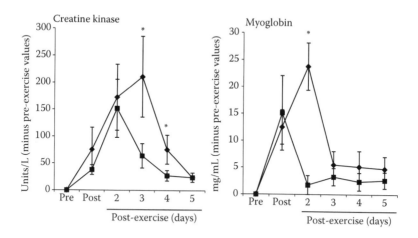

FIGURE 12.2 Consumption of black currant anthocyanins modulates the plasma markers of muscle damage. The volunteers ($n = 10$) underwent repetitive eccentric quadriceps contractions designed to cause muscle damage after the ingestion of either a placebo (♦) or an anthocyanin-rich New Zealand black currant extract (■) (240 mg anthocyanins. The degree of muscle damage was assessed by measuring plasma creatine kinase and myoglobin levels. The results are expressed as minus pre-exercise values, creatine kinase (U/L), and myoglobin (mg/mL). *Represents the statistical significance ($P < 0.05$) between treatments at specified times post-exercise.

function (Connolly et al. 2006, Bowtell et al. 2011). In a preliminary study performed by us, we found that the ingestion of a New Zealand black currant extract rich in anthocyanins (240 mg) 1 h prior to a series of repetitive quadriceps muscle contractions had no effect on the initial oxidative stress or inflammatory response but appeared to accelerate its resolution and initiate repair and adaptive processes that promote the recovery of muscle function. A decrease in both plasma myoglobin and creatine kinase (known plasma markers of muscle damage) was observed on day 2 or 3, respectively; post-eccentric exercise-induced muscle damage in the black currant anthocyanin treatment group was compared to placebo where a decline was not observed until day 3 or 4, respectively, post-muscle damage (Figure 12.2). However, no performance parameters were examined in this study and so, it is unclear whether a single dose of anthocyanins prior to the intense muscle damaging exercise accelerates the recovery of muscle function, although subjective analysis using the visual analog scale of perceived muscle soreness did support this possibility (Figure 12.3). The support for berry-derived anthocyanins attenuating muscle damage comes from studies exploring the effect of blueberries on muscle damage. Preliminary studies by our group showed that blueberry-derived anthocyanins, in particular malvidin glycosides, ameliorated the skeletal muscle cell damage in *in vitro* studies (Hurst et al. 2010). Furthermore, a dietary intervention study in humans performed by colleagues at Massey University (McLeay et al. 2012) involving 300 repetitive eccentric quadriceps contractions (designed to cause muscle damage) and blueberry fruit

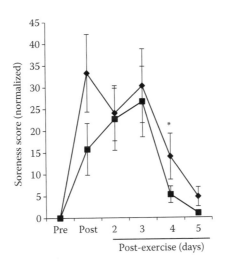

FIGURE 12.3 Consumption of black currant anthocyanins influences the perceived muscle soreness. The volunteers ($n = 10$) underwent repetitive eccentric quadriceps contractions designed to cause muscle damage after the ingestion of either a placebo (♦) or an anthocyanin-rich New Zealand black currant extract (■) (240 mg anthocyanins. DOMS, using a visual analog scale (VAS), was assessed prior to and then at specified times (post-5 days). The results are expressed as perceived soreness normalized to pre-exercise measurements by VAS. *Represents the statistical significance ($P < 0.05$) at a specified recovery time post-exercise.

intervention (consumed before and after muscle damaging exercise) showed an initial increase in oxidative stress and inflammatory parameters post-damage in both blueberry-fed and placebo groups. The recovery in isometric strength was accelerated in the blueberry treatment group, and there was a significant increase in plasma antioxidant capacity observed on day-4 post-muscle damage, suggesting that blueberry-derived polyphenolics may facilitate the adaptive processes that result in a performance recovery. However, no direct relationship between the increase in antioxidant capability and muscle recovery was revealed. The results from these studies would indicate that the consumption of fruits rich in anthocyanins probably do not suppress exercise-induced inflammatory responses but moreover, they facilitate the resolution of inflammation, assisting the repair and recovery. Although the underlying mechanisms and the efficacy of individual identified anthocyanins have yet to be determined, anthocyanin-rich foods may be useful in both promoting accelerated recovery between intensive exercise sessions and limiting the risk of developing overtraining.

12.3   CONCLUSIONS

Exercise maintains and builds cardiac, physical, and immune fitness essential for health and well-being. The relationships between health, regular exercise, and the consumption of plant-derived phytochemicals such as anthocyanins are only now being slowly revealed but evidence to date suggests that any health benefits are partly derived through the initiation of adaptive cell and tissue events leading to the modulation of the innate immunity with associated protection and aided recovery from stress. It is recognized that exercise of an appropriate frequency, duration, and intensity is the key to reap any optimized benefits and the same general notion is likely to be true for the consumption of plant-derived phytochemicals such as anthocyanins. Therefore, the future studies should aim to discover the particular type(s) of anthocyanins responsible for assisting the benefits of exercise with an additional focus on evaluating the dose and timing of consumption according to the particular exercise regime and/or fitness and the physiological makeup of individuals. There is great promise suggested for anthocyanins aiding the benefits of exercise and these future studies should reveal exciting opportunities for efficacious anthocyanin-derived functional foods.

REFERENCES

Armstrong RB. Mechanisms of exercise-induced delayed onset muscular soreness: A brief review. *Med. Sci. Sports Exerc.* 1984. 16(6): 529–538.

Bandyopadhyay A, Bhattacharjee I, Sousana PK. Physiological perspective of endurance overtraining—A comprehensive update. *Al Ameen J. Med. Sci.* 2012. 5(1): 7–20.

Beutler BA. TLRs and innate immunity. *Blood* 2009. 113(7): 1399–1407.

Bianchi L, Lazze C, Pizzala R, Stivala LA, Savio M, Prosperi E. Anthocyanins protect against oxidative damage in cell cultures. Biologically active phytochemicals in foods. *Book Series: R. Soc. Chem. Spec. Publ.* 2001. 269: 311–318.

Birringer M. Hormetics: Dietary triggers of an adaptive stress response. *Pharm. Res.* 2011. 28: 2680–2694.

Bitsch R, Netzel M, Frank T, Strass G, Bitsch I. Bioavailability and the biokinetics of anthocyanins from red grape juice and red wine. *J. Biomed. Biotechnol.* 2004. 5: 293–298.

Bowtell JL, Sumners DP, Dyer A, Fox P, Mileva KN. Montmorency cherry juice reduces muscle damage caused by intensive strength exercise. *Med. Sci. Sports Exerc.* 2011. 43(8): 1544–1551.

Boysen P, Eide DM, Storset K. Natural killer cells in free-living *Mus musculus* have a primed phenotype. *Mol. Ecol.* 2011. 20(23): 5103–5110.

Chang W-H, Chen CM, Hu SP, Kan N-W, Chiu C-C, Liu J-F. Effect of purple sweet potato leaves consumption on the modulation of the immune response in basketball players during the training period. *Asia Pac. J. Clin. Nutr.* 2007. 16(4): 609–615.

Cho BO, Ryu HW, Jin CH, Choi DS, Kang SY, Kim DS, Byun MW, Jeong IY. Blackberry extract attenuates oxidative stress through up-regulation of Nrf2-dependent antioxidant enzymes in carbon tetrachloride-treated rats. *J. Agric. Food Chem.* 2001. 59: 11442–11448.

Connolly DAJ, McHugh MP, Padilla-Zakour OI. Efficacy of a tart cherry juice blend in preventing the symptoms of muscle damage. *Br. J. Sports* 2006. 40(8): 679–683.

Connolly DAJ, Sayers SP, McHugh MP. Treatment and prevention of delayed onset muscle soreness. *J. Stren. Condit. Res.* 2003. 17(1): 197–208.

Fiander H, Schneider H. Dietary ortho phenols that induce glutathione S-transferase and increase the resistance of cells to hydrogen peroxide are potential cancer chemopreventives that act by two mechanisms: The alleviation of oxidative stress and detoxification of mutagenic xenobiotic. *Cancer Lett.* 2000. 156: 117–124.

Franco AA, Odom RS, Rando TA. Regulation of antioxidant enzyme gene expression in response to oxidative stress and during differentiation of muscle skeletal muscle. *Free Rad. Biol. Med.* 1999. 27(9–10): 1122–1132.

García JJ, Bote E, Hinchado MD, Ortega E. A single session of intense exercise improves the inflammatory response in health sedentary women. *J. Physiol. Biochem.* 2011. 67(1): 87–94.

Getz GS. Bridging the innate and adaptive immune systems. *J. Lipid Res.* 2005. 46: 619–621.

Gleeson M, Bishop NC, Stensel DJ, Lindley MR, Mastana SS, Nimmo MA. The anti-inflammatory effects of exercise: Mechanisms and implications for the prevention and treatment of disease. *Nature Rev. Immunol.* 2011. 11: 607–615.

He J, Giusti MM. Anthocyanins: Natural colorants with health-promoting properties. *Annu. Rev. Food Sci. Technol.* 2010. 1: 163–187.

Hou DX, Yanagita T, Uto T, Masuzaki S, Fujii M. Anthocyanins inhibit cyclooxygenase-2 expression in LPS-evoked macrophages: Structure-activity relationship and molecular mechanism involved. *Biochem. Pharmacol.* 2005. 70: 417–425.

Hurst RD, Wells RW, Hurst SM et al. Blueberry fruit polyphenolics suppress oxidative stress-induced skeletal muscle cell damage *in vitro*. *Mol. Nutr. Food.* 2010. 54(3): 353–363.

Jackson MJ. Reactive oxygen species and redox-regulation of skeletal muscle adaptation to exercise. *Philos. Trans. R. Soc. Lond. B Biol. Sci.* 2005. 360(1464): 2285–2291.

Kakanis MW, Peake J, Brenu EW, Simmonds M, Gray B, Hooper SL, Marshall-Gradisnik SM. The open window of susceptibility to infection after acute exercise in healthy young male elite athletes. *Exerc. Immunol. Rev.* 2010. 16(suppl.): 119–137.

Karlsen A, Rettersol L, Laake P, Paur I, Kjolsrud-Bohn S, Sandvik L, Blomhoff R. Anthocyanins inhibit nuclear factor-κB activation in monocytes and reduce plasma concentrations of pro-inflammatory mediators in healthy adults. *J. Nutr.* 2007. 137: 1951–1954.

Kasapis C, Thompson PD. The effects of physical activity on serum C-reactive protein and inflammatory markers: A systemic review. *J. Am. Col. Cardiol.* 2005. 45(10): 1563–1569.

Kensler TW, Nobunao W, Biswal S. Cell survival responses to environmental stresses via the Keap1–Nrf2–ARE pathway. *Annu. Rev. Pharmacol. Toxicol.* 2007. 47: 89–116.

Kohchi C, Inagawa H, Nishizawa T, Soma G-I. ROS and innate immunity. *Anti Cancer Res.* 2009. 29: 817–822.

Kohut ML, Arntson BA, Lee W, Rozeboom K, Yoon KJ, Cunnick JE, McElhaney J. Moderate exercise improves antibody response to influenza immunization in older adults. *Vaccine* 2004. 22(17–18): 2298–2306.

Lev-yadun S, Gould KS. Role of anthocyanins in plant defence, in *Anthocyanins Biosynthesis Functions and Applications.* K. Gould et al., eds. 2009. Springer, New York, NY, pp. 21–48.

Li J. Antioxidant signalling in skeletal muscle: A brief review. *Exp. Gerontol.* 2007. 42(7): 582–593.

Li J, Gomez-Cabrera M-C, Vina J. Exercise and hormesis. Activation of cellular antioxidant signalling pathways. *Ann. N.Y. Acad. Sci.* 2006. 1067: 425–435.

Long EO. Ready for prime time: NK cell priming by dendritic cells. *Immunity* 2007. 26(4): 385–387.

Lyall KA, Hurst SM, Cooney J, Jensen D, Lo K, Hurst RD, Stevenson LM. Short-term black-currant extract consumption modulates exercise-induced oxidative stress and lipopoly-saccharide-stimulated inflammatory responses. *Am. J. Physiol. Integr. Comp. Physiol.* 2009. 297(1): R70–R81.

McAnulty LS, Nieman DC, Dumke CL, Shooter LA, Henson DA, Utter AC, Milne G, McAnulty SR. Effect of blueberry ingestion on natural killer cell counts, oxidative stress, and inflammation prior to and after 2.5 h of running. *Appl. Physiol. Nutr. Metab.* 2011. 36: 976–984.

McGhie TK, Walton MC. The bioavailability and absorption of anthocyanins: Towards a better understanding. *Mol. Nutr. Food Res.* 2007. 51: 702–713.

McLeay Y, Barnes MJ, Mundel T, Hurst SM, Hurst RD, Stannard SR. Effect of New Zealand blueberry consumption on recovery from eccentric exercise-induced muscle damage. *J. Inter. Soc. Sports Nutr.* 2012. 9(1): 19–31.

Medzhitov R, Janeway C. Innate immunity. *N. Engl. J. Med.* 2000. 343: 338–344.

Miguel MG. Anthocyanins: Antioxidant and/or anti-inflammatory activities. *J. Appl. Pharm. Sci.* 2011. 01(06): 07–15.

Miyazaki K, Makino K, Iwadate E, Deguchi Y, Ishikawa F. Anthocyanins from purple sweet potato *Ipomoea batatas* cultivar *Ayamurasaki* suppress the development of atheroscle-rotic lesions and both enhancement of oxidative stress and soluble vascular cell adhesion molecular-1 in apolipoprotein E-deficient mice. *J. Agric. Food Chem.* 2008. 56(23): 11485–11492.

Morillas-Ruiz JM, Villegas Garcia JA, Lopez FJ, Vidal-Guevara ML, Zafrilla P. Effects of poly-phenolic antioxidants on exercise-induced oxidative stress. *Clin. Nutr.* 2006. 25: 444–453.

Nieman DC. The effect of exercise on immune function. *Bull Rheum. Dis.* 1994. 43: 5–9.

Nieman DC, Konrad M, Henson DA, Kennerly K, Shanely RA, Wallner-Liebmann SJ. Variance in the acute inflammatory response to prolonged cycling is linked to exercise intensity. *J. Interferon Cytokine* 2012. 32(1): 12–17.

Nieman DC, Miller AR, Henson DA, Warren BJ, Gusewitch G, Johnson RL, Davis JM, Butterworth DE, Nehlsen-Cannarella SL. Effects of high- vs. moderate-intensity exer-cise on natural killer cell activity. *Med. Sci. Sports Exerc.* 1993. 25(10): 1126–1134.

Nizamutdinova IT, Kim YM, Chung JI, Shin SC, Jeong YK, Seo HG, Lee JH, Chang KC, Kim HJ. Anthocyanins from black soybean seed coats preferentially inhibits TNF-alpha-mediated induction of VCAM-1 over ICAM-1 through the regulation of GATAs and IRF-1. *J. Agric. Food Chem.* 2009. 57(16): 7324–7330.

Ortega E, Collazos ME, Maynar M, Barriga C, De la Fuente M. Stimulation of the phagocytic function of neutrophils in sedentary men after acute moderate exercise. *Euro. J. Appl. Physiol. Occ. Physiol.* 1993. 66(1): 60–64.

Peake JM. Exercise-induced alterations in neutrophil degranulation and respiratory burst activity: Possible mechanisms of action. *Exerc. Immunol. Rev.* 2002. 8: 49–100.

Peake J, Wilson G, Hordern M, Suzuki K, Yamaya K, Nosaka K, Mackinnon L, Coombes JS. Changes in neutrophil surface receptor expression, degranulation, and respiratory burst activity after moderate- and high-intensity exercise. *J. Appl. Physiol.* 2004. 97: 612–618.

Peternelj T-T, Coombes JS. Antioxidant supplementation during exercise training: Beneficial or detrimental? *Sport Med.* 2011. 41(12): 1043–1069.

Petersen AMW, Pedersen BK. The anti-inflammatory effect of exercise. *J. Appl. Physiol.* 2005. 98(4): 1154–1162.

Pilaczynska-Szczesniak L, Skarpanska-Steinborn A, Deskur E, Basta P, Horoszkiewicz-Hassan M. The influence of chokeberry juice supplementation on the reduction of oxidative stress resulting from an incremental rowing ergometer exercise. *Int. J. Sport Nutr. Exerc. Metab.* 2005. 15(1): 48–58.

Radak Z, Chung HY, Koltai E, Taylor AW, Goto S. Exercise, oxidative stress and hormesis (Review). *Ageing Res. Rev.* 2007. 7(1): 34–42.

Romeo J, Warnberg J, Pozo T, Marcos A. 3rd International Immunonutrition Workshop. Session 6: Role of physical activity on immune function. Physical activity, immunity and infection. *Proc. Nutr. Soc.* 2010. 69: 390–399.

Rowe CA, Nantz MP, Nieves C, West RL, Percival SS. Regular consumption of Concord grape juice benefits human immunity. *J. Med. Food* 2011. 14(1–2): 69–78.

Ryan GB, Majno G. Acute inflammation: A review. *Am. J. Pathol.* 1977. 86(1): 183–276.

Shih P-H, Yeh C-T, Yen G-C. Anthocyanins induce the activation of phase II enzymes through the antioxidant response element pathway against oxidative stress-induced apoptosis. *J. Agric. Food Chem.* 2007. 55: 9427–9435.

Starkie R, Ostrowski SR, Jauffred S, Febbraio M, Pedersen BK. Exercise and IL-6 infusion inhibit endotoxin-induced TNFα production in humans. *FASEB J.* 2003. 17: 884–886.

Swain S, Rohn TT, Quinn MT. Neutrophils priming in host defense: Role of oxidants as priming agents. *Antiox. Redox Sig.* 2002. 4(1): 69–83.

Teixeira VH, Valente HF, Casal SI, Marques AF, Moreira PA. Antioxidants do not prevent post exercise peroxidation and may delay muscle recovery. *Med. Sci. Sports Exerc.* 2009. 41(9): 1752–1760.

Thimmulappa RK, Lee H, Rangasamy T, Reddy SP, Yamamoto M, Kensler TW, Biswal S. Nfr2 is a critical regulator of the innate immune response and survival during experimental sepsis. *Clin. Invest.* 2006. 116(4): 984–995.

Tiball JG. Inflammatory processes in muscle injury and repair. *Am. J. Physiol. Regu. Physiol.* 2004. 288(2): R345–R353.

Turner MK. Anthocyanins increase antioxidant enzyme activity in HT-29 adenocarcinoma cells. 2009. MSc thesis. Athens, Georgia, USA.

Vari R, D'Archivio M, Filesi C, Cartenuto S, Scazzocchio B, Santangelo C, Giovannini C, Masella R. Protocatechuic acid induces antioxidant/detoxifying enzyme expression through JNK-mediated Nrf2 activation in murine macrophages. *J. Nutr. Biochem.* 2011. 22: 409–417.

Walsh NP, Gleeson M, Pyne DB, Nieman DC, Dhabhar FS, Shephard RJ, Oliver SJ, Bermon S, Kajeniene A. Position statement part two: Maintaining immune health. *Exerc. Immunol. Rev.* 2011. 17: 64–103.

Walsh NP, Gleeson M, Shephard RJ et al. Position statement part one: Immune function and exercise. *Exerc. Immunol. Rev.* 2011. 17: 6–63.

Wang J, Mazza G. Effects of anthocyanins and other phenolic compounds on the production of tumour necrosis factor α in LPS/IFN-γ-activated RAW 264.7 macrophages. *J. Agric. Food Chem.* 2002. 50(15): 4183–4189.

Wood JA, Vieira VJ, Keylock KT. Exercise, inflammation, and innate immunity. *Neurol. Clin.* 2006. 24: 585–599.

Yakeu G, Butcher L, Isa S, Webb R, Roberts AW, Thomas AW, Backx K, James PE, Morris K. Low-intensity exercise enhances expression of markers of alternative activation in circulating leukocytes: Roles of PPARγ and Th2 cytokines. *Atherosclerosis* 2010. 212: 668–673.

Yang M, Koo SI, Song WO, Chun OK. Food matrix affecting anthocyanin bioavailability: Review. *Curr. Med. Chem.* 2011. 18(2): 291–300.

Zafra-Stone S, Yasmin T, Bagchi M, Chatterjee A, Vinson JA, Bagchi D. Berry anthocyanins as novel antioxidants in human health and diseases prevention. *Mol. Nutr. Food Res.* 2007. 51: 675–683.

Zhu W, Jia QJ, Wang Y, Zhang YH, Xia M. The anthocyanin cyanidin-3-*O*-beta-glucoside, a flavonoid, increases hepatic glutathione synthesis and protects hepatocytes against reactive oxygen species during hyperglycaemia: Involvement of a cAMP-PKA-dependent signalling pathway. *Free Rad. Biol. Med.* 2012. 52(2): 314–327.

Index